Physical Chemistry

Physical Chemistry
is one of a series of textbooks published in
cooperation with E. K. Georg Landsberger.

Physical Chemistry

Eugene J. Rosenbaum
DREXEL UNIVERSITY

APPLETON–CENTURY–CROFTS
EDUCATIONAL DIVISION
NEW YORK   MEREDITH CORPORATION

Copyright © 1970 by

MEREDITH CORPORATION

All rights reserved

This book, or parts thereof, must not be used or reproduced in any manner without written permission. For information address the publisher, Appleton-Century-Crofts, Educational Division, Meredith Corporation, 440 Park Avenue South, New York, N. Y. 10016.

6129-1

Library of Congress Card Number: 72-100040

PRINTED IN THE UNITED STATES OF AMERICA

390-75998-8

To Ruth and Judy
to whom this book owes much

Preface

Anyone who has taught physical chemistry in recent years is aware that the range of topics presently considered as part of this field has grown to such an extent that it cannot be covered adequately in the time normally available in the curriculum. The author of a textbook of physical chemistry must decide between a book that approaches an abridged encyclopedia by touching on every part of this subject, and one that presents fewer topics more completely.

I have written this book from the point of view that the main function of a textbook is to aid students in the process of learning, and that students can learn better from a treatment in depth of a limited number of topics than from an encyclopedic treatment. The process of selection is not an easy one and unanimity of opinion on a particular choice of subject matter is not to be expected. I have scrutinized every topic included to assess its contribution to the understanding of physical chemistry and to the development of the student's background in other branches of chemistry. Many interesting and significant topics, such as nuclear chemistry and polymers, have been deliberately omitted to allow more space for the treatment of those topics that are included. I assume that an instructor who is particularly interested in one of the omitted topics will have no serious difficulty in supplementing the material in this book.

I have paid careful attention to the order in which the various topics are introduced. Although the existing textbooks vary widely in this respect, they can be generally classified as starting with (and usually emphasizing) either the thermodynamic or the molecular basis of physical chemistry. In my opinion these are equally important and it is unfortunate that one must be introduced before the other. My decision to develop the thermodynamic approach first was based on the thought that it is preferable to provide the student with a solid foundation for the organization of equilibrium concepts and relations. Also, the addition of molecular theory to a structure of thermodynamics seems easier than the reverse process.

After the introductory material in Chapter 1, basic thermodynamics is developed in Chapters 2, 3, and 4 to the first logical stopping-place — the treatment of the chemical equilibria of ideal systems. Then the molecular approach is introduced in Chapter 5 by way of the kinetic theory of gases with its simplified molecular models. This is followed in Chapters 6 and 7 by a treatment of atomic structure and energy levels based on wave mechanics, and an introduction to the theory of chemical bonding. Chapter 8, which deals with methods for the determination of molecular structure, includes an elementary treatment of molecular spectroscopy and rotational and vibrational energy levels. The statistical bridge between molecular theory and thermodynamic functions is developed for ideal gases in Chapter 9 and is extended in Chapter 10 to an elementary treatment of the effect of weak intermolecular forces. (This chapter can be omitted without serious detriment to the continuity of the remaining material.)

In the rest of the book either a thermodynamic or a molecular approach is employed, whichever seems more appropriate. Thus the student becomes familiar with the power and limitations of each approach and the fruitful way they supplement each other. In Chapter 11 the subject of reaction kinetics is presented with emphasis on gas-phase reactions. The structure and properties of crystals and liquids are treated in Chapters 12 and 13. Chapter 14 is concerned with phase equilibria in 1-component systems and the dynamics of phase transitions. The treatment is extended in Chapter 15 to multicomponent systems, both ideal and nonideal. Finally, in Chapter 16 the equilibrium and dynamic aspects of the electrolytic solution are treated.

In writing this book I have tried to give the chemistry student a realization that physical chemistry is more than a collection of individual topics, presented in a more or less arbitrary sequence. He should recognize that it is a coherent structure with logical relations between its component parts. I have also tried to aid the student by indicating a distribution of emphasis in the text. Portions of the text set off by a special symbol can be omitted in a first reading. At the end of most chapters supplementary material is included which should prove helpful to students (or classes) with better-than-average background.

No effort has been spared to make the development of the topics in this book as clear as possible and thus make this a book that students can read with understanding. That this very difficult goal has been reached in all parts is improbable. I would appreciate receiving any critical comments or suggestions for improving the clarity of the presentation.

This book has benefited from the comments of colleagues and students at Drexel University. In particular I would like to express my thanks to Professor Herman B. Wagner who read the entire manuscript.

E.J.R.

Contents

Preface   vii

1 Definitions: Properties of Gases   1

2 The First Law of Thermodynamics   27

3 The Second Law of Thermodynamics   59

4 Free Energy Change and Chemical Equilibrium   99

5 Kinetic Theory of Gases   123

6 Atomic Structure and Quantum Theory   157

7 Chemical Bonding and Valence   215

8 Determination of Molecular Structure   267

Contents

9 Statistical Mechanics 335

10 Statistical Mechanics of Real Gases 373

11 Reaction Kinetics 393

12 Solids 445

13 Liquids 491

14 Phase Transitions and Equilibria for Pure Substances 513

15 Phase Equilibria in Multicomponent Systems 539

16 Electrochemistry 595

Glossary of Symbols 667

Index 671

Physical Chemistry

1

Definitions: Properties of Gases

Introduction

Physical chemistry deals with the application of physical concepts and techniques to the study of the behavior of matter. Its goal is the description, understanding, and prediction of this behavior under all circumstances. In pursuit of this goal, two broad approaches have been developed which start from different foundations, use different concepts, and even use different mathematical tools. One of these approaches is thermodynamics; the other is molecular theory in the broadest sense of this term.

Each of these approaches has its own strengths and capabilities, its own weaknesses and deficiencies. Neither one by itself can give us an adequate understanding of the behavior of matter. Together, mutually reinforced by numerous cross-connections, they provide us with a breadth and depth of comprehension which would have seemed fantastic only fifty years ago.

Thermodynamics is a theoretical structure, built on a foundation of a few definitions and three general laws, which is preeminently suited for the study of all types of equilibria, both physical and chemical. Many useful relationships between the properties of matter, such as temperature, pressure, and volume, have been derived by rigorous mathematical reasoning. A conclusion that is based on sound thermodynamic arguments is one of the firmest conclusions in all physical science.

Molecular theory, which was foreshadowed in the vague intuitions of a few ancient Greeks and Romans and the equally vague discussions in the early days of physics in the seventeenth and eighteenth centuries, is based on concepts of the structure of matter. These concepts first became a part of chemistry with the work of John Dalton in the early nineteenth century. In the latter half of this century, acceptance of the ideas of Avogadro on the molecular nature of gases led to our present ideas on atomic and molecular weights.

During the same period, physicists were actively building the kinetic theory of gases. Subsequently the introduction of the quantum theory, the development of our current concepts of atomic and molecular structure, and the applications of quantum mechanics paved the way for a comprehensive and successful treatment of phenomena on the atomic and molecular scale.

If we could start with a few kinds of elementary particles and by purely theoretical methods calculate all of the properties of matter, no other approach to the understanding of the behavior of matter would be needed. Although much has been done in this direction, we remain in the primitive stages of such a program and formidable difficulties must still be overcome. In this situation thermodynamics, which does not make use of any theory of the structure of matter, is indispensable. But thermodynamics cannot predict a numerical value for any property of matter, and it does not reveal anything about the rate at which changes occur. This information must come either from measurements in the laboratory or from the use of molecular theory to its practical limit.

It should be clear at this point that both thermodynamics and molecular theory are essential parts of physical chemistry. Unfortunately, we cannot consider both at the same time. After some introductory material — mostly in the nature of a review — we first take up the subject of thermodynamics and its application to physical and chemical equilibria. This is followed by a presentation of the basic concepts of molecular theory. After that, both approaches are used together.

Some Definitions

The meaning of words can be a problem in a discourse on any topic. Especially in a discussion of a scientific topic, we must be able to express our ideas precisely. This requires a careful definition of terms and symbols. Some of the words are the same as those in common usage, but here they have a technical meaning.

In the physical sciences a *system* is any part of the universe on which we choose to focus our attention. To a physicist, it may consist of a single atom or molecule. To the astronomer, a system might be the sun and its planets, or a galaxy containing billions of stars. For most of our work in chemistry, the systems under consideration contain an amount of matter conveniently handled in the laboratory.

It is essential that a system have a boundary, either real or imagined, so well specified that there is no doubt as to what is in the system and what is not. Everything outside the system will be called the *surroundings*. An *open system* is one in which there can be a transfer of matter or energy to or from the surroundings. For a *closed system* the transfer of matter is excluded, but there can be a transfer of energy. An *isolated system* does not exchange either matter or energy with its surroundings. Such a system is an idealization, of course,

Definitions: Properties of Gases

because no system is completely independent of its surroundings. A *homogeneous system* is one that has the same value of each of its properties in all parts of the system.

The properties of a system which are important for physical chemistry are those that can be measured quantitatively. Of the many properties included in this category, a few are more useful than the rest for an exact description of a system. These are the following: mass, pressure, temperature, and chemical composition. In most cases we specify the mass and composition of a system by a statement of the number of moles of each component in the system.

The properties of a system can be divided into two classes, extensive and intensive. An *extensive property* is one whose value depends on the amount of matter in the system; an *intensive property* is independent of the amount of matter. For example, if we have two identical systems separated by a partition and then remove the partition, the volume of the combined systems is twice the volume of one of the individual systems. In contrast, the temperature and pressure of the combined systems are the same as the original temperatures and pressures. Clearly, mass and volume are extensive properties, while temperature and pressure are intensive properties. The ratio of any two extensive properties is an intensive property. Thus the density, which is the ratio of mass to volume, is an intensive property.

The *state of a system* is a condition which can be precisely specified by giving the values of enough properties to avoid any ambiguity in the description of the system. It has been found from experience that the specification of a relatively small number of properties will suffice for the determination of the state. If we consider, for example, a homogeneous system with a definite composition, fixing the values of two intensive properties fixes the values of all other intensive properties, and thus determines the state of the system. This means that two investigators in different locations who make a variety of measurements on systems of the same composition find that when they bring their respective systems to a condition in which two intensive properties are the same in both systems, the measured values of all other intensive properties, such as specific heat, viscosity, or dielectric constant, are the same in both systems within the respective experimental errors. The two systems are then in the same state.

In most cases of interest to chemists, the temperature and the pressure are the intensive properties which are specified.

Equilibrium

The state of a system can be specified in a rigorous sense only if the system is at *equilibrium*. The concept of equilibrium is basic for much of this book. It is easy to define this term for a simple mechanical system such as the one

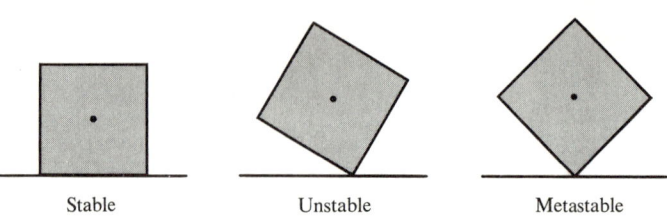

Figure 1–1. Equilibrium in a gravitational field.

illustrated in Figure 1–1. It is difficult, however, to give a definition which is broad enough to cover all cases in physical chemistry. For the present we say that a system is at equilibrium if its properties do not change with time and if there is no net transfer of matter or energy within the system or between the system and its surroundings.

In some cases care is needed to distinguish between a system at equilibrium and one in which changes are taking place very slowly. Also, it is possible for a system to be at equilibrium with respect to some properties and not in equilibrium with respect to others. Both of these points are exemplified by a system consisting of a mixture of hydrogen and oxygen in a glass bulb at room temperature and atmospheric pressure. This system is clearly at equilibrium with respect to temperature and pressure; it is not in equilibrium with respect to chemical composition. In spite of the fact that no change in composition will be observed even after many weeks have elapsed, there is good reason to believe that a chemical reaction is occurring at an unobservable rate. The introduction of a small piece of spongy platinum into the system will markedly accelerate the rate of the reaction with dramatic effects on the system.

States of Aggregation: Phase

It is a familiar fact that matter can exist in three states of aggregation, solid, liquid, and gas, and that transitions between these states can occur. At a later point we are greatly concerned with these transitions and with the internal structures associated with these states of aggregation. For the present, it is sufficient to limit our consideration to simple definitions in terms of observable behavior.

A *solid* is characterized by a definite shape which is maintained in the absence of a container. A *liquid* takes the shape of any container and if the container is partially filled a surface will be formed. A *gas* rapidly occupies any space which is made available to it. Liquids and gases flow under an external force and collectively are often called *fluids*. Liquids and solids collectively are often called *condensed states*.

Definitions: Properties of Gases

For the description of some systems, the terms liquid state and solid state are not sufficiently definite. Consequently it is necessary to introduce another word, phase, which unfortunately has several meanings. In the present context, a *phase* is defined as a system or part of a system which has uniform properties throughout, and a boundary which separates it from the rest of the system. A homogeneous system, such as a solution of a salt in water, constitutes a single phase. If there are several parts of a system with identical properties, they are considered to form one phase. For example, a system containing chopped ice, undissolved salt, a saturated salt solution, mercury, and water vapor consists of two solid phases, two liquid phases, and a gas phase.

A system that contains more than one phase is said to be *heterogeneous*. A system can never have more than one gas phase because all gases are completely miscible in all proportions.

Behavior of Gases

The study of gases is simpler than the study of solids or liquids. For this reason, the first significant work on the behavior of matter was done on gases. In 1662 Boyle studied the relation between the pressure and the volume of a definite amount of air at constant temperature. He trapped a fixed amount of air in the closed arm of a U-tube containing mercury, and measured the volume of the air at a pressure which was the sum of atmospheric pressure and that produced by the column of mercury (Fig. 1–2). By adding or removing mercury, he changed the pressure and measured the corresponding volume. Results similar to his were obtained later by other investigators, using gases other than air. These results can be summarized in the following statement of *Boyle's law:* For a constant amount of gas at a constant temperature, the

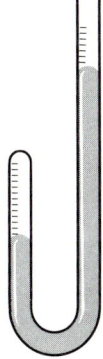

Figure 1–2. Boyle's U-tube experiment.

volume is inversely proportional to the pressure. In mathematical terms, Boyle's law is expressed by the following equation:

$$PV = k \quad \text{(constant temperature and amount of gas)}$$

It seems clear to us now that the next step should have been the investigation of the effect of temperature change. Actually, more than a hundred years passed before this step was taken. Before temperature could be treated as a variable, a method for its measurement had to be developed. This required the invention of a practical and reliable thermometer and the adoption of a temperature scale.

In 1742 Celsius suggested the use of the freezing point and the boiling point of water as reference temperatures for the establishment of a temperature scale. He assigned the values 0° and 100°, respectively, to these reference temperatures and divided the interval between them into 100 degrees. This temperature scale, which was formerly known as the centigrade scale, is now preferably called the Celsius scale.

Toward the end of the eighteenth century, temperature measurements had become sufficiently reliable that meaningful results could be obtained for the relation between the volume of a gas and its temperature. The earliest work was done by Charles (1787); this was followed by more quantitative measurements by Gay-Lussac (1802). *Gay-Lussac's law* may be stated in the following way: At a constant pressure, the volume of a fixed amount of gas is a linear function of the temperature. Expressed in symbols,

$$V_t = a + bt \quad \text{(constant pressure and amount of gas)}$$

in which V_t is the volume at temperature t on the Celsius scale, and a and b are constants. If V_0 represents the volume at 0°C, the *temperature coefficient of expansion*, α, is defined by

$$\alpha = \frac{1}{V_0}\left(\frac{V_t - V_0}{t}\right)$$

On solving this equation for V_t, we obtain

$$V_t = V_0 + V_0\alpha t = V_0(1 + \alpha t)$$

Measurements on a variety of gases show that α is nearly the same for all gases and that its numerical value is approximately $\frac{1}{273}$ per degree.

When the volumes of a fixed amount of a gas at constant pressure are found at two temperatures, t_1 and t_2, the ratio of the volumes is

$$\frac{V_2}{V_1} = \frac{1 + \frac{t_2}{273}}{1 + \frac{t_1}{273}} = \frac{273 + t_2}{273 + t_1}$$

Definitions: Properties of Gases

This suggests that a simpler relation would result if we defined a new temperature scale with the same degree as the Celsius scale, but with the zero shifted downward by 273°. This new scale, which we use almost exclusively, is often called the *absolute temperature scale*. Temperatures on this scale are said to be in degrees Kelvin in honor of Lord Kelvin who did much to clarify the concept of temperature. By international agreement the freezing point of water on the absolute temperature scale is taken to be 273.15°K. Thus a temperature of T°K has the following relation to t in °C: $T°K = t°C + 273.15$.

We can now write Gay-Lussac's law in the following forms:

$$\frac{V_2}{V_1} = \frac{T_2}{T_1}$$

and

$$V_T = k'T \quad \text{(constant pressure and amount of gas)}$$

The equation above implies that the volume of a gas would become zero at 0°K. This, of course, never occurs because all gases become liquid or solid well before this temperature could be reached.

We now have two empirical laws which describe the behavior of gases. If we could combine them into one, we would achieve a further simplification. To do this, let us assume that we have a system consisting of a certain quantity of gas in an initial state with volume V_1, pressure P_1, and temperature T_1. We then change the state of the gas to a final volume V_2, pressure P_2, and temperature T_2. Let us carry out this change in two steps. First we change the pressure to P_2, keeping the temperature constant. The resulting volume, V_r, as given by Boyle's law is

$$V_r = \frac{P_1 V_1}{P_2}$$

Now let us change the temperature to T_2, keeping the pressure constant. The final volume is given by Gay-Lussac's law:

$$V_2 = \frac{V_r T_2}{T_1}$$
$$= \frac{P_1 V_1}{P_2} \frac{T_2}{T_1}$$

Separating the initial and final values, we have

$$\frac{P_2 V_2}{T_2} = \frac{P_1 V_1}{T_1}$$

It follows that no matter how we change the state of this amount of gas, the ratio PV/T always remains constant.

We have thus obtained a combination of the empirical laws of Boyle and Gay-Lussac which we can write as follows:

$$PV = KT$$

From this equation we can predict that the pressure of a definite amount of gas is directly proportional to the absolute temperature at constant volume. This prediction is found to be verified by measurements.

Our combined gas law is still not completely satisfactory, however, because the value of K depends on the amount of gas in the system. Since V is an extensive property, its value at constant P and T is proportional to the number of moles in the system. Then K must also be proportional to the number of moles because P and T are intensive properties. We can express this by writing $K = nR$, in which n is the number of moles of gas in the system and R is independent of all variables and therefore is a universal constant. We have thus obtained the *general gas law* in its final form:

$$PV = nRT \tag{1-1}$$

Let us now assume that our system consists of a mixture of gases with n_1 moles of the first gas, n_2 moles of the second gas, and so on, in a container with volume V and a total pressure P_{tot}. We define the *partial pressure* of each gas in the system as the pressure that the same number of moles of this gas would exert if it were present alone in the same container and at the same temperature as the mixture. Thus the partial pressure, P_i, of component i is given by

$$P_i = \frac{RT}{V} n_i \qquad (i = 1, 2, 3 \cdots) \tag{1-2}$$

By laboriously adding gases to a container and measuring the resulting pressures, Dalton found that the total pressure of a mixture of gases is the sum of the partial pressures. This statement, known as *Dalton's law of partial pressures*, can be represented in the following ways:

$$\begin{aligned} P_{tot} &= P_1 + P_2 + P_3 \cdots \\ &= \Sigma P_i = \frac{RT}{V} \Sigma n_i = \frac{RT}{V} n_{tot} \end{aligned} \tag{1-3}$$

Dalton's law of partial pressures is thus seen to be completely consistent with the general gas law.

When we divide Equation 1–2 by Equation 1–3 we obtain

$$\frac{P_i}{P_{tot}} = \frac{n_i}{n_{tot}} = X_i \tag{1-4}$$

The quantity X_i is called the *mole fraction* of component i. This quantity is a measure of composition that we use frequently. The mole fraction is clearly an intensive property.

Evaluation and Units of R

As the techniques of measurement improved and experimental errors became smaller, it became apparent that no two gases behaved exactly alike and that no gas followed the general gas law within experimental error. It was observed, however, that for every gas the deviation from this law becomes smaller, the higher the temperature and the lower the pressure. We can imagine a gas which follows this law exactly, and we call such a hypothetical gas an ideal gas. Correspondingly, we refer to the general gas law (Eq. 1-1) as the *ideal gas law*.

A *real gas* — one we can work with in the laboratory — follows the ideal gas law approximately at ordinary pressures and temperatures, and the be-

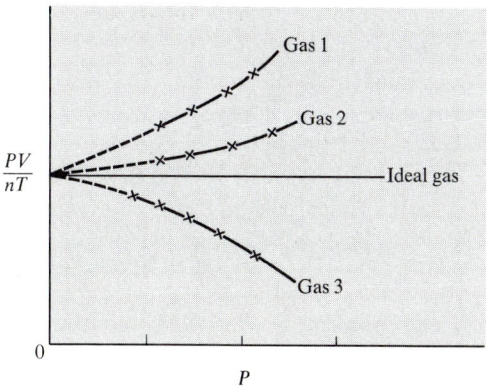

Figure 1-3. Determination of the value of R (schematic).

havior of the real gas approaches that of the ideal gas as the temperature is raised or the pressure is lowered. We return shortly to a consideration of deviations of real gases from ideality.

If no real gas follows the ideal gas law exactly, how can a precise value be found for the gas constant R? This is done by making measurements of volume on known amounts of a gas at a series of pressures. The value of the ratio PV/nT is plotted against the pressure (Fig. 1-3), and the resulting graph line is extrapolated to zero pressure. When this procedure is carried out on a variety of gases, it is found that all of the intercepts at zero pressure are the same within quite small errors of measurement. This common intercept is the value of the gas constant R.

The gas constant is one of the most important quantities in physical chemistry. Its numerical value depends on the units in which it is expressed. We always express n in moles and T in degrees Kelvin. This leaves the units and dimensions of the PV product to be considered. By definition, pressure is force per unit area; $P = f/A$. Volume has the dimensions of length^3 or area $\times$ length. Consequently,

$$\frac{f}{A} \times A \times L = f \times L$$

Since the product of a force and a length has the dimensions of work, or energy, this is also true of the PV product. We thus conclude that no matter in what units R is expressed, its physical significance is energy per mole degree.

Pressures are often expressed in terms of the *atmosphere* as a unit, but we will also use two other pressure units. One of these is the *torr* which is defined as the pressure corresponding to a column of mercury 1 mm in height. The other is the cgs unit of pressure, the dyne/cm^2. These units have the following relations to the atmosphere:

$$1 \text{ atm} = 760 \text{ torr}$$
$$= 1.013 \times 10^6 \text{ dynes/cm}^2$$

The liter is a convenient unit of volume. We also use the milliliter (ml) which is essentially the same as the cubic centimeter (cc or cm^3).

With the atmosphere and the liter as units, the value of R from data of the kind shown in Figure 1-3 is 0.08205 liter-atm per mole degree. It follows that the volume of 1 mole of ideal gas at 1 atm and 0°C is given by

$$V = \frac{0.08205 \times 273.15}{1} = 22.414 \text{ liter}$$

When the volume is expressed in cc, $R = 82.05$ cc atm/mole degree. In cgs units,

$$R = \frac{1.013 \times 10^6 \text{ dynes/cm}^2 \times 22{,}414 \text{ cm}^3}{273.15 \times 1 \text{ mole}} = 8.314 \times 10^7 \text{ ergs/mole deg}$$
$$= 8.314 \text{ joules/mole deg}$$

The *calorie* is a widely used unit of thermal energy which was originally defined as the amount of heat required to raise the temperature of one gram of water by 1°C. Careful measurements revealed that this amount of heat does not remain constant as the initial temperature of the water varies. Today an average value is adopted which is based on the joule:

$$1 \text{ calorie} = 4.184 \text{ joules}$$

It follows that $R = 1.987$ cal/mole deg. Except for very accurate calculations this can be rounded off to 2 cal/mole deg. Other units for R are sometimes used but the ones mentioned are sufficient for our purposes.

Equations of State: Phase Diagrams

The ideal gas law is the simplest example of an *equation of state*. An equation of state is a mathematical relation between the variables which determine the

Definitions: Properties of Gases

state of a system — in the present case P, V, T, and n. When the values of any three of these properties are given the value of the fourth is fixed and can be obtained by a simple calculation provided the value of R is known.

When we divide both sides of the ideal gas equation of state by n we have

$$\frac{PV}{n} = PV_m = RT$$

where V_m is the molar volume. We now have a relation connecting three intensive properties. From a mathematical point of view any one of the three may be considered as a dependent variable which can be expressed as a function of two independent variables.

If we consider a 3-dimensional graph with P, V_m, and T plotted along the x, y, and z coordinates, the state of our system with definite values of P, V_m, and T is represented by a point. A mathematical relation of the form $V_m = f(P,T)$ is represented by a surface (Fig. 1–4). Such a graph is often called a *phase diagram*.

Because 3-dimensional graphs are awkward to draw it is simpler to use 2-dimensional cross-sections obtained by assuming that one of the variables

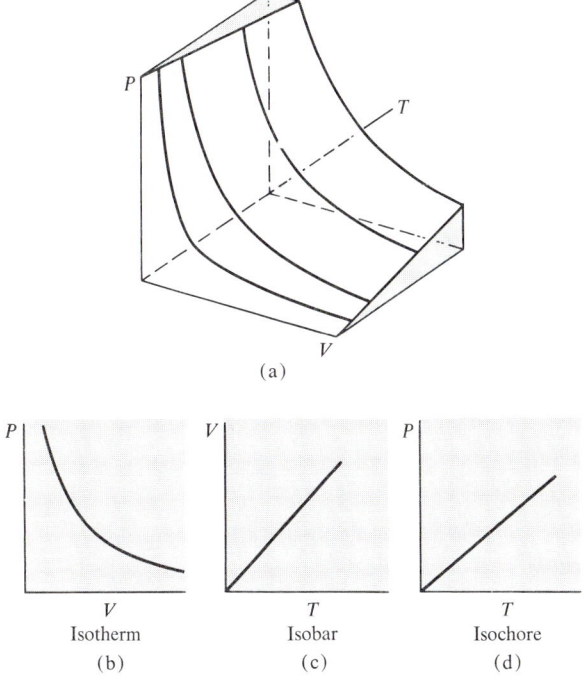

Figure 1–4. (a) P–V–T surface (phase diagram) for an ideal gas. (b), (c), (d) Cross-sections through the 3-dimensional phase diagram.

is held constant. A graph line which represents a series of states with constant temperature is called an *isotherm*. Lines representing states at constant pressure or volume are called *isobars* and *isochores*, respectively. We will make frequent use of 2-dimensional phase diagrams in future work.

Molecular Weights of Gases

One useful application of the ideal gas law is the determination of the molecular weight of an unknown substance which either is a gas or can be vaporized without decomposition. Representing a known weight of the unknown substance by w and its molecular weight by M, we have

$$n = \frac{w}{M}; \quad PV = \frac{w}{M} RT;$$

$$M = \frac{w}{V} \frac{RT}{P} = \rho \frac{RT}{P} \tag{1-5}$$

in which $\rho = w/V$ is the density of the gas at measured T and P. Depending on the nature of the sample, different techniques are used to make the required measurements.

The molecular weights obtained in this way cannot be highly accurate since they depend on the ideal gas law. In spite of this they are very useful and have played an important role in the development of chemistry, because even approximate molecular weights are adequate for the purpose of determining the molecular formula of a substance. Suppose, for example, that the chemical analysis of a compound provides an empirical formula $(CH_2O)_x$.

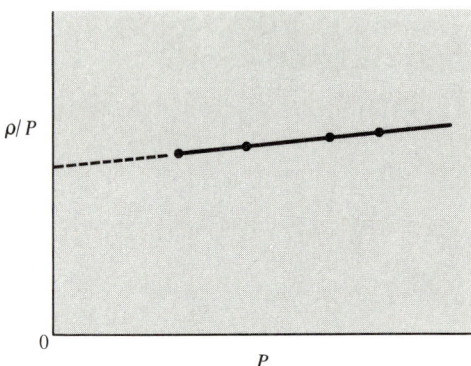

Figure 1-5. Illustration of the method of limiting densities for the determination of accurate molecular weights. The value of $\left(\frac{\rho}{P}\right)_{P=0}$ is obtained by extrapolation of the graph line.

Definitions: Properties of Gases

All that is required to fix the value of x is a molecular weight determination which can clearly distinguish between multiples of 30. For this purpose, an uncertainty of $\pm 5\%$ in the molecular weight would not be important.

If an accurate value of the molecular weight is desired, the deviation from the ideal gas law must be eliminated. This can be accomplished by using the method of limiting densities. In carrying out this method, the gas density is measured at a series of moderately low pressures and constant temperature. The ratio ρ/P is plotted against P (Fig. 1-5), the graph line is extrapolated to $P = 0$, and the value of $(\rho/P)_0$ so obtained is multiplied by RT (Eq. 1-5). The main application of this procedure has been the determination of atomic weights. For example, an accurate molecular weight of CH_4 combined with an accurate molecular weight of hydrogen yields a reliable value for the atomic weight of carbon.

Real Gases

Many careful measurements of P, V, and T have been made for a variety of gases over wide ranges of pressures and temperatures. At elevated pressures and moderate temperatures, substantial deviations from the ideal gas law are found, as illustrated in Figure 1-6. In this graph the ratio PV/nRT, which is called the *compressibility factor* Z, is plotted against P for several typical gases and temperatures. It will be noted that for some gases the values of Z are less than the ideal value of 1 at intermediate pressures but become greater than 1 at higher pressures. For hydrogen and helium at ordinary temperatures, Z is greater than 1 at all pressures.

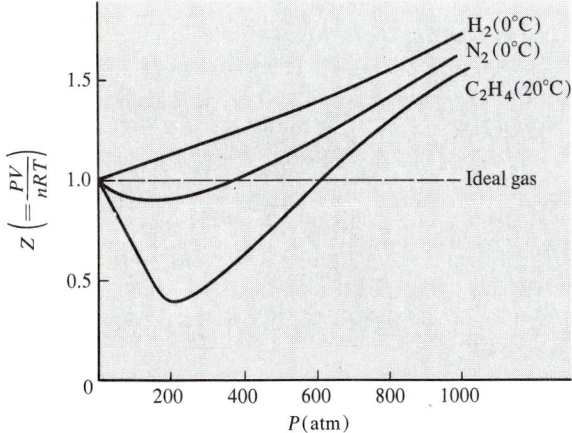

Figure 1-6. Compressibility factor as a function of the pressure.

The P,V,T behavior of a real gas can be described by tables or graphs of Z as a function of P and T. This is useful for some purposes but in general it is awkward because each gas must have its own set of tables or graphs. Much effort has been spent in devising empirical equations of state which would be applicable to all gases by assigning values of a few parameters to individual gases. In general, the larger the number of parameters used, the more reliable is the equation of state. The price paid for this reliability is the large amount of computational labor involved.

Of the dozens of equations of state which have been proposed for real gases, we mention only two which have particular interest. The first of these is the following equation which is known as the *van der Waals equation of state*:

$$\left(P + \frac{a}{V_m^2}\right)(V_m - b) = RT \tag{1-6}$$

in which the van der Waals constants, a and b, have values that are characteristic for each kind of gas.

Van der Waals arrived at his equation of state by a plausible but qualitative line of reasoning which we only sketch here. He noticed that there must be at least two causes for deviations from the ideal gas law, one producing positive deviations (that is, values of Z greater than 1) and one producing negative deviations. He concluded that at least two molecular parameters must be used to account for these types of deviations.

On the basis of a simple molecular model for a gas (which we consider in detail at a later point) he asserted that the volume V which appears in the ideal gas law should be the volume which is available for the motion of the molecules; in other words, the total volume minus the volume occupied by the molecules themselves. For a gas at low pressure, the molar volume is relatively large and the volume occupied by the molecules can be considered as negligible. When the pressure is high, the molar volume is relatively small and the volume occupied by the molecules is a significant fraction of the total volume. If the total volume is used in the ideal gas equation of state, positive deviations in the PV product will occur — which is the case for all gases at sufficiently high pressures. To correct for this positive deviation, van der Waals subtracted from the total molar volume a quantity, symbolized by b, related to the volume of Avogadro's number of molecules. At this stage the equation of state is

$$P(V_m - b) = RT$$

The negative deviations remain to be accounted for. To do this, van der Waals pointed out that the molecules exert attractive forces on each other

Definitions: Properties of Gases

and that these forces would be stronger at relatively small intermolecular distances. The pressure of a gas is the result of impacts of molecules on the walls of the container. Considering a molecule about to make an impact on a wall, we can conclude that the force it will exert is lessened by the attractive forces between it and its neighboring molecules. The attractive force on the impacting molecule is proportional to the number of molecules per unit volume and is thus inversely proportional to the molar volume. The number of molecules hitting the wall of the container in unit time is also proportional to the molecular concentration and inversely proportional to the molar volume. Since the total effect is equal to the product of the attractive force on an impacting molecule and the number of molecules making impacts in unit time, it is inversely proportional to the square of the molar volume. Therefore the reduction of the pressure from the value it would have if there were no intermolecular forces is given by a/V_m^2, in which the constant of proportionality is known as the van der Waals constant a. To correct the ideal gas law for the effect of intermolecular forces, a/V_m^2 must then be added to the observed pressure, thus yielding Equation 1-6. It should be noted that the discussion just given is not a rigorous derivation of the van der Waals equation.

From Equation 1-6 it can be seen that as V_m increases, b becomes negligible with respect to V_m, and a/V_m^2 becomes negligible with respect to P. The van der Waals equation then reduces to the ideal gas law. As the pressure is raised or the temperature is lowered, V_m is decreased. The correction to the pressure usually becomes significant first, but at high enough pressures the correction to the volume also becomes significant. In the case of hydrogen and helium, the value of a is relatively small. This accounts for the fact that at ordinary temperatures no negative deviations from the ideal gas law are observed with these gases.

It should be clear that the units of the van der Waals constants depend on the units used for pressure and volume. Because the atmosphere and the liter are widely used, the representative values given in Table 1-1 are given in terms of these units.

The van der Waals equation is not an outstandingly good equation of state, which is implied by the fact that the constants actually vary with temperature. The inadequacy of this equation is particularly evident at high pressures. To cover a broad range of temperature and pressure a larger number of adjustable parameters must be used.

The other equation of state we present here is known as the *virial equation of state*. One form of this equation is as follows:

$$Z = \frac{PV_m}{RT} = 1 + a(T)P + b(T)P^2 + \cdots$$

TABLE 1-1   VAN DER WAALS CONSTANTS*

Gas	a (liter^2-atm/mole^2)	b (liter/mole)
H_2	0.245	0.0267
He	0.034	0.0236
N_2	1.39	0.0394
Cl_2	6.49	0.0562
NH_3	4.17	0.0370
CO_2	3.59	0.0427
CH_4 (methane)	2.25	0.0428
C_2H_6 (ethane)	5.49	0.0638

American Institute of Physics Handbook, McGraw-Hill, New York, 1963.

This equation expresses the compressibility factor as a power series in the pressure with coefficients which are functions of the temperature. An alternative form is a power series in reciprocal molar volumes:

$$Z = 1 + \frac{B(T)}{V_m} + \frac{C(T)}{V_m^2} + \cdots \qquad (1\text{-}7)$$

In principle, any type of P,V,T behavior can be represented by an equation of this form if enough terms in the series are used. This equation is worth mentioning here because some of the coefficients B, C, and so on, which are known as the virial coefficients, can be calculated theoretically. We return to this point in Chapter 10.

Critical State and Liquefaction

It is a familiar fact that gases condense to liquids if the temperature is lowered sufficiently. Long ago it was observed that some gases, such as ammonia and sulfur dioxide, can be condensed at laboratory temperatures by raising the pressure. Other gases, such as hydrogen, oxygen, and nitrogen, could not be condensed in this way and these were referred to as permanent gases. The investigation of Andrews on carbon dioxide in 1869 contributed much to the understanding of this distinction and the relation between the liquid and gaseous states of aggregation.

Andrews' results are summarized in the phase diagram shown in Figure 1-7. The isotherm T_1 represents the data obtained at a relatively high temperature. Its form is very close to that of a rectangular hyperbola and shows that at this temperature CO_2 follows Boyle's law rather closely. The next isotherm T_2 shows a marked deviation from ideal behavior at a lower temperature.

Now consider the isotherm T_4 at a relatively low temperature. As the pressure on the system is raised from a low initial value the volume at first decreases roughly in accordance with Boyle's law. At a certain pressure P_4 the volume continues to decrease while the pressure remains constant. At volume V_2 another discontinuity is apparent and the large slope of the isotherm shows that the rate of change of pressure with volume becomes very large. Observation of the system shows that when the pressure reaches the value P_4 the first evidence of a liquid phase appears. As the volume decreases more of the gas is transferred to the liquid phase while the pressure remains constant.

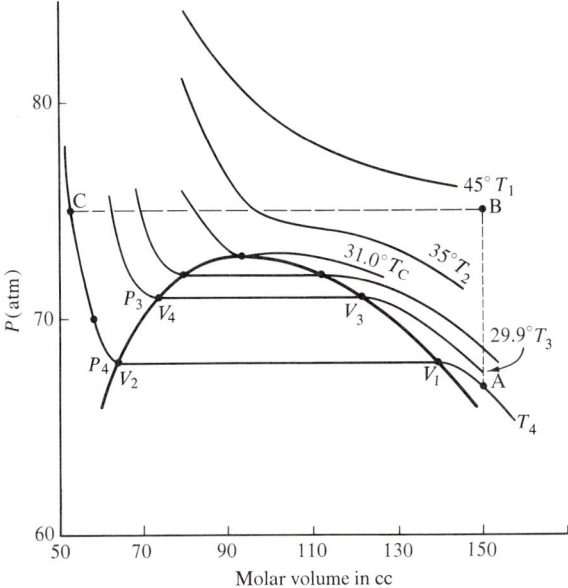

Figure 1-7. Phase diagram of CO_2.

Finally at V_2 all of the CO_2 is in the liquid phase. To decrease the volume of the liquid requires a very high pressure. As long as both liquid and gas phases are present and the system is at equilibrium, the pressure will be P_4. This equilibrium pressure is known as the *vapor pressure*.

If the system is at a somewhat higher temperature T_3, the vapor pressure is higher and the volume range in which the two phases can coexist is decreased. As the temperature is raised, the coexistence range approaches zero. The isotherm labelled T_c, which is tangent to the boundary line of the dome-shaped coexistence region, is called the critical isotherm and the corresponding temperature is the *critical temperature*. If the temperature of a pure gas is above its critical temperature, the gas cannot be liquified by pressure alone.[1] A gas whose temperature is below its critical temperature is usually called a vapor.

The pressure in the system when the critical temperature is reached is called the *critical pressure;* it is the highest possible value of the vapor pressure. The molar volume at the critical temperature and the critical pressure is the *critical volume.*

Each substance has a unique set of critical constants (Table 1-2) that characterizes its critical state. The critical temperature is relatively easy to

TABLE 1-2 CRITICAL CONSTANTS*

Substance	T_c (°K)	P_c (atm)	V_c (l/mole)	$P_c V_c / R T_c$
Helium	5.3	2.26	0.0578	0.300
Hydrogen	33.3	12.80	0.0650	0.305
Nitrogen	126.2	33.5	0.0901	0.291
Oxygen	154.8	50.1	0.0780	0.307
Methane	190.7	45.8	0.0990	0.290
Ethane	305.5	48.2	0.148	0.284
Ethylene (C_2H_4)	282.4	50.0	0.124	0.268
Acetylene (C_2H_2)	309	61.6	0.113	0.274
Benzene (C_6H_6)	562	48.6	0.260	0.274
Carbon dioxide	304.2	72.9	0.0940	0.274
Ammonia	405.6	111.5	0.072	0.242
Water	647.4	218.3	0.056	0.231

*K. A. Kobe and R. E. Lynn, *Chem. Rev.* **52**, 117 (1953).

measure. This can be accomplished by sealing off enough sample in a heavy-walled glass tube (to withstand rather high critical pressures) to half fill the tube at a temperature slightly below the critical temperature, and then slowly raising the temperature. A surface (meniscus) between the liquid and the vapor is visible by reflected light so long as both phases are present. In the critical state only one phase is present and the meniscus disappears. The temperature at which this occurs is read on a suitable thermometer. The critical pressure can be measured directly by connecting the sample tube to a high pressure manometer.

The measurement of the critical volume is more difficult and an indirect method is usually employed to obtain this quantity. In a system consisting of a liquid and its vapor in equilibrium at a definite temperature, the density of the liquid is greater than the density of the vapor. These densities are known as orthobaric densities. As the temperature is raised the liquid expands and its density is lowered. At the same time the vapor pressure and the vapor

---

[1] If the system consists of a mixture, its behavior may be more complex than that of a single substance. This fact is discussed by J. G. Roof, *J. Chem. Ed.* **34**, 492 (1957).

Values of some Physical Constants

Acceleration of gravity	g	980.665 cm sec^{-2}
Avogadro's number	N	6.02252 x 10^{23} molecules mole^{-1}
Boltzmann constant	k	1.38054 x 10^{-16} erg molecule^{-1} deg^{-1}
Electron rest mass	m_e	9.1091 x 10^{-28} g
Elementary charge	ϵ	1.60210 x 10^{-19} coulomb
		4.80298 x 10^{-10} esu
Faraday constant	$\mathcal{F}$	9.64870 x 10^4 coulomb equiv^{-1}
Gas constant	R	8.3143 x 10^7 erg mole^{-1} deg^{-1}
		1.9872 cal mole^{-1} deg^{-1}
		82.056 cm^3 atm mole^{-1} deg^{-1}
Planck's constant	h	6.6256 x 10^{-27} erg sec
Proton rest mass	m_p	1.67252 x 10^{-24} g
Radius of first Bohr orbit	a_o	5.29167 x 10^{-9} cm
Vacuum speed of light	c	2.997925 x 10^{10} cm sec^{-1}

Conversion Factors

1 atmosphere = 1.01325 x 10^6 dyne cm^{-2}
1 calorie = 4.18400 x 10^7 erg
1 electron volt = 1.60210 x 10^{-12} erg = 23,060.9 cal mole^{-1}
ln x = 2.30259 log$_{10}$ x

Conversion Factors for Energy

	Erg molecule^{-1}	Joules mole^{-1}	Kcal mole^{-1}	Electron volts	Centimeters^{-1}
erg molecule^{-1}	1	6.024 x 10^{16}	1.4397 x 10^{13}	6.243 x 10^{11}	5.036 x 10^{15}
joule mole^{-1}	1.660 x 10^{-17}	1	2.389 x 10^{-4}	1.036 x 10^{-5}	8.300 x 10^{-2}
kcal mole^{-1}	6.946 x 10^{-14}	4.184 x 10^3	1	4.336 x 10^{-2}	3.498 x 10^2
eV	1.602 x 10^{-12}	9.649 x 10^4	23.060	1	8.067 x 10^3
cm^{-1}	1.986 x 10^{-16}	1.196 x 10	2.859 x 10^{-3}	1.240 x 10^{-4}	1

density are raised. At the critical temperature the two densities become identical.

Cailletet and Mathias found empirically that the average of the orthobaric liquid and vapor densities is a linear function of the temperature (Fig. 1-8):

$$\frac{\rho_l + \rho_v}{2} = a + bt$$

When the orthobaric densities are known for at least two temperatures (which may be well below the critical temperature) the constants in this equation can

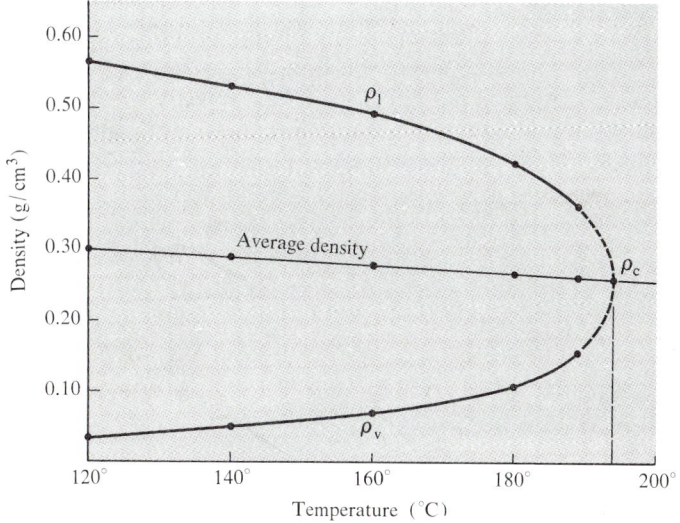

Figure 1-8. Law of Cailletet and Mathias illustrated by use of the orthobaric densities of diethyl ether. $\rho_c = 0.263$ g/cm³, $t_c = 193.8°C$.

be evaluated. Then by substituting the critical temperature in the equation above the critical density can be computed. The molecular weight divided by the critical density yields the critical volume.

The orthobaric densities at a particular temperature can be determined by introducing a known weight w of sample into a graduated tube, equilibrating the system at that temperature, and obtaining the liquid and vapor volumes from the position of the meniscus. When this has been done for two different amounts of sample, a pair of simultaneous equations can be set up and solved for the desired densities:

$$w_1 = v_{l_1}\rho_l + v_{v_1}\rho_v$$
$$w_2 = v_{l_2}\rho_l + v_{v_2}\rho_v$$

Continuity of Gaseous and Liquid States

Consider a system consisting of a fixed amount (1 mole, for example) of a real gas in a cylinder closed off by a piston with temperature T_4 and volume V. By applying an external pressure on the piston, keeping the temperature constant, the system will follow the isotherm T_4 in Figure 1–7, as we have seen before. When the volume reaches V_1 condensation will start and when it reaches V_2 the last bubble of vapor will have disappeared. At point C the system is entirely liquid.

It is possible to start at point A and arrive at point C without going through the discontinuities represented by the limits of the coexistence region. Starting at point A, we can raise the temperature and pressure at constant volume to reach point B. When the temperature is lowered at constant pressure the system reaches point C. At no time did a meniscus appear. We can say that the transition to the liquid phase occurred when the temperature dropped below its critical value. However, the properties of the system just above and below the critical temperature are nearly the same and no discontinuity can be observed at the critical temperature. Clearly, the distinction between a gas at high pressure and a liquid cannot easily be made if only one phase is present. In this case the term fluid is useful to include both.

Critical Constants and the van der Waals Equation

We have just seen that the gaseous and liquid states are closely related. This fact suggests that it might be possible to find an equation of state which would be applicable to all fluids. To a limited extent the van der Waals equation has this capability. After expansion of the parentheses, it is apparent that this equation is a cubic with the molar volume as the variable:

$$V_m^3 - \left(b + \frac{RT}{P}\right)V_m^2 + \frac{a}{P}V_m - \frac{ab}{P} = 0 \qquad (1\text{–}8)$$

A series of isotherms which are graphs of this equation is shown in Figure 1–9. For values of $T < T_c$ there are three real roots; for $T > T_c$ there are one real and two imaginary roots. The similarity between Figure 1–9 and Figure 1–7 is striking. The main difference is the fact that the experimental isotherm has a straight segment representing constant pressure in the coexistence range while the van der Waals isotherm has a maximum and a minimum in this range.

There is some evidence for the reality of the portions of the van der Waals isotherm designated by A–B and C–D. Under special conditions, mainly the absence of dust particles, the formation of a new phase can be delayed. When this delay occurs we have a condition of metastable equilibrium, analogous to supersaturation of a solution. As soon as a small amount of a new phase is formed a rapid transition to a stable equilibrium occurs. The central portion of the isotherm (B–C) has no physical significance because it implies that the volume of a system would increase under increasing pressure.

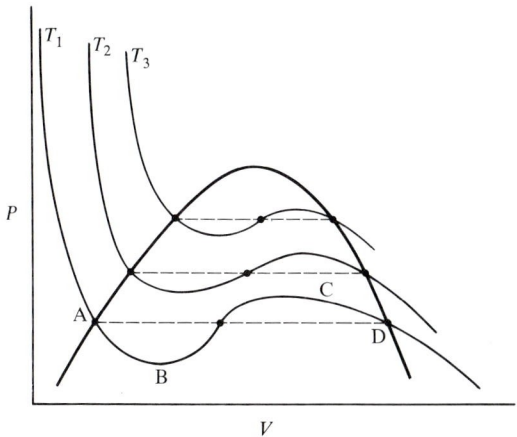

Figure 1–9. Isotherms of the van der Waals equation.

The three real roots of the van der Waals equation approach each other as the temperature is raised, and at the critical temperature they become coincident at the critical volume. In this case $V_m - V_c = 0$, and consequently

$$(V_m - V_c)^3 = 0 = V_m^3 - 3V_m^2 V_c + 3V_m V_c^2 - V_c^3$$

By comparing the coefficients of the various powers of V_m in this equation with those in the general equation 1–8, we obtain the following relations:

$$3V_c = b + \frac{RT_c}{P_c}; \quad 3V_c^2 = \frac{a}{P_c}; \quad V_c^3 = \frac{ab}{P_c} \qquad (1\text{--}9)$$

These three equations can be solved simultaneously to give the critical constants in terms of the van der Waals constants:

$$V_c = 3b; \quad P_c = \frac{a}{27b^2}; \quad T_c = \frac{8a}{27Rb}$$

Conversely, we can solve these equations to express the van der Waals constants in terms of the critical constants:

$$b = \frac{V_c}{3}; \quad a = 3P_c V_c^2$$

We can also obtain

$$\frac{P_c V_c}{RT_c} = \frac{3}{8} = 0.375 \tag{1-10}$$

and this value should be independent of the nature of the substance. Values of this expression calculated from experimental critical data are listed in Table 1–2. It can be seen that they are approximately constant and are not widely different from the value calculated above. The differences, however, are far beyond experimental error and this is a consequence of the fact, mentioned above, that the van der Waals equation is not a reliable equation of state at high pressures and near the critical state.

Corresponding States

If we substitute in the van der Waals equation the values of a and b in terms of the critical constants we obtain:

$$\left[P + 3P_c\left(\frac{V_c}{V_m}\right)^2\right]\left(V_m - \frac{V_c}{3}\right) = RT$$

From Equation (1–10), $P_c V_c = 3/8\ RT_c$.
On dividing the first of these equations by the second, the result is

$$\left[\frac{P}{P_c} + 3\left(\frac{V_c}{V_m}\right)^2\right]\left(\frac{V_m}{V_c} - \frac{1}{3}\right) = \frac{8}{3}\frac{T}{T_c}$$

We define the *reduced pressure* $\pi = P/P_c$, the *reduced volume* $\phi = V_m/V_c$, and the *reduced temperature* $\theta = T/T_c$. In terms of these new variables we obtain the van der Waals *reduced equation of state:*

$$(\pi + 3/\phi^2)(3\phi - 1) = 8\theta$$

This equation does not contain any parameters characteristic of individual gases, although values of the critical constants are implied. According to this equation, if any two gases are in states in which two of the reduced variables are the same, the third variable should also be the same. Gases in such states are said to be in *corresponding states*.

Deviations from ideality should depend only on the values of the reduced variables and it should be possible to express the compressibility factor Z in terms of these variables. To do this we can make use of the fact, which is independent of any equation of state, that $P_c V_c / RT_c$ is approximately constant. Thus we have

$$Z = \frac{P_c V_c}{RT_c}\frac{\pi \phi}{\theta} = c\frac{\pi \phi}{\theta} \tag{1-11}$$

Definitions: Properties of Gases

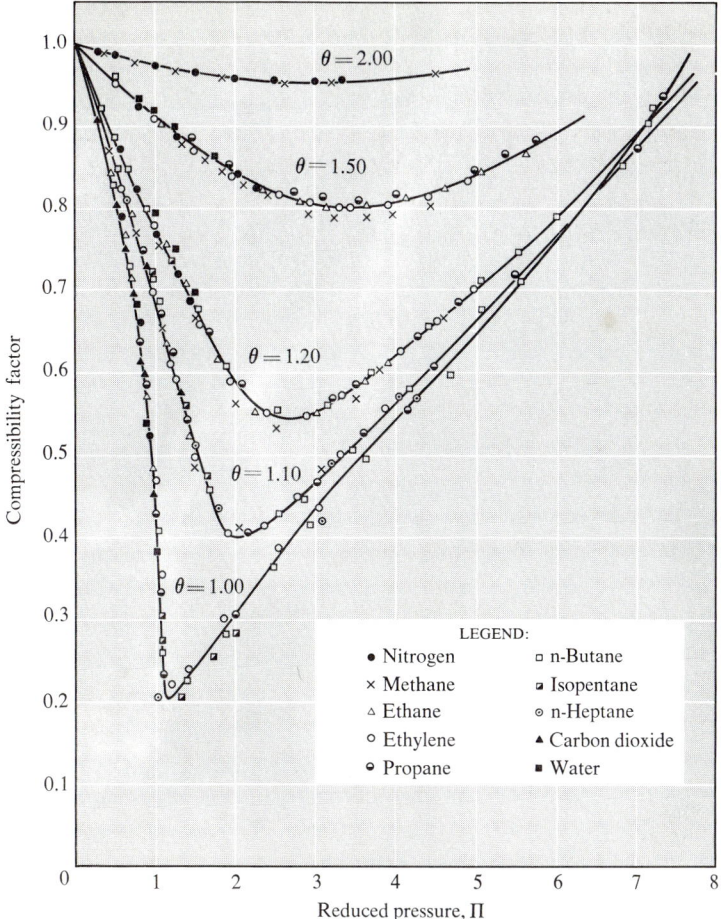

Figure 1-10. Compressibility factor as function of reduced state variables. Gouq-Jen Su, *Ind. Eng. Chem.*, **38,** 803 (1946).

In Figure 1-10 Z is plotted as a function of reduced pressure for a variety of substances at a number of reduced temperatures. It is apparent that the experimental points fall fairly close to the lines representing Equation 1-11. The values of Z obtained from such a graph are not accurate to within experimental error, but they are good enough to be useful in many cases.

It should be pointed out that although attention here has been focussed on the van der Waals equation of state, any equation of state with two parameters can be expressed in reduced form.

Supplementary References

R. W. Hakala, "Dimensional Analysis and the Law of Corresponding States." *J. Chem. Ed.* **41,** 380 (1964).

R. G. Neville, "The Discovery of Boyle's Law." *J. Chem. Ed.* **39,** 356 (1962).

Problems

1. Calculate the pressure exerted by 10.0 g of oxygen in a 5-liter vessel at 100°C. **Ans.** 1.91 atm

2. A gas occupies a volume of 300 ml at a pressure of 300 torr and a temperature of 25.0°C. What will its volume be when the pressure is 1.45 atm and the temperature is 50.0°C? **Ans.** 88.5 ml

3. 1.234 g of an ideal gas in a bulb with a volume of 550 ml exerts a pressure of 2.00 atm at a temperature of 20.0°C. How much of this gas does the bulb hold when the pressure is 0.80 atm and the temperature is 5.0°C?
Ans. 0.520 g

4. The density of a mixture of CH_4 and C_2H_6 at 100°C and a pressure of 700 torr is 0.694 g/liter. What is the composition of this mixture (a) by volume; (b) by weight; (c) in mole %? What is the partial pressure of each of the components?

5. A 1-liter bulb containing H_2 at a pressure of 2.50 atm is connected to a 2-liter bulb containing N_2 at a pressure of 1.50 atm, both at the same temperature. Calculate (a) the total pressure of the system; (b) the partial pressure of each gas; (c) the mole fraction of H_2.
Ans. (a) 1.83 atm; (b) P_{H_2} = 0.83 atm; (c) 0.45

6. Calculate the density of air that is saturated with water at 25.0°C at a total pressure of 500 torr. Assume that the composition of dry air is 80 mole % N_2 and 20 mole % O_2. The vapor pressure of water at 25.0°C is 23.8 torr.

7. 0.1160 g of a mixture of H_2 and N_2 collected over water at 50.0°C occupies a volume of 275 ml when the total pressure is 1.00 atm. Calculate the composition of the dry mixture in mole %. The vapor pressure of water at 50.0°C is 92.5 torr. **Ans.** X_{H_2} = 0.587

8. 2.040 g of a hydrocarbon containing 7.80 wt % hydrogen and 92.20 wt % carbon occupies a volume of 800 ml at 100°C and a pressure of 1 atm. What is its molecular formula?

Definitions: Properties of Gases

9. A 500-ml bulb has a weight of 36.542 g when evacuated and a weight of 37.380 g when filled with a gas at a pressure of 750 torr and a temperature of 20.0°C. Calculate the molecular weight of the gas. **Ans. 41**

10. 1.00 g of hydrogen is introduced into a 10-liter flask containing air at a temperature of 50°C and an initial pressure of 1.00 atm. (a) What is the average molecular weight of the hydrogen-air mixture? (b) What is the corresponding density? (c) What is the final pressure? (Assume that air is a mixture of 80 mole % nitrogen and 20 mole % oxygen.)

11. A flask contains 1.25 g of pure CH_4 at 1.00 atm and 20°C. How many grams of C_2H_6 must be added to the flask in order for the density of the mixture to be the same as the density of air at 1.00 atm and 20°C? (Average mol. wt. of air = 29.)

12. When 1.50 g of gaseous substance A is introduced into an initially evacuated flask held at 25°C, the pressure is observed to be 1.00 atm. When 3.00 g of gaseous substance B is added to the 1.50 g of A in the flask the pressure is observed to be 1.50 atm. Assuming that A and B are ideal gases, calculate the ratio of their molecular weights M_A/M_B.

13. A flask A of unknown volume containing O_2 at 5.00 atm was connected to a 2-liter flask B containing helium at 3.00 atm. Chemical analysis showed that the mole fraction of oxygen in the resulting mixture was 0.20. Calculate the volume of flask A. **Ans. 300 ml**

14. A mixture of CH_4 and C_3H_8 occupied a volume of 100 ml at 25°C and 380 torr. Enough O_2 was added to make the pressure 560 torr when the volume was 400 ml. The mixture was then ignited and after the temperature returned to 25°C the volume was 300 ml at 487 torr. Find the composition of the mixture in volume %. (Assume that the combustion was complete. Ignore the volume of liquid water formed and take the vapor pressure of water at 25°C to be 24 torr.)

15. The density of CH_4 at 0°C was measured at a series of pressures, with the following results:

P (atm)	0.2500	0.5000	0.7500	1.0000
ρ (g/liter)	0.17893	0.35808	0.53745	0.71707

Assuming the atomic weight of H to be 1.008, find the atomic weight of C. **Ans. 12.00**

16. Using the van der Waals equation calculate the pressure exerted by 20.0 g of CO_2 in a 1-liter vessel at 25°C and compare with the ideal gas value.

17. The orthobaric densities of CCl_4 liquid and vapor at a series of temperatures have the following values:

t (°C)	150	250	270	280
ρ_l (g/ml)	1.3215	0.9980	0.8666	0.7634
ρ_v (g/ml)	0.0304	0.1754	0.2710	0.3597

Given that the critical temperature of CCl_4 is 283.1°C, find the critical volume of CCl_4. **Ans.** 276 ml/mole

18. Obtain the critical constants in terms of the van der Waals constants by solving Equations 1–9 simultaneously. (Hint: Start by dividing the third equation by the second one.)

19. Calculate values of the critical constants of hydrogen from the van der Waals constants and compare with the values in Table 1–2. What is the reason for the observed differences?

2

The First Law of Thermodynamics

Introduction

Thermodynamics is concerned with the relationships between changes in the states of systems and the associated exchange of work, heat, and other forms of energy with the surroundings. One of the important problems treated by thermodynamic methods is the conversion of heat into work. Our major interest in thermodynamics, however, is the development of criteria for equilibrium, both physical and chemical. On the way to this objective we will obtain many useful results. We start with a consideration of the concepts of work and heat.

Work

Our definition of work is taken directly from the subject of mechanics. In simplest terms, when a constant force f is exerted on an object which is displaced through a distance L in the direction of the force, the work done by the force is equal to fL. When the force is expressed in dynes and the displacement in centimeters the unit of work is the dyne-cm, or erg.

We recognize that work can be done against a variety of opposing forces: gravitational, electrostatic, magnetic, and others. A type of work which is particularly important here is the work done by a system when it expands against an external pressure. This type of work, which is called expansion work or PV work, can be illustrated by a system consisting of a gas in a cylinder closed by a piston (Fig. 2-1). To focus attention on the essential points, we can make the simplifying assumptions that the piston has a negligible mass and that friction between it and the cylinder can be neglected. We assume that the piston, whose area is A, moves a distance L against the force provided by the constant external pressure, P_{ext}. Then the work done by the system on the surroundings in this expansion is

$$w = fL = P_{ext}AL = P_{ext}\Delta V = P_{ext}(V_2 - V_1).$$

ΔV is the change in the volume of the system. In this book the symbol ΔX always represents a finite, measurable change in the property X.

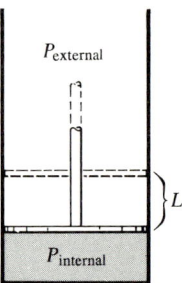

Figure 2-1. Diagram of system consisting of a gas in a cylinder closed by a piston. Work done by the gas in expanding equals the product of external pressure, the area of piston, and distance of travel.

We have seen before that the PV product has the dimensions of work and energy and the present result is consistent with this fact. To avoid any ambiguity we adopt the convention that work done by a system is always taken as positive and work done on a system by the surroundings during a compression is negative.

Heat

The concept of heat was a troublesome topic in the early days of physics and chemistry. Heat was thought to be a substance which could flow from one part of a body to another or from one body to another. Early in the nineteenth century Rumford observed that large amounts of heat were produced when cannons were bored. He could not see how indefinitely large amounts of a substance could be formed by this mechanical operation and he suggested that heat must be a form of energy. The relation between work and heat was clarified and made quantitative by Joule. He dissipated measured amounts of work as friction in turning paddle wheels in a tank of water, and found that the ratio of the amount of heat produced in the water (measured by the temperature rise of the water) to the work done was always the same.

Because work can be quantitatively transformed into heat, heat must be a form of energy. We define heat as that form of energy which is transferred as a result of a temperature difference.

The First Law of Thermodynamics

As soon as heat was recognized to be a form of energy the stage was set for the statement of one of the most basic laws in all science, the law of conservation of energy: Energy can be neither created nor destroyed; it can only be transferred from one form to another.[1] We are familiar with the fact that there are many forms of energy. For the present we include only heat and expansion work in our treatment of thermodynamics.

Mathematical Prelude

Before proceeding with our consideration of thermodynamics it seems desirable at this point to summarize some mathematical results which we need. These results may be found in any book on calculus.

If we have a dependent variable F which is a function of two independent variables, x and y, the total differential of F is

$$dF = \left(\frac{\partial F}{\partial x}\right)_y dx + \left(\frac{\partial F}{\partial y}\right)_x dy$$

where $\left(\frac{\partial F}{\partial x}\right)_y$ means the derivative of F with respect to x, keeping y constant. If F is held constant, $dF = 0$ and

$$\left(\frac{\partial y}{\partial x}\right)_F = -\frac{\left(\frac{\partial F}{\partial x}\right)_y}{\left(\frac{\partial F}{\partial y}\right)_x} \qquad (2\text{-}1)$$

When successive partial differentiations are carried out the order of differentiation is unimportant:

$$\left[\frac{\partial}{\partial x}\left(\frac{\partial F}{\partial y}\right)_x\right]_y = \left[\frac{\partial}{\partial y}\left(\frac{\partial F}{\partial x}\right)_y\right]_x = \frac{\partial^2 F}{\partial x \partial y} \qquad (2\text{-}2)$$

As an example of the application of these abstract equations to a specific function, let us make use of the equation of state. Consider the volume as a dependent variable which is some function of the pressure and temperature. The total differential of V is

$$dV = \left(\frac{\partial V}{\partial P}\right)_T dP + \left(\frac{\partial V}{\partial T}\right)_P dT$$

---

[1] Today we know that the law of conservation of energy is part of a still more general statement, the law of conservation of energy and mass. Einstein concluded on the basis of his theory of relativity that mass and energy are equivalent, and this conclusion has been completely confirmed by many precise measurements. The interconversion of mass and energy, however, is not detectable under the conditions under which ordinary chemical reactions are carried out. Consequently, the separate, more restricted laws of conservation of energy and of matter are adequate for our purpose here.

For the special case of a system composed of a mole of an ideal gas, $V = \dfrac{RT}{P}$ and

$$dV = -\frac{RT}{P^2} dP + \frac{R}{P} dT \tag{2-3}$$

At constant volume $dV = 0$, and it follows that $\left(\dfrac{\partial P}{\partial T}\right)_V = \dfrac{R/P}{RT/P^2}$. Thus,

$$\left(\frac{\partial P}{\partial T}\right)_V = \frac{P}{T}$$

We can find the same result by obtaining $\left(\dfrac{\partial P}{\partial T}\right)_V$ directly from the equation of state.

By obtaining the second partial derivatives we can verify the statement that the order of partial differentiation is unimportant.

$$\left[\frac{\partial}{\partial T}\left(\frac{\partial V}{\partial P}\right)_T\right]_P = \left[\frac{\partial}{\partial T}\left(-\frac{RT}{P^2}\right)\right]_P = -\frac{R}{P^2} = \left[\frac{\partial}{\partial P}\left(\frac{\partial V}{\partial T}\right)_P\right]_T$$

Now let us turn to the problem of the integration of a differential equation involving two independent variables. A general form of such an equation is

$$dG = M(x,y)\, dx + N(x,y)\, dy \tag{2-4}$$

in which M and N are two different functions of x and y. Is there a function G which can be found by integration of this equation? The function G can be found only if dG is an exact differential. The condition for dG to be an exact differential (often called the Euler criterion for exactness) is the following equation:

$$\left(\frac{\partial M}{\partial y}\right)_x = \left(\frac{\partial N}{\partial x}\right)_y \tag{2-5}$$

If this condition is not satisfied, that is, if dG is an inexact differential, there is no function G and Equation 2-4 cannot be integrated.

If dG is an exact differential and it is integrated between limits, the value of the definite integral depends only on the limits of integration. If the limits are reversed the value of the definite integral is the negative of the original value. It follows that the value of the cyclic integral of an exact differential is always zero:

$$\oint_a^b dG = \int_a^b dG + \int_b^a dG = 0$$

If dG is an inexact differential and a particular functional relationship between x and y is known, M and N can be expressed in terms of only one variable and the integration can be carried out. In this case, however, for fixed values of the limits of integration the definite integral depends on the relation-

The First Law of Thermodynamics

ship between x and y. An integral of this type is called a line integral because a definite path of integration is implied by the relationship $y = f(x)$.

◈◈◈◈◈◈◈

To illustrate what is meant by a line integral let us take the very simple differential equation:

$$dG = y\, dx$$

Since in this equation $M = y$ and $N = 0$, the Euler criterion for exactness is clearly not satisfied and dG is an inexact differential. If we do not know a functional relation between y and x, dG cannot be integrated. In a graphical representation (Fig. 2–2) if the point x_1, y_1 represents the initial values of the

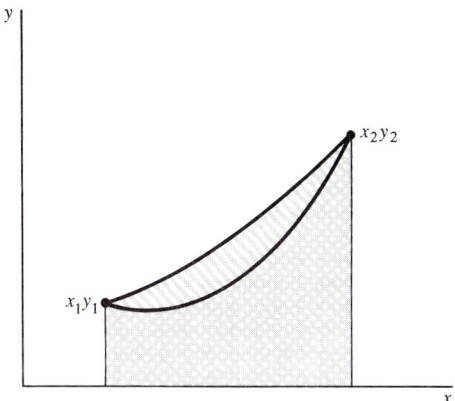

Figure 2–2. Graphical representation of two line integrals. $\int_{x_1 y_1}^{x_2 y_2} y\, dx$.

variables and the point x_2, y_2 represents the final values, the value of the line integral $\int_{G_1}^{G_2} dG$ is represented by the area under a line connecting x_1, y_1 and x_2, y_2; but each different line gives a different value of the area, and thus the integral has no definite value. If, however, a path of integration is specified, for example $y = x^2$, the value of the definite integral can now be found either graphically, as the area under a certain curve, or analytically, since the definite integral is equal to

$$\int_{x_1}^{x_2} x^2\, dx$$

which in turn is equal to

$$\frac{x_2^3 - x_1^3}{3}$$

To emphasize the distinction between exact and inexact differentials, we now consider an exact differential. Let us calculate the volume change when the temperature and pressure of 1 mole of ideal gas are changed from initial values T_1, P_1 to final values T_2, P_2 (Fig. 2-3). For this purpose we use Equation 2-3:

$$dV = \frac{R}{P} dT - \frac{RT}{P^2} dP$$

(It should be noted that here P is the pressure of the gas and *not* the external pressure.)

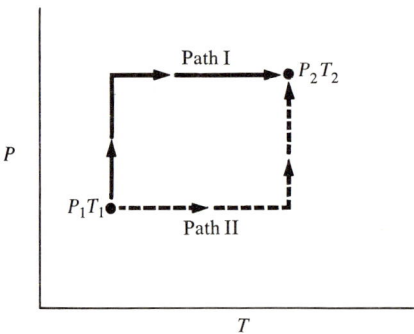

Figure 2-3. Graphical representation of a change of state from $P_1 T_1$ to $P_2 T_2$.

First we obtain the integral of dV along path I. In the isothermal portion of this expansion $T = T_1$ and $dT = 0$; in the constant pressure portion $P = P_2$ and $dP = 0$. We have then for all of path I,

$$dV = \frac{R}{P_2} dT - \frac{RT_1}{P^2} dP$$

Since the first term on the right depends on T only and the second term depends on P only, we can carry out the integration as follows:

$$\Delta V = \int_{V_1}^{V_2} dV = \frac{R}{P_2} \int_{T_1}^{T_2} dT - RT_1 \int_{P_1}^{P_2} \frac{dP}{P^2}$$

$$= \frac{R}{P_2} (T_2 - T_1) + RT_1 \left(\frac{1}{P_2} - \frac{1}{P_1} \right)$$

$$= R \left(\frac{T_2}{P_2} - \frac{T_1}{P_1} \right)$$

Similarly, along path II we have

$$\Delta V = \frac{R}{P_1} \int_{T_1}^{T_2} dT - RT_2 \int_{P_1}^{P_2} \frac{dP}{P^2}$$

$$= R \left(\frac{T_2}{P_2} - \frac{T_1}{P_1} \right)$$

The First Law of Thermodynamics

We have thus found the same volume increase along both paths and we would obtain the same result along any other path. We can obtain the same volume increase directly from the ideal gas law by subtraction; thus

$$\Delta V = V_2 - V_1 = \frac{RT_2}{P_2} - \frac{RT_1}{P_1}$$

Because ΔV depends *only* on the initial and final states and not on the path of integration, it is clear that dV is an exact differential. It can be shown similarly that dP and dT are also exact differentials.

Differentials of Work and Heat

Let us consider again the work done by a gas when it expands against an external pressure. For an expansion in which the volume increase is infinitesimal, the work done by the system is given by

$$dw = P_{\text{ext}}\, dV$$

This equation cannot be integrated because there is no functional relationship between the external pressure and the volume of a gas. Thus we see that dw is an inexact differential. To call attention to the inexactness of certain differentials we use the special symbol $đ$; for example, an infinitesimal amount of work is represented by $đw$.

Since $đw$ is an inexact differential, the amount of work done in a finite expansion from V_1 to V_2 is not a definite quantity but instead depends on how the expansion is carried out. To show this graphically, let us refer to Figure 2–4 in which the coordinates are the external pressure and the volume of the gas. The work done in an expansion from P_1,V_1 to P_2,V_2 is equal to the

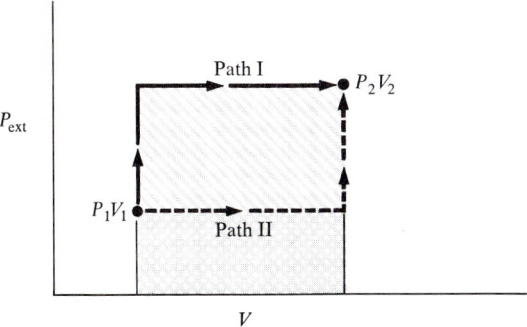

Figure 2–4. Graphical demonstration that the work done in the expansion of a gas from a given initial volume to a given final volume depends on the path.

area under a line, the path of integration, connecting these points. It is clear that the work done by the gas will be different for every different path.

For example, along path I the external pressure is increased from P_1 to P_2 at constant volume, and then the volume is increased at constant pressure. The work done in the first step is zero and the work done in the second step is

$$w_\mathrm{I} = \int_{V_1}^{V_2} P_2\, dV = P_2\, \Delta V$$

Along path II the system is expanded at the constant external pressure P_1 and then the pressure is raised at constant volume to P_2. In this case $w_\mathrm{II} = P_1 \Delta V$ and we see that $w_\mathrm{I} \neq w_\mathrm{II}$.

The heat transferred to or from a system which changes from an initial state to a final state is also not a definite quantity, but depends on the way in which the change occurs. This is exemplified by Joule's paddle wheel experiment. The temperature of a given amount of water can be raised from an initial value T_1 to a final value T_2 by transfer of a certain amount of heat from the surroundings to the water. The same change in the state of the water can be carried out without the transfer of any heat by turning the paddle wheels and so doing work on the system. Alternatively, some heat may be absorbed by the water and some work done to raise the temperature from T_1 to T_2.

We represent a finite amount of heat transferred to a system by $+q$. From the discussion above it can be concluded that an infinitesimal amount of heat is an inexact differential, and therefore we represent it by dq.

First Law of Thermodynamics: Energy

The first law of thermodynamics is the law of conservation of energy. One form of this law is the following statement: The energy of an isolated system is constant; thus $\Delta E = 0$ for any process occurring in the system. If the system under consideration is closed, but not isolated, it can interact with its surroundings in a variety of ways. We assume for the present that this interaction consists only in doing expansion work and transfer of heat. It is not difficult to generalize this treatment by the inclusion of gravitational, electrical, magnetic, and other types of work, but we defer such a generalization to a later time. We also assume that our system as a whole is stationary, and thus it has no kinetic energy.

Let us consider a closed system in a definite initial state characterized by definite values of V and T. Keeping V constant, allow the system to absorb an amount of heat q_V from the surroundings. The first law tells us that this absorbed heat must be equal to the increase in the energy of the system because no work is done:

$$q_V = E_2 - E_1 = \Delta E \qquad (\Delta V = 0) \qquad (2\text{--}6)$$

The First Law of Thermodynamics

If the system expands against an external pressure some of the heat absorbed must go into expansion work:

$$q = \Delta E + w \quad (\Delta V \neq 0)$$

When we carry out any process which returns the system to the initial state, the energy of the system must be the same as it was before the cyclical process was carried out and the net work done by the system must be equal to the net heat absorbed. If this were not so it would be possible to carry out a series of cyclical processes at the end of which the system would be in its initial state but its energy would be indefinitely high (or low). This is contrary to all experience. Therefore

$$\Delta E_{1 \to 2} + \Delta E_{2 \to 1} = 0$$

We can now express the first law of thermodynamics by a number of equivalent statements:

1. $\Delta E = q - w$. For a differential process $dE = đq - đw$. Although $đq$ and $đw$ are both inexact differentials their difference is an exact differential.

2. $\Delta E_{1 \to 2} = \int_{E_1}^{E_2} dE$. The change in the energy of a closed system depends only on the initial and final states of the system and not on the manner in which the change occurs.

3. $\oint dE = \oint (đq - đw) = 0$.

A property of a system whose differential is exact, such as the energy, is said to be a *function of the state*. Such properties, whose values depend only on the state of the system, are particularly important in thermodynamics because it is not necessary to be concerned with the details of a process in order to obtain the resulting changes in the value of the property. We soon meet other functions of the state.

It should be noted that we have not mentioned the absolute value of the energy of a system. There is no way to measure the energy of a system directly. Changes in the energy can be inferred from the interaction of the system with its surroundings. Because only changes in energy can be considered, a zero of energy is not defined in thermodynamics.

Since energy is a function of the state of a system, it is a function of the independent variables which determine the state. For a homogeneous system V and T are useful variables. In this case $E = f(V,T)$, and

$$dE = \left(\frac{\partial E}{\partial V}\right)_T dV + \left(\frac{\partial E}{\partial T}\right)_V dT \quad (2\text{-}7)$$

Classification of Processes

Even with our restriction on the forms of energy treated, the first law of thermodynamics is so general and the variety of possible processes with which it deals is so great that it is desirable to focus our attention on simple processes for which certain variables are held constant. In the laboratory, too, it is common to carry out experiments or measurements in which some variables are held constant. There is no loss of generality in concentrating on simple processes because any actual process, however complicated, is equivalent to a sequence of simple processes which has the same initial and final state as the actual process.

Simple processes may be classified as:
1. *Isothermal* ($\Delta T = 0$).
2. *Isobaric* or *isopiestic* ($\Delta P = 0$).
3. *Isochoric* ($\Delta V = 0$).
4. *Adiabatic* ($q = 0$).

In the last case there is no heat transfer between the system and its surroundings. We assume that our system is completely enclosed by walls which are perfect thermal insulators.

Processes may be carried out either reversibly or irreversibly. A *thermodynamically reversible process* is defined as a process whose direction can be reversed by an infinitesimal change in one of the properties of the system. No actual process is reversible, but reversibility is a limit which actual processes can be made to approach more or less closely by a choice of experimental conditions.

An example of a reversible process is the expansion of a gas in a cylinder closed by a piston when the external pressure is very close to the gas pressure. Ignoring the weight of the piston and friction between the piston and the cylinder, we can say that if the gas pressure is infinitesimally larger than the external pressure the system will expand. If the gas pressure is lower by an infinitesimal amount the system will contract.

During a reversible expansion, as the pressure of the gas decreases the external pressure must decrease at the same rate in order to maintain the infinitesimal difference and continue the expansion. Of course, an infinite time interval would be required for a finite expansion or compression with an infinitesimal driving force. For this reason reversible processes are sometimes called quasistatic processes. The time required to carry out a process, however, is of no concern here because time is not a thermodynamic variable.

The basic significance of the reversible process lies in the fact that the system at all stages of the process can be considered to be in equilibrium. In contrast, if the external pressure on the piston were suddenly cut in half there would be a lowering of the pressure of the gas near the piston relative to the remainder of the gas, turbulence would occur, and the temperature would not be constant throughout the gas. The system would not be in a well-defined state and thermodynamics would not be applicable.

The work that is done by a system during a reversible isothermal expansion is the *maximum work* which can be done for a given isothermal change in volume

The First Law of Thermodynamics

because the work is done against the largest opposing force which will permit an expansion to occur. Conversely, the work done on a system in a reversible isothermal compression is the minimum work.

Another way of classifying processes is to describe them as chemical or physical. If the composition of the system is changed by the process, it is a chemical process; otherwise it is physical.

Enthalpy

We have seen that when a process takes place in a closed system of constant volume no work is done and the heat transferred to or from the surroundings, q_V, is equal to the change in the energy of the system (Eq. 2–6). This amount of heat, therefore, depends only on the initial and final states.

Many chemical processes are studied under conditions in which the pressure of the system, rather than its volume, is held constant. What can be said about the heat transfer under these conditions? To answer this question, let us start with an infinitesimal process. The first law tells us that

$$đq = dE + đw$$

Assuming that all the work done is expansion work we have

$$đq = dE + P_{ext}\, dV$$

Now we assume that the expansion is reversible. In this case the difference between the external pressure and the pressure of the system is infinitesimal and we can replace P_{ext} by P, the pressure of the system:

$$đq = dE + P\, dV$$

If we further assume that the pressure of the system remains constant during the process, we can integrate between the limits E_1 and V_1 for the initial state and E_2 and V_2 for the final state:

$$q_P = \int_{E_1}^{E_2} dE + P \int_{V_1}^{V_2} dV = E_2 - E_1 + PV_2 - PV_1$$
$$= (E_2 + PV_2) - (E_1 + PV_1)$$

This combination of E, P, and V occurs so frequently in discussing constant pressure processes that it is convenient to give the combination a name, the *enthalpy*, and a special symbol, H. The definition of enthalpy is

$$H = E + PV \tag{2-8}$$

It follows that

$$q_P = H_2 - H_1 = \Delta H = \Delta E + \Delta(PV) = \Delta E + P\, \Delta V \quad \text{(constant } P\text{)} \tag{2-9}$$

Because the enthalpy is composed of functions of the state it is also a function of the state. Therefore, although we assumed a reversible process in arriving at the concept of enthalpy, the enthalpy change depends only on the initial and final states and does not depend on the reversibility of the process or on the constancy of the pressure. Only for an isobaric process, however, does the heat transferred equal the enthalpy change.

In dealing with changes of enthalpy, P and T are useful independent variables. Accordingly, we have:

$$dH = \left(\frac{\partial H}{\partial P}\right)_T dP + \left(\frac{\partial H}{\partial T}\right)_P dT \tag{2-10}$$

From the definition of enthalpy (Eq. 2–8) we can also write the general differential as:

$$dH = dE + P\,dV + V\,dP \tag{2-11}$$

Heat Capacity

By adding known amounts of heat to a homogeneous system and measuring the resulting temperature rise, we can calculate the average heat capacity C of the system, defined as the ratio of q to ΔT. A more precise definition of heat capacity is:

$$C = \lim_{\Delta T \to 0} \frac{q}{\Delta T} = \frac{đq}{dT}$$

We have seen that q in general depends on the details of the process and that $đq$ is an inexact differential. Consequently we must specify our process more definitely. We do this by focusing on two special cases: constant V and constant P.

At constant volume $q_V = \Delta E$ and $đq_V = dE$. The *heat capacity at constant volume* is defined by

$$C_V = \frac{đq_V}{dT} = \left(\frac{\partial E}{\partial T}\right)_V$$

For an isobaric process $q_P = \Delta H$ and $đq_P = dH$. The *heat capacity at constant pressure* is defined by

$$C_P = \frac{đq_P}{dT} = \left(\frac{\partial H}{\partial T}\right)_P = \left(\frac{\partial E}{\partial T}\right)_P + P\left(\frac{\partial V}{\partial T}\right)_P$$

We can now obtain a general expression for the difference between C_P and C_V:

$$C_P - C_V = \left(\frac{\partial E}{\partial T}\right)_P + P\left(\frac{\partial V}{\partial T}\right)_P - \left(\frac{\partial E}{\partial T}\right)_V \tag{2-12}$$

The First Law of Thermodynamics

We have seen that

$$dE = \left(\frac{\partial E}{\partial T}\right)_V dT + \left(\frac{\partial E}{\partial V}\right)_T dV$$

Performing the mathematically inelegant (but correct) operation of dividing both sides of this equation by dT and imposing the condition of constant P, we have

$$\left(\frac{\partial E}{\partial T}\right)_P = \left(\frac{\partial E}{\partial T}\right)_V + \left(\frac{\partial E}{\partial V}\right)_T \left(\frac{\partial V}{\partial T}\right)_P \qquad - \textcircled{2}$$

On substitution of this expression in Equation 2–12 there results

$$C_P - C_V = \left(\frac{\partial E}{\partial T}\right)_V + \left(\frac{\partial E}{\partial V}\right)_T \left(\frac{\partial V}{\partial T}\right)_P + P\left(\frac{\partial V}{\partial T}\right)_P - \left(\frac{\partial E}{\partial T}\right)_V$$

$$= \left[\left(\frac{\partial E}{\partial V}\right)_T + P\right]\left(\frac{\partial V}{\partial T}\right)_P \qquad (2\text{–}13)$$

The partial derivative $\left(\dfrac{\partial E}{\partial V}\right)_T$ has the dimensions of pressure. In fact, it is often called the internal pressure. For condensed phases it has a large value but the temperature coefficient of expansion is so small in this case that the difference between C_P and C_V is usually neglected. We evaluate the internal pressure of ideal gases in the next section.

The heat capacity of a substance is an important property and we have many occasions to refer to it. Because energy and enthalpy are extensive properties, the heat capacity is also an extensive property. The heat capacity per gram of substance is called the specific heat. For our purpose the heat capacity per mole is much more useful and we use the symbols C_P and C_V to represent molar heat capacities with units of calories per mole degree.

Thermodynamics cannot supply the value of any heat capacity (or any other property). We see later (Chapter 9) that it is possible to calculate the heat capacities of a number of substances on the basis of molecular theory. The calculated values of C_P for monatomic and diatomic ideal gases are $\frac{5}{2}R$ and $\frac{7}{2}R$, respectively. Most heat capacities, however, are obtained from calorimetric measurements. Because heat capacities, particularly those of polyatomic gases, vary with the temperature, it is customary to represent the measured values for these gases by empirical equations whose coefficients are adjusted to fit the data. One common form of such an empirical equation is a power series in terms of the temperature:

$$C_P = a + bT + cT^2 + \cdots \qquad (2\text{–}14)$$

The number of terms used depends on the precision of the data. Another form used by some investigators is:

$$C_P = a' + b'T + \frac{c'}{T^2} + \cdots$$

The values of the coefficients for a number of typical gases are given in Table 2-1. Values for C_P only are quoted because, as is shown later, the values of C_V can be obtained from those for C_P. These empirical equations are de-

TABLE 2-1 COEFFICIENTS IN EMPIRICAL EQUATION
$Cp = a + bT + cT^2$ cal/mole degree
FOR MOLAR HEAT CAPACITY OF GASES AT CONSTANT PRESSURE*

(Valid in the range 300-1500°K)

Gas	a	$b \times 10^3$	$c \times 10^7$
H_2	6.947	−0.200	4.81
N_2	6.524	1.250	−0.01
O_2	6.148	3.102	−9.23
CH_4	3.381	18.044	−43.00
C_2H_6	2.195	38.282	−110.0
C_2H_4	2.706	29.160	−90.59
NH_3	6.189	7.887	−7.28
H_2O	7.256	2.298	2.83
CO_2	6.214	10.396	−35.45
HCl	6.732	0.433	3.70

*H. M. Spencer and J. L. Justice, *J. Am. Chem. Soc.* **56**, 2311 (1934). H. M. Spencer and G. N. Flannagan, *J. Am. Chem. Soc.* **64**, 2511 (1942). H. M. Spencer, *J. Am. Chem. Soc.* **67**, 1859 (1945).

veloped to represent the data obtained over the temperature range of the measurements. It is important to note that such an equation should never be used outside the indicated temperature range of validity.

Energy and Enthalpy Changes of an Ideal Gas

Because the ideal gas plays an important role in our work we now consider the simplifications which result from ideal behavior. We wish to know how the energy of an ideal gas changes during an isothermal expansion. To be more specific, we can rewrite Equation 2-7, which is completely general, in the following form:

$$dE = \left(\frac{\partial E}{\partial V}\right)_T dV + C_V\, dT \qquad (2\text{-}15)$$

Our question then is, what is the value of $\left(\dfrac{\partial E}{\partial V}\right)_T$? The clue for this evaluation was provided by another famous experiment carried out by Joule. A schematic diagram of his apparatus is shown in Figure 2-5. He joined two flasks together through a stopcock. A gas at elevated pressure was introduced into one of the flasks and the other one was evacuated. The combination was submerged in water and after thermal equilibrium was reached at a measured temperature the stopcock was opened. Joule could not detect any temperature change in the water. This implied that no heat was transferred to or from the gas, and thus $q = 0$. Since no work was done on the water, $w = 0$, and therefore $\Delta E = 0$. Joule concluded that the energy of the gas was unchanged by the expansion and thus $\left(\dfrac{\partial E}{\partial V}\right)_T = 0$.

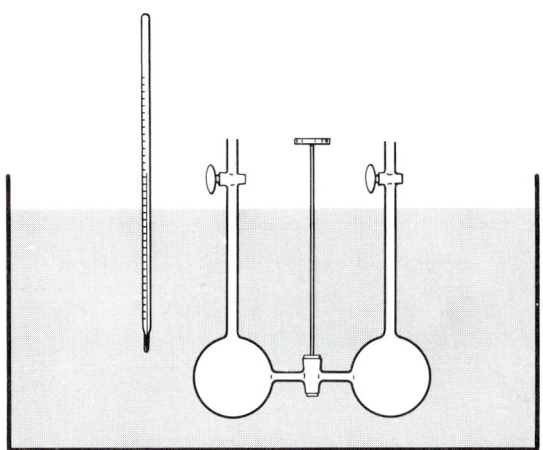

Figure 2–5. Schematic representation of Joule's apparatus for his expansion experiment.

We know now that Joule's apparatus was too crude to permit the measurement of small thermal effects. When the measurements are repeated with modern techniques, a heat transfer is observed ($q \neq 0$). The observed values, however, are quite small at conditions for which deviations from the ideal gas law are small, and as PV_m/RT approaches 1, $\left(\dfrac{\partial E}{\partial V}\right)_T$ approaches 0. We thus have obtained an additional criterion for the ideality of a gas, namely, the energy of an ideal gas is a function of the temperature only. It follows that for any isothermal expansion or compression of an ideal gas, $\Delta E = 0$ and therefore $q = w$.

We can now consider the evaluation of Equation 2–13 for the case of an ideal gas. Since $\left(\dfrac{\partial E}{\partial V}\right)_T = 0$, this equation becomes

$$C_P - C_V = P\left(\dfrac{\partial V}{\partial T}\right)_P = P\dfrac{R}{P}$$
$$= R \qquad (2\text{–}16)$$

This is an adequate approximation for real gases at pressures not much above 1 atmosphere.

Because the first term on the right-hand side of Equation 2–15 is zero, it follows that when the temperature of n moles of ideal gas is changed

$$dE = nC_V \, dT \quad \text{and} \quad \Delta E = n \int_{T_1}^{T_2} C_V \, dT \qquad (2\text{–}17)$$

These expressions are valid whether the volume (or the pressure) changes or not, because the energy change depends only on the temperature change. But only when the volume is constant does ΔE equal q.

The enthalpy of an ideal gas also is a function of the temperature only because the energy and the PV product of an ideal gas are functions of the temperature only. It follows that in this case

$$\left(\frac{\partial H}{\partial P}\right)_T = \left(\frac{\partial H}{\partial V}\right)_T = 0 \qquad (2\text{–}18)$$

and $\Delta H = 0$ for any isothermal expansion or compression of an ideal gas. It also follows that when the temperature of n moles of ideal gas is changed

$$dH = nC_P \, dT \quad \text{and} \quad \Delta H = n \int_{T_1}^{T_2} C_P \, dT$$

Only when the pressure is constant does ΔH equal q. When we introduce Equation 2–14 into the equation above we obtain

$$q_P = \Delta H = n \int_{T_1}^{T_2} (a + bT + cT^2) \, dT$$

Applications of the First Law to Simple Physical Processes

We follow the practice of first illustrating the use of our thermodynamic relations with simple physical processes and then going on to consider the chemical processes which are our main interest. It is important to be quite clear about the meaning of every symbol used and the restrictions, if any, that are implied in a particular equation.

Let us start with expansions and compressions of gaseous systems. In general, $dw = P_{\text{ext}} \, dV$. We have already seen that nothing can be said about the general case because this equation cannot be integrated. When, however, certain restrictions are imposed on the process under consideration, this equation can be integrated. The following are the special cases for which this can be done.

1. *Expansions against constant external pressure.* In this case

$$w = P_{ext} \int_{V_1}^{V_2} dV = P_{ext} \Delta V$$

If the gas is ideal we can use its equation of state to express V in terms of n, P, and T, and thus we obtain

$$\boxed{w = P_{ext}\, nR \left(\frac{T_2}{P_2} - \frac{T_1}{P_1}\right)} \qquad (2\text{-}19)$$

in which P_1 and P_2 represent the initial and final pressures of the gas.

2. *Isothermal expansions.* If the temperature of the gas as well as the external pressure is constant, the work done is given by the equation in the preceding paragraph with $T_1 = T_2$. The only other isothermal expansion which can be treated is the reversible isothermal expansion. In this case, as mentioned before, the external pressure is essentially the same as the pressure of the gas at all stages of the expansion and the work done by the gas has a maximum value. Thus we have

$$w_{max} = \int_{V_1}^{V_2} P\, dV$$

We can substitute for P its value as a function of V from the equation of state of the gas and then carry out the indicated integration. If the gas is ideal, $P = nRT/V$ and it follows that

$$w_{max} = nRT \int_{V_1}^{V_2} \frac{dV}{V}$$

$$= nRT \ln \frac{V_2}{V_1} = 2.303\, nRT \log \frac{V_2}{V_1} \qquad (2\text{-}20)$$

Since we have seen that $\Delta E = 0$ for this isothermal expansion, and consequently $w_{max} = q_{rev}$, we obtain

$$\boxed{q_{rev} = nRT \ln \frac{V_2}{V_1}}$$

From Boyle's law, at constant temperature $V_2/V_1 = P_1/P_2$, and we can write the equations above with the pressure as the independent variable in the following way:

$$w_{max} = q_{rev} = nRT \ln \frac{P_1}{P_2}$$

3. *Adiabatic expansions.* In this case $q = 0$ and therefore $w = -\Delta E$. The work done by the gas must come from its energy and consequently its temperature will drop. Conversely, in an adiabatic compression the tempera-

ture of the gas must rise. The only irreversible adiabatic expansion which can be treated is one in which P_{ext} is constant. If the gas is ideal, w is given by Equation 2–19 and ΔE is given by Equation 2–17. Assuming that C_V is constant during the expansion, we can write the following equation:

$$P_{ext} R \left(\frac{T_2}{P_2} - \frac{T_1}{P_1} \right) = -C_V(T_2 - T_1)$$

When the values of C_V, the initial temperature and pressure of the gas, its final pressure, and the external pressure are given, this equation can be solved for the unknown final temperature, and then the work done in the expansion can be calculated.

Now we turn to the reversible adiabatic expansion of an ideal gas. Let us choose T and V as independent variables. Since in this case $dw = -dE$ and for an ideal gas $dw = nRT\, dV/V$ and $dE = nC_V\, dT$, it follows that

$$nRT \frac{dV}{V} = -nC_V\, dT$$

We can separate the variables in this equation by dividing both sides by nT; then it becomes

$$R \frac{dV}{V} = -C_V \frac{dT}{T}$$

On integrating between limits, assuming that C_V is constant, we obtain

$$R \ln \frac{V_2}{V_1} = -C_V \ln \frac{T_2}{T_1} \tag{2-21}$$

For an expansion, the left-hand side of this equation is always positive; therefore the final temperature is always less than the initial temperature. It should be noted that the final temperature for a given volume ratio does not depend on the number of moles of gas in the system.

The relationship between the volume ratio and the temperature ratio can be expressed in a number of equivalent ways; for example,

$$\log \frac{T_2}{T_1} = -\frac{R}{C_V} \log \frac{V_2}{V_1} = \frac{R}{C_V} \log \frac{V_1}{V_2}$$

and thus

$$\frac{T_2}{T_1} = \left(\frac{V_1}{V_2} \right)^{R/C_V}$$

Since

$$T_2 V_2^{R/C_V} = T_1 V_1^{R/C_V}$$

it follows that

$$TV^{R/C_V} = \text{constant}$$

The First Law of Thermodynamics

If we now choose T and P as our independent variables, from the two expressions for dH in Equations 2–11 and 2–10, we have in general

$$dH = dE + P\,dV + V\,dP = \left(\frac{\partial H}{\partial T}\right)_P dT + \left(\frac{\partial H}{\partial P}\right)_T dP$$

For the reversible adiabatic expansion of an ideal gas, $dE + P\,dV = 0$ because $dq = 0$ and also $\left(\frac{\partial H}{\partial P}\right)_T = 0$. Therefore

$$V\,dP = \left(\frac{\partial H}{\partial T}\right)_P dT$$

and thus

$$nRT\,\frac{dP}{P} = nC_P\,dT$$

Dividing by nT and integrating between limits, we obtain

$$\log \frac{T_2}{T_1} = \frac{R}{C_P} \log \frac{P_2}{P_1} \qquad (2\text{--}22)$$

For an expansion from V_1, P_1 to V_2, P_2

$$\log \frac{T_2}{T_1} = \frac{R}{C_V} \log \frac{V_1}{V_2} = \frac{R}{C_P} \log \frac{P_2}{P_1}$$

and

$$\log \frac{P_2}{P_1} = \frac{C_P}{C_V} \log \frac{V_1}{V_2}$$

A widely used symbol for the ratio of C_P to C_V is γ. With its use, on taking antilogarithms we have

$$\frac{P_2}{P_1} = \left(\frac{V_1}{V_2}\right)^\gamma$$

Since $P_2 V_2^\gamma = P_1 V_1^\gamma$, it follows that for a reversible adiabatic expansion of an ideal gas

$$PV^\gamma = \text{constant}$$

4. *Joule–Thomson expansion.* The Joule–Thomson expansion is a special type of adiabatic expansion. We have seen that Joule was unable to detect a temperature change when he expanded a gas under conditions in which no work was done. Subsequently, he and William Thomson (later Lord Kelvin) devised a different type of experiment that led to significant results. A schematic diagram of their apparatus is shown in Figure 2–6. A gas at a moderately high pressure P_1 was allowed to stream through a porous plug into a region of lower pressure

P_2. The entire apparatus was thermally insulated, and the gas temperatures on the high and low pressure sides of the plug were measured. Joule and Thomson found that for nearly all gases the temperature dropped on passage through the porous plug but for a few, notably hydrogen and helium, the temperature rose.

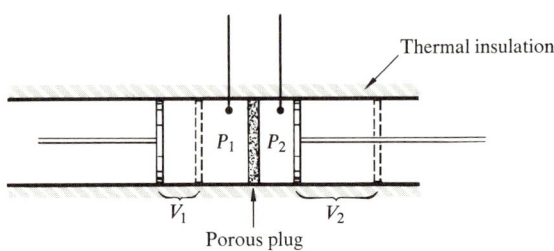

Figure 2–6. Schematic diagram of the Joule-Thomson porous-plug experiment.

This observed change in temperature is evidence of the nonideality of gases. To see why this is so, let us analyze this experiment. Assume for simplicity that a piston maintains the pressure P_1 while a mole of gas passes through the plug, and that the resulting volume change is V_1. On the low pressure side of the plug a piston maintains the pressure P_2 and the resulting volume change is V_2. The work done on the gas by the high-pressure piston is P_1V_1 and the work done by the gas on the low-pressure piston is P_2V_2. Since no heat transfer takes place ($q = 0$), the net work done by the gas during this process is equal to the change in its energy; thus we have

$$w = P_2V_2 - P_1V_1 = E_1 - E_2$$

It follows that

$$E_2 + P_2V_2 - (E_1 + P_1V_1) = \Delta H = 0$$

We see that this process is not only adiabatic but also isoenthalpic.

The rate of change of the temperature with pressure at constant enthalpy is known as the Joule–Thomson coefficient, represented by μ and given by the following equations:

$$\mu = \lim_{\Delta P \to 0} \left(\frac{\Delta T}{\Delta P}\right)_H = \left(\frac{\partial T}{\partial P}\right)_H$$

Making use of Equation 2–1, we obtain

$$\mu = -\frac{\left(\frac{\partial H}{\partial P}\right)_T}{\left(\frac{\partial H}{\partial T}\right)_P} = -\frac{1}{C_P}\left(\frac{\partial H}{\partial P}\right)_T \tag{2–23}$$

The First Law of Thermodynamics

For an ideal gas $\mu = 0$ because $\left(\dfrac{\partial H}{\partial P}\right)_T = 0$ (Eq. 2–18), and therefore no temperature change would be observed for such a gas. For a real gas μ is a function of the temperature. For each gas μ has a negative value above, and a positive value below, a characteristic temperature which is called the inversion temperature. Only for hydrogen and helium is the inversion temperature below room temperature.

The Joule–Thomson expansion is of great technical importance in the liquefaction of gases. For gases whose temperatures are below their inversion temperatures, it provides a method of cooling that does not require heat transfer to a colder substance. By using part of the cooled gas to precool feed gas at a high pressure, the temperature is lowered sufficiently to cause liquefaction. Because the temperature drop depends on deviations from ideality, the process is most efficient when the initial temperature is relatively low.

Another way of cooling a gas, one that does not depend on gas imperfection, is carried out by allowing the gas to do mechanical work in an adiabatic expansion. Modern liquefaction plants use both of these cooling methods, each in the temperature range in which its efficiency is the greater.

5. *Phase changes.* Phase changes are important examples of physical processes. For the present we concentrate on the reversible vaporization of a pure liquid. Our system consists of a liquid in equilibrium with its vapor at a definite temperature, T. Because the vapor pressure depends on the temperature only, the pressure is fixed at a value P. We can picture our 2-phase system in the cylinder-piston combination we have discussed before with $P_{ext} = P$.

When we add reversibly a small amount of heat to this system T and P remain constant and a certain amount of the liquid is transferred to the vapor phase. This amount of heat is equal to the enthalpy change of the system because the heat is added at constant pressure. The enthalpy change per mole, which is called the *molar heat of vaporization*, is designated by ΔH_v.

The work done in the reversible vaporization of a mole of liquid is

$$w = P(V_{vapor} - V_{liquid})$$

Because the volume of a liquid is a small fraction of the corresponding vapor volume, it can be neglected in many cases. If the vapor is assumed to behave ideally we have

$$w \sim PV_{vapor} = RT$$

and

$$\Delta E_v = \Delta H_v - RT$$

When a liquid vaporizes into an evacuated space no work is done and the heat absorbed is ΔE_v.

For a transition of one condensed phase into another, the volume change is so small that the work involved is usually negligible, and $\Delta H \sim \Delta E$.

Thermochemistry

It is a familiar fact that chemical reactions are accompanied by evolution or, less commonly, absorption of heat. Thermochemistry, which is the study of this heat effect, is basically an application of the first law of thermodynamics to chemical processes.

For a chemical process the initial state of the system under consideration consists of the reactants in specified states, and the final state consists of the products. This means that reactions are assumed to go to completion. Because chemical reactions are usually studied at constant pressure, the *heat of reaction*, unless otherwise specified, is understood to be the enthalpy change ($q_P = \Delta H$) resulting from the reaction. For an exothermic reaction ΔH is negative, and for an endothermic reaction ΔH is positive.

In the chemical equations used to describe reactions a chemical formula represents a mole (or formula weight) of the substance involved. All variables which have a significant effect on the reaction must be stated explicitly. This is usually done in the following manner:

$$\text{C (graphite)} + 2Cl_2(g, 1 \text{ atm}) = CCl_4(l); \Delta H_{298°K} = -33.19 \text{ kcal}$$

Because carbon can exist in an amorphous form and in two crystalline forms with different properties, graphite and diamond, it is necessary to state which form is used. That Cl_2 is a gas and CCl_4 is a liquid under the given conditions is made explicit. The temperature subscript on ΔH implies that the process is isothermal. Normally it is not necessary to give the pressures for condensed phases because their properties are not affected significantly by moderate pressure changes.

From the fact that an enthalpy change depends only on the initial and final states of the system, it can be inferred that the heat of a reaction does not depend on any temperature changes which might occur while the reaction is proceeding, provided that the products are brought to the specified temperature. Also, it makes no difference whether the reaction takes place in one step or in a series of steps. The following reactions illustrate this point:

$$\text{S(rhombic)} + O_2(g, 1 \text{ atm}) = SO_2(g, 1 \text{ atm}); \Delta H_{298} = -70.96 \text{ kcal}$$
$$SO_2(g, 1 \text{ atm}) + \tfrac{1}{2}O_2(g, 1 \text{ atm}) = SO_3(g, 1 \text{ atm}); \Delta H_{298} = -23.49 \text{ kcal}$$
$$\overline{\text{S(rhombic)} + \tfrac{3}{2}O_2(g, 1 \text{ atm}) = SO_3(g, 1 \text{ atm}); \Delta H_{298} = -94.45 \text{ kcal}}$$

That heats of successive reactions can be added in this way to give the heat of the over-all reaction was discovered empirically by Hess in 1840 before the first law of thermodynamics was conceived. *Hess' law of constant heat summation*, as it is called, is now seen to be simply a direct consequence of the first law. We often make use of the fact that chemical equations, when properly interpreted, can be added or subtracted like algebraic equations.

The heats of some reactions are measured in a bomb calorimeter in which the reaction occurs at constant volume. In this case the heat effect, q_V, is equal to ΔE. We can easily obtain ΔH from ΔE by using the definition of enthalpy (T is constant):

$$\Delta H = \Delta E + \Delta(PV) = \Delta E + \Delta(n_g RT) = \Delta E + RT\, \Delta n_g$$

In this expression Δn_g is the difference between the number of moles of gaseous products and the number of moles of gaseous reactants. If $\Delta n_g = 0$, or if the reaction involves only condensed phases (for which the volume change is negligible), $\Delta H = \Delta E$.

A very large number of chemical reactions are known. A tabulation of all of the known heats of reaction would take up much space and would be cumbersome to use. Fortunately, the same amount of information is provided by a table of heats of formation of compounds. The *heat of formation* of a compound is defined as the difference between the enthalpy of a mole of the compound and the enthalpies of the elements contained in the compound.

To be more definite than this, the compound and the elements are taken to be in their most stable states[2] at 1 atm pressure and a reference temperature which is usually 25°C. The enthalpy change is then called the *standard heat of formation* and is symbolized by ΔH_f°. The heats of reaction given above for the formation of SO_2 and SO_3, respectively, from the elements are standard heats of formation. The values for a number of compounds are given in Table 2-2.

TABLE 2-2 STANDARD HEATS (ENTHALPIES) OF FORMATION*
AT 25°C IN KCAL/MOLE

Compound	ΔH_f°
$H_2O(g)$	−57.798
$H_2O(l)$	−68.317
$CO_2(g)$	−94.052
$HCl(g)$	−22.063
$SO_2(g)$	−70.96
$SO_3(g)$	−94.45
$NH_3(g)$	−11.04
$CH_4(g)$	−17.889
$C_2H_6(g)$	−20.236
$C_2H_4(g)$	12.496
$C_2H_2(g)$	54.194
$CH_3OH(l)$	−57.02
$C_6H_6(l)$	11.718

*F. D. Rossini, D. D. Wagman, W. H. Evans, S. Levine and I. Jaffe, *Natl. Bur. Standards Circular* 500, U. S. Government Printing Office, Washington, D. C. 1952. Many other values of thermodynamic quantities are listed in this volume.

---

[2] We have not yet arrived at a criterion of stability but this should cause no difficulty.

The heat of formation is a measure of the enthalpy of a compound relative to the enthalpies of its elements which are given a value of zero by this convention. Consequently the heat of a reaction can be calculated from the tabulated values of standard heats of formation by addition and subtraction, considering the heats of formation of the elements in their standard states as zero:

$$\Delta H° = \Sigma \Delta H_f° \text{ (products)} - \Sigma \Delta H_f° \text{ (reactants)}$$

There are some reactions for the formation of compounds from elements which are simple enough and fast enough that their heats can be determined directly in a calorimeter. Such reactions, however, are the exception rather than the rule and the heats of formation of most compounds are found indirectly. Let us take methanol as an example:

$$C(\text{graphite}) + 2H_2(g, 1\text{ atm}) + \tfrac{1}{2}O_2(g, 1\text{ atm}) = CH_3OH(l)$$
$$\Delta H_f° = ?$$

This is a perfectly good chemical equation. The only difficulty is that no one has ever carried out this reaction with methanol as the only product, and it is safe to predict that no one ever will. How then can the heat of formation of methanol be found?

One reaction which can be readily carried for all organic compounds (and many others) is combustion, the complete reaction with O_2 to give in most cases CO_2 and H_2O. *Heats of combustion* can be measured relatively easily. For methanol we have:

$$CH_3OH(l) + \tfrac{3}{2}O_2(g, 1\text{ atm}) = CO_2(g, 1\text{ atm}) + 2H_2O(l); \Delta H_{298}° = -173.67 \text{ kcal}$$

The heats of formation of CO_2 and H_2O are well known. The equations for the reverse reactions are:

$$CO_2(g, 1\text{ atm}) = C(\text{graphite}) + O_2(g, 1\text{ atm}); \Delta H_{298}° = 94.05 \text{ kcal}$$
$$2H_2O(l) = 2H_2(g, 1\text{ atm}) + O_2(g, 1\text{ atm}); \Delta H_{298}° = 136.64 \text{ kcal}$$

Adding all three equations, we have

$$CH_3OH(l) + \tfrac{3}{2}O_2(g, 1\text{ atm}) + CO_2(g, 1\text{ atm}) + 2H_2O(l) =$$
$$CO_2(g, 1\text{ atm}) + 2H_2O(l) + C(\text{graphite}) + 2H_2(g, 1\text{ atm}) + 2O_2(g, 1\text{ atm})$$

Subtracting from each side those quantities which appear on both sides, we are left with:

$$CH_3OH(l) = C(\text{graphite}) + 2H_2(g, 1\text{ atm}) + \tfrac{1}{2}O_2(g, 1\text{ atm}); \Delta H_{298}° = 57.02 \text{ kcal}$$

This equation represents just the reverse of the formation reaction and therefore the desired heat of formation is the negative of the value we have found.

Another way of arriving at the same result is to write the enthalpy change for the combustion reaction in terms of heats of formation.

$$CH_3OH(l) + \tfrac{3}{2}O_2(g, 1\text{ atm}) = CO_2(g, 1\text{ atm}) + 2H_2O(l)$$
$$\begin{array}{cccc} x & 0 & -94.05 & -136.64 \end{array}$$

$$\Delta H^\circ_{298} = -173.67 = -(94.05 + 136.64) - x$$
$$x = \Delta H^\circ_f = -57.02 \text{ kcal}$$

The heats of combustion of some representative compounds are listed in Table 2–3.

TABLE 2-3 STANDARD HEATS OF COMBUSTION* AT 25°C IN KCAL/MOLE

$CH_4(g)$	212.80
$C_2H_6(g)$	372.82
$C_2H_4(g)$	337.23
$C_3H_6(g)$ (propylene)	491.99
$C_2H_2(g)$	310.62
$C_3H_4(g)$ (methyl acetylene)	463.11
$C_6H_{12}(l)$ (cyclohexane)	944.79
$C_6H_6(l)$	780.98
$C_7H_8(l)$ (toluene)	934.50
$CH_3OH(l)$	173.67

*American Institute of Physics Handbook.

Heats of solution are treated in a way that is quite similar to the treatment of heats of reaction. Here, too, it is necessary to be specific about the initial and final states. For example, the dissolving of one mole of gaseous HCl in four moles of water can be represented by the following equation:

$$HCl(g) + 4H_2O = HCl(H_2O, X = 0.20); \Delta H_{298} = -14.63 \text{ kcal}$$

In this equation, X is the mole fraction of HCl in the solution. The enthalpy change per mole of solute in forming a solution of specified composition from its pure components is called the *integral heat of solution* (Fig. 2–7). When a solute such as HCl is dissolved in so much water that the addition of more water would produce no additional heat effect (infinite dilution), the process is represented by the equation

$$HCl(g) + aq = HCl(aq); = -17.96 \text{ kcal}$$

The *integral heat of dilution* is the enthalpy change when a solution containing 1 mole of solute at a specified concentration is diluted to make a solution at another specified concentration. This quantity is the difference between the integral heats of solution at the two concentrations. For example, the heat of

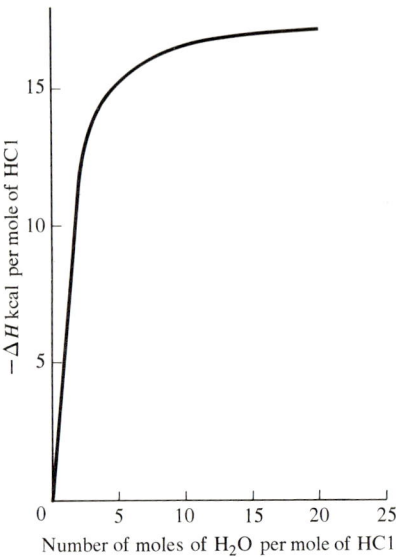

Figure 2–7. Integral heat of solution of HCl at 25°C in water.

dilution of the HCl solution mentioned above by a large amount of water is given by -17.96 kcal $- (-14.63$ kcal), which equals -3.33 kcal. The equation for this process is

$$\text{HCl}(\text{H}_2\text{O}, X = 0.20) + \text{aq} = \text{HCl(aq)}; \Delta H = -3.33 \text{ kcal}$$

In some cases the differential heat of solution is a useful concept. This quantity, $\Delta \bar{H}_1$, is defined as the rate of change of the integral heat of solution with change in the number of moles of solute, keeping the number of moles of solvent constant; thus

$$\Delta \bar{H}_1 = \left(\frac{\partial \Delta H}{\partial n_1} \right)_{n_2}$$

in which n_1 and n_2 are the numbers of moles of solute and solvent, respectively. The differential heat of solution can be considered as the enthalpy change produced when 1 mole of solute is added to so large a volume of solution that the concentration of the solution is essentially unchanged.

Temperature Dependence of the Heat of Reaction

We have seen that heats of reaction are tabulated at a reference temperature which is usually 25°C. This fact, of course, does not mean that all reactions are studied at just this temperature. It suggests that we need a method for calculating the heat of reaction at one temperature from data obtained at another temperature. To see how this is done let us use the following schematic representation of the general reaction $A + B \rightarrow C + D$ at two temperatures and constant pressure, and indicate the associated enthalpy changes:

$$\begin{array}{ccc} & \Delta H_2 & \\ & A + B \rightarrow C + D & T_2 \\ \Delta H_a \downarrow & \uparrow \Delta H_b & \\ & A + B \rightarrow C + D & T_1 \\ & \Delta H_1 & \end{array}$$

In this diagram, ΔH_1 and ΔH_2 are the heats of reaction at temperatures T_1 and T_2, respectively. ΔH_a is the amount of heat transferred in changing the temperature of the reactants from T_2 to T_1. It is equal to $[C_P(A) + C_P(B)](T_1 - T_2)$ in which $C_P(A)$ and $C_P(B)$ are average values of the heat capacities for the temperature range. Similarly, ΔH_b, the amount of heat transferred in changing the temperature of the products from T_1 to T_2, is given by $[C_P(C) + C_P(D)](T_2 - T_1)$. Assuming that ΔH_1 is known (T_1 might be 25°C, for example), we can express the value of ΔH_2 as follows:

$$\Delta H_2 = \Delta H_a + \Delta H_1 + \Delta H_b$$

The sum of ΔH_a and ΔH_b is given by

$$\Delta H_a + \Delta H_b = [C_P(C) + C_P(D) - C_P(A) - C_P(B)](T_2 - T_1)$$

Representing the quantity in brackets by ΔC_P, we have

$$\Delta H_a + \Delta H_b = \Delta C_P \Delta T$$

and it follows that

$$\Delta H_2 = \Delta H_1 + \Delta C_P \Delta T \qquad (2\text{-}24)$$

This equation provides an adequate approximation in most cases for temperature ranges up to about 100°.

For a more accurate treatment we can express the enthalpy change for the reaction in the following way:

$$\Delta H = H_C + H_D - H_A - H_B$$

When both sides of this equation are differentiated with respect to temperature at constant pressure, the result is

$$\left(\frac{\partial \Delta H}{\partial T}\right)_P = \left(\frac{\partial H_C}{\partial T}\right)_P + \left(\frac{\partial H_D}{\partial T}\right)_P - \left(\frac{\partial H_A}{\partial T}\right)_P - \left(\frac{\partial H_B}{\partial T}\right)_P$$
$$= C_P(C) + C_P(D) - C_P(A) - C_P(B)$$
$$= \Delta C_P \qquad (2\text{--}25)$$

This equation is known as the *Kirchhoff equation*.

When we integrate Equation 2–25 between temperature limits, we obtain

$$\Delta H_2 = \Delta H_1 + \int_{T_1}^{T_2} \Delta C_P \, dT \qquad (2\text{--}26)$$

This equation reduces to Equation 2–24 if ΔC_P can be taken as constant over the temperature range. For a large temperature range, the heat capacities of all substances involved must be known as a function of the temperature. When these substances are gases we can add and subtract the empirical heat capacity equations described above (Eq. 2–14) to obtain an expression for ΔC_P of the following form:

$$\Delta C_P = \Delta a + \Delta b T + \Delta c T^2$$

in which

$$\Delta a = a_C + a_D - a_A - a_B$$

Δb and Δc are calculated in an analogous way. Introducing this expression into Equation 2–26, we obtain

$$\Delta H_2 = \Delta H_1 + \int_{T_1}^{T_2} (\Delta a + \Delta b T + \Delta c T^2) \, dT$$

Carrying out the indicated integrations leads to

$$\Delta H_2 = \Delta H_1 + \Delta a \, \Delta T + \frac{\Delta b}{2}(T_2^2 - T_1^2) + \frac{\Delta c}{3}(T_2^3 - T_1^3)$$

Alternatively we can find an expression for the heat of reaction as a function of the temperature by indefinite integration:

$$\Delta H_T = \Delta H_o + \Delta a T + \frac{\Delta b}{2} T^2 + \frac{\Delta c}{3} T^3 \qquad (2\text{--}27)$$

In this case a constant of integration, symbolized by ΔH_o, must be evaluated from a known heat of reaction at some temperature.

In this discussion we have implicitly assumed that the states of aggregation of the reactants and products are unchanged in the temperature range considered. If this is not the case, the heat of phase change must be included in the calculation.

The First Law of Thermodynamics

It should be noted that the Kirchhoff equation is completely general and is not limited to chemical processes. As an example we can calculate the heat of vaporization of water at 150°C. ΔH_v for water at 100°C is 9700 cal/mole. The heat capacities of liquid and vapor can be approximated as 18 and 9 cal/mole, respectively, and $\Delta C_P = -9$ cal/mole. Then

$$\Delta H_{v\ 150°C} = 9700 - (9 \times 50) = 9250 \text{ cal/mole}$$

Supplementary References

I. Klotz, *Chemical Thermodynamics*, W. A. Benjamin, New York, 1964.
F. T. Wall, *Chemical Thermodynamics*, 2nd edition, W. H. Freeman, San Francisco, 1965.
M. W. Zemansky, *Heat and Thermodynamics*, 5th edition, McGraw-Hill, New York, 1968.

G. T. Armstrong, "The Calorimeter and Its Influence on the Development of Chemistry." *J. Chem. Ed.*, **41**, 297 (1954).
D. Kivelson and I. Oppenheim, "Work in Irreversible Expansions." *J. Chem. Ed.*, **43**, 233 (1966).

Problems

1. 20 liters of CO_2 are formed by adding a solution of HCl to solid Na_2CO_3 while the system is under a constant pressure of one atmosphere. Calculate the work done by the gas in liter-atm, cal, and ergs. **Ans. 484 cal**

2. Show that Equation 2–3,

$$dV = \frac{R}{P} dT - \frac{RT}{P^2} dP$$

satisfies the Euler criterion for the exactness of differentials.

3. 10.0 liters of O_2 at 1.00 atm and 25.0°C are heated to 75.0°C. Calculate the amount of heat absorbed, ΔH, and ΔE if this process occurs at (a) constant pressure; (b) constant volume.   **Ans.** (a) $q_p = \Delta H = 143$ cal, $\Delta E = 102$ cal; (b) $q_v = \Delta E = 102$ cal, $\Delta H = 143$ cal

4. A mixture of gases contains 50% He, 30% N_2, and 20% O_2 by volume. How much heat is required to raise the temperature of 200 g of this mixture from 100°C to 150°C at (a) constant pressure; (b) constant volume?

5. The value of C_P for a certain gas is given by the following empirical equation: $C_P = 4.82 + 0.0876\ T$. (a) Calculate the amount of heat absorbed

when the temperature of 1 mole of this gas is raised from 25°C to 200°C at constant pressure. (b) For the same process, calculate the value of ΔE and ΔH.
Ans. (a) $q_p = 6.76$ kcal; (b) $\Delta H = q_p$, $\Delta E = 6.41$ kcal

6. Show that

$$C_P - C_V = \left[V - \left(\frac{\partial H}{\partial P}\right)_T\right]\left(\frac{\partial P}{\partial T}\right)_V$$

7. 5 moles of an ideal gas at 20°C are expanded isothermally from an initial pressure of 4 atm to a final pressure of 1 atm against a constant external pressure of 1 atm. Calculate ΔE, q, w, ΔH. Express your answers in calories.
Ans. $\Delta E = \Delta H = 0$; $q = w = 2200$ cal

8. 5 moles of an ideal gas at 20°C are expanded isothermally and reversibly from an initial pressure of 4 atm to a final pressure of 1 atm. Calculate ΔE, q, w, and ΔH.

9. (a) 7.00 g of N_2 at 25°C are expanded isothermally from an initial pressure of 5.00 atm to a final pressure of 2.00 atm against a constant external pressure of 1.00 atm. Calculate w, q, ΔE, and ΔH in calories. (b) The same amount of N_2 is expanded isothermally between the same initial and final volumes, but this time the expansion is carried out reversibly. Calculate w, q, and ΔH.
Ans. (a) $\Delta H = \Delta E = 0$; $q = w = 44.5$ cal. (b) $q = w = 136$ cal

10. An ideal gas does 1000 cal of work in carrying out a reversible isothermal expansion from an initial volume of 1 liter at 25°C and 10 atm pressure. What is the final volume?

11. Derive an expression for the heat absorbed in an isothermal reversible expansion from V_1 to V_2 of n moles of gas obeying the van der Waals equation of state.

12. 2.0 moles of ideal diatomic gas at 27°C and 5.0 atm pressure are expanded adiabatically to a final pressure of 2.0 atm against a constant external pressure of 1.0 atm. Calculate the final temperature, q, w, ΔE, and ΔH.
Ans. $T_f = 270°K$, $q = 0$, $w = -\Delta E = 300$ cal, $\Delta H = -420$ cal

13. 2.0 moles of ideal diatomic gas at 27°C and 5.0 atm are expanded adiabatically and reversibly to a final pressure of 2.0 atm. Calculate the final temperature, q, w, ΔE, and ΔH. Compare the values obtained with those of the preceding problem.

14. 20 g of N_2 at 27°C are compressed reversibly and adiabatically from 20 liters to 10 liters. Calculate the final temperature, w, ΔE, and ΔH.
Ans. $T_f = 227°K$, $w = -\Delta E = 261$ cal, $\Delta H = -365$ cal

The First Law of Thermodynamics

15. 8 g of O_2 at 27.0°C and a pressure of 10.0 atm are expanded adiabatically and reversibly until the final temperature is −118°C. What is the final pressure?

16. Show that for the reversible adiabatic change in volume of an ideal gas

$$TP^{-\frac{R}{C_P}} = \text{constant}$$

17. Show that the work done by an ideal gas in a reversible adiabatic expansion from P_i and V_i to P_f and V_f is given by

$$w = \frac{P_iV_i - P_fV_f}{\gamma - 1}, \quad \text{where} \quad \gamma = \frac{C_P}{C_V}$$

18. The standard heat of combustion at 25°C for ethanol is 326.7 kcal. Calculate the amount of heat evolved when 10.0 g of ethanol are completely burned in a combustion bomb with a fixed volume, and the products are brought to 25.0°C. **Ans.** −70.9 kcal

19. Calculate the heat of the following reaction at 25°C:

$$C_2H_2(g) + 2H_2(g) \rightarrow C_2H_6(g)$$

20. The heat of combustion of acetaldehyde (CH_3CHO) at 25°C is 285.0 kcal/mole. The standard heat of formation of ethylene oxide ($H_2C\!-\!\!-\!CH_2$, with O bridging) is −12.2 kcal/mole. Calculate ΔH at 25°C for the isomerization reaction:

$$CH_3CHO \rightarrow H_2C\!-\!\!-\!CH_2 \text{ (epoxide)}$$

21. From the heat of combustion at 25°C for benzene calculate its standard heat of formation at this temperature. **Ans.** 11.7 kcal

22. The integral heat of solution of NaOH(s) in n moles of water in kcal/mole is −6.90 for $n = 3$, −9.02 for $n = 5$, and −9.85 for $n = 7$. (a) Calculate the heat evolved when 0.30 mole of NaOH(s) is dissolved in enough water to make $X_{NaOH} = 0.125$. (b) Calculate the heat of dilution when 2 moles of water are added to a solution containing 1 mole of NaOH with $X_{NaOH} = 0.25$.

23. The reaction represented by

$$Na_2SO_4(s) + 10H_2O(l) \rightarrow Na_2SO_4 \cdot 10H_2O(s)$$

is called hydration. Calculate the heat of hydration of $Na_2SO_4(s)$ from the following data:

$$Na_2SO_4(s) + aq \rightarrow Na_2SO_4(aq) \quad \Delta H = -0.56 \text{ kcal}$$
$$Na_2SO_4 \cdot 10\,H_2O + aq \rightarrow Na_2SO_4(aq) \quad \Delta H = 18.85 \text{ kcal}$$

Ans. -19.41 kcal

24. Using the coefficients for C_P in Table 2–1, obtain an expression for ΔH as a function of the temperature for the reaction:

$$N_2(g) + 3H_2(g) \rightarrow 2NH_3(g)$$

25. Using the coefficients for C_P in Table 2–1, calculate ΔH at $1000°K$ for the reaction:

$$C_2H_6(g) + \tfrac{7}{2}O_2(g) \rightarrow 2CO_2(g) + 3H_2O(g)$$

26. Calculate the heat evolved when 9 g of supercooled water in a thermostat at $-15°C$ freeze at this temperature. The heat of fusion of ice at $273°K$ is 1440 cal/mole; the specific heats of ice and water are 0.49 and 1.0 cal deg^{-1}g^{-1}, respectively. **Ans.** -651 cal

3

The Second Law of Thermodynamics

Introduction

We have seen that many important and useful results can be obtained from the first law of thermodynamics, but we have not yet come to the main problem. We want to be able to predict whether or not, under given circumstances, a certain chemical reaction can be carried out. We want to be able to predict the composition of a system when chemical equilibrium has been reached. The fundamental basis for these predictions is now developed.

It is a curious fact of history that this development was started by a French army officer named Carnot who was interested in the factors which limit the efficiency of steam engines. The operation of such an engine results in the conversion of heat into work. It is remarkable that Carnot made his contribution before heat was generally recognized to be a form of energy and before the first law had been proposed. Although we are not concerned with the technical aspects of the efficiency of engines, consideration of this subject leads to criteria for the establishment of equilibria in general, and chemical equilibrium in particular.

The outstanding characteristic of a system at equilibrium is that nothing seems to happen. No spontaneous process can occur in a system at equilibrium and no work can be obtained from it. All spontaneous processes are thermodynamically irreversible. The free expansion of a gas into a vacuum, the melting of a piece of ice immersed in warm water, the dissolving of a crystal of cupric nitrate in water — each of these is an example of a spontaneous and thermodynamically irreversible process. When the concept of a reversible process was introduced in Chapter 2 it was pointed out that the only kind of a process that can occur in a finite time is an irreversible process.

All spontaneous, and therefore irreversible, processes are observed to take place in a direction which leads to equilibrium. In order to reach conclusions about equilibria a quantitative measure of irreversibility is needed. This is provided by the second law of thermodynamics.

It is an observable fact that every form of energy — mechanical, electrical, magnetic, and so forth — can be quantitatively converted into heat. The converse of this statement, namely, that heat can be quantitatively converted into some other form of energy, is observed not to be true. Under some circumstances no amount of heat can be converted. Consider, for example, a steamship which runs out of fuel in the middle of the ocean. Surrounding this ship is a practically infinite source of heat, the thermal energy of the water in the ocean. We know, however, that none of this tremendous reservoir of energy is available for propelling the ship.

This limitation on the conversion of heat into mechanical energy is not imposed by the first law of thermodynamics which merely states that the loss of energy by the water must be equal to the work done in propelling the ship. The study of many examples of this kind led Clausius and Kelvin to the conclusion that another general principle must be involved and this principle is the second law of thermodynamics. No exception to this law has been observed.

Statement of the Second Law of Thermodynamics

The second law can be stated in a number of different ways all of which are equivalent to each other, although this equivalence is not immediately obvious. The following two statements of the second law are due to Clausius and Kelvin, respectively:

1. It is impossible for a cyclic process to transfer heat from a body at a lower temperature to one at higher temperature without the simultaneous conversion of work into heat.

2. It is impossible for a cyclic process to convert heat into work without the simultaneous transfer of heat from a body at a higher temperature to one at a lower temperature.

The first statement implies that when work is used for the continuous transfer of heat from a body at a lower temperature to one at a higher temperature, some work must be wasted by direct conversion into heat. The second statement implies that in the conversion of heat into work some of the heat in a body at a higher temperature must be wasted by direct transfer to a body at a lower temperature.

The second statement also implies that it is impossible to convert heat into work by an isothermal cyclic process. To see what this means let us consider the reversible isothermal expansion of a mole of ideal gas from V_1 to V_2. We have found that the work done by the system in this expansion is $RT \ln V_2/V_1$ (Eq. 2–20). Because $\Delta E = 0$, it follows that $q = w$, and the heat absorbed has been completely converted into work. This, however, is not a cyclic process. In order to complete the cycle, a reversible isothermal compression from V_2 to V_1 must be carried out. At the end of this step the system is back in its original state, no net heat has been absorbed, and significantly, no net work has been done. This is true for all isothermal cyclic processes.

Carnot Cycle

In order to have a system do a net amount of work a more complicated cycle must be used. The simplest of these cycles is the one devised by Carnot and named in his honor. The Carnot cycle consists of four steps which are assumed to be carried out reversibly in order to obtain the maximum amount of work (p. 36). These steps are: (a) an isothermal expansion of the system at a temperature T_2; (b) an adiabatic expansion which results in a lowering of the temperature to a value T_1; (c) an isothermal compression at temperature T_1; and (d) an adiabatic compression which raises the temperature to the initial value, T_2, and leaves the system in the initial state.

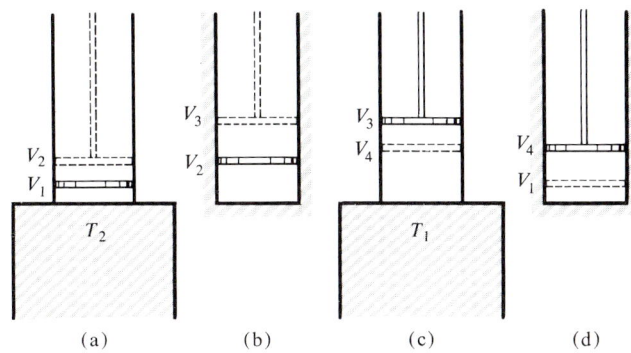

Figure 3–1. Schematic diagram of the Carnot cycle. (a) Isothermal expansion at temperature T_2. (b) Adiabatic expansion. (c) Isothermal compression at temperature T_1. (d) Adiabatic compression to original volume.

In step (a) we assume that the system, represented again by 1 mole of ideal gas in a cylinder closed by a piston (Fig. 3–1a), is in thermal contact with an indefinitely large heat reservoir at a temperature T_2. The transfer of some heat from the reservoir to the system during the isothermal expansion leaves the temperature of the reservoir unchanged. The heat absorbed by the system in this step is completely converted into work. In step (b) the system is disconnected from the reservoir and is thermally insulated; the work done in this adiabatic expansion comes from the energy of the system.

Next we start to bring the system back to its initial state. In step (c) thermal contact is made between the system and a heat reservoir at temperature T_1, and the system is compressed at this temperature to a volume such that after the adiabatic compression in step (d) the system has been returned to its initial state.

The values of P and T as functions of V for the Carnot cycle are graphed in Figure 3-2(a) and (b).

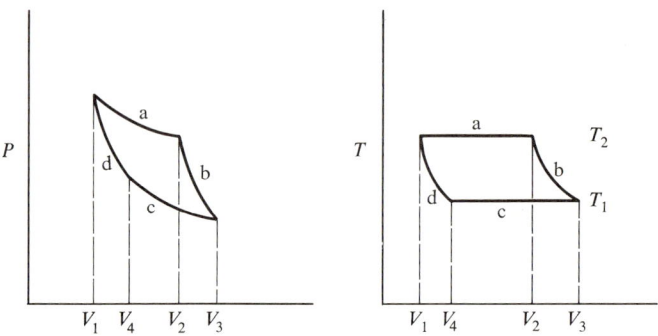

Figure 3-2. Graphs of P vs V and T vs V for Carnot cycle.

Let us now calculate the work done and the heat absorbed in each step when 1 mole of ideal gas is carried through the Carnot cycle.

a. Isothermal expansion from V_1 to V_2

$$q_a = w_a = RT_2 \ln V_2/V_1 \tag{3-1}$$

b. Adiabatic expansion from V_2 to V_3

$$q_b = 0; \quad w_b = -\Delta E = C_V(T_2 - T_1)$$

c. Isothermal compression from V_3 to V_4

$$q_c = w_c = RT_1 \ln V_4/V_3 \tag{3-2}$$

d. Adiabatic compression from V_4 to V_1

$$q_d = 0; \quad w_d = C_V(T_1 - T_2)$$

Efficiency of the Carnot Cycle

The efficiency of a process is defined as the ratio of the net work done by the system to the heat absorbed from the reservoir at the higher temperature. In our case the heat absorbed is q_a. The work done in step (b) is equal and opposite in sign to that done in step (d), and thus the net work done is $w_a + w_c$.

The Second Law of Thermodynamics

It follows that the efficiency can be expressed by

$$\text{Eff} = \frac{w_a + w_c}{q_a} \tag{3-3}$$

$$= \frac{RT_2 \ln V_2/V_1 + RT_1 \ln V_4/V_3}{RT_2 \ln V_2/V_1}$$

To proceed further we need a relation between the ratios V_2/V_1 and V_4/V_3. Because steps (b) and (d) are reversible adiabatic processes between the same two temperatures, we have, using Equation 2–21,

$$\ln \frac{V_3}{V_2} = \frac{C_V}{R} \ln \frac{T_2}{T_1} = \ln \frac{V_4}{V_1}$$

It follows that

$$\frac{V_3}{V_2} = \frac{V_4}{V_1} \quad \text{and} \quad \frac{V_2}{V_1} = \frac{V_3}{V_4}$$

Consequently,

$$\text{Eff} = \frac{T_2 \ln V_2/V_1 - T_1 \ln V_2/V_1}{T_2 \ln V_2/V_1}$$

$$= \frac{T_2 - T_1}{T_2} \tag{3-4}$$

It is clear from this relationship that even in the idealized cycle for which we neglected all frictional losses and used reversible processes to obtain maximum work, only a fraction of the heat absorbed by the system can be converted into work. This fraction depends on the temperature difference between the heat reservoirs. Without a temperature difference the efficiency is zero and no work can be done. When work is done in a cyclic process some heat is transferred from the reservoir at a higher temperature to the one at a lower temperature, in agreement with the second statement of the second law given above.

One might suppose that the expression we have obtained for the efficiency is limited to the special case for which we derived it, namely, an ideal gas for the system and a Carnot cycle for the process. We now use the second law to show that this is not the case and that Equation 3–4 gives the maximum efficiency for any system and any cyclic process carried out between two temperatures.

First, let us note that the Carnot cycle can be operated equally well in the reverse direction. Starting with an isothermal expansion at temperature T_1 (the reverse of step c) we can carry our system around the cycle to the initial state. As a result of this reversed cycle, an amount of heat, q_c, has been absorbed by the system at the lower temperature T_1 and an amount of heat, q_a, has been transferred to the reservoir at the higher temperature. At the same time a net amount of work equal to $w_a + w_c$ has been done on the system. In carrying out this cycle the system has acted as a refrigerator or, what amounts to the same function, a heat pump. By the expenditure of work which is supplied by the surroundings, a certain amount of heat has been "pumped" from a lower temperature to a higher one. This process is consistent with the first statement of the second law given above.

Now let us assume that in addition to our ideal gas system, which we call system I, we have another which can contain any substance which is not an ideal gas, called system II. We imagine that these systems are coupled together in such a way that the work done *by* system II is done *on* system I. In other words, system II acts as an engine and system I acts as a refrigerator. Then the net work done by the coupled systems on the surroundings ($w_I + w_{II}$) is zero. We assume that both systems carry out Carnot cycles between the same two temperatures, T_1 and T_2.

Finally, let us further assume that the cyclic process for system II has a higher efficiency than that we have calculated for system I. Because $Eff_{II} > Eff_I$, we would then have the following inequality

$$\frac{w_{II}}{q_{aII}} > \frac{-w_I}{-q_{aI}}$$

in which both w_I and q_{aI} are negative because system I is carrying out a reverse cycle. Since, by hypothesis, w_{II} is equal in absolute magnitude to w_I, it follows that $q_{aI} > q_{aII}$. Remembering that $\Delta E = 0$ for a complete cycle in each system, we see that

$$w_{II} = q_{aII} - q_{cII}; \quad -w_I = -(q_{aI} - q_{cI})$$

and

$$q_{cII} = q_{aII} - w_{II}; \quad q_{cI} = q_{aI} - w_I$$

Because $w_{II} = w_I$ and $q_{aI} > q_{aII}$, it follows that $q_{cI} > q_{cII}$.

In view of the fact that q_{aI} represents the amount of heat transferred by system I in one complete cycle to the surroundings at temperature T_2, and q_{aII} represents the amount of heat absorbed at this temperature, we conclude that the coupled systems would transfer a net amount of heat $q_{aI} - q_{aII}$ to the surroundings at T_2. Likewise, because q_{cI} and q_{cII} represent the amounts of heat absorbed by system I and evolved by system II, respectively, at temperature T_1, we find that a net amount of heat $q_{cI} - q_{cII}$ would be absorbed from the surroundings by the coupled systems at T_1.

The over-all effect of the cycle for the coupled systems would be to transfer some heat from a lower temperature to a higher one without any work being converted into heat. Such a process is not possible according to the second law. Therefore our assumption that system II has a higher efficiency than system I must be invalid.

The Second Law of Thermodynamics

We can demonstrate in a similar way that the assumption that system II can have a lower efficiency than system I is also in disagreement with the second law. We must therefore conclude that the efficiency we have calculated for the Carnot cycle is independent of the nature of the substance in the system.

By completely analogous reasoning, we can also demonstrate that the assumption that *any* cyclic process carried out reversibly between two temperatures has a higher (or lower) efficiency than the Carnot cycle leads to disagreement with the second law. Consequently the efficiency of *any* reversible cyclic process involving *any* substance is given by Equation 3–4.

Thermodynamic Temperature Scale

The temperature scale we have been using in our treatment of thermodynamics so far is the ideal gas temperature scale which is defined in terms of the ideal gas equation of state in the form $T = \frac{PV}{nR}$. Kelvin recognized that Equation 3–3 provides the basis for the definition of a temperature scale that is independent of the nature of any substance.

Before Kelvin's definition is presented, let us make a slight change in our notation. Since the symbols q_a and q_c in Equations 3–1 and 3–2 represent amounts of heat transferred at temperatures T_2 and T_1, respectively, it is convenient to replace them by q_2 and q_1. Then $w_a + w_c = q_2 + q_1$, and Equation 3–3 becomes

$$\text{Eff} = \frac{q_2 + q_1}{q_2}$$

From Equation 3–4 we have

$$\frac{q_2 + q_1}{q_2} = \frac{T_2 - T_1}{T_2}$$

It follows that

$$-\frac{q_1}{q_2} = \frac{T_1}{T_2} \tag{3-5}$$

(Note that q_1 is actually a negative quantity because it represents an amount of heat lost by the system at temperature T_1.)

Kelvin proposed that a temperature scale be so defined that the ratio of two temperatures should be equal to the ratio of the amounts of heat absorbed and evolved in a Carnot cycle between these two temperatures. Fortunately, the temperature scale defined in this way is identical with the ideal gas temperature we have been using and which we call the Kelvin scale. We now know, however, that this scale is not dependent on the particular properties of an ideal gas.

Treatment of General Cycle

The processes we have discussed so far have been cycles operating reversibly between two fixed temperatures. For such processes it follows from Equation 3-5 that

$$\frac{q_1}{T_1} + \frac{q_2}{T_2} = 0 \tag{3-6}$$

Now we see what happens to this relationship when the limitation to two heat reservoirs is dropped. First, let us note that a Carnot cycle operating between two temperatures, T_2 and T_1, is equivalent to a series of smaller Carnot cycles involving heat exchange with a number of heat reservoirs with temperatures between T_2 and T_1. This equivalence is shown in Figure 3-3. The net

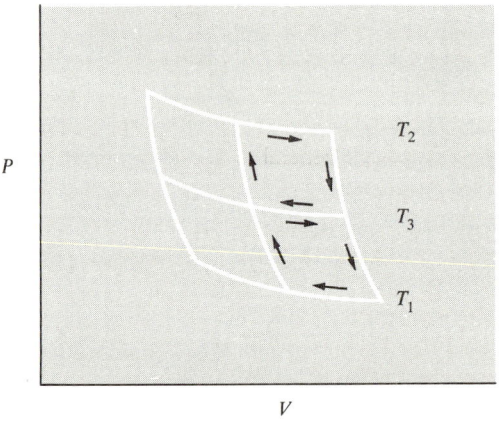

Figure 3-3. Equivalence of large Carnot cycle to sum of small Carnot cycles.

amount of work done in each of the small Carnot cycles is given by the area enclosed by its graph in the PV plane, and the sum of all of the areas of the small cycles is equal to the area of the single large Carnot cycle.

In addition, when all of the small Carnot cycles are completed, each interior line in the graph has been traversed twice, once in each direction. This means that each of the heat reservoirs at the intermediate temperatures has absorbed and given up the same amount of heat. It follows that the net amount of heat absorbed from the reservoir at T_2 and given up to the reservoir at T_1 is the same as for the large Carnot cycle. Thus the general form of Equation 3-6 which is applicable to the heat transfers for all of the heat reservoirs is

The Second Law of Thermodynamics

$$\sum \frac{q_i}{T_i} = 0$$

where q_i is the amount of heat transferred at temperature T_i.

In a similar way, any reversible cycle can be approximated by a sum of Carnot cycles (Fig. 3-4). The smaller the Carnot cycles and the larger the number of intermediate temperatures, the better is the approximation. In the

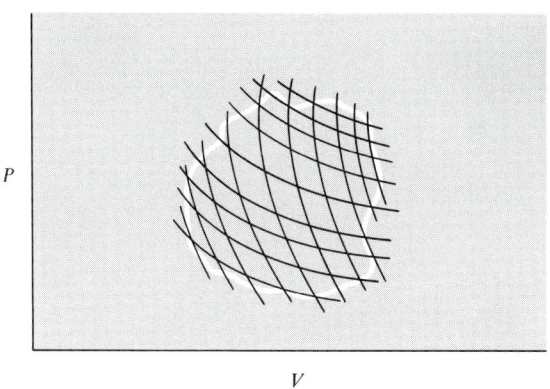

Figure 3-4. Equivalence of a general cycle and a sum of small Carnot cycles.

limit, the amount of heat transferred at each step in each cycle can be considered to become infinitesimal, and the sum written above becomes a cyclic integral; thus,

$$\oint \frac{dq_{rev}}{T} = 0 \qquad (3\text{-}7)$$

in which the symbol dq_{rev} represents an infinitesimal amount of heat transferred reversibly at temperature T.

Finally, we consider what happens when the requirement of reversibility in the cycle is dropped. If there is any irreversibility at any stage of a cycle, the net work done in the cycle is less than the maximum done in the reversible cycle operating between the same two temperatures. Consequently the efficiency of an irreversible cycle is always less than the efficiency of the corresponding reversible cycle. It follows that

$$\frac{w_{net}}{q_2} < \frac{T_2 - T_1}{T_2}$$

and

$$\frac{q_2 + q_1}{q_2} < \frac{T_2 - T_1}{T_2}$$

Then

$$\frac{q_1}{q_2} < -\frac{T_1}{T_2}$$

On adding T_1/T_2 to both sides of this inequality, we obtain

$$\frac{q_1}{q_2} + \frac{T_1}{T_2} < 0$$

and consequently

$$\frac{q_1}{T_1} + \frac{q_2}{T_2} < 0$$

In the limiting case of infinitesimal heat transfer, this inequality becomes

$$\oint \frac{dq_{\text{irr}}}{T} < 0 \tag{3-8}$$

Entropy

Let us return to Equation 3–7 which we have shown to be a general equation applicable to any reversible cycle:

$$\oint \frac{dq_{\text{rev}}}{T} = 0$$

On page 30, we saw that the necessary and sufficient condition for a differential to be exact is the vanishing of its cyclic integral. We also saw that a property whose differential is exact is a function of the state. Clausius in 1850 recognized that Equation 3–7 implies that dq_{rev}/T is an exact differential and that the integral of this differential is a function which represents a property with fundamental significance. He named this property the *entropy* and showed that entropy change provides the quantitative measure of irreversibility needed for the study of equilibrium.

The entropy of a system, symbolized by S, is a property of the system. Like volume, mass, and energy, it is an extensive property. For the present we are not concerned with absolute values of the entropy, but only with entropy changes during processes. For an infinitesimal process

$$dS = \frac{dq_{\text{rev}}}{T}$$

We first consider a process carried out reversibly between two states of a system designated as A and B. Because dq_{rev}/T is an exact differential and $\oint_A^B dS = 0$, it follows that

The Second Law of Thermodynamics

$$\int_A^B dS = -\int_B^A dS \qquad (3\text{-}9)$$

Thus

$$S_B - S_A = \Delta S_{AB} = -\Delta S_{BA}$$

The heat which is absorbed (or lost) by the system during the process $A \to B$ must be transferred reversibly from (or to) the surroundings. Consequently in any process carried out reversibly the entropy gained by the system must be lost by the surroundings (and conversely), and the sum of the entropy changes of the system and the surroundings must be zero. In symbols,

$$\Delta S_{AB} \text{ (system)} = -\Delta S_{AB} \text{ (surroundings)}$$

and

$$\boxed{\Delta S_{\text{total}} = \Delta S_{\text{sys}} + \Delta S_{\text{sur}} = 0}$$

Now let us consider what happens when the process $A \to B$ is carried out irreversibly. No matter what the nature of this process might be, we can assume that the reverse process $B \to A$ is carried out reversibly. Then for the system in a complete cycle from A to B and then back to A, we have

$$\oint dq_{\text{irr}}/T = \int_A^B dq_{\text{irr}}/T + \int_B^A dq_{\text{rev}}/T$$

Making use of Equation 3-9, we can rewrite this equation in the following way:

$$\oint dq_{\text{irr}}/T = \int_A^B dq_{\text{irr}}/T - \int_A^B dq_{\text{rev}}/T$$

Because $\oint dq_{\text{irr}}/T < 0$ (Eq. 3-8), it follows that

$$\int_A^B \frac{dq_{\text{irr}}}{T} < \int_A^B \frac{dq_{\text{rev}}}{T}$$

and therefore

$$\int_A^B \frac{dq_{\text{irr}}}{T} < \Delta S_{AB}$$

If the system under consideration is assumed to be isolated from its surroundings, no heat transfer can occur and

$$\int_A^B dq_{\text{irr}}/T = 0$$

In this case

$$\Delta S_{AB} > 0$$

Because by hypothesis no change has occurred in the surroundings, the inequality above is equivalent to

$$\Delta S_{tot} > 0$$

This important and completely general result implies that whenever any irreversible process takes place in an isolated system the total entropy change is always positive.

Now let us suppose that there is a transfer of heat between a system which we call system I and its surroundings. We can always imagine that system I and its surroundings form a larger system, called system II, which is isolated from *its* surroundings. Then any spontaneous and irreversible process which occurs in system II is accompanied by a positive entropy change. Consequently, no matter what the interaction between system I and its surroundings in this irreversible process might be, the entropy of system II always increases. Since the entropy change of system II is the sum of the entropy changes within this system, it follows that

$$\Delta S_{\text{system II}} = \Delta S_{\text{system I}} + \Delta S_{\text{surroundings I}} > 0$$

Clausius assumed that the entire universe could be considered as an isolated system in which all naturally-occurring processes are irreversible. This is the basis for his often-quoted statement: "The energy of the universe is constant; the entropy of the universe always tends toward a maximum."

The conclusions we have reached in this section can be summarized by the following statements:

1. The entropy of a system is a function of the state.
2. The entropy change when a system undergoes a process depends on the initial and final states only and not on the nature of the process.
3. In a reversible process, entropy is transferred between the system and its surroundings and the total entropy change is zero.
4. In a spontaneous (and therefore irreversible) process, the total entropy change is always positive. In other words, entropy is always produced in an irreversible process.
5. The entropy change for an actual irreversible process between given initial and final states can be found only by calculating the entropy change for a reversible process (or series of processes) between the same initial and final states.

Entropy Changes in Simple Processes

The statements above are somewhat abstract. In order to show their significance we illustrate their application by finding the entropy changes for the following simple systems and processes: (1) Heat transfer between solids at

The Second Law of Thermodynamics

different temperatures. (2) Changes in temperature and volume (or pressure) of an ideal gas. (3) Mixing of two ideal gases.

1. *Heat transfer between solids at different temperatures.* Consider a system which consists of two large masses of metal, one of which is at a higher temperature (T_h) than the other (T_l). Imagine that the two masses are touched together for a moment and then they are separated. We know in this case what the spontaneous process is: a small amount of heat, q, is transferred from the hotter mass to the colder one, while the temperatures are essentially unchanged. What is the entropy change? To answer this question, according to statement (5) above we must devise a reversible path from the initial state to the final state.

We can accomplish our objective by imagining that a certain amount of gas at the temperature T_h is contained in a cylinder in thermal contact with the hotter mass. By undergoing a reversible isothermal expansion, the gas absorbs q calories from this mass. The gas is then thermally insulated and a reversible adiabatic expansion is carried out which lowers the gas temperature from T_h to T_l. Finally the gas is placed in thermal contact with the colder mass, and by a reversible isothermal compression q calories are transferred from the gas to this mass.

Now we can calculate the entropy changes for these processes. In the first step the hotter mass lost q calories reversibly at a constant temperature T_h. Consequently, $\Delta S_h = -q/T_h$. The adiabatic step involved no change in entropy, but in the second isothermal step the colder mass gained q calories at temperature T_l. Thus the entropy change for the system of two masses is

$$\Delta S_{\text{sys}} = \Delta S_h + \Delta S_l = -\frac{q}{T_h} + \frac{q}{T_l}$$

The corresponding entropy change for the gas in the two isothermal steps is

$$\Delta S_{\text{gas}} = \frac{q}{T_h} - \frac{q}{T_l}$$

It follows that the total entropy change for these reversible processes is

$$\Delta S_{\text{tot}} = \Delta S_{\text{sys}} + \Delta S_{\text{gas}} = 0$$

This result is consistent with statement 3.

Let us return to the irreversible transfer of q calories from the hotter mass to the colder one. The initial and final states of the system are the same for this case as for the reversible transfer we have just considered and therefore the value of ΔS_{sys} is the same (statement 2). But now there is no interaction with the surroundings. Consequently,

$$\Delta S_{\text{tot}} = \Delta S_{\text{sys}} = \frac{q}{T_l} - \frac{q}{T_h} > 0$$

This result is consistent with statement 4. If we had not known the direction in which the spontaneous transfer of heat occurred, we could have predicted this direction from the requirement that $\Delta S_{tot} > 0$.

2. *Changes in temperature and volume* (or *pressure*) *of an ideal gas.* The expression of the first law for any infinitesimal reversible process is:

$$dq_{rev} = dE + dw_{max}$$

When the only work done is PV work this expression becomes

$$dq_{rev} = dE + P\,dV$$

If the system consists of 1 mole of ideal gas,

$$dq_{rev} = C_V\,dT + \frac{RT}{V}\,dV$$

When we divide both sides of this equation by T we obtain

$$\frac{dq_{rev}}{T} = dS = C_V \frac{dT}{T} + R \frac{dV}{V} \tag{3-10}$$

which is equivalent to

$$\boxed{dS = C_V\,d\ln T + R\,d\ln V} \tag{3-11}$$

In Chapter 2 it was pointed out that dq_{rev} is an inexact differential. Because each of the terms in Equation 3–10 is a function of one variable only, Euler's criterion for the exactness of differentials (Eq. 2–5) is satisfied for dq_{rev}/T. From a mathematical point of view we can say that $1/T$ is an integrating factor for dq_{rev}. This statement is, in fact, one form of the second law.

For a process in which n moles of ideal gas undergo a change in temperature and volume from T_1, V_1 to T_2, V_2, the entropy change for the gas is found by integrating Equation 3–11 between these limits; thus,

$$\Delta S = n \int_{T_1}^{T_2} C_V\,d\ln T + nR \int_{V_1}^{V_2} d\ln V \tag{3-12}$$

With the assumption that C_V is independent of the temperature, this equation becomes

$$\boxed{\Delta S = n\left(C_V \ln \frac{T_2}{T_1} + R \ln \frac{V_2}{V_1}\right)} \tag{3-13}$$

The Second Law of Thermodynamics

We see from Equation 3-13 that the entropy of the gas increases when its temperature is raised or its volume is increased, and that the entropy change is proportional to the number of moles of gas in the system.

As an example of the use of Equation 3-13 let us calculate the entropy change when 2 moles of a monatomic ideal gas are heated from 27°C to 127°C and the volume is doubled. Since C_V for a monatomic gas has the value $\frac{3}{2}R$, we obtain

$$\Delta S = 2(\tfrac{3}{2}R \ln \tfrac{400}{300} + R \ln 2)$$
$$= 2(0.87 + 1.38)$$
$$= 4.50 \text{ cal/deg}$$

Notice that nothing need be said about the details of this process. Whether it is carried out reversibly or irreversibly, and whether the temperature and volume changes occur at the same time or not, the entropy change for the gas has the same value because entropy is a function of the state.

If temperature and pressure are used as independent variables instead of temperature and volume, we can substitute P_1T_2/P_2T_1 for V_2/V_1 in Equation 3-13 and thus obtain

$$\Delta S = nC_V \ln \frac{T_2}{T_1} + nR \ln \frac{P_1 T_2}{P_2 T_1}$$
$$= n\left[(C_V + R) \ln \frac{T_2}{T_1} + R \ln \frac{P_1}{P_2}\right]$$

Since $C_V + R = C_P$, this equation becomes

$$\Delta S = n\left(C_P \ln \frac{T_2}{T_1} + R \ln \frac{P_1}{P_2}\right) \tag{3-14}$$

From Equations 3-13 and 3-14 it can be seen that the entropy change resulting from a temperature change at constant volume depends on C_V, while the corresponding entropy change at constant pressure depends on C_P.

We now make use of Equations 3-13 and 3-14 to find the total entropy changes for expansions (or compressions) of an ideal gas. Let us first consider the case of an isothermal expansion. The change in the molar entropy of the gas for an isothermal change in volume from V_1 to V_2 is

$$\Delta S_\text{gas} = R \ln \frac{V_2}{V_1}$$

If this expansion is carried out reversibly, the heat absorbed by the gas is given by

$$q_\text{rev} = RT \ln \frac{V_2}{V_1}$$

This amount of heat has been transferred to the gas by the surroundings at temperature T. Therefore the entropy change of the surroundings, ΔS_sur, is equal to $-q_{rev}/T$, which is equal in magnitude and opposite in sign to ΔS_{gas}. Thus we have

$$\Delta S_{tot} = \Delta S_{gas} + \Delta S_{sur} = 0$$

Now let us say that the isothermal expansion from V_1 to V_2 is carried out in the most irreversible way possible; that is, the gas expands into a vacuum. In this case no work is done by the gas and no heat is transferred from the surroundings. Since ΔS_{gas} is the same as for the reversible expansion and $\Delta S_{sur} = 0$, it follows that

$$\Delta S_{tot} = R \ln \frac{V_2}{V_1}$$

Thus we see that in this case ΔS_{tot} is necessarily positive and it has the largest magnitude possible for the given volume limits.

If the same expansion is carried out with an intermediate degree of irreversibility, the work done by the gas is less than in the reversible expansion, and both the amount of heat lost by the surroundings, $-q_{sur}$, and the entropy change ($\Delta S_{sur} = -q_{sur}/T$) have correspondingly smaller magnitudes. Therefore we conclude that

$$\Delta S_{tot} = R \ln \frac{V_2}{V_1} - \frac{q_{sur}}{T} > 0$$

Thus ΔS_{tot} for this irreversible expansion is a positive quantity whose magnitude is between 0 and $R \ln V_2/V_1$.

Let us next turn to the adiabatic expansion of an ideal gas from V_1 to V_2. If this expansion is carried out reversibly, $q_{rev} = 0$ and consequently $\Delta S_{gas} = 0$. Since there is no heat transfer between the gas and the surroundings in any adiabatic expansion, the entropy of the surroundings is unchanged. It follows that

$$\Delta S_{tot} = \Delta S_{gas} = 0$$

Because $\Delta S = 0$ for a reversible adiabatic expansion, from Equation 3–13 we have

$$R \ln \frac{V_2}{V_1} = - C_V \ln \frac{T_2}{T_1}$$

and from Equation 3–14 we have

$$R \ln \frac{P_2}{P_1} = C_P \ln \frac{T_2}{T_1}$$

These are the equations we derived previously (Eqs. 2–21 and 2–22) from a different point of view.

If the adiabatic expansion from V_1 to V_2 is carried out irreversibly, less work is done by the gas than in the reversible case and the drop in the energy

The Second Law of Thermodynamics

of the gas is also smaller. Consequently the final temperature T_2' is not as low as T_2 for the reversible process, and it follows that

$$\Delta S_{\text{tot}} = \Delta S_{\text{gas}} = R \ln \frac{V_2}{V_1} + C_V \ln \frac{T_2'}{T_1} > 0$$

Again we see that ΔS_{tot} for an irreversible process is a positive quantity.

3. *Mixing of two ideal gases.* Suppose that n_A moles of an ideal gas A are contained in one bulb and n_B moles of ideal gas B are contained in another, with the two gases at the same temperature and pressure P. Let us imagine that the two bulbs are connected by a tube which is closed off by a stopcock. When the stopcock is opened we know what spontaneous process will take place: the two gases diffuse into each other until a uniform mixture is reached at equilibrium. Each gas will occupy the total available volume at partial pressures we represent by P_A and P_B.

The entropy change which occurs when the stopcock is opened is called the entropy of mixing, and is the sum of the entropy changes for each gas. Since the temperature and the total volume of the system remain constant during this process and there is no interaction with the surroundings outside the bulbs, the entropy change is due to the pressure change for each gas as given by Equation 3–14. Thus the entropy of mixing is given by the following equation:

For ideal gases
$$\Delta S_{\text{tot}} = \Delta S_{\text{mixing}} = n_A R \ln \frac{P}{P_A} + n_B R \ln \frac{P}{P_B} \qquad (3\text{--}15)$$

Because P_A and P_B are less than P, the entropy change for the spontaneous and irreversible process of mixing is positive.

It is sometimes convenient to express Equation 3–15 in a somewhat different form. From Dalton's law of partial pressures (Eq. 1–4), P_A/P is equal to the mole fraction X_A, and $n_A = X_A n_{\text{tot}}$. When these substitutions are made Equation 3–15 becomes

$$\Delta S_{\text{mixing}} = -n_{\text{tot}} R (X_A \ln X_A + X_B \ln X_B) \qquad (3\text{--}16)$$

Entropy as a Function of T and V

In the last section we found the entropy changes associated with a series of special processes. Now we derive a number of general relationships, applicable to any homogeneous system, in which the temperature and volume are taken as independent variables.

Because $dq_{rev} = T\,dS$, we can combine the first and second laws in the following way:

$$dE = T\,dS - P\,dV \tag{3-17}$$

On solving this equation for dS, we find

$$dS = \frac{dE}{T} + \frac{P}{T}dV \tag{3-18}$$

We consider the energy to be a function of T and V. From the general differentiation of $E(T,V)$ we have

$$dE = \left(\frac{\partial E}{\partial T}\right)_V dT + \left(\frac{\partial E}{\partial V}\right)_T dV$$

When this expression for dE is substituted in Equation 3–18 we obtain

$$dS = \frac{1}{T}\left[\left(\frac{\partial E}{\partial T}\right)_V dT + \left(\frac{\partial E}{\partial V}\right)_T dV\right] + \frac{P}{T}dV$$

When the terms in this equation are regrouped, it becomes

$$dS = \frac{1}{T}\left(\frac{\partial E}{\partial T}\right)_V dT + \frac{1}{T}\left[\left(\frac{\partial E}{\partial V}\right)_T + P\right]dV \tag{3-19}$$

From the general differentiation of $S(T,V)$ we have

$$dS = \left(\frac{\partial S}{\partial T}\right)_V dT + \left(\frac{\partial S}{\partial V}\right)_T dV \tag{3-20}$$

On equating the coefficients of dT in Equations 3–19 and 3–20, we find

$$\left(\frac{\partial S}{\partial T}\right)_V = \frac{1}{T}\left(\frac{\partial E}{\partial T}\right)_V \tag{3-21}$$

From the definition of C_V this equation can be written in the following way:

$$\boxed{\left(\frac{\partial S}{\partial T}\right)_V = \frac{C_V}{T} \quad \text{(for 1 mole)}} \tag{3-22}$$

In a similar way, from the coefficients of dV in Equations 3–19 and 3–20 we find

$$\left(\frac{\partial S}{\partial V}\right)_T = \frac{1}{T}\left[\left(\frac{\partial E}{\partial V}\right)_T + P\right] \tag{3-23}$$

When Equation 3–21 is differentiated with respect to V at constant T, the result is

$$\frac{\partial^2 S}{\partial V \partial T} = \frac{1}{T}\frac{\partial^2 E}{\partial V \partial T} \tag{3-24}$$

The Second Law of Thermodynamics

Now let us differentiate Equation 3–23 with respect to T at a constant V. We obtain

$$\frac{\partial^2 S}{\partial T \partial V} = \frac{1}{T}\frac{\partial}{\partial T}\left[\left(\frac{\partial E}{\partial V}\right)_T + P\right] - \frac{1}{T^2}\left[\left(\frac{\partial E}{\partial V}\right)_T + P\right]$$

Using Equation 3–23 to substitute the bracketed part of the second term on the right-hand side, we obtain

$$\frac{\partial^2 S}{\partial T \partial V} = \frac{1}{T}\frac{\partial^2 E}{\partial T \partial V} + \frac{1}{T}\left(\frac{\partial P}{\partial T}\right)_V - \frac{1}{T}\left(\frac{\partial S}{\partial V}\right)_T \qquad (3\text{--}25)$$

Since the order of differentiation is unimportant (Eq. 2–2), we can equate the two expressions for the second derivative of S with respect to T and V in Equations 3–24 and 3–25. We thus obtain the useful relation

$$\boxed{\left(\frac{\partial S}{\partial V}\right)_T = \left(\frac{\partial P}{\partial T}\right)_V} \qquad (3\text{--}26)$$

Equation 3–26 provides us with a completely general relation for the isothermal change of entropy with volume for any homogeneous system. This relation depends only on the equation of state of the system.

For a finite isothermal change in volume we can now write the following equation:

$$\Delta S = \int_{V_1}^{V_2} \left(\frac{\partial P}{\partial T}\right)_V dV$$

Equation 3–12 is a special case of the equation above, since for an ideal gas $\left(\frac{\partial P}{\partial T}\right)_V = nR/V$.

By integrating Equation 3–22 we obtain the following general equation for the molar entropy change resulting from a temperature change at constant volume:

$$\Delta S = \int_{T_1}^{T_2} \frac{C_V}{T} dT$$

If the temperature range under consideration is so small that C_V can be treated as constant in this range, we have

$$\Delta S = C_V \int_{T_1}^{T_2} d\ln T = C_V \ln \frac{T_2}{T_1}$$

In the general case, however, C_V must be known as a function of the temperature. As previously mentioned (p. 39), the heat capacity of gases which is usually recorded is C_P rather than C_V, and C_P is often expressed as an empirical function of the temperature in the form of some type of power series such as $C_P = a + bT + cT^2$. In this case we can use the fact that $C_V = C_P - R$

(Eq. 2–16) is an adequate approximation for most purposes, and thus obtain the following equation for the entropy change:

$$\Delta S = \int_{T_1}^{T_2} \left(\frac{a-R}{T} + b + cT \right) dT$$

On carrying out the indicated integrations, we have

$$\Delta S = (a - R) \ln \frac{T_2}{T_1} + b(T_2 - T_1) + \frac{c}{2}(T_2^2 - T_1^2)$$

For solids and liquids the difference between C_P and C_V is usually negligible. The entropy change resulting from a temperature change of a condensed phase is often found by plotting experimental values of C_P/T against the temperature and carrying out a graphical integration between the desired temperature limits.

Let us now return to Equation 3–23 which we rewrite in the following form:

$$\left(\frac{\partial E}{\partial V} \right)_T = T \left(\frac{\partial S}{\partial V} \right)_T - P$$

With the use of Equation 3–26 we substitute $\left(\frac{\partial P}{\partial T} \right)_V$ for $\left(\frac{\partial S}{\partial V} \right)_T$, and thus obtain

$$\left(\frac{\partial E}{\partial V} \right)_T = T \left(\frac{\partial P}{\partial T} \right)_V - P \qquad (3\text{–}27)$$

This is a general expression for the isothermal change of molar energy with volume in terms of the equation of state of the system. It is sometimes called the thermodynamic equation of state. In order to apply it to an ideal gas we need the value of $\left(\frac{\partial P}{\partial T} \right)_V$ which, as we found above, is equal to R/V. Consequently,

$$\left(\frac{\partial E}{\partial V} \right)_T = T \frac{R}{V} - P$$

Thus we see that the statement that the molar energy of an ideal gas is independent of the volume is not a part of the definition of an ideal gas. Rather, this statement is a conclusion which follows directly from the ideal gas equation of state.

As another application of Equation 3–27, we can use it to reduce Equation 2–13 for $C_P - C_V$ to a more convenient form which depends only on the equation of state:

$$C_P - C_V = \left[\left(\frac{\partial E}{\partial V} \right)_T + P \right] \left(\frac{\partial V}{\partial T} \right)_P$$

$$= T \left(\frac{\partial P}{\partial T} \right)_V \left(\frac{\partial V}{\partial T} \right)_P \qquad (3\text{–}28)$$

Entropy as a Function of T and P

We now carry out a treatment, analogous to that in the preceding section, of entropy changes when T and P are chosen as independent variables. This treatment yields some additional useful relations.

From the definition of enthalpy, $H = E + PV$, by a general differentiation we obtain

$$dH = dE + P\,dV + V\,dP$$

We can substitute dE from Equation 3–17 to obtain

$$dH = T\,dS + V\,dP \tag{3-29}$$

It follows that

$$dS = \frac{dH}{T} - \frac{V}{T}dP \tag{3-30}$$

Considering H as a function of T and P, we have

$$dH = \left(\frac{\partial H}{\partial T}\right)_P dT + \left(\frac{\partial H}{\partial P}\right)_T dP$$

Substitution of this expression for dH in Equation 3–30 yields

$$dS = \frac{1}{T}\left[\left(\frac{\partial H}{\partial T}\right)_P dT + \left(\frac{\partial H}{\partial P}\right)_T dP\right] - \frac{V}{T}dP$$

$$= \frac{1}{T}\left(\frac{\partial H}{\partial T}\right)_P dT + \frac{1}{T}\left[\left(\frac{\partial H}{\partial P}\right)_T - V\right]dP$$

When S is differentiated as a function of T and P, the result is

$$dS = \left(\frac{\partial S}{\partial T}\right)_P dT + \left(\frac{\partial S}{\partial P}\right)_T dP$$

On equating the coefficients of dT and dP in the two equations above, we obtain

$$\boxed{\left(\frac{\partial S}{\partial T}\right)_P = \frac{1}{T}\left(\frac{\partial H}{\partial T}\right)_P} \tag{3-31}$$

and

$$\left(\frac{\partial S}{\partial P}\right)_T = \frac{1}{T}\left[\left(\frac{\partial H}{\partial P}\right)_T - V\right] \tag{3-32}$$

Differentiation of Equation 3–31 with respect to P at constant T yields

$$\frac{\partial^2 S}{\partial P\,\partial T} = \frac{1}{T}\frac{\partial^2 H}{\partial P\,\partial T} \tag{3-33}$$

Likewise, differentiation of Equation 3–32 with respect to T at constant P leads to

$$\frac{\partial^2 S}{\partial T \partial P} = \frac{1}{T}\left[\frac{\partial^2 H}{\partial T \partial P} - \left(\frac{\partial V}{\partial T}\right)_P - \left(\frac{\partial S}{\partial P}\right)_T\right] \quad (3\text{–}34)$$

By equating the right-hand sides of Equations 3–33 and 3–34 we obtain

$$\boxed{\left(\frac{\partial S}{\partial P}\right)_T = -\left(\frac{\partial V}{\partial T}\right)_P} \quad (3\text{–}35)$$

The general equation for the change of entropy resulting from an isothermal pressure change for a homogeneous system is

$$\Delta S = \int_{P_1}^{P_2} \left(\frac{\partial S}{\partial P}\right)_T dP$$

Making use of Equation 3–35, we can express this entropy change in terms of the equation of state, as follows:

$$\Delta S = -\int_{P_1}^{P_2} \left(\frac{\partial V}{\partial T}\right)_P dP$$

The corresponding relation for molar entropy change with temperature at constant pressure from Equation 3–31 is

$$\Delta S = \int_{T_1}^{T_2} \frac{1}{T}\left(\frac{\partial H}{\partial T}\right)_P dT$$

which can also be written as

$$\Delta S = \int_{T_1}^{T_2} \frac{C_P}{T} dT \quad (3\text{–}36)$$

We can apply some of the relations obtained in this section to a further consideration of the Joule–Thomson coefficient which is defined by Equation 2–23 as follows:

$$\mu = \left(\frac{\partial T}{\partial P}\right)_H = -\frac{1}{C_P}\left(\frac{\partial H}{\partial P}\right)_T$$

From Equations 3–32 and 3–35 we obtain

$$\left(\frac{\partial H}{\partial P}\right)_T = -T\left(\frac{\partial V}{\partial T}\right)_P + V$$

It follows that

$$\boxed{\mu = \frac{T\left(\frac{\partial V}{\partial T}\right)_P - V}{C_P}}$$

Whether the Joule–Thomson coefficient is positive or negative clearly depends on the relative values of $T\left(\frac{\partial V}{\partial T}\right)_P$ and V which, in turn, depend on the equation of state.

◈◈◈◈◈◈◈

Entropy Changes Resulting from Phase Changes

For a phase change carried out reversibly at constant pressure, $q_{\text{rev}} = \Delta H$, the heat absorbed per mole in the phase change at the equilibrium temperature T. The corresponding molar entropy change for the substance undergoing the phase change is given by

$$\Delta S = \frac{\Delta H}{T} \qquad (3\text{-}37)$$

For example, the heat of fusion of ice is 1440 cal/mole. Consequently the molar entropy of fusion of water is $1440/273 = 5.26$ cal/deg.

The entropy of a substance always increases on melting or vaporization because the heat of fusion or vaporization is always positive. A useful empirical generalization known as Trouton's rule states that the molar entropy of vaporization of most liquids that do not contain hydrogen bonds is about 21 cal/deg. Benzene is such a liquid. Its heat of vaporization is 7450 cal/mole at its normal boiling point of 353°K. Thus its molar entropy of vaporization is $7450/353 = 21.1$ cal/deg. In contrast, the molar heat of vaporization of water is 9700 cal/mole and its molar entropy of vaporization is 26.0 cal/deg.

Entropy Change as a Criterion of Equilibrium

A system for which no spontaneous process is possible is at equilibrium. We have seen that when an irreversible process occurs in a system the total entropy always increases. Consequently we conclude that when a system is in such a state that for all possible processes $\Delta S_{\text{tot}} = 0$, the system is at equilibrium.

We have now arrived at a very general criterion of equilibrium. In fact, it is so general that it is not very useful because it requires knowledge not only of the possible processes in the system, but also of what the interactions with the surroundings might be. It would be much more useful to have a criterion which depends on the state of the system only. We now show that several such criteria can be found, but only at the expense of a limitation on the kind of process which can be employed.

For this purpose, let us return to the combination of the first and second laws (Eq. 3–18) which we now write in the following more general form:

$$dS \geqslant \frac{dE}{T} + \frac{P_{\text{ext}}}{T} dV \qquad (3\text{–}38)$$

The equality sign is applicable to a reversible process; in this case the external pressure is the same as the system pressure, P. The inequality is applicable to any irreversible process.

If we impose as restrictions on the process under consideration the requirement that the energy and volume of the system remain constant, Equation 3–38 reduces to

$$dS_{E,V} \geqslant 0$$

Since the entropy of the system will increase when an irreversible process is possible, it follows that when the entropy is at a maximum and thus $dS_{E,V} = 0$ for all possible processes at constant energy and volume, the system is at equilibrium. A system at constant energy and volume, however, is an isolated system because PV work and energy in any form including heat cannot be exchanged between the system and its surroundings. Although this criterion depends only on the state of the system, it clearly is not a practical one.

Another possible set of restrictions is the requirement that the entropy and volume of the system be constant. Rewriting Equation 3–38 in the form

$$dE \leqslant T\,dS - P_{\text{ext}}\,dV$$

we see that $dS = dV = 0$ implies that $dE \leqslant 0$. In this case the system will decrease its energy spontaneously until a minimum is reached at equilibrium, when $dE_{S,V} = 0$ for all possible processes. This is the criterion for equilibrium in mechanics, in which entropy, volume, temperature, and heat are not treated as variables, but it is of no use in chemistry.

Free Energy

We now consider how criteria of equilibrium can be found that are more useful to the chemist than those discussed in the preceding section. We accomplish this by imposing restrictions on the behavior of the system that are more practical than those mentioned above. Again we start from the combination of the first and second laws, this time in the following form:

$$đw \leqslant -dE + T\,dS$$

Now we take into consideration for the first time the fact that the work done in a general process may include types of work other than PV work. One such type of work which is ordinarily unimportant in chemistry is the work done by an external magnetic field on a system that has a magnetic moment. A type which is of major importance, however, is the electrical work done by a chemical reaction carried out in an electrochemical cell. We treat this subject in detail in Chapter 16.

The symbol $đw$ in the equation above represents the total work of all types that may be done in an infinitesimal process. If this process is carried out isothermally, $d(TS) = T\,dS$ and we have

$$đw \leqslant -d(E - TS) \qquad (3\text{–}39)$$

It is convenient to express the quantity $(E - TS)$ by a single symbol, and thus we are led to introduce a new thermodynamic function, A, defined by

$$A = E - TS \qquad (3\text{–}40)$$

This function is a function of the state because it is composed of functions of the state, and therefore its differential is exact. From integration of Equation 3–39 for a reversible isothermal process, the work done is given by

$$w_{\max} = -\Delta(E - TS) = -\Delta E + T\Delta S = -\Delta A$$

In an irreversible process with the same initial and final states, the work done is less than $w_{\max}$, but ΔA has the same value in both cases because A is a function of the state.

The function A is called the *work function*, the *work content*, or the *Helmholtz free energy*. The name free energy was given to this function by Helmholtz because the change in A represents the maximum amount of energy that is free or available to be converted into work. This amount may be more, or less, than ΔE, depending on the sign of ΔS.

Since $d(E - TS) = dA$, we can rewrite Equation 3–39 in the following way:

$$đw + dA \leqslant 0$$

Processes of interest in chemistry, such as chemical reactions and phase changes, are usually carried out in such a way that the only work done is PV work. If the volume of the system under consideration remains constant, no PV work can be done and it follows that in this case

$$dA_{T,V} \leqslant 0$$

This means that if our system can carry out an irreversible process at constant temperature and volume without doing any work, it will do so spontaneously. The Helmholtz free energy will decrease as a result of a decrease in the energy and an increase in the entropy of the system. The process will continue until the Helmholtz free energy reaches a minimum value when the system is at equilibrium. In this state $dA_{T,V} = 0$ for any process, and this is the criterion of equilibrium at constant temperature and volume.

Although we arrived at the Helmholtz free energy function A by considering an isothermal process, we can differentiate this function without any restriction to obtain

$$dA = dE - T\,dS - S\,dT$$

For a reversible process in which the only work done is PV work, we can substitute for dE its equal, $T\,dS - P\,dV$. Then we have

$$dA = -P\,dV - S\,dT$$

From this equation we can obtain several other relationships. When we divide both sides by dV and impose the condition of constant temperature, the result is

$$\left(\frac{\partial A}{\partial V}\right)_T = -P \tag{3-41}$$

Similarly, when we divide by dT and impose the condition of constant volume we obtain

$$\left(\frac{\partial A}{\partial T}\right)_V = -S$$

Because dA is an exact differential, we can employ the Euler criterion of exactness (Eq. 2-5) to obtain

$$\left(\frac{\partial S}{\partial V}\right)_T = \left(\frac{\partial P}{\partial T}\right)_V$$

This is a relationship which we found previously by another method (Eq. 3-26).

We now develop the criterion for equilibrium appropriate for the conditions which are particularly important in chemistry, namely, constancy of temperature and pressure. In order to do this we define the *net work* done in a process as the difference between the total work and the PV work. In a process carried out at constant pressure, the volume of the system will usually change and thus some PV work will be done against the external pressure, which is often atmospheric pressure. Then the net work is given by

$$đw_{net} = đw_{tot} - P_{ext}\,dV \tag{3-42}$$

The Second Law of Thermodynamics

When we subtract $P_{ext}\,dV$ from both sides of Equation 3-39, we obtain

$$\dbar w_{net} \leqslant -dE + T\,dS - P_{ext}\,dV \tag{3-43}$$

($P_{ext} = P$ for a reversible process, for which the equal sign is applicable.)

We can write Equation 3-43 in a simpler form by introducing the enthalpy, H. Recalling the definition of H as $E + PV$, we see that for a process carried out at constant pressure, $dH = dE + P\,dV$. Thus Equation 3-43 becomes

$$\dbar w_{net} \leqslant -dH + T\,dS \tag{3-44}$$

If the process is carried out at constant temperature as well as constant pressure,

$$-dH + T\,dS = -d(H - TS)$$

This suggests that it is useful to define another new thermodynamic function by the equation

$$G = H - TS \tag{3-45}$$

Then Equation 3-44 takes the following form:

$$\dbar w_{net} \leqslant -dG$$

or

$$\dbar w_{net} + dG \leqslant 0 \tag{3-46}$$

The function G is called the *Gibbs free energy*,[1] or just the free energy when no ambiguity is likely. (It should be noted that some authors use F, rather than G, as a symbol for the Gibbs free energy.)

If the process under consideration is carried out in such a way that no net work is done, Equation 3-46 reduces to

$$dG_{T,P} \leqslant 0$$

This equation implies that if the system under consideration can carry out any irreversible process at constant temperature and pressure, it will do so spontaneously and thus lower its Gibbs free energy until this function reaches a minimum value. When this value is reached, $dG_{T,P} = 0$ and the system is at equilibrium.

Now let us summarize what we have accomplished in this section. We saw earlier that the most general criterion for equilibrium is $dS_{tot} = 0$ for any process whatever, but that the use of this criterion requires knowledge of the changes in the surroundings as well as those in the system. By limiting the processes to those occurring at constant energy and volume of the system, the criterion $dS_{E,V} = 0$, which depends only on the state of the system, is

---

[1] A more logical name would be free enthalpy, for it can be seen that G bears the same relationship to H that A bears to E.

obtained. This, however, is not useful because energy is not a convenient independent variable and it is not usually held constant. When the processes are limited to those occurring at constant temperature and volume, $dA_{T,V} = 0$ is the criterion of equilibrium. Finally, and most importantly, for processes occurring at constant temperature and pressure, $dG_{T,P} = 0$ is the criterion of equilibrium, and this is the one we will use in most of the subsequent applications of thermodynamics.

We can obtain some additional relations from Equation 3–45 which we can write in the following form:

$$G = E + PV - TS$$

The general differentiation of this equation leads to

$$dG = dE + P\,dV + V\,dP - T\,dS - S\,dT$$

We have seen previously (Eq. 3–17) that for a reversible process in which the only work done is PV work, the combination of the first and second laws is expressed by

$$T\,dS = dE + P\,dV$$

It follows that

$$dG = V\,dP - S\,dT \qquad (3\text{–}47)$$

Again employing the Euler criterion for the exactness of differentials, from this equation we obtain

$$\left(\frac{\partial S}{\partial P}\right)_T = -\left(\frac{\partial V}{\partial T}\right)_P$$

which is the same as Equation 3–35.

Free Energy Change with Pressure and Temperature

Although the free energy change is useful as a criterion of equilibrium for a system whose temperature and pressure are constant, we need to calculate the changes in the free energy as a function of these variables. From Equation 3–47 we see that the free energy change with pressure at constant temperature is given by

$$dG = V\,dP$$

in which V is the volume of the system under consideration. With the constancy of T shown explicitly, this equation is

$$\left(\frac{\partial G}{\partial P}\right)_T = V \qquad (3\text{–}48)$$

The Second Law of Thermodynamics

For the special case of an ideal gas, at constant temperature

$$dG = \frac{nRT}{P} dP$$

$$= nRT \, d \ln P$$

When this equation is integrated between the limits P_1 and P_2, we have

$$\Delta G = nRT \ln \frac{P_2}{P_1}$$

which is equal to

$$\Delta G = nRT \ln \frac{V_1}{V_2}$$

If, for example, the pressure of 3 moles of ideal gas at 25°C is decreased to one-third its initial value, and thus $V_2 = 3V_1$, the free energy change for this process is

$$\Delta G = 3 \times 1.99 \times 298 \ln \tfrac{1}{3}$$
$$= -1960 \text{ cal}$$

Since $G = H - TS$ and $A = E - TS$, it follows that $G = A + PV$ and for any isothermal process

$$\Delta G = \Delta A + \Delta(PV)$$

For the special case of an isothermal expansion of an ideal gas, $\Delta G = \Delta A$, because $\Delta(PV) = 0$.

From Equation 3–47 the change in free energy with temperature at constant pressure is given by

$$\left(\frac{\partial G}{\partial T}\right)_P = -S \tag{3-49}$$

From the definition of G, $-S = \dfrac{G - H}{T}$, and the equation above can be written as

$$\left(\frac{\partial G}{\partial T}\right)_P = \frac{G - H}{T}$$

These are not, however, generally useful relationships because they imply knowledge of either the absolute value of the entropy or the free energy and enthalpy as a function of the temperature. If, on the other hand, we have an isothermal process for which the free energy change at temperature T is ΔG, we can express ΔG as a function of the temperature. From

$$\left(\frac{\partial \Delta G}{\partial T}\right)_P = -\Delta S = \frac{\Delta G - \Delta H}{T}$$

we obtain

$$\Delta G = \Delta H + T\left(\frac{\partial \Delta G}{\partial T}\right)_P \qquad (3\text{--}50)$$

This equation is known as the *Gibbs–Helmholtz equation*.

To find the value of ΔG at one temperature when its value is known at some other temperature, it is often more convenient to make use of the temperature dependence of $\frac{\Delta G}{T}$, rather than the temperature dependence of ΔG itself. When we differentiate $\frac{\Delta G}{T}$ with respect to the temperature at constant pressure, the result is

$$\left(\frac{\partial \frac{\Delta G}{T}}{\partial T}\right)_P = \frac{\partial}{\partial T}\left(\frac{\Delta H}{T} - \Delta S\right)_P$$

$$= \frac{T\left(\frac{\partial \Delta H}{\partial T}\right)_P - \Delta H}{T^2} - \left(\frac{\partial \Delta S}{\partial T}\right)_P$$

$$= \frac{\Delta C_P}{T} - \frac{\Delta H}{T^2} - \frac{\Delta C_P}{T}$$

$$= -\frac{\Delta H}{T^2} \qquad (3\text{--}51)$$

Since $d\left(\frac{1}{T}\right) = -\frac{1}{T^2}\,dT$, a variant of this equation is

$$\left(\frac{\partial \frac{\Delta G}{T}}{\partial \left(\frac{1}{T}\right)}\right)_P = \Delta H$$

Equilibrium Between Two Phases

As a first example of the usefulness of the free energy concept let us apply it to a system consisting of a pure substance in two phases at equilibrium. To be specific, we consider a liquid and a vapor phase in equilibrium at a constant temperature T. Because the vapor pressure of a pure liquid depends on the temperature only, the pressure in the system is also constant. If some liquid is evaporated reversibly (with $P_{\text{ext}} = P$) or if some vapor is condensed reversibly, the volume of the system will change but not its free energy, since $\Delta G_{T,P} = 0$. Thus at equilibrium the free energy per mole of liquid is equal to the free energy per mole of vapor ($G_{\text{liq}} = G_{\text{vap}}$).

The Second Law of Thermodynamics

Now let us assume that the temperature is changed by an infinitesimal amount. When equilibrium is reestablished the pressure is also changed infinitesimally. Under these conditions,

$$G_{liq} + dG_{liq} = G_{vap} + dG_{vap}$$

Consequently,

$$dG_{liq} = dG_{vap}$$

Replacing dG by its equal from Equation 3–47, we have

$$V_{liq}\, dP - S_{liq}\, dT = V_{vap}\, dP - S_{vap}\, dT$$

Rearrangement of this equation results in

$$(V_{vap} - V_{liq})\, dP = (S_{vap} - S_{liq})\, dT$$

which can be written as

$$\Delta V\, dP = \Delta S\, dT$$

It follows that

$$\frac{dP}{dT} = \frac{\Delta S}{\Delta V}$$

For a reversible vaporization, $\Delta S = \Delta H_v/T$ (Eq. 3-37) and consequently,

$$\boxed{\frac{dP}{dT} = \frac{\Delta H_v}{T\,\Delta V}} \qquad (3\text{--}52)$$

This is an exact equation which is known as the *Clapeyron equation*. If the appropriate value of ΔH is used it is applicable to any kind of phase change.

For liquid-vapor equilibria we can obtain a very useful approximation to the Clapeyron equation. Because the molar volume of a liquid is small compared to the volume of the corresponding vapor, we can replace ΔV by V_{vap}. With the assumption that the vapor behaves like an ideal gas, $V_{vap} = RT/P$ and we have

$$\frac{dP}{dT} = \frac{\Delta H_v P}{RT^2}$$

When we divide both sides by P the result is

$$\boxed{\frac{d \ln P}{dT} = \frac{\Delta H_v}{RT^2}} \qquad (3\text{--}53)$$

This equation is called the *Clausius–Clapeyron equation*. Integrating it from an initial state (P_1, T_1) to a final state (P_2, T_2) and assuming that ΔH_v remains constant over this temperature range, we obtain

$$\ln \frac{P_2}{P_1} = \frac{\Delta H_v}{R}\left(\frac{1}{T_1} - \frac{1}{T_2}\right) \qquad (3\text{--}54)$$

The integrated Clausius–Clapeyron equation involves five variables. If any four of these are known, the fifth can be readily found.

As a typical example, let us consider the following problem: Given that the value of ΔH_v for benzene is 7450 cal/mole and that its normal boiling point (that is, the temperature at which its vapor pressure equals 760 torr) is 80°C, what is the vapor pressure at 70°C? We can represent the desired pressure by P_2 and solve Equation 3–54 for this pressure. When we substitute numerical values and change to common logarithms, we find

$$\log P_2 = \log 760 + \frac{7450}{2.30 \times 1.99}\left(\frac{1}{353} - \frac{1}{343}\right)$$
$$= 2.881 - 0.138 = 2.743$$
$$P_2 = 553 \text{ torr}$$

We can also integrate Equation 3–53 indefinitely to obtain the following expression for the vapor pressure as a function of the temperature:

$$\ln P = \frac{1}{R}\int \frac{\Delta H_v}{T^2}\, dT + K$$

With the assumption that ΔH_v is constant this becomes

$$\log P = -\frac{1}{4.58}\frac{\Delta H_v}{T} + K \tag{3–55}$$

The constant of integration can be evaluated when either the heat of vaporization and the vapor pressure at one temperature, or the vapor pressures at two temperatures are known. From the form of Equation 3–55 it follows that a graph of $\log P$ versus $1/T$ should be a straight line whose slope is equal to $-\Delta H_v/4.58$. For temperature ranges that are not too large, the graphs of experimental data are indeed straight lines, as shown in Figure 3–5, from which the molar heats of vaporization can be calculated. This is, in fact, the most common method for obtaining heats of vaporization.

Equations of the same form as Equation 3–55 are found to represent the sublimation pressures of solids as functions of the temperature. For example, the measured values of the sublimation pressure of benzophenone, expressed in torr, are represented by the equation, $\log P = -4966/T + 11.76$. Consequently $\Delta H_s/4.58 = 4966$, and the heat of sublimation of benzophenone is given by

$$\Delta H_s = 4.58 \times 4966 = 22.7 \text{ kcal/mole}$$

For the case of equilibrium between the solid and liquid phases of a substance, we must return to the Clapeyron equation, 3–52. In this case T is the melting point (or freezing point) when the pressure on the system is P. Since the melting point is a function of P, we can use Equation 3–52 to calculate the value of the melting point at some pressure if its value at some other pressure

The Second Law of Thermodynamics

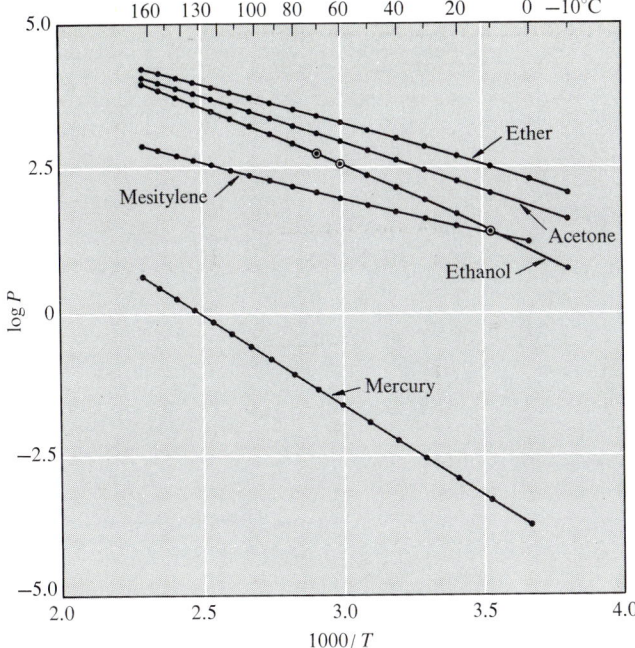

Figure 3-5. Graph of log P vs $1/T$ for several liquids.

is known as well as its heat of fusion, ΔH_f, and the molar volumes of the solid and liquid phases. Because neither ΔH_f nor ΔV changes much with change in temperature, it is a good approximation to replace the differential coefficient dP/dT by the ratio of finite differences, $\Delta P/\Delta T$. Then the change in melting point resulting from a change in pressure ΔP is given by

$$\Delta T = \frac{T \Delta V}{\Delta H_f} \Delta P$$

For nearly all substances, $\Delta V (= V_{liq} - V_{sol})$ is positive and consequently the melting point is raised with increasing pressure. Water is an outstanding exception to this generalization and the fact that the melting point of ice is lowered with increasing pressure is responsible for many phenomena, such as the flow of glaciers.

Systems of Variable Composition

The systems we have discussed so far in this chapter have consisted of a fixed amount of some substance. Because we want to apply our results to systems in which chemical reactions take place and whose compositions are therefore variable, we need to consider a more general type of system. Let us assume that we have a multicomponent open system; that is, a system for which not only heat and work but also several kinds of matter can be exchanged with the surroundings. The numbers of moles of the various substances in the system are to be treated as variables.

From the definition of free energy we can see that it is an extensive property because enthalpy and entropy are extensive properties. The free energy of a system is a function not only of the temperature and pressure, but also of the number of moles of each substance in the system. We can express this function in the following way: $G = f(T,P,n_1, \ldots n_k)$ where n_1, and so forth, represent the numbers of moles of each of the k components in the system. If we assume for simplicity that our system contains only two components, the total differential of G is

$$dG = \left(\frac{\partial G}{\partial T}\right)_{P,n_1,n_2} dT + \left(\frac{\partial G}{\partial P}\right)_{T,n_1,n_2} dP + \left(\frac{\partial G}{\partial n_1}\right)_{T,P,n_2} dn_1 + \left(\frac{\partial G}{\partial n_2}\right)_{T,P,n_1} dn_2 \quad (3\text{–}56)$$

The first two terms of this equation represent the change in G with temperature and pressure for a system of constant composition, and they are essentially the same as Equations 3–49 and 3–48, respectively. The last two terms introduce something new, namely, the partial derivative of a thermodynamic function with respect to the number of moles of a component. This partial derivative has the mathematical form $\left(\frac{\partial Y}{\partial n_i}\right)_{T,P,n_j}$, in which Y represents any extensive thermodynamic function (or variable) such as volume, entropy, or free energy, and n_j represents the numbers of moles of all substances in the system other than i.

The partial derivative, often symbolized by $\overline{Y}_i$, represents the change in Y per mole of component i when an infinitesimal amount of this component is added to a system of definite composition. An alternative way of looking at such a partial derivative is to consider it as the change in Y when 1 mole of component i is added to a system which is so large that this addition has a negligible effect on the composition.

Depending on the nature of Y, the partial derivative denoted by $\overline{Y}_i$ may be called the partial molar volume, entropy, or free energy. The partial molar free energy is a quantity which is so important in chemistry that it is usually given a special name, the chemical potential, and a special symbol, μ_i.

$$\mu_i = \left(\frac{\partial G}{\partial n_i}\right)_{T,P,n_j} \quad (3\text{–}57)$$

The chemical potential is an intensive property because it is a molar quantity. For a pure substance it is simply the free energy per mole of this substance.

With the use of the symbol μ for the chemical potential and Equations 3–48 and 3–49, we can rewrite Equation 3–56 in the following form:

$$dG = -S\,dT + V\,dP + \mu_1\,dn_1 + \mu_2\,dn_2 \quad (3\text{–}58)$$

The Second Law of Thermodynamics

in which S and V are the entropy and volume, respectively, of the system which contains n_1 moles of component 1 and n_2 moles of component 2. For an infinitesimal process which takes place at constant T and P, this equation reduces to

$$dG = \mu_1 \, dn_1 + \mu_2 \, dn_2 \tag{3-59}$$

In order to express the free energy of this system in terms of chemical potentials and the number of moles of each component, we need the integrated form of Equation 3-59. We can obtain this for the special case in which the composition of the system, and thus the chemical potentials, remain constant by imagining that we synthesize our system by adding infinitesimal amounts of the two components in the molar ratio n_1/n_2. This process is continued until n_1 moles of component 1 and n_2 moles of component 2 have been added. The addition of these infinitesimal amounts is equivalent to the mathematical operation of integration. The result that we obtain in this way is

$$G = \mu_1 n_1 + \mu_2 n_2 \tag{3-60}$$

Although we have obtained the equation above by consideration of a special case, the free energy of a system is clearly independent of the particular way the system is formed. Thus this equation is a general one and we can differentiate it without any restriction for the following result:

$$dG = \mu_1 \, dn_1 + n_1 \, d\mu_1 + \mu_2 \, dn_2 + n_2 \, d\mu_2$$

In view of Equation of 3-59, it follows that

$$n_1 \, d\mu_1 + n_2 \, d\mu_2 = 0 \tag{3-61}$$

and

$$\boxed{d\mu_1 = -\frac{n_2}{n_1} \, d\mu_2}$$

Equation 3-61 gives the relationship between a change in the chemical potential of one component and the change in chemical potential of the other component. This equation (or its generalized form applicable to any number of components) is known as the *Gibbs–Duhem equation.* We make use of it later in Chapter 15.

Standard States

It was pointed out earlier that there is no basis in thermodynamics for assigning absolute values to the energy or enthalpy of a system, and also that absolute values are not needed because only changes in these quantities are of interest. The same statements apply to the free energy of a system. Since we are concerned with free energy changes, it is convenient to define a standard state for the system under consideration and to refer the free energy and chemical potentials in any other state to the values (unknown in absolute magnitude) of these quantities in the standard state.

As a simple example of the choice of a standard state, let us consider the isothermal change in chemical potential of an ideal gas with pressure. Since the chemical potential in this case is the free energy per mole, from Equation 3–48 we have

$$\left(\frac{\partial \mu}{\partial P}\right)_T = V_m$$

Keeping the constancy of temperature in mind, we can write this equation in the following form:

$$d\mu = V_m \, dP \tag{3-62}$$

Substituting for V_m its equal from the ideal gas law, we obtain

$$d\mu = RT \, d \ln P$$

On integration between the pressure limits P_1 and P_2 this equation becomes

$$\mu_2 - \mu_1 = RT \ln \frac{P_2}{P_1}$$

Now let us choose a pressure of 1 atmosphere as the standard state for our gas. For the change in chemical potential in an isothermal pressure change from 1 atmosphere to P atmospheres, we can write

$$\mu_P - \mu^\circ = RT \ln \frac{P}{1}$$

where μ° represents the chemical potential of the gas in its standard state at temperature T. Thus we can express μ_P at any pressure in the following way:

$$\mu_P = \mu^\circ + RT \ln P \tag{3-63}$$

In this equation it should be understood that a division of P by 1 atmosphere is implied, and thus the quantity whose logarithm is taken is actually dimensionless — as it must be.

For other types of systems, such as solutions, other choices of standard states will be made and these are discussed at the appropriate places.

Deviations from Ideality

The simple expression we have obtained above for the chemical potential as a function of the pressure depends, of course, on the ideal gas law. If it is not an adequate approximation to treat the gas under consideration as ideal, then, in principle, we could solve the equation of state to find the molar volume as a function of the pressure and integrate $V_m\, dP$ term by term. In general, this procedure would lead to awkward expressions which are difficult to manipulate.

Many years ago G. N. Lewis adopted another approach to this problem. He defined a function he called the *fugacity* (f) which reduces to the pressure when the gas behaves ideally and which, when substituted into the equation for the chemical potential, makes this equation exact by definition. This device makes it possible to retain the simple mathematical form found above for the ideal gas. It shifts the difficulty produced by deviations from ideality to the problem of finding the relation between the fugacity and pressure of a real gas.

The fugacity is defined in the following way:

$$d\mu = RT\, d\ln f; \quad f \to P \text{ as } P \to 0$$

With the use of Equation 3-62 we obtain

$$d\ln f = \frac{V_m}{RT}\, dP$$

When $d\ln P$ is subtracted from both sides of this equation, it becomes

$$d\ln f - d\ln P = \frac{V_m}{RT}\, dP - d\ln P$$

This can be written in the following form:

$$d\ln \frac{f}{P} = \left(\frac{V_m}{RT} - \frac{1}{P}\right) dP$$

In order to obtain the change in the ratio f/P in going from state 1 to state 2, we integrate this equation between the limits P_1 and P_2. We can express the result in the following way:

$$\ln \frac{f_2}{P_2} - \ln \frac{f_1}{P_1} = \int_{P_1}^{P_2} \left(\frac{V_m}{RT} - \frac{1}{P}\right) dP$$

Now let us allow P_1 to approach 0. Then by definition f_1 approaches P_1, and $\ln f_1/P_1$ approaches 0. We thus have derived the following expression for the ratio f/P, where P is any value of the pressure:

$$\ln \frac{f}{P} = \int_0^P \left(\frac{V_m}{RT} - \frac{1}{P}\right) dP$$

The indicated integration is usually carried out graphically from experimental P,V,T data.

The fugacity has the dimensions of pressure. The appropriate standard state for a real gas is a state of unit fugacity. Consequently for any gaseous component i the chemical potential is given by

$$\mu_i = \mu_i^\circ + RT \ln f_i \tag{3-64}$$

In dealing with a system containing a solution or other condensed phase, the fugacity of a component can be considered as the corrected vapor pressure of the component. In such cases it is convenient to express chemical potentials in terms of the ratio of the fugacity of a component in the system to the fugacity of this component in its standard state. This dimensionless ratio is called the activity (symbol a) of the component.

The best choice of standard state depends on the nature of the system under consideration. For a pure substance in a liquid or a solid phase, the standard state is taken to be the substance under a pressure of 1 atmosphere. Thus by definition, the activity of such a liquid or solid phase is 1. Then the general expression for the chemical potential is

$$\mu = \mu^\circ + RT \ln a \tag{3-65}$$

Supplementary References

J. D. Fast, *Entropy*, McGraw-Hill, New York, 1962.

Problems

1. A 50 g mass of copper at a temperature of 120°C is placed in contact with a 100 g mass of copper at a temperature of 30°C in a thermally insulated container. Calculate q and ΔS_{tot} for the resulting process. Use a value of 0.10 cal/g deg for the specific heat of copper.

Ans. $q = 300$ cal, $\Delta S_{tot} = 0.11$ cal/deg

2. 1 mole of ideal gas is expanded isothermally at 25°C until its volume is tripled. Find the values of ΔS_{gas} and ΔS_{tot} under the following conditions: (a) The expansion is carried out reversibly. (b) The expansion is carried out irreversibly and 200 cal less are absorbed than in case (a). (c) The expansion is a free expansion ($P_{ext} = 0$).

Ans. (a) $\Delta S_{gas} = 2.19$ cal/deg, $\Delta S_{tot} = 0$;
(b) $\Delta S_{gas} = 2.19$ cal/deg, $\Delta S_{tot} = 0.67$ cal/deg;
(c) $\Delta S_{gas} = 2.19$ cal/deg, $\Delta S_{tot} = 2.19$ cal/deg

3. Calculate ΔS for 5.0 moles of helium heated from a temperature of 100°C to 300°C at (a) constant volume; (b) constant pressure.

Ans. (a) 6.4 cal/deg; (b) 10.7 cal/deg

The Second Law of Thermodynamics

4. Calculate the entropy change when 10.0 liters of neon at 25°C and 1 atm are heated at constant pressure to 125°C.

5. Calculate ΔS for 3 moles of diatomic ideal gas which is heated and compressed from 25°C and 1 atm to 125°C and 5 atm. **Ans.** $-$ 3.60 cal/deg

6. 10 grams of neon initially at a pressure of 5 atm and a temperature of 200°C expand adiabatically to a pressure of 2 atm. Calculate the total entropy change for the following ways of carrying out this expansion: (a) The expansion is carried out reversibly. (b) The expansion occurs against a constant external pressure of 2 atm. (c) The expansion is a free expansion.

7. (a) 1 mole of ideal monatomic gas at 25°C, occupying a volume of 3 liters, is expanded adiabatically and reversibly to a pressure of 1 atm. It is then compressed isothermally and reversibly at the new temperature until the volume is again 3 liters. Calculate q, w, ΔE, ΔH, and ΔS. (b) The same gas in the same initial state is expanded isothermally and reversibly to a pressure of 1 atm. It is then compressed adiabatically and reversibly until the volume is again 3 liters. Calculate q, w, ΔE, ΔH, and ΔS.

8. 1 mole of helium is mixed with 2 moles of neon, both at the same temperature and pressure. Calculate ΔS for this process if the total volume remains constant. **Ans.** 3.8 cal/deg

9. 1 mole of helium at 25°C is mixed with 2 moles of neon at the same pressure but at a temperature of 125°C, in such a way that the total volume remains constant and there is no heat exchange with the surroundings. Calculate ΔS_{tot} for this process.

10. Using the heat capacity data in Table 2–1, calculate ΔS for 2 moles of nitrogen heated at constant pressure from 25°C to 1000°C. **Ans.** 21.3 cal/deg

11. Show that $C_P - C_V = \alpha^2 TV/\kappa$, in which $\alpha = \frac{1}{V}\left(\frac{\partial V}{\partial T}\right)_P$ and $\kappa = -\frac{1}{V}\left(\frac{\partial V}{\partial P}\right)_T \cdot \frac{1}{V}\left(\frac{\partial U}{\partial P}\right)_T$
(Hint: Use Eqs. 3–28 and 2–1.)

12. 10 g of ice are heated to become vapor at 100°C and 1 atm. Calculate ΔS for the system. (Take the heat of fusion of ice at 0°C as 80 cal/g, the heat of vaporization of water at 100°C as 540 cal/g, and the average specific heat of liquid water as 1 cal/g deg.) **Ans.** 20.5 cal/deg

13. 5.0 g of ice at 0°C are added to 30 g of water at 50°C in a thermally insulated container. (a) What is the final temperature? (b) What is the total entropy change? (Use the physical constants for water given in the preceding problem.)

14. Calculate ΔG for the conversion of 3 moles of liquid benzene at 80°C to vapor at the same temperature and a pressure of 500 torr. Consider the vapor as an ideal gas. **Ans.** -885 cal

15. It is possible to supercool water without freezing. 100 g of water are supercooled to $-10°C$ in a thermostat held at this temperature, and then crystallization takes place. Calculate ΔS_{water}, ΔS_{tot}, and ΔG_{water} for this process. (Take C_P for ice as 9 cal/mole.)

16. Calculate the free energy change in the freezing of 100 g of water at $-10°C$, given that the vapor pressures of water and ice at $-10°C$ are 2.15 torr and 1.95 torr, respectively. (Compare the answer with that of the preceding problem.)

17. At 100°C and 1 atm the specific volume of water vapor is 1674 cm³/g and the value of dP/dT is 27.12 torr/deg. Calculate ΔH_v in calories/mole. **Ans.** 9750 cal/mole

18. The vapor pressure of toluene is 59.1 torr at 40.0°C and 289.7 torr at 80.0°C. Calculate the heat of vaporization.

19. A certain liquid has a normal boiling point of 65.0°C. Using Trouton's rule: (a) Estimate the vapor pressure at 52°C. (b) Estimate the boiling point at a pressure of 200 torr. (c) Obtain an equation for the vapor pressure of this liquid in torr as a function of the temperature.
Ans. (a) 500 torr; (b) 300°K; (c) $\log P = 7.47 - 1550/T$

20. The vapor pressure of a liquid which obeys Trouton's rule rises by 52 torr between temperatures 1 degree below and 1 degree above the normal boiling point. Find the value of the normal boiling point and the molar heat of vaporization of the liquid.

4

Free Energy Change and Chemical Equilibrium

Free Energy Change in a Chemical Reaction

We are now in a position to apply the general results obtained from the second law in the last chapter to chemical reactions. At first we restrict our attention to homogeneous systems in which all reactants and products are ideal gases.

As always, it is necessary to be quite specific in describing the process under consideration. The chemical equation we use to represent a reaction must include all relevant information about the initial and final states of the reacting system. The reaction is assumed to be reversible in the chemical sense of the word; that is, we assume that the products can react under suitable conditions to form the original reactants, although there are many cases where chemical reversibility has never been observed.

The usage of the word reversible to describe a chemical reaction must be carefully distinguished from its meaning in the term thermodynamically reversible. It is unfortunate that the same word is used with two different meanings, but the context will always show which meaning is intended. A chemically reversible reaction which occurs spontaneously can clearly be seen to be a thermodynamically irreversible process.

Let us consider a general gas reaction carried out at the constant temperature T and constant total pressure P:

$$aA(P_A) + bB(P_B) = eE(P_E) + fF(P_F) \qquad (4\text{-}1)$$

This equation represents a reaction in which a moles of A and b moles of B, each at the partial pressure specified, disappear from the system and are replaced by e moles of E and f moles of F at their specified partial pressures.

When this reaction has proceeded to a certain extent, the numbers of moles of the substances formed or reacted are related by the coefficients in the chemical equation. If we consider the special case in which infinitesimal numbers of moles of A and B, $-dn_A$ and $-db_B$, react to form corresponding quantities of products, dn_E and dn_F, then it follows that the ratios of these four differentials to the corresponding coefficients a, b, e, and f, are all equal. Thus

$$-\frac{dn_A}{a} = -\frac{dn_B}{b} = \frac{dn_E}{e} = \frac{dn_F}{f}$$

If we express the common value of these ratios by $d\xi$, then $dn_A = -a\, d\xi$, $dn_E = e\, d\xi$, and so on. The quantity ξ is a measure of the extent to which the reaction has occurred.

We now find the free energy change when the reaction has proceeded at constant T and P to an extent $d\xi$. By a slight extension of Equation 3–59, we can express this change in the following way:

$$dG = \mu_E\, dn_E + \mu_F\, dn_F + \mu_A\, dn_A + \mu_B\, dn_B$$

On replacing each of the differentials dn_E, and so forth, by its equal in terms of $d\xi$, we obtain

$$dG = (e\mu_E + f\mu_F - a\mu_A - b\mu_B)\, d\xi$$

In this equation the chemical potentials of the reactants and products have the values corresponding to the initial and final states of the reacting system, respectively. Since these chemical potentials are not functions of ξ, we can integrate this equation from $\xi = 0$ to $\xi = 1$ (which implies that the reaction has gone to completion). We thus obtain for the free energy change of the reaction as described by Equation 4–1.

$$\Delta G = e\mu_E + f\mu_F - a\mu_A - b\mu_B \tag{4-2}$$

When each of the μ_i is replaced by its equal, $\mu_i^\circ + RT \ln P_i$, (Eq. 3–63) this equation becomes

$$\Delta G = e(\mu_E^\circ + RT \ln P_E) + f(\mu_F^\circ + RT \ln P_F) - a(\mu_A^\circ + RT \ln P_A)$$
$$- b(\mu_B^\circ + RT \ln P_B)$$

Collecting all of the terms related to standard states, we have

$$\Delta G = e\mu_E^\circ + f\mu_F^\circ - a\mu_A^\circ - b\mu_B^\circ + RT(e \ln P_E + f \ln P_F - a \ln P_A - b \ln P_B)$$

The first four terms on the right give the standard free energy change, ΔG°, which is the change in free energy due to the reaction if each of the substances involved is in its standard state. When the logarithmic terms are also combined, the result is

$$\Delta G = \Delta G^\circ + RT \ln \frac{P_E^e P_F^f}{P_A^a P_B^b} \tag{4-3}$$

The ratio of products of partial pressures in this equation is often called the reaction quotient, represented by Q. In terms of this symbol, the free energy change of the reaction is given by

$$\Delta G = \Delta G^\circ + RT \ln Q \tag{4-4}$$

Free Energy Change and Chemical Equilibrium

In contrast to the reaction quotient, which is a function of the partial pressures specified in the chemical equation, $\Delta G°$ is a function of the temperature only, since it involves the chemical potentials in the standard state.

Whether or not the reaction represented by our chemical equation can take place spontaneously depends on the sign of ΔG. If ΔG has a positive value, the reaction as written is forbidden by the second law but the reaction in the reverse direction can occur spontaneously. If ΔG is negative the reaction is possible. We cannot tell, however, whether it will proceed at a finite rate because thermodynamics tells us nothing about reaction rates. In Chapter 11 we consider the factors on which reaction rates depend.

Chemical Equilibrium

If ΔG for a chemical reaction is zero at constant T and P, the process is thermodynamically reversible and the reacting system is at equilibrium. In this case,

$$\Delta G° = -RT \ln Q_{eq}$$

where Q_{eq} is the value of the reaction quotient when the set of partial pressures is such that equilibrium exists. Since $\Delta G°$ at a definite temperature is a constant, as mentioned above, the right-hand side of this equation must also be a constant. To emphasize the fact that Q_{eq} is independent of the value of any of the partial pressures, we replace it by another symbol, K_{eq}. Thus,

$$\Delta G° = -RT \ln K_{eq} \qquad (4\text{--}5)$$

and

$$K_{eq} = e^{-\frac{\Delta G°}{RT}}$$

In view of Equation 4–3, K_{eq} is also given by

$$K_{eq} = \left(\frac{P_E^e P_F^f}{P_A^a P_B^b}\right)_{eq} \qquad (4\text{--}6)$$

We have now arrived at a most important result. Every chemical reaction carried out at constant temperature and total pressure is characterized by a constant, K_{eq}, which can be calculated either from thermodynamic data which yield the standard free energy change of the reaction, or from the partial pressures which exist at equilibrium. We will soon see that knowledge of the value of the equilibrium constant and the initial composition of the reacting system makes it possible to calculate the composition of the system at equilibrium.

The value of K_{eq} has significance only with respect to a specific chemical equation describing the reaction under consideration because the value of $\Delta G°$

depends on the coefficients used in the chemical equation. If, for example, all of the coefficients in Equation 4–1 are divided by two, $\Delta G°$ must also be divided by two, and the corresponding equilibrium constant is the square root of the constant defined above. If the equation is written in the reverse direction, the sign of $\Delta G°$ is changed and the related equilibrium constant is the reciprocal of the constant defined above. Of course, the state of the reacting system at equilibrium cannot depend on the way we write the chemical equation.

If our reaction is carried out in the usual way by introducing the reactants at definite partial pressures into a closed container, keeping the temperature and total pressure constant, the partial pressures of the reactants will decrease while those of the products will increase until equilibrium is reached. After this time all partial pressures will have constant values.

A convenient way of looking at the approach to equilibrium is provided by combining Equations 4–4 and 4–5 in the following way:

$$\Delta G = -RT \ln K_{eq} + RT \ln Q$$
$$= RT \ln \frac{Q}{K_{eq}}$$

If the value of Q for a particular state of the reacting system is less than the value of K_{eq}, ΔG is negative and the reaction can occur spontaneously. During the course of the reaction Q will increase until it reaches K_{eq}. Then $\Delta G = 0$ and no further change will be observed. If the initial value of Q is larger than K_{eq}, the reaction as represented by the chemical equation will not occur.

The equilibrium constant is defined in Equation 4–6 in terms of partial pressures and is appropriately symbolized by K_P. In some cases it is convenient to express the composition of a system in terms of molar concentrations, c, rather than partial pressures. Making use of the relation between these quantities for an ideal gas,

$$P_i = n_i \frac{RT}{V} = c_i RT$$

we obtain the relation between K_P and the equilibrium constant in terms of concentrations, K_c, as follows:

$$K_P = \frac{(c_E RT)^e (c_F RT)^f}{(c_A RT)^a (c_B RT)^b}$$
$$= \frac{c_E^e c_F^f}{c_A^a c_B^b} (RT)^{e+f-a-b}$$
$$= K_c (RT)^{\Delta n_g}$$

where Δn_g ($=e+f-a-b$) is the change in the number of moles of gas when the reaction as written is carried to completion. (The units of R must, of course, be consistent with the units used for K_P and K_c.)

Free Energy Change and Chemical Equilibrium

Similarly, we can express the equilibrium constant in terms of mole fractions, K_X, by use of Dalton's law, $P_i = X_i P$:

$$K_P = \frac{(X_E P)^e (X_F P)^f}{(X_A P)^a (X_B P)^b} = \frac{X_E^e X_F^f}{X_A^a X_B^b} P^{\Delta n_g}$$

(2) $\quad K_X = K_P P^{-\Delta n_g}$

Since $P_i = n_i \dfrac{P}{n_t}$, where n_t is the total number of moles in the system at equilibrium, we can also express the equilibrium constant, K_n, in the following way:

(3) $\quad K_n = K_P \left(\dfrac{P}{n_t}\right)^{-\Delta n_g}$

If the equilibrium partial pressures are so high that gas ideality cannot be assumed, Equation 4–6 can be made exact by substituting for each partial pressure the corresponding fugacity:

(4) $\quad K_f = \dfrac{f_E^e f_F^f}{f_A^a f_B^b}$

The most general form of the equilibrium constant is that expressed in terms of activities because, as we will see later, this form is applicable to systems that contain condensed phases as well as gases.

◈◈◈◈◈◈

At equilibrium all properties of the reaction system have constant values. As far as thermodynamics is concerned, the reaction (or process, in general) has stopped. We know from other considerations, however, that at equilibrium both forward and reverse reactions are taking place, and that each component of the system is being formed at the same rate at which it is being removed by reaction. The existence of an equilibrium constant and its relation to the composition of the system were actually found from a study of the rates of some reversible reactions before the thermodynamic approach was developed. Although the subject of reaction rates is treated in some detail in Chapter 11, we anticipate a few of the results to be obtained there in order to show the relationship of the two ways of looking at the equilibrium state.

Let us consider the reversible reaction

$$A + B \rightleftharpoons E + F$$

In the simplest cases it is found that at a constant temperature the rate of such a reaction is proportional to the concentrations of the reactants:

$$-\frac{d[A]}{dt} = k_f [A][B] \qquad (4-7)$$

where [A] and [B] represent the concentrations of substances A and B, respectively, $-d[A]/dt$ represents the rate at which the concentration of A is reduced by the reaction, and k_f (a function of the temperature only) is the rate constant of the forward reaction. Similarly, the rate of the reverse reaction is given by

$$-\frac{d[E]}{dt} = k_r[E][F]$$

in which k_r is the rate constant of the reverse reaction.

When we introduce definite concentrations of A and B into a reaction vessel, the rate of the forward reaction is initially at a maximum and then decreases as A and B are consumed. Conversely, the initial rate of the reverse reaction is zero because E and F are not present, but this rate increases as E and F are formed. Ultimately, the two rates will become equal at equilibrium. Then we have

$$k_f[A]_{eq}[B]_{eq} = k_r[E]_{eq}[F]_{eq}$$

It follows that

$$\left(\frac{[E][F]}{[A][B]}\right)_{eq} = \frac{k_f}{k_r} = K_{eq} \qquad (4\text{-}8)$$

From this point of view the equilibrium constant is the ratio of the rate constants for the forward and reverse reactions.

The above relation of K_{eq} to the composition of the system has the same form as the one we derived on a thermodynamic basis. The thermodynamic derivation is rigorous, however, and it is as general as the laws of thermodynamics. The derivation on the reaction-rate basis depends on a very simple form of the rate law which, as we see later, is actually followed by very few reactions. In general we can say that the state of equilibrium must be independent of the form of rate laws.

Equation 4–8 is sometimes called the Law of Mass Action, although this name is more appropriately given to the rate law, Equation 4–7. The name has historical significance only. Guldberg and Waage who first carried out this derivation in 1864 used the term active mass to designate what we call concentration.

Le Chatelier's Rule

Before the quantitative aspects of chemical equilibrium were fully developed, several investigators, principally Le Chatelier, formulated qualitative rules to predict the effects on a system at equilibrium when some of the variables are changed. Such a rule can be helpful as a guide when quantitative results are not needed or as a rough check on the results of calculation.

Le Chatelier's rule may be stated in the following way: If a system at equilibrium is subjected to a change, the system will react in such a way as to oppose or reduce the change if this is possible. If, for example, the pressure on a system is suddenly increased, a net reaction will occur in the direction in which the number of gas molecules is reduced, thus lowering the pressure. If some heat is added to the system to raise its temperature, the equilibrium will shift in the endothermic direction. If one of the components of the system is added, a reaction occurs to reduce the amount of this component.

Equilibrium for Gas Reactions with $\Delta n_g = 0$

In several respects those reactions in which the total number of gas molecules remains constant are easier to deal with than the others. When $\Delta n_g = 0$, it can be seen that $K_P = K_c = K_X = K_n$. It can also be seen that if we have a set of equilibrium partial pressures, multiplication of each of the partial pressures by the same factor does not change the value of K. It follows that the relative amounts of the components present at equilibrium are independent of the total pressure or volume of the system, and they are not changed by the addition of an inert gas. It also follows that the value of the equilibrium constant is independent of the units in which pressures or concentrations are expressed.

We can take as an example of a reaction with $\Delta n_g = 0$ the so-called water–gas reaction which was studied carefully early in this century. The equation for this reaction is

$$H_2(g) + CO_2(g) \rightleftharpoons H_2O(g) + CO(g)$$

When various proportions of H_2 and CO_2 were added to a reaction vessel and brought to equilibrium at 1259°K, the composition of the system was found[1] to be given by the values listed below:

Equilibrium Data at 1259°K for
$$H_2 + CO_2 \rightleftharpoons CO + H_2O$$

Composition in mole percent			
H_2	CO_2	$CO = H_2O$	K
46.93	7.15	22.96	1.58
22.95	21.22	27.90	1.60
12.67	34.43	26.45	1.60
6.85	47.50	22.82	1.60

(Notice that CO and H_2O are formed in equimolar quantities.) The value of K can be seen to be reasonably constant.

[1]Hahn, Z. physik. Chem. **44**, 513 (1903).

When the equilibrium constant is known, the composition of the system at equilibrium can be calculated from any given initial composition. Let us suppose that a moles of H_2 and b moles of CO_2 are allowed to reach equilibrium at 1259°K. If x represents the number of moles of CO at equilibrium, then there are also x moles of H_2O, $a - x$ moles of H_2, and $b - x$ moles of CO_2. It follows that

$$\frac{x^2}{(a-x)(b-x)} = 1.60 \tag{4-9}$$

The equation above is a quadratic equation in x. One of its two solutions will determine the equilibrium composition completely, while the other solution will have an unreasonable value and must be ignored. As an example, let us calculate the composition of the system when 3 moles of H_2 and 2 moles of CO_2 are allowed to reach equilibrium at 1259°K. In this case we have

$$\frac{x^2}{(3-x)(2-x)} = 1.60$$

This equation has the two solutions $x = 12$ and $x = 1.33$. The first solution is clearly inapplicable to our problem. From the second solution we find that at equilibrium the system contains 1.33 moles each of CO and H_2O, 1.67 moles of H_2, and 0.67 moles of CO_2.

For the special case in which $a = b$ in Equation 4–9, this equation reduces to the following form:

$$\left(\frac{x}{a-x}\right)^2 = K$$

and

$$\frac{x}{a-x} = \pm\sqrt{K}$$

The choice of the negative sign must be rejected because the left-hand side of this equation is necessarily positive. Then the value of x in terms of K is given by

$$x = \frac{a\sqrt{K}}{1+\sqrt{K}}$$

Gas Reactions with $\Delta n_g \neq 0$

Most gas reactions involve a change in the total number of moles of gas. We select the dissociation reaction as representative of reactions of this type. It is often important to determine the extent to which a certain compound will dissociate under specified conditions when the equilibrium constant is known.

Free Energy Change and Chemical Equilibrium

To see how the extent, or degree, of dissociation can be calculated, let us consider a reaction represented by $A \rightleftharpoons B + C$. When a moles of A are introduced into a reaction vessel and x moles of A have dissociated at equilibrium with total pressure P, the number of moles of B and C is each equal to x. The total number of moles in the system is then $a - x + 2x = a + x$. Because the partial pressure of each gas is equal to its mole fraction times the total pressure, we can express K_P in the following way:

$$K_P = \frac{P_B P_C}{P_A} = \frac{\left(\dfrac{x}{a+x} P\right)^2}{\left(\dfrac{a-x}{a+x}\right) P} = \frac{x^2}{a^2 - x^2} P \qquad (4\text{-}10)$$

Notice that although K_P is independent of P for reactions of ideal gases, P appears on the right-hand side of this equation. This fact has the following consequences:

1. At a given temperature the composition of the system at equilibrium depends on the value of P.
2. The numerical value of K_P depends on the unit of pressure, although K_P is fundamentally dimensionless. This apparent inconsistency arises from the fact, mentioned on page 94, that it is customary not to indicate that P_i in the equation $\mu_i = \mu_i^\circ + RT \ln P_i$ is actually the ratio of a partial pressure to the unit pressure. Accordingly, K_P by convention is given the units of $P^{\Delta n_g}$, and it is always necessary to indicate clearly the pressure units which are used when $\Delta n_g \neq 0$.

For a specific example of a dissociation reaction let us take the reaction represented by the equation $PCl_5 \rightleftharpoons PCl_3 + Cl_2$. At 229°C the value of K_P with P in atmospheres is found to be 0.460. When, for example, 2 moles of PCl_5 are heated at this temperature until equilibrium is reached at a total pressure of 1 atm, the composition of the system can be calculated from the following equation:

$$0.460 = \frac{x^2}{4 - x^2}$$

From the solution of this equation, $x = 1.12$, we find that the system at equilibrium consists of 1.12 moles each of PCl_3 and Cl_2, and 0.88 moles of PCl_5. Thus we conclude that $1.12/2$, or 56%, of the initial amount of PCl_5 has dissociated.

If the pressure on the system is raised to 10 atm, keeping the temperature constant, the equilibrium equation becomes

$$0.460 = \frac{10x^2}{4 - x^2}$$

and its solution is $x = 0.42$. From this result we see that raising the pressure of the system from 1 to 10 atm lowers the degree of dissociation from 56% to 21%.

On the basis of Le Chatelier's rule it can be predicted that the dissociation of a compound is partially suppressed when some of the dissociation product is initially present. To evaluate this effect, let us use the reaction discussed above, $A \rightleftarrows B + C$, as an example. When a moles of A are introduced into a reaction vessel that already contains b moles of B, at equilibrium there are $a - x$ moles of A, $b + x$ moles of B, and x moles of C. The total number of moles in the system is then $a + b + x$. The equation for K_P in this case is

$$K_P = \frac{\frac{(b+x)xP}{(a+b+x)^2}}{\frac{a-x}{a+b+x}} = \frac{(b+x)xP}{(a-x)(a+b+x)}$$

Now let us apply this equation to the PCl_5 dissociation reaction. Suppose that 2 moles of PCl_5 are dissociated in the presence of 1 mole of Cl_2 at 229°C and a total pressure of 1 atm. The equation above then becomes

$$0.460 = \frac{(1+x)x}{(2-x)(3+x)}$$

The solution of this equation is $x = 0.96$. It follows that the presence of the Cl_2 lowers the degree of dissociation from 56% to 48%.

Because the degree of dissociation α, defined by x/a, is usually the quantity of interest in the study of dissociation reactions, it is convenient to express the equilibrium equations in terms of α. Thus by dividing the numerator and denominator of Equation 4–10 by a^2, this equation takes the following form:

$$K_P = \frac{\alpha^2}{1 - \alpha^2} P$$

When the number of moles in a reacting system changes at constant temperature and pressure during the approach to equilibrium, the volume and the density of the system also change. As an example, let us take the reaction discussed above, $A \rightleftarrows B + C$. When a moles of compound A are introduced into the reaction vessel, the initial volume is given by $V_o = \dfrac{RT}{P} a$, and the volume at equilibrium is $V_{eq} = \dfrac{RT}{P}(a + x)$. It follows that

$$\frac{V_{eq}}{V_o} = \frac{a + x}{a} = 1 + \alpha$$

Consequently, the value of α, and of K_P also, can be found from volume measurements.

In a similar way, α can be calculated from density measurements. Since the density is inversely proportional to the volume, we have

$$\frac{\rho_{eq}}{\rho_o} = \frac{1}{1 + \alpha}$$

and

$$\alpha = \frac{\rho_o}{\rho_{eq}} - 1$$

When ρ_o is replaced by its equal from the ideal gas law, the result is

$$\alpha = \frac{PM_A}{RT\rho_{eq}} - 1$$

where M_A is the molecular weight of compound A.

These equations can be readily generalized for applicability to any type of dissociation reaction. If the dissociation of 1 mole of A leads to the formation of m moles of products, the ratio of the total number of moles at equilibrium to the initial number of moles is $1 - \alpha + m\alpha$. It follows that

$$\frac{\rho_o}{\rho_{eq}} = 1 - \alpha + m\alpha$$

and

$$\alpha = \frac{1}{m - 1}\left(\frac{PM_A}{RT\rho_{eq}} - 1\right)$$

Heterogeneous Equilibria Involving Pure Condensed Phases

If any of the reacting components in a system at equilibrium is present as a pure liquid or solid, that is, the reacting system is heterogeneous, the free energy change of the reaction is given by an equation of the same form as the homogeneous systems previously considered (Eq. 4-2). The general form of this equation is:

$$\Delta G = \sum_i \lambda_i \mu_i$$

where the subscript i represents any of the reactants or products and λ_i is the coefficient of component i in the related chemical equation. λ_i has a positive value for products and a negative one for reactants.

For a pure substance in a condensed phase, however, the chemical potential is very nearly independent of the pressure on the system. At any definite temperature, such a pure substance in a condensed phase can be considered to be in its standard state at that temperature. It follows that in the expression for K_P there are no factors for substances in condensed phases. Only the partial pressures of the gaseous components are included.

An example of a heterogeneous reaction is that of solid carbon with oxygen to form carbon monoxide:

$$C(s) + \tfrac{1}{2}O_2 \rightleftarrows CO$$

The expression for K_P related to this chemical equation is

$$K_P = \frac{P_{CO}}{P_{O_2}^{1/2}}$$

Whenever O_2 and CO are at equilibrium in the presence of solid carbon, the ratio of their partial pressures is determined by the value of K_P.

Standard Free Energy Change

We have seen that when the equilibrium constant for a reaction has been determined a simple calculation yields the corresponding standard free energy change. The experimental determination of an equilibrium constant, however, is in general not easy. A more common procedure is to obtain the standard free energy change, which can be done in several different ways, and then from this value calculate the equilibrium constant. It is possible to obtain an equilibrium constant from calorimetric data (heats of reaction and heat capacities) without making any measurements on the system at equilibrium. In fact, it is possible to calculate the equilibrium constant for a reaction which has never been carried out!

Because the free energy is a thermodynamic function, $\Delta G°$ for a reaction can be obtained by simple arithmetic if the equation for this reaction can be considered as the sum or difference of equations for other reactions whose $\Delta G°$ values are known at the temperature of interest. There is a close similarity here with calculations involving $\Delta H°$.

Suppose, for example, that K_P at 298°K is needed for the reaction

$$S(\text{rhombic}) + \tfrac{3}{2}O_2 = SO_3(g)$$

and that this is difficult to determine directly. The value of $\Delta G°_{298}$ for this reaction can be obtained from the known $\Delta G°$ values of the following two reactions:

$$\begin{aligned}
\text{S(rhombic)} + \text{O}_2 &= \text{SO}_2(g); \Delta G° = -71.79 \text{ kcal} \\
\text{SO}_2(g) + \tfrac{1}{2}\text{O}_2 &= \text{SO}_3(g); \Delta G° = -16.71 \text{ kcal} \\
\hline
\text{S(rhombic)} + \tfrac{3}{2}\text{O}_2 &= \text{SO}_3(g); \Delta G° = -88.50 \text{ kcal}
\end{aligned}$$

It is clear that a table of $\Delta G°$ values for a variety of reactions could be helpful. Because such a table for a large number of chemical reactions would be bulky, it is more convenient to tabulate the standard free energies of formation of compounds $\Delta G_f°$. The standard free energy of formation of a compound is defined as the free energy change when 1 mole of the compound in its standard state is formed from elements in their standard states. By convention the free energy of formation of the elements is set equal to zero. The standard free energy of formation is quite analogous to the standard heat of formation defined on page 49. The standard free energies of formation of many compounds have been determined and tabulated for 298°K (Table 4-1). From

TABLE 4-1 STANDARD FREE ENERGY
OF FORMATION AT 25°C IN KCAL/MOLE

$H_2O(g)$	−54.636
$H_2O(l)$	−56.690
$CO_2(g)$	−94.260
$HCl(g)$	−22.769
$SO_2(g)$	−71.79
$SO_3(g)$	−88.52
$NH_3(g)$	−3.976
$CH_4(g)$	−12.140
$C_2H_6(g)$	−7.860
$C_2H_4(g)$	16.282
$C_2H_2(g)$	50.000
$CH_3OH(l)$	−39.73
$C_6H_6(l)$	29.756

such data it is easy to calculate the standard free energy change for any reaction which involves these compounds by use of the following equation:

$$\Delta G° = \Sigma \Delta G_f°(\text{products}) - \Sigma \Delta G_f°(\text{reactants})$$

Another way of obtaining $\Delta G°$ values makes use of the basic relationship: $\Delta G° = \Delta H° - T\Delta S°$. We have seen in Chapter 2 how $\Delta H°$ for a reaction can be calculated from tabulated values of $\Delta H_f°$ for compounds. All that is needed is the value of $\Delta S°$ for the reaction. It would be possible to define a standard entropy of formation analogous to $\Delta H_f°$ and $\Delta G_f°$. This is not done however; instead another approach to entropy change is employed. This difference in treatment results from the fact that there is a difference between entropy on the one hand, and energy, enthalpy, and free energy, on the other. It was pointed out earlier that in thermodynamics we do not know the absolute values of E, H, or G, but only differences in these properties resulting from processes. In contrast, by use of the third law of thermodynamics we can calculate an absolute value of the entropy of a substance. From such entropy values, known as third-law entropies, $\Delta S°$ for a reaction can be calculated.

Third Law of Thermodynamics: Third-Law Entropies

For a pure substance the entropy change with temperature at constant pressure is given by (Eq. 3-31)

$$\left(\frac{\partial S}{\partial T}\right)_P = \frac{C_P}{T}$$

and

$$\Delta S = S_{T_2} - S_{T_1} = \int_{T_1}^{T_2} \frac{C_P}{T} dT \qquad (3\text{-}36)$$

Let us take as our system 1 mole of pure substance in the crystalline state and consider what happens as T_1 approaches $0°K$.

It is an experimental fact that as the temperature is lowered the heat capacity of any crystalline substance decreases and seems to approach a value of zero as a limit. In 1912 Debye derived a theoretical equation for the heat capacity of a crystal which fits measurements very well in the lowest temperature

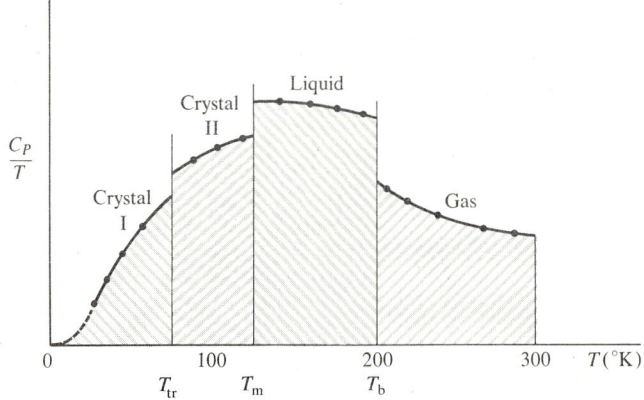

Figure 4–1. Schematic illustration of the calculation of the third-law entropy of a gas at 298°K by graphical integration. T_{tr}, T_m, and T_b are respectively the temperature of transition between two crystalline forms, the melting point of Crystal II, and the normal boiling point. The graph line is extrapolated from 21°K to 0°K by use of the Debye T^3 law.

range in which heat capacity measurements can be made without extreme difficulty ($\sim$10 to 20°K). In this range the Debye equation is simply $C_P = \alpha T^3$ where α is a constant for each kind of crystal. It follows that not only C_P but also C_P/T ($= \alpha T^2$) approaches zero as $T \rightarrow 0$°K (Fig. 4–1).

On the basis of these and related considerations, Planck in 1913 proposed a generalization which has become known as the *Third Law of Thermodynamics*: For an ideal [2] single crystal of a pure substance at 0°K, the value of the entropy is zero. (It can be shown that this is closely related to the statement that it is impossible to reach a temperature of 0°K.)

Since $S_0 = 0$ according to the third law, we conclude that the entropy of a crystalline substance at temperature T is given by

$$S_T = \int_0^T \frac{C_P}{T} dT$$

The value of this integral is usually found by graphical integration. When heat capacity measurements have been made at a series of temperatures down to about 10 or 15°K, the constant α in the Debye "T^3 equation" is evaluated. This equation is used to extrapolate the graph of C_P/T versus temperature to 0°K. The entropy at temperature T is then obtained from the area under the C_P/T curve from 0°K to T.

When the crystal is melted, the resulting liquid has a heat capacity which is different from that of the crystal. Consequently there is a discontinuity in the C_P/T curve at the melting point. In order to obtain the entropy of a substance in the liquid phase at a temperature T, the entropy change for raising the temperature of the liquid from the melting point to T is obtained by graphical integration. Then this quantity is added to the entropy of fusion ($= \Delta H_f/T_{mp}$) and the entropy of the crystal at T_{mp}.

At the normal boiling point there is another discontinuity in the C_P/T graph because the heat capacity of the vapor is different from that of the liquid. The entropy of the vapor at temperature T is the sum of five terms which may be represented in the following way:

$$S_T \text{(vapor)} = \Delta S_{0 \rightarrow T_{mp}} \text{(crystal)} + \frac{\Delta H_f}{T_{mp}} + \Delta S_{T_{mp} \rightarrow T_{bp}} \text{(liquid)}$$
$$+ \frac{\Delta H_v}{T_{bp}} + \Delta S_{T_{bp} \rightarrow T} \text{(vapor)}$$

Many substances have different crystalline forms which exist in different temperature ranges. In such cases a calculation of the entropy by use of the third law must include all entropies of transition from one form to another obtained from the heat of transition and the transition temperature.

As an example of the evaluation of a third-law entropy we calculate the

[2] What is implied by the word ideal is not clear at this point. It is discussed in Chapter 9.

entropy of benzene vapor at the normal boiling point from the data of Oliver, Eaton, and Huffman.[3] The lowest temperature at which these measurements of heat capacity were made is 13°K. From the value of C_P for crystalline benzene at this temperature, 0.685 cal/mole deg, the value of α in the Debye T^3 equation is

$$\alpha = \frac{0.685}{(13)^3} = 3.12 \times 10^{-2}$$

With the use of this value of α, the graph of C_P/T (Fig. 4–1) can be extrapolated to 0°K. From the area under the graph line the entropy of crystalline benzene at its melting point, 278.7°K, is found to be 30.79 cal/mole deg.

The heat of fusion of benzene is 2358 cal/mole and therefore the entropy of fusion is 2358/278.7 = 8.46 cal/mole deg. The integral of C_P/T for liquid benzene from the melting point to the boiling point at 1 atm, 353.3°K, has a value of 7.93 cal/mole deg. The heat of vaporization at the boiling point, 7350 cal/mole, yields a value of the entropy of vaporization of 20.80 cal/mole deg. Thus the third-law entropy of benzene vapor at 353.3°K and 1 atm pressure is 30.79 + 8.46 + 7.93 + 20.80 = 67.98 cal/mole deg.

A selection of third-law entropies at 298°K is given in Table 4–2. It should be noted that elements as well as compounds are included.

TABLE 4-2 THIRD-LAW ENTROPIES AT 25°C IN CAL/MOLE DEG

Compounds		Elements	
$H_2O(g)$	45.11	$H_2(g)$	31.21
$H_2O(l)$	16.72	$O_2(g)$	49.00
$CO_2(g)$	51.06	$N_2(g)$	45.77
$HCl(g)$	44.62	$Cl_2(g)$	53.29
$SO_2(g)$	59.40		
$SO_3(g)$	61.24		
$NH_3(g)$	46.01		
$CH_4(g)$	44.50		
$C_2H_6(g)$	54.85		
$C_2H_4(g)$	52.45		
$C_2H_2(g)$	48.00		
$CH_3OH(l)$	30.3		
$C_6H_6(l)$	41.30		

The third law is of considerable significance in the treatment of entropy from the molecular point of view, and we return to it in Chapter 9. For the present it is sufficient to say that entropy data such as those in Table 4–2 make it possible to calculate values of ΔS°_{298} for chemical reactions. When these

[3] Oliver, Eaton, and Huffman, *J. Am. Chem. Soc.* **70**, 1502 (1948).

Change of Equilibrium Constant with Temperature

To obtain the value of ΔG°_{298} (and therefore K_P at 298°K) by itself is of very limited usefulness because it would be only a coincidence if 298°K happened to be the temperature of interest. Clearly we need the ability to calculate K_P at any temperature at which a reaction might be carried out.

We already have the relationships which make this calculation possible. There are several alternative approaches which are essentially the same. One way is to start from the equation

$$\ln K_P = -\frac{\Delta G^\circ}{RT}$$

On differentiating both sides of this equation with respect to T, we obtain

$$\frac{d \ln K_P}{dT} = -\frac{1}{R}\frac{d\frac{\Delta G^\circ}{T}}{dT} \qquad (4\text{-}11)$$

(The differentials in this equation are written as total differentials because K_P and ΔG° are functions of T only.) We previously found (Eq. 3-51) that the differential coefficient on the right-hand side is equal to $-\Delta H^\circ/T^2$. Consequently Equation 4-11 can be written in the following form:

$$d \ln K_P = \frac{\Delta H^\circ}{RT^2} dT$$

in which ΔH° is the heat of the reaction at temperature T.

We can integrate both sides of this equation either between limits or indefinitely. If we know the equilibrium constant at one temperature, T_1 (usually 298°K), and want to obtain its value at another temperature, T_2, we integrate between these temperature limits. We can express the result in the following way:

$$\ln K_P(T_2) = \ln K_P(T_1) + \frac{1}{R}\int_{T_1}^{T_2} \frac{\Delta H^\circ}{T^2} dT \qquad (4\text{-}12)$$

If the temperature range is so small that ΔH° can be considered as constant over this range, this equation becomes

$$\ln K_P(T_2) = \ln K_P(T_1) + \frac{\Delta H^\circ}{R}\left(\frac{T_2 - T_1}{T_1 T_2}\right)$$

It is apparent that the equilibrium constant for an exothermic reaction decreases with rising temperature, in agreement with Le Chatelier's rule.

If the temperature range is too large for this approximation to be adequate, we must know $\Delta H°$ as a function of the temperature in order to carry out this integration. In Chapter 2 we obtained this function (Eq. 2-27) by integrating $d\,\Delta H° = \Delta C_P\,dT$, in which ΔC_P is obtained from empirical heat capacity equations for gases. This integration yielded

$$\Delta H_T° = \Delta H_o° + \Delta a\,T + \frac{\Delta b}{2}T^2 + \frac{\Delta c}{3}T^3$$

where $\Delta H_o°$ is a constant of integration. When this function is substituted for $\Delta H°$ in Equation 4-12, the result is

$$\ln K_P(T_2) = \ln K_P(T_1) + \frac{1}{R}\int_{T_1}^{T_2}\left(\frac{\Delta H_o°}{T^2} + \frac{\Delta a}{T} + \frac{\Delta b}{2} + \frac{\Delta c}{3}T\right)dT$$

On carrying out the indicated integrations we have

$$\ln K_P(T_2) = \ln K_P(T_1) + \frac{\Delta H_o°}{R}\frac{T_2 - T_1}{T_1 T_2} + \frac{\Delta a}{R}\ln\frac{T_2}{T_1} + \frac{\Delta b}{2R}(T_2 - T_1)$$
$$+ \frac{\Delta c}{6R}(T_2^2 - T_1^2)$$

When values are desired for the equilibrium constant of a reaction at a series of temperatures, these integrations are carried out indefinitely and the constant of integration is calculated from the value of K_P (or $\Delta G°$) at one temperature. In the case that $\Delta H°$ can be considered as constant, the equation for K_P is

$$\ln K_P = -\frac{\Delta H°}{RT} + \text{constant}$$

In the general case this equation is

$$\ln K_P = \frac{1}{R}\left(-\frac{\Delta H_o°}{T} + \Delta a\ln T + \frac{\Delta b}{2}T + \frac{\Delta c}{6}T^2\right) + \text{constant}$$

The results of this section can be summarized as follows: From the heat of reaction and the equilibrium constant (or standard free energy change) at one temperature and the heat capacities of all reactants and products as functions of the temperature, we can calculate the equilibrium constant at any temperature within the range of the heat capacity data.

Free Energy Function

We have seen that a complete solution for the problem of expressing the equilibrium constant for a reaction as a function of the temperature is provided by the use of standard heats of formation and third-law entropies, together with

Free Energy Change and Chemical Equilibrium

heat capacity data. This solution is not the most convenient one, however, because two integrations over a number of terms are needed for a large temperature range.

Another approach which is widely adopted for the tabulation of thermodynamic data for pure substances involves the use of the free energy function which is defined by $(G° - H_0°)/T$. There are two reasons for the choice of this function:

1. The rate of change of the free energy function with temperature is relatively small. Consequently, for nearly all purposes it is sufficient to tabulate values of this function at a few temperatures and to use linear interpolation for intermediate temperatures.

2. It is possible to calculate the free energy function for gases by a theoretical method based on statistical mechanics. This method is developed in Chapter 9.

For a pure crystalline substance to which the third law is applicable, the free energy function can be calculated from heat capacity data alone. From

$$S_T° = \int_0^T \frac{C_P}{T} dT$$

and

$$H_T° = H_0° + \int_0^T C_P \, dT$$

we can express $G_T°$ in the following way:

$$G_T° = H_T° - TS_T°$$

$$= H_0° + \int_0^T C_P \, dT - T \int_0^T \frac{C_P}{T} dT$$

It follows that

$$\frac{G° - H_0°}{T} = \frac{1}{T}\int_0^T C_P \, dT - \int_0^T \frac{C_P}{T} dT$$

If the crystalline substance undergoes a phase transition at a temperature between 0°K and T, the heat of transition must be included in the enthalpy and entropy terms. For substances that are liquid in the standard state at temperature T, the heat of fusion, and, for gases, the heat of vaporization also must be included to obtain the free energy function.

For a chemical reaction at temperature T the change in the free energy function can be expressed as follows:

$$\Delta\left(\frac{G° - H_0°}{T}\right) = \sum\left(\frac{G° - H_0°}{T}\right)_{\text{products}} - \sum\left(\frac{G° - H_0°}{T}\right)_{\text{reactants}}$$

$$= \frac{\Delta G°}{T} - \frac{\Delta H_0°}{T}$$

When the relationship between $\Delta G°$ and $\ln K_P$ is introduced, this equation becomes

$$\Delta\left(\frac{G° - H_0°}{T}\right) = -R \ln K_P - \frac{\Delta H_0°}{T}$$

In order to use the equation above to calculate an equilibrium constant from tabulated values of the free energy function it is necessary to evaluate $\Delta H_0°$ for the reaction. This can be done in several different ways, depending on the available data:

1. If the equilibrium constant is known at some temperature T, $\Delta H_0°$ can be found from a slight modification of the equation above:

$$\Delta H_0° = -RT \ln K_P - T \Delta\left(\frac{G° - H_0°}{T}\right)$$

2. If the heat of reaction is known at some temperature T and equations for C_P as a function of the temperature have been developed for all reactants and products, $\Delta H_0°$ can be calculated from the following equation:

$$\Delta H_0° = \Delta H_T° - \int_0^T \Delta C_P \, dT$$

3. The method used most frequently is to adopt the convention that the enthalpies of all elements, considered to be in their standard states at temperature T, are equal to zero at $0°K$ On this basis, the enthalpy change for the formation of 1 mole of compound from its elements evaluated at $0°K$ can be

TABLE 4-3 FREE-ENERGY FUNCTION OF GASES*

Substance	$-\frac{G° - H_0°}{T}$ (cal/mole deg)			$\Delta H_{f(0)}°$ (kcal/mole)
	298°K	500°K	1000°K	
H_2O	37.17	41.29	47.01	−57.11
CO_2	43.56	47.67	54.11	−93.97
CO	40.25	43.86	48.77	−27.20
SO_2	50.82	55.38	62.28	−70.36
SO_3	51.89	57.14	66.08	−93.06
CH_4	36.46	40.75	47.65	−15.99
C_2H_4	43.98	48.74	57.29	14.52
H_2	24.42	27.95	32.74	—
O_2	42.06	45.68	50.70	—
N_2	38.82	42.42	47.31	—
Cl_2	45.93	49.85	55.43	—

*G. N. Lewis, M. Randall, K. S. Pitzer, and L. Brewer, *Thermodynamics*, McGraw-Hill, New York, 1961. This reference includes extensive tables of the free-energy function.

Free Energy Change and Chemical Equilibrium

considered to be the standard heat of formation at $0°K$, $\Delta H°_{f(0)}$, in close analogy to the standard heat of formation, $\Delta H°_{f298}$, defined on page 49. Tables of the free energy function are usually accompanied by values of $\Delta H°_{f(0)}$ for the compounds included in the tabulation. In this case, the calculation of $\Delta H°_0$ for a reaction is reduced to

$$\Delta H°_0 = \Sigma \Delta H°_{f(0)} \text{ (products)} - \Sigma \Delta H°_{f(0)} \text{ (reactants)}$$

As an example of the use of the free energy function let us calculate K_P at $1000°K$ for the reaction $CO + \tfrac{1}{2}O_2 = CO_2$. For this reaction,

$$R \ln K_P = -\frac{\Delta G°}{1000}$$

$$= -\left(\frac{G° - H°_0}{1000}\right)_{CO_2} + \frac{1}{2}\left(\frac{G° - H°_0}{1000}\right)_{O_2} + \left(\frac{G° - H°_0}{1000}\right)_{CO} - \frac{\Delta H°_{f(0)}(CO_2) - \Delta H°_{f(0)}(CO)}{1000}$$

On inserting the appropriate values from Table 4–3 we find

$$R \ln K_P = 54.11 - 25.35 - 48.77 - \frac{-93{,}970 + 27{,}200}{1000}$$

$$= 46.76 \text{ cal/deg}$$

It follows that

$$\log K_P = 10.22$$

and

$$K_P = 1.7 \times 10^{10}$$

◈◈◈◈◈◈◈

Problems

1. Hydrogen and nitrogen react according to the equation $3H_2 + N_2 = 2NH_3$. When a system containing hydrogen, nitrogen, and ammonia reached equilibrium at $400°C$, the partial pressures of the 3 gases were found to be: $P_{H_2} = 12.7$ atm, $P_{N_2} = 17.9$ atm, and $P_{NH_3} = 2.45$ atm. Calculate (a) K_P and (b) $\Delta G°_{400°}$ for this reaction.   Ans. (a) 1.64×10^{-4}; (b) 11.7 kcal

2. At $1105°K$, the value of K_P for the reaction $SO_2 + \tfrac{1}{2}O_2 = SO_3$ is 0.63. (a) Calculate the standard free energy change for this reaction at $1105°C$. (b) Calculate the free energy change at $1105°C$ for the reaction:

$$SO_2(1 \text{ atm}) + \tfrac{1}{2}O_2(25 \text{ atm}) = SO_3(2 \text{ atm})$$

(c) What is the significance of the signs of the answers in parts (a) and (b)?
   Ans. (a) 1.01 kcal; (b) -1.00 kcal

3. In a study of the water-gas reaction, $CO_2(g) + H_2(g) = CO(g) + H_2O(g)$, a mixture of CO_2 and H_2 initially containing 42.4 mole % H_2 was brought to equilibrium in a closed vessel at 986°C. The system was then found to contain 15.2 mole % H_2. Calculate K_P and $\Delta G°$ for this reaction at 986°C.
   Ans. $K_P = 1.60$, $\Delta G° = -1.17$ kcal

4. For the reaction $H_2(g) + I_2(g) = 2HI(g)$, $K_P = 54.8$ at 417°C. (a) When a mixture initially containing 8 moles of HI, 2 moles of H_2, and 0.5 mole of I_2 is heated to 417°C, will more HI be formed? (2) What is the maximum amount of HI that can be formed when 1 g of H_2 and 100 g of I_2 are heated to 417°C in a 2-liter vessel? (c) A mixture of 1 mole each of H_2, I_2, and HI was brought to equilibrium at 417°C. What was the composition of the system?

5. When a mixture of 1.0 mole of SO_2 and 0.5 mole of O_2 was brought to equilibrium at 1000°K and 1 atm, chemical analysis showed that 46% of the SO_2 had reacted to form SO_3 according to the equation

$$SO_2(g) + \tfrac{1}{2}O_2(g) = SO_3(g).$$

(a) Calculate K_P for this reaction at 1000°K. (b) The pressure of the system was then raised isothermally to 5 atm. What was the composition of the system?
   Ans. (a) 1.85; (b) $X_{SO_2} = 0.32$, $X_{SO_3} = 0.52$

6. Sulfuryl chloride dissociates according to the equation $SO_2Cl_2(g) = SO_2 + Cl_2$ and the value of K_P at 112°C is 2.74 atm. 5.0 moles of sulfuryl chloride are brought to equilibrium at a temperature of 112°C and a total pressure of 1 atm. Find (a) the composition of the system in mole % and (b) the degree of dissociation.

7. H_2S dissociates according to the equation $2H_2S(g) = 2H_2(g) + S_2(g)$. At 1125°C and a total pressure of 1 atm, the degree of dissociation of H_2S is 0.305. Calculate K_P for this reaction at 1125°C.   Ans. 0.25

8. HCl and O_2 react according to the equation

$$4HCl(g) + O_2(g) = 2H_2O(g) + 2Cl_2(g)$$

When 1.00 mole of HCl and 0.48 mole of O_2 were added to a vessel held at 386°C and 1 atm until equilibrium was reached, 0.40 mole of Cl_2 was formed. Calculate K_P.

9. $COCl_2$ gas dissociates according to the equation $COCl_2 = CO + Cl_2$. When $COCl_2$ is heated to 724°K at 1 atm, the density of the gas mixture at equilibrium is 1.16 g/liter. Calculate (a) the degree of dissociation, (b) K_P, and (c) $\Delta G°$ for this reaction at 724°K.   Ans. (a) 0.43; (b) 0.23; (c) 2.1 kcal

10. N_2O_4 dissociates according to the equation $N_2O_4(g) = 2NO_2(g)$. When 0.578g of N_2O_4 was introduced into a 1-liter flask maintained at 35°C, the equi-

librium pressure was 0.238 atm. Calculate (a) the degree of dissociation and (b) K_P at this temperature. **Ans.** (a) 0.50; (b) 0.32 atm

11. PCl_5 dissociates according to the equation $PCl_5(g) = PCl_3(g) + Cl_2(g)$. When 0.030 mole of PCl_5 was brought to equilibrium at 229°C and 1 atm, the volume of the system was 2.09 liters. Calculate (a) the degree of dissociation and (b) K_P.

12. The value of K_P in atmospheres for the dissociation reaction $PCl_5 = PCl_3 + Cl_2$ is 1.78 at 250°C. Calculate the degree of dissociation when 0.20 mole of PCl_5 is brought to equilibrium in a 3-liter vessel at 250°C.

13. Solid NH_4HS dissociates according to the equation

$$NH_4HS(s) = NH_3(g) + H_2S(g)$$

The dissociation pressure of solid NH_4HS is 0.66 atm at 25°C. (a) Calculate K_P for this reaction. (b) What fraction of the solid will dissociate when 0.1 mole of NH_4HS is introduced into a 1-liter flask at 0°C? (c) What fraction will dissociate when 0.1 mole of NH_4HS is introduced into a 1-liter flask that contains NH_3 at 0.2 atm and 25°C? **Ans.** (a) 0.11; (b) 0.13; (c) 0.10

14. Ammonium carbamate dissociates according to the equation

$$NH_4CO_2NH_2(s) = CO_2(g) + 2NH_3(g)$$

The dissociation pressure at 30°C is 125 torr. (a) Calculate K_P. (b) What is the minimum weight of ammonium carbamate which must be added to a 3-liter vessel to establish equilibrium between the solid compound and its dissociation products at 30°C?

15. For the reaction

$$C_2H_4(g) + H_2(g) = C_2H_6(g)$$

(a) Calculate K_P at 25°C from the data in Table 4–1. (b) Calculate K_P from the data in Tables 2–2 and 4–2. (c) Calculate K_c.

16. Using tabulated data, calculate ΔG at 298°K for the reaction

$$C_2H_2(10 \text{ atm}) + 2H_2(0.1 \text{ atm}) = C_2H_6(5 \text{ atm})$$

Ans. 55.6 kcal

17. For the reaction $2SO_2 + O_2 = 2SO_3$, calculate K_P at 100°C, assuming that ΔC_P for the reaction is essentially zero. Use data in Tables 2–2 and 4–2.

18. The density of an equilibrium mixture of N_2O_4 and NO_2 at 1 atm is 3.62 g/liter at 15°C and 1.84 g/liter at 75°C. What is the heat of reaction for $N_2O_4 = 2NO_2$? **Ans.** 17.5 kcal

19. Calculate the equilibrium partial pressure of HCl when 3 moles of H_2 and 2 moles of Cl_2 are heated to 727°C at a total pressure of 10 atm. Use the following C_P values:

$$HCl, 6.73 + 0.43 \times 10^{-3}T \text{ cal/mole deg}$$
$$H_2, 6.95 - 0.20 \times 10^{-3}T \text{ cal/mole deg}$$
$$Cl_2, 7.58 + 2.42 \times 10^{-3}T \text{ cal/mole deg}$$

20. Mercuric oxide dissociates according to the equation

$$HgO(s) = Hg(g) + \tfrac{1}{2}O_2(g)$$

The dissociation pressure is 387 torr at 420°C and 810 torr at 450°C. For this reaction, calculate (a) ΔH and (b) ΔS at 450°C.

Ans. (a) 36.6 kcal; (b) 48.8 cal/deg

21. Using the data in Table 4–3, calculate the equilibrium constant at 1000°K for the reaction

$$SO_2(g) + \tfrac{1}{2}O_2(g) = SO_3(g)$$

5

Kinetic Theory of Gases

Introduction

Up to this point in our treatment of physical chemistry we have not made use of any ideas concerning the structure of matter. (Although we have referred to the terms moles and molecular weight, these quantities can be defined in a way that does not depend on the concept of molecule.) We have obtained results of great significance and generality, but we have not been able to calculate any of the properties of matter on a purely theoretical basis. If physical chemistry were limited to thermodynamics, many important aspects of the behavior of matter would be omitted and we would be left with a feeling of incompleteness.

The desire to interpret and understand the behavior of matter in terms of the existence and interaction of elementary particles can be traced back to the days of the Greek philosophers and underlies much research in physics and chemistry today. For many centuries the ideas introduced to satisfy this desire remained vague and entirely speculative. By the eighteenth century, however, a picture of the structure of gases was developed by Bernoulli which resembles the one we accept today.

A major advance in the development of a theory of matter was the publication early in the nineteenth century of Dalton's Atomic Theory which accounted for the laws of definite and multiple proportions for chemical compounds by assuming the existence of atoms. An initial confusion between the concepts of atom and molecule caused some difficulty which was eventually resolved by the use of Avogadro's hypothesis, but the power of Dalton's ideas in explaining some of the basic facts of chemistry was one of the important factors in the evolution of this science. Also during the nineteenth century physicists, notably Maxwell, Clausius, and Boltzmann, developed a detailed theory of the behavior of gases, the kinetic theory, which has become an essential part of physical chemistry.

In discussing the kinetic theory we start with the simplest model of a gas and we use this model to derive the values of measurable properties of gases which can be compared with experimental data. A model in the present context is basically a set of assumptions which is thought to be adequate to account for certain properties and which is simple enough for us to explore its implications by means of mathematics. When our model turns out to be inadequate, we modify it or replace it with another. It is not believed to be a "true-to-life" picture of a real system.

Kinetic-Theory Model of an Ideal Gas

The choice of a model is guided by consideration of the properties observed to be common to all gases. Because gases interpenetrate and diffuse relatively rapidly, are easily compressible, and rapidly fill all available space, we can make the following assumptions:

1. A gas consists of a large number of particles (called molecules) whose volume is negligible and which contains all of the mass of the gas.
2. The molecules are in rapid motion which is completely random and independent of direction (isotropic).
3. No forces act on a molecule except at the instant of a collision.
4. All collisions are completely elastic; that is, there is no transfer of molecular kinetic energy into other forms of energy.
5. The laws of classical mechanics, in particular Newton's second law of motion, are applicable to the molecules.

Gas Pressure

We now use our model to account for the existence of the pressure of a gas as a result of a very high rate of molecular collisions with the walls of the gas container. We assume that we have a gas consisting of n molecules, each of mass m, in a cubical container of edge-length l. (The choice of this shape for the container is a matter of convenience and involves no loss of generality because the pressure of a gas is independent of the shape of the container.) We also assume that the walls of the container are smooth on a molecular scale and that they are oriented perpendicular to the x, y, and z axes of a Cartesian coordinate system.

Let us focus on a molecule which is about to strike a wall located in the y-z plane. This molecule has a *velocity* **v**, a vector with components v_x, v_y, and v_z along the coordinate axes. (We use the word *speed* for the magnitude of v with no sense of direction.) For the velocity vector we have the general equation:

$$v^2 = v_x^2 + v_y^2 + v_z^2$$

Kinetic Theory of Gases

A collision with a wall changes only the velocity component that is perpendicular to the wall. Because the collision is elastic and the wall is assumed stationary, only the sign of this velocity component is changed. The resulting change in momentum, Δp_x, is

$$m[v_x - (-v_x)] = 2mv_x$$

After traversing a distance l in the $-x$ direction, the molecule will strike the other wall that is perpendicular to the x axis. The time interval between successive collisions on the two walls is l/v_x, and the reciprocal of this quantity is the number of collisions with both walls per unit time. The momentum change per unit time for collisions of one molecule with both walls is the product of the momentum change per collision and the number of collisions per unit time. Consequently,

$$\frac{\Delta p_x}{\Delta t} = 2mv_x \cdot \frac{v_x}{l} = \frac{2mv_x^2}{l}$$

For collisions at all six walls we have

$$\frac{\Delta p}{\Delta t} = \frac{2m}{l}(v_x^2 + v_y^2 + v_z^2) = \frac{2mv^2}{l}$$

To obtain the total momentum change per unit time for all n molecules in the container, we add the contributions of each molecule; thus,

$$\frac{\Delta p_{tot}}{\Delta t} = \sum_{i=1}^{i=n} \frac{2mv_i^2}{l} = \frac{2m}{l} \Sigma v_i^2$$

Introducing the mean square speed defined by

$$\overline{v^2} = \frac{\Sigma v_i^2}{n}$$

we have

$$\frac{\Delta p_{tot}}{\Delta t} = \frac{2mn\overline{v^2}}{l}$$

The change in momentum of the molecules per unit time or, more precisely, the rate of change of momentum, is by Newton's second law the force exerted by the walls on the molecules. By Newton's third law, this force is numerically equal to the force exerted on the walls by the molecules. Recalling the definition of pressure as force per unit area, we obtain the basic relationship of the kinetic theory:

$$P = \frac{f}{A} = \frac{2mn\overline{v^2}}{6l^2 \cdot l} = \frac{1}{3}\frac{nm\overline{v^2}}{V} \qquad (5\text{--}1)$$

With m in grams, $\overline{v^2}$ in (cm/sec)² and the volume V in cm³, this equation yields the pressure in dynes/cm².

Equation 5–1 can be written in the following form:

$$PV = \tfrac{2}{3}n(\tfrac{1}{2}m\overline{v^2}) = \tfrac{2}{3}n\bar{\epsilon} \tag{5-2}$$

where $\bar{\epsilon}$ is the average kinetic energy of translation per molecule. With Avogadro's number ($\mathfrak{N}$) of molecules under consideration, this becomes

$$PV_m = \tfrac{2}{3}\mathfrak{N}\bar{\epsilon} = \tfrac{2}{3}E_K$$

where V_m is the molar volume and E_K, the translational kinetic energy of a mole of ideal gas, is equal to $\tfrac{1}{2}\mathfrak{N}m\overline{v^2}$.

For this theoretical equation to be consistent with the ideal gas law, we must have

$$PV_m = \tfrac{2}{3}E_K = RT$$

and consequently

$$E_K = \tfrac{3}{2}RT \tag{5-3}$$

We have thus arrived at the important conclusion that the translational energy of an ideal gas is directly proportional to the absolute temperature. We can express this result in terms of the average molecular kinetic energy:

$$\bar{\epsilon} = \frac{3RT}{2\mathfrak{N}} = \frac{3}{2}kT$$

The constant $k\left(=\dfrac{R}{\mathfrak{N}}\right)$, which is known as Boltzmann's constant, has the value 1.38×10^{-16} ergs/molecule degree.

The average molecular kinetic energy can be expressed as a sum of contributions associated with each of the velocity components as follows:

$$\bar{\epsilon} = \tfrac{1}{2}m\overline{v^2} = \tfrac{1}{2}m\overline{v_x^2} + \tfrac{1}{2}m\overline{v_y^2} + \tfrac{1}{2}m\overline{v_z^2} \tag{5-4}$$

Because $\bar{\epsilon} = \tfrac{3}{2}kT$ and the molecular motion is isotropic, we conclude that the average molecular kinetic energy associated with any one of the components of the velocity is $\tfrac{1}{2}kT$. This is a particular example of a more general theoretical conclusion known as the law of equipartition of energy which is discussed later in this chapter.

From Equation 5–3 it follows that all ideal gases at the same temperature have the same molar (or average molecular) kinetic energy of translation. Therefore for two different gases at the same temperature, with $M\,(=\mathfrak{N}m)$ representing the molecular weight, we have

$$\tfrac{1}{2}M_1\overline{v_1^2} = \tfrac{1}{2}M_2\overline{v_2^2}$$

and

$$\frac{\sqrt{\overline{v_1^2}}}{\sqrt{\overline{v_2^2}}} = \sqrt{\frac{M_2}{M_1}}$$

Kinetic Theory of Gases

We can express this equation in terms of the root-mean-square speed, defined by $v_{rms} = \sqrt{\overline{v^2}}$, as follows:

$$\frac{v_{rms(1)}}{v_{rms(2)}} = \sqrt{\frac{M_2}{M_1}}$$

From Equation 5-3 we can obtain the following value of v_{rms} in terms of the properties of the gas:

$$v_{rms} = \sqrt{\frac{3RT}{M}} \tag{5-5}$$

For example, the value of v_{rms} for hydrogen at 27°C is given by

$$v_{rms} = \left(\frac{3 \times 8.32 \times 10^7 \times 300}{2}\right)^{\frac{1}{2}}$$

$$= 1.93 \times 10^5 \text{ cm/sec} \sim 4300 \text{ miles/hr}$$

Since the density $\rho = PM/RT$ from the ideal gas law, an alternative expression for v_{rms} is the following:

$$v_{rms} = \sqrt{\frac{3P}{\rho}}$$

The magnitude of v_{rms} is of the same order as the speed of sound in the same gas and at the same temperature. It should be noted that v_{rms} is *not* the same as the average molecular speed. We will soon see what the difference is between these two quantities.

It is easy to show that Equation 5-2 is consistent with Dalton's law of partial pressures (Eq. 1-3). For a mixture of ideal gases at a temperature T and a total pressure P_{tot} in a container of volume V, we have

$$P_{tot} = \frac{2\bar{\epsilon}}{3V} n_{tot}$$
$$= \frac{2\bar{\epsilon}}{3V} n_1 + \frac{2\bar{\epsilon}}{3V} n_2 + \cdots$$
$$= P_1 + P_2 + \cdots$$

Distribution of Molecular Speeds

According to our model of an ideal gas, the molecular motion is completely random and chaotic in both direction and speed. The molecules continuously collide with each other and exchange momentum among themselves. At any instant some molecules undergo collisions which leave them with very small speeds, while others acquire speeds far above the average value. Because the observed properties of an isolated sample of gas at equilibrium do not change with time, we conclude that the distribution of molecular speeds (that is, the fraction of the total number of molecules with speeds in any definite range) must be constant even though the speeds of individual molecules may be changing by large amounts.

One of the major problems of the kinetic theory is the determination of the law according to which molecular speeds are distributed. From a mathematical point of view the problem is the derivation of the distribution function $f(v)$ defined by the statement that the fraction of molecules dn_v/n having a speed between v and $v + dv$ is given by

$$\frac{dn_v}{n} = f(v)\, dv$$

On considering the physical significance of this equation we notice several peculiar aspects which we will meet elsewhere. Mathematically, dv is an infinitesimal quantity which can be allowed to approach zero just as closely as we wish. When we interpret dv as a range of speeds, however, we require that it be small relative to v but we cannot allow it to become too small for then there may be no molecule whose speed is in the specified range. Mathematically, n is a continuous variable and its differential can approach zero as a limit. Physically, n is the total number of molecules and it is necessarily an integer, as is dn. The treatment of n as a continuous variable is a convenient approximation that is thoroughly justified by the fact that n is a number of the order of magnitude of Avogadro's number. In consistency with our restriction of the magnitude of dv, we require that dn be large relative to 1 but at the same time "practically" infinitesimal with respect to n. These restrictions introduce no complications.

It might seem that the derivation of the distribution function for 10^{23} molecules with random speeds and directions is a hopeless task, but Maxwell solved the problem over a hundred years ago. Although subsequently several other derivations have been developed (and we treat one of these at a later point) we now follow rather closely Maxwell's original treatment because it involves some concepts which have proved to be very significant. It also shows how purely mathematical considerations can in some cases lead to a physically useful result.

We start our derivation of the Maxwell distribution function by looking at a closely related problem. Let us again express the components of the molecular velocity vector in terms of a Cartesian coordinate system. What is the fraction of the total number of molecules under consideration whose x component of the velocity is in the range v_x to $v_x + dv_x$? For brevity it is usual to employ a less precise expression: What is the fraction of molecules, dn_x/n, whose x component of the velocity is v_x?

Our first conclusion is that this fraction must have the same value for $-v_x$ that it has for v_x because the molecular motion is isotropic and we cannot

Kinetic Theory of Gases

distinguish the $+x$ direction from $-x$. Consequently whatever the mathematical form of the distribution function might be, it must be a function of v_x^2. We can express this conclusion in the following way:

$$\frac{dn_x}{n} = f(v_x^2)\,dv_x$$

Similar equations apply to the fractions of molecules with velocity components v_y and v_z.

Now we ask what fraction of the molecules will simultaneously have the velocity components v_x, v_y, and v_z. An analogy may be helpful here. Suppose that we wish to know the fraction of the population of this country which consists of blue-eyed girls between the ages of 19 and 20. If there is no correlation among eye color, sex, and age (that is, if these properties are statistically independent) then the desired fraction is the product of the fraction of the population with blue eyes, the fraction which is female, and the fraction in the particular age group.

In the molecular case, because the motion is assumed to be completely random and the perpendicular velocity components are statistically independent, the fraction of molecules which simultaneously have the velocity components v_x, v_y, and v_z is the product of factors for each component. Hence,

$$\frac{dn_{xyz}}{n} = f(v_x^2)\,f(v_y^2)\,f(v_z^2)\,dv_x\,dv_y\,dv_z \tag{5-6}$$

Next, we note that the orientation of our coordinate system is completely arbitrary because the molecular motion is isotropic. The distribution of molecular velocities does not depend on the arbitrary coordinate system we use. Therefore we can express the distribution function for the total velocity in a form that is independent of a coordinate system. The fraction of molecules with a velocity $\mathbf{v}$ is given by

$$\frac{dn_v}{n} = \phi(v^2)\,dv \tag{5-7}$$

Expressing v^2 in terms of the velocity components, we have

$$\phi(v^2) = \phi(v_x^2 + v_y^2 + v_z^2)$$

Because Equations 5–6 and 5–7 describe the same molecular motion from two different points of view, the two distribution functions must have identical values. Consequently,

$$\phi(v_x^2 + v_y^2 + v_z^2) = f(v_x^2)\,f(v_y^2)\,f(v_z^2) \tag{5-8}$$

This relationship is sufficient to determine the forms of the desired distribution functions. To show this we consider the following simpler problem in only two independent variables which we call p and q:

$$\phi(r) = f(p)\,f(q)$$

in which $r = p + q$, and $f(p)$ and $f(q)$ each is a function of one variable only. Partial differentiation of both sides of this equation with respect to p yields

$$\left(\frac{\partial \phi(r)}{\partial p}\right)_q = f(q)\frac{df(p)}{dp}$$

We also have the identity

$$\left(\frac{\partial \phi(r)}{\partial p}\right)_q = \frac{d\phi(r)}{dr}\left(\frac{\partial r}{\partial p}\right)_q$$

Similarly, differentiation with respect to q yields

$$\left(\frac{\partial \phi(r)}{\partial q}\right)_p = f(p)\frac{df(q)}{dq} = \frac{d\phi(r)}{dr}\left(\frac{\partial r}{\partial q}\right)_p$$

Because $\left(\frac{\partial r}{\partial p}\right)_q = \left(\frac{\partial r}{\partial q}\right)_p = 1$ from the definition of r, we have

$$\frac{d\phi(r)}{dr}\left(\frac{\partial r}{\partial p}\right)_q = \frac{d\phi(r)}{dr}\left(\frac{\partial r}{\partial q}\right)_p$$

and it follows that

$$f(q)\frac{df(p)}{dp} = f(p)\frac{df(q)}{dq}$$

When we separate the variables in this equation by dividing both sides by $f(p)\,f(q)$, we have

$$\frac{1}{f(p)}\frac{df(p)}{dp} = \frac{1}{f(q)}\frac{df(q)}{dq}$$

This is an equation in which the function is unknown, and the equality must hold for all values of the variables p and q. This can be true only if each side is equal to a quantity that is independent of p and q. In other words, if we have an equation in which the two sides depend on different variables and if the equation is to be satisfied for all values of the variables, the two sides of the equation must each be equal to the same constant.

Representing this constant by α and recognizing that

$$\frac{df(p)}{f(p)} = d \ln f(p)$$

we have

$$d \ln f(p) = \alpha \, dp$$

Integrating both sides of this equation, we obtain

$$\ln f(p) = \alpha p + \ln A$$

and consequently

$$f(p) = Ae^{\alpha p}$$

where $\ln A$ is a constant of integration.

It follows that

$$\phi(p+q) = Ae^{\alpha p} \cdot Ae^{\alpha q} = A^2 e^{\alpha(p+q)}$$

Only the exponential function satisfies the equation $\phi(p+q) = f(p)f(q)$ for all values of p and q.

We now apply this conclusion to our 3-dimensional molecular problem. The desired form of the distribution function $f(v_x^2)$ is the exponential function $Ae^{-\alpha v_x^2}$. Accordingly we can write Equation 5-8 in the following ways:

$$\phi(v^2) = (Ae^{-\alpha v_x^2})(Ae^{-\alpha v_y^2})(Ae^{-\alpha v_z^2})$$
$$= A^3 e^{-\alpha(v_x^2 + v_y^2 + v_z^2)}$$
$$= A^3 e^{-\alpha v^2}$$

Notice that the exponents are written as negative quantities. This choice of sign is necessary because a positive exponent would imply that the fraction of molecules with velocity v increases without limit as v increases. Such an unrealistic situation must be excluded.

Our next step is to evaluate the constants A and α. To evaluate A we can focus on any one of the velocity components because they all behave similarly. The fraction of molecules with values of v_x between $-\infty$ and ∞ is 1 because this fraction includes all of the molecules. Consequently when we integrate the distribution function over the range of v_x we obtain

$$A \int_{-\infty}^{\infty} e^{-\alpha v_x^2} \, dv_x = 1 \tag{5-9}$$

Making use of the fact that this integral is a standard form[1] which has the value $\sqrt{\dfrac{\pi}{\alpha}}$, we find that

$$A\sqrt{\frac{\pi}{\alpha}} = 1 \quad \text{and} \quad A = \sqrt{\frac{\alpha}{\pi}}$$

Because the integrand approaches zero as v_x^2 becomes large the contribution to the integral from very large values of v_x^2 is negligible. The use of the limits $-\infty$ and ∞ is a mathematical convenience and does not imply that molecular velocities are believed to approach infinitely large values.

The evaluation of α is not quite so simple. In order to accomplish this evaluation we must relate α to some known property of an ideal gas. A property of a gas is calculated as an average over values for the individual mole-

---

[1] A collection of some useful integrals needed in later work is given in Table 5-1.

TABLE 5-1   SOME USEFUL DEFINITE INTEGRALS*

$$\int_0^\infty e^{-ax^2}\,dx = \frac{1}{2}\left(\frac{\pi}{a}\right)^{\frac{1}{2}} \qquad \int_0^\infty x e^{-ax^2}\,dx = \frac{1}{2a}$$

$$\int_0^\infty x^2 e^{-ax^2}\,dx = \frac{1}{4a}\left(\frac{\pi}{a}\right)^{\frac{1}{2}} \qquad \int_0^\infty x^3 e^{-ax^2}\,dx = \frac{1}{2a^2}$$

$$\int_0^\infty x^4 e^{-ax^2}\,dx = \frac{3}{8a^2}\left(\frac{\pi}{a}\right)^{\frac{1}{2}} \qquad \int_0^\infty x^5 e^{-ax^2}\,dx = \frac{1}{a^3}$$

$$\int_0^\infty x^{2n} e^{-ax^2}\,dx = \frac{1\cdot 3\cdot 5\cdots(2n-1)}{2^{n+1}a^n}\left(\frac{\pi}{a}\right)^{\frac{1}{2}} \qquad \int_0^\infty x^n e^{-ax}\,dx = \frac{n!}{a^{n+1}}$$

$$\int_0^\infty x^{2n+1} e^{-ax^2}\,dx = \frac{n!}{2}\frac{1}{a^{n+1}}$$

*The integral of any even function taken between the limits $-\infty$ and $+\infty$ is twice the integral from 0 to ∞. The integral of any odd function between the limits $-\infty$ and $+\infty$ is equal to zero.

cules in the gas. The way that a distribution function is used to calculate such an average property is a generalization of the elementary method for obtaining weighted averages.

Suppose, for example, we wish to find the average examination grade for a class of students. We can multiply each grade (g_i) by the number of students receiving that grade (n_i), add all of the products, and divide this sum by the number of students in the class:

$$\bar{g} = \frac{\Sigma g_i n_i}{\Sigma n_i} = \Sigma g_i \frac{n_i}{n} \tag{5-10}$$

In these equations i is an index which designates a particular grade and n_i/n is the fraction of students who received the grade g_i.

Now let us suppose that we have a continuous distribution (both n and g are continuous variables) defined by a distribution function $dn = f(g)\,dg$. Then the sums in Equation 5-10 can be replaced by integrals, and we can express the average value of g in the following way:

$$\bar{g} = \frac{\int g\,dn}{\int dn} = \frac{\int g f(g)\,dg}{\int f(g)\,dg}$$

The integrals are taken over the range of the variable g and in this case the denominator is equal to n. By the same procedure it is possible to use a dis-

Kinetic Theory of Gases

tribution function for finding the average value of any function of the variable g, such as g^2.

We now apply this procedure to find the value of α. We use the distribution function $\sqrt{\dfrac{\alpha}{\pi}}\, e^{-\alpha v_x^2}$ to calculate $\overline{v_x^2}$, the average value of v_x^2. Since the integral of this function is equal to 1 (Eq. 5–9),

$$\overline{v_x^2} = \sqrt{\frac{\alpha}{\pi}} \int_{-\infty}^{\infty} v_x^2\, e^{-\alpha v_x^2}\, dv_x \tag{5-11}$$

The value of the integral is $\dfrac{1}{2\alpha}\sqrt{\dfrac{\pi}{\alpha}}$ and therefore $\overline{v_x^2} = \dfrac{1}{2\alpha}$. Because $\bar{\epsilon}_x$, the average value of the molecular kinetic energy associated with v_x, is equal to $\dfrac{m}{2}\overline{v_x^2}$, we have

$$\bar{\epsilon}_x = \frac{m}{4\alpha}$$

From Equation 5–4 we see that $\bar{\epsilon}_x = \dfrac{kT}{2}$. Consequently

$$\frac{m}{4\alpha} = \frac{kT}{2}$$

and

$$\alpha = \frac{m}{2kT}$$

Having evaluated the constants, we can write the complete distribution law for the x component of the velocity:

$$\frac{dn_x}{n} = \left(\frac{m}{2\pi kT}\right)^{\frac{1}{2}} e^{-\frac{mv_x^2}{2kT}}\, dv_x \tag{5-12}$$

A graph of this function is given in Figure 5–1. Notice that it is symmetrical about $v_x = 0$. It follows that $\overline{v_x} = 0$.

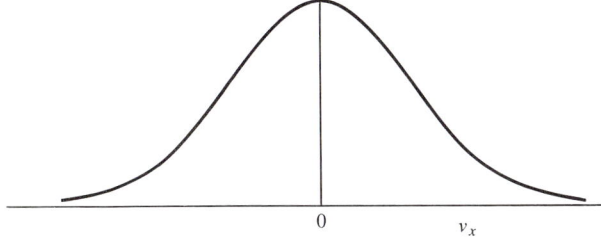

Figure 5–1. Graph of $\dfrac{1}{n}\dfrac{dn_x}{dv_x}$ vs v_x, the distribution function for one component of molecular velocity.

It is interesting to observe that this function has the same mathematical form as the Gauss error function which gives the relative frequency of a random error in a large number of repetitive measurements as a function of the magnitude of the error. The same function can be derived by consideration of the 1-dimensional "random walk" problem; in this case the function gives the probability of being at a certain distance from the origin after having taken a large number of random steps, either forward or backward. The common feature in all of these problems is the concept of complete randomness.

With consideration of all 3 components of the velocity we can write Maxwell's distribution law for molecular velocities in the following way:

$$\frac{dn_{xyz}}{n} = \left(\frac{m}{2\pi kT}\right)^{\frac{3}{2}} e^{-\frac{mv^2}{2kT}} dv_x \, dv_y \, dv_z \qquad (5\text{--}13)$$

For most purposes we are not concerned with the directional aspects of molecular velocities but only with their magnitudes; that is, the molecular speeds. To obtain the distribution law for molecular speeds we imagine that

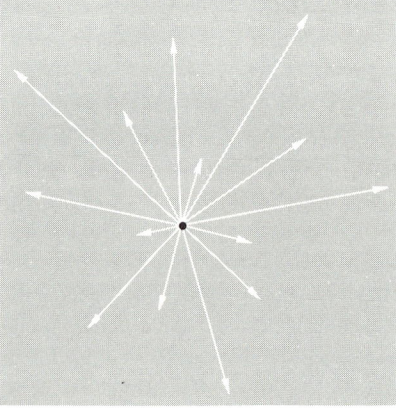

Figure 5–2. Schematic representation of molecular velocities in velocity space.

we could plot the velocities of all n molecules at one instant in a Cartesian coordinate system with the velocity components referred to a common origin as coordinates, as in Figure 5–2. (This three-dimensional space is often called

Kinetic Theory of Gases

velocity space.) Then the number of velocity vectors (dn_{xyz}) which terminate in an element of volume in velocity space $(dv_x\, dv_y\, dv_z)$ is given by

$$n \left(\frac{m}{2\pi kT}\right)^{\frac{3}{2}} e^{-\frac{mv^2}{2kT}} dv_x\, dv_y\, dv_z$$

We can describe the distribution of the velocity vectors in velocity space just as well by using spherical polar coordinates as by using Cartesian coordinates. We specify a particular set of polar coordinates (v, θ, and ϕ) by relating them to our original Cartesian coordinates (Fig. 5–3). Assuming that the axis

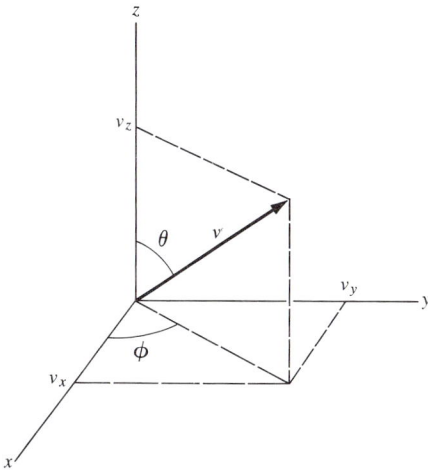

Figure 5–3. Relation between polar and Cartesian coordinate systems.

of polar coordinates is the same as the z axis, the relationship between the two sets of coordinates is: $v_z = v \cos \theta$; $v_x = v \sin \theta \cos \phi$; $v_y = v \sin \theta \sin \phi$. (We will find other cases for which this transformation from Cartesian to polar coordinates is useful.)

The element of volume in this set of spherical polar coordinates is $v^2\, dv\, \sin \theta\, d\theta\, d\phi$. We can rewrite Equation 5–13 in terms of these coordinates:

$$\frac{dn_{v\theta\phi}}{n} = \left(\frac{m}{2\pi kT}\right)^{\frac{3}{2}} e^{-\frac{mv^2}{2kT}} v^2\, dv\, \sin \theta\, d\theta\, d\phi$$

Now we ask the following question: What is the fraction of the molecules (dn_v/n) whose velocity vectors have the length v without regard to direction? To include the vectors at all angles we integrate over the angular coordinates. Noting that the angle ϕ varies from 0 to 2π, θ varies from 0 to π, and that

$$\int_0^{2\pi} d\phi = 2\pi; \quad \int_0^{\pi} \sin \theta\, d\theta = 2$$

we have

$$\frac{dn_v}{n} = \left(\frac{m}{2\pi kT}\right)^{\frac{3}{2}} e^{-\frac{mv^2}{2kT}} v^2 \, dv \int_0^\pi \sin\theta \, d\theta \int_0^{2\pi} d\phi$$

$$= 4\pi \left(\frac{m}{2\pi kT}\right)^{\frac{3}{2}} e^{-\frac{mv^2}{2kT}} v^2 \, dv \qquad (5\text{-}14)$$

This is Maxwell's distribution law for molecular *speeds*. It is represented graphically in Figure 5-4. We see that the distribution function depends on the temperature in two different ways. The most important temperature dependence comes from the exponential factor while the dependence on $T^{-\frac{3}{2}}$ in the pre-exponential factor is relatively minor. With rising temperature the maximum of the distribution function shifts to higher speed values and the spread of molecular speeds increases, as illustrated by the graph.

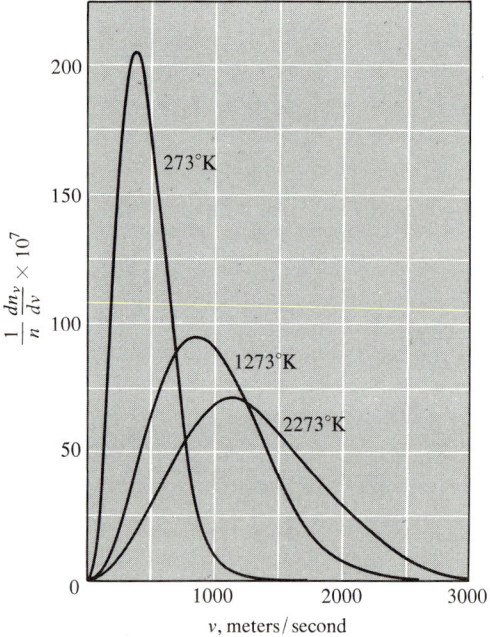

Figure 5-4. Graph of Maxwell's distribution function for molecular speeds at three temperatures.

For many years after Maxwell derived his distribution law there was no way to test it directly by experimental measurements of molecular velocities. More recently a number of ways for a direct test have been developed, and they all confirm Maxwell's law within their respective experimental errors.

Kinetic Theory of Gases

Having obtained the Maxwell distribution law for molecular speeds, we can use it to find the average values of several quantities related to speed. Let us start with the calculation of the mean-square speed. In the same way we found the average of $\overline{v_x^2}$ (Eq. 5-11), we now obtain the value of $\overline{v^2}$ from the following equations:

$$\overline{v^2} = \int_0^\infty \frac{v^2 \, dn_v}{n} = \int_0^\infty v^2 \, 4\pi \left(\frac{m}{2\pi kT}\right)^{\frac{3}{2}} e^{-\frac{mv^2}{2kT}} v^2 \, dv$$

$$= 4\pi \left(\frac{m}{2\pi kT}\right)^{\frac{3}{2}} \int_0^\infty e^{-\frac{mv^2}{2kT}} v^4 \, dv$$

Note that the lower limit of the speed is zero. From Table 5-1 we find that the value of the definite integral is $\frac{3}{8}\pi^{\frac{1}{2}} \left(\frac{m}{2kT}\right)^{-\frac{5}{2}}$. Consequently,

$$\overline{v^2} = 4\pi \left(\frac{m}{2\pi kT}\right)^{\frac{3}{2}} \frac{3}{8} \pi^{\frac{1}{2}} \left(\frac{m}{2kT}\right)^{-\frac{5}{2}}$$

$$= \frac{3kT}{m}$$

and

$$(\overline{v^2})^{\frac{1}{2}} = v_{\rm rms} = \left(\frac{3kT}{m}\right)^{\frac{1}{2}} = \left(\frac{3RT}{M}\right)^{\frac{1}{2}}$$

This is just the result we found previously in Equation 5-5, a fact which should not be too surprising because we implicitly used this value of $v_{\rm rms}$ in evaluating the constants in the distribution function. It is clear, however, that our derivation is internally consistent.

By the method just employed for the mean-square speed, we obtain the mean speed $\bar{v}$:

$$\bar{v} = 4\pi \left(\frac{m}{2\pi kT}\right)^{\frac{3}{2}} \int_0^\infty e^{-\frac{mv^2}{2kT}} v^3 \, dv$$

$$= 4\pi \left(\frac{m}{2\pi kT}\right)^{\frac{3}{2}} \cdot 2 \left(\frac{kT}{m}\right)^2$$

$$= \left(\frac{8kT}{\pi m}\right)^{\frac{1}{2}} \tag{5-15}$$

It was mentioned earlier that $v_{\rm rms}$ and $\bar{v}$ are different quantities; now we can compare them. Their ratio is given by

$$\frac{v_{\rm rms}}{\bar{v}} = \frac{\left(\frac{3kT}{m}\right)^{\frac{1}{2}}}{\left(\frac{8kT}{\pi m}\right)^{\frac{1}{2}}} = \left(\frac{3\pi}{8}\right)^{\frac{1}{2}} = 1.085$$

Although the numerical difference between these speeds is not great, v_{rms} should be used in relations involving temperature or average molecular kinetic energy while $\bar{v}$ appears in equations connected with molecular collisions.

Distribution of Molecular Kinetic Energies

We can easily transform the distribution law for molecular speeds into a distribution law for molecular kinetic energies by introducing the molecular kinetic energy as the variable. From the differentiation of $\epsilon = \frac{1}{2}mv^2$ we obtain $dv = (2m\epsilon)^{-\frac{1}{2}} d\epsilon$. Replacement of v and dv in the Maxwell distribution function by their equals leads to

$$\frac{dn_\epsilon}{n} = 4\pi \left(\frac{m}{2\pi kT}\right)^{\frac{3}{2}} e^{-\frac{\epsilon}{kT}} \left(\frac{2\epsilon}{m}\right) (2m\epsilon)^{-\frac{1}{2}} d\epsilon$$

$$= 2\pi^{-\frac{1}{2}}(kT)^{-\frac{3}{2}} e^{-\frac{\epsilon}{kT}} \epsilon^{\frac{1}{2}} d\epsilon \tag{5-16}$$

A graph of this equation is given in Figure 5–5. We will make use of the distribution law for molecular kinetic energy in Chapter 11 in connection with a treatment of the reaction kinetics of gases.

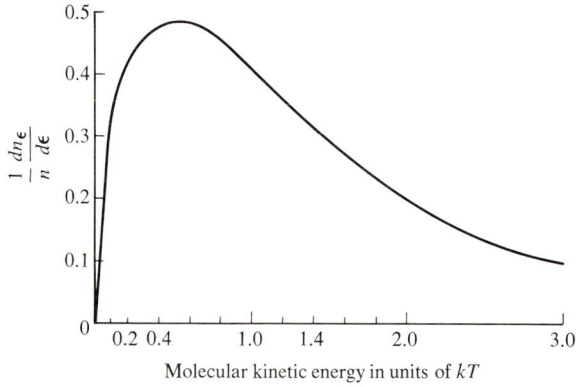

Figure 5–5. Distribution function for molecular kinetic energy.

Collisions with a Wall: Effusion

For many reasons it is important to know the rate at which molecules strike a surface. We can carry out the calculation of this rate in the following way: Let us assume that we have n molecules contained in a volume V and let us represent the molecular concentration (n/V) by N. Imagine that we have a

Kinetic Theory of Gases

unit area of wall located in the y, z plane. (The shape of this area is unimportant but for definiteness we assume that it is a circle.)

First we calculate the number of molecules with the x-component of velocity v_x that strike this unit area in 1 second. This number is equal to the number of molecules (dn_x) with the velocity component v_x which are contained in a cylindrical volume of unit cross-section and a length numerically

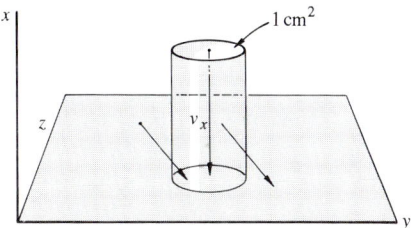

Figure 5-6. Cylinder of volume v_x cm³ containing molecules that strike a unit area of the y, z plane in 1 second.

equal to v_x (Fig. 5–6).[2] Because the distribution of molecules in space is uniform, the total number of molecules contained in the cylindrical volume of v_x cm³ is Nv_x. To find dn_x we multiply this total number by the fraction of the molecules with velocity component v_x (Eq. 5–12); thus we have

$$dn_x = Nv_x \left(\frac{m}{2\pi kT}\right)^{\frac{1}{2}} e^{-\frac{mv_x^2}{2kT}} dv_x$$

This represents the number of molecules with velocity component v_x that strikes a unit area in 1 second.

In order to obtain the total number of molecules striking a unit area each second, we must integrate this expression over all *positive* values of v_x. This implies that the lower limit of v_x is zero. Therefore the desired rate is given by

$$n_x = N\left(\frac{m}{2\pi kT}\right)^{\frac{1}{2}} \int_0^\infty e^{-\frac{mv_x^2}{2kT}} v_x\, dv_x$$

$$= N\left(\frac{m}{2\pi kT}\right)^{\frac{1}{2}} \frac{kT}{m}$$

$$= N\left(\frac{kT}{2\pi m}\right)^{\frac{1}{2}}$$

Because $\left(\dfrac{kT}{2\pi m}\right)^{\frac{1}{2}} = \dfrac{\bar{v}}{4}$, the equation for n_x reduces to the following simple form:

$$n_x = \tfrac{1}{4}N\bar{v} \qquad (5\text{-}17)$$

[2]All molecules with velocity component v_x which are initially at a distance of v_x cm or less from the unit area can reach this area in 1 second if their y- and z-components are sufficiently small. Some of the molecules move in directions which take them outside the cylinder. Because of the randomness of molecular motion, however, an equal number, initially outside the cylinder, move in such directions that they strike the unit area by the end of the second.

We can use the method in the preceding paragraphs to calculate the pressure of the gas. The change in momentum when a molecule with the x-component of velocity v_x strikes the wall mentioned above is $2mv_x$. The total change in momentum of all molecules striking a unit area in one second is equal to the pressure exerted by the gas. Consequently,

$$P = 2mN \left(\frac{m}{2\pi kT}\right)^{\frac{1}{2}} \int_0^\infty v_x^2 e^{-\frac{mv^2}{2kT}} dv_x$$

$$= 2mN \left(\frac{m}{2\pi kT}\right)^{\frac{1}{2}} \frac{\pi^{\frac{1}{2}}}{4} \left(\frac{2kT}{m}\right)^{\frac{3}{2}}$$

$$= NkT = \tfrac{2}{3}N\bar{\epsilon}$$

We have thus derived the ideal gas equation of state.

If there is a small hole of area A in the wall of the container, we would expect that all molecules that hit the hole will escape from the container. For this to be the case the hole must be so small and the wall must be so thin that the molecules undergo no collisions in passing through the hole. This type of flow is called effusion or molecular flow. The rate of effusion in molecules per second $\left(\dfrac{dn}{dt}\right)$ is given by $n_x A$. Thus we have

$$\frac{dn}{dt} = NA \left(\frac{kT}{2\pi m}\right)^{\frac{1}{2}}$$

Since $N = P/kT$, we have

$$\frac{dn}{dt} = \frac{PA}{kT}\left(\frac{kT}{2\pi m}\right)^{\frac{1}{2}} = \frac{PA}{(2\pi mkT)^{\frac{1}{2}}} \qquad (5\text{--}18)$$

The number of grams of gas effusing per second is found by multiplying this equation by m.

The most interesting feature of Equation 5–18 is the dependence of the rate of effusion on $m^{-\frac{1}{2}}$. This is the theoretical explanation of the law of effusion found empirically by Graham which states that the rate of effusion of a gas is inversely proportional to the square root of its molecular weight. This relationship makes it possible to use the rate of effusion to determine an unknown molecular weight. It also makes possible the partial separation of

Kinetic Theory of Gases

molecules of differing molecular weights. This phenomenon has been used on a large scale to concentrate the $U^{235}F_6$ in a mixture of isotopic uranium hexafluorides.

Mean Free Path and Collision Number

In discussing collisions of molecules with surfaces and the distribution of molecular speeds, we assumed that the molecules have negligible volumes. The only properties we have ascribed to our molecular model up to this point are mass and velocity.

When we consider collisions of the molecules among themselves we need to define another molecular property, the molecular size. To specify the size with only one parameter for simplicity, we assume that a molecule can be represented as a rigid sphere of diameter σ. (We will discuss more realistic models later.) The diameter of the rigid sphere model of a molecule is often called the *collision diameter* because if two like spherical molecules come so close together that the distance between their centers is σ, a collision has occurred.

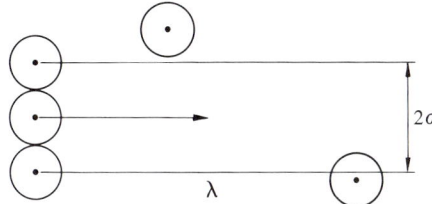

Figure 5-7. Exclusive volume $\pi\sigma^2\lambda$ cm^3 swept out by molecule traversing a path λ cm long.

A useful concept in the treatment of molecular motion is the mean free path (λ) which is defined as the average distance a molecule travels between successive collisions with other molecules. To evaluate this quantity, let us picture a container in which there is a molecular concentration of N molecules/cm^3 and in which all molecules are assumed to be stationary except one. If this molecule moves a distance λ between collisions, it sweeps out a cylindrical volume $\pi\sigma^2\lambda$ in which no other molecule has its center (Fig. 5-7). (The area $\pi\sigma^2$ is sometimes called the collision cross-section of the molecule.) On the average, each molecule must have this "exclusive" volume and since there are N molecules/cm^3,

$$N\sigma^2\pi\lambda = 1$$
$$\lambda = \frac{1}{N\pi\sigma^2} \text{ cm}$$

This derivation is oversimplified because we assumed that only one molecule is moving. When consideration is given to the fact that all molecules are moving, a correction factor of $2^{-\frac{1}{2}}$ for relative motion is obtained. Then the corrected expression for the mean free path is

$$\lambda = \frac{1}{\sqrt{2}N\pi\sigma^2} = \frac{kT}{\sqrt{2}\pi\sigma^2 P} \qquad (5-19)$$

Thus we see that for a particular gas the mean free path is inversely proportional to the gas pressure. This fact is of importance in work in vacuum systems. At low pressures the mean free path may be of the same order of magnitude as (or larger than) the dimensions of the apparatus.

Every mean free path is terminated by a bimolecular collision. A molecule moving with an average speed $\bar{v}$ will move through $\bar{v}/\lambda$ mean free paths in 1 second and consequently will make this many collisions with other molecules. Since there are N molecules per cm³, the total number of bimolecular collisions per second per cm³ is

$$Z = \frac{1}{2}N\frac{\bar{v}}{\lambda} = \frac{1}{\sqrt{2}}N^2\pi\sigma^2\bar{v} = 2N^2\sigma^2\left(\frac{\pi kT}{m}\right)^{\frac{1}{2}} \qquad (5-20)$$

Notice the factor $\frac{1}{2}$ in the above equation. This is introduced to take into account the fact that a collision of molecule 1 with molecule 2 is the same event as a collision of molecule 2 with molecule 1.

We now have derived expressions for the mean free path and the collision number (Z) in terms of the molecular collision diameter. To evaluate these expressions we need a method for obtaining numerical values of σ. One method is provided by a study of the transport properties of a gas. These properties have a significance which is broader than just the determination of σ and we now proceed to their treatment.

Transport Properties of Gases

Up to this point our discussion of the kinetic theory has been limited to gases in equilibrium and without motion of the gas as a whole; that is, the average molecular velocity (not speed!) is zero. Now we remove these restrictions. We assume that the gas under consideration is in a nonuniform state in which some quantity, such as momentum or energy, is transported through the gas as a result of molecular motion.

One of the transport properties is viscosity, which is a characteristic property of all fluids — liquids as well as gases. In a general way, viscosity is a measure of resistance to flow. A force must be applied to a fluid to cause it to flow, and in the absence of external forces flow will stop.

Suppose that we have two parallel surfaces with a fluid between them

(Fig. 5-8), and assume that one of the surfaces S_1 is made to move in its own plane (the x,y plane) at a constant speed. After a steady state is reached it is found that the fluid is flowing in such a way that the velocity of a volume element is a simple function of the distance z of this volume element from S_1. This type of flow is known as laminar flow.

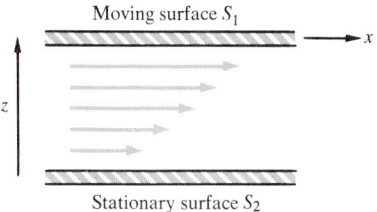

Figure 5-8. Distribution of velocity in a fluid located between a moving surface and a stationary one; the velocity gradient, $\dfrac{dv_x}{dz}$, is pictured as constant.

It is also found that work must be done against a "drag" force to keep S_1 moving at a constant speed. At the same time a force is observed to act on S_2 in the direction of the motion.

On the basis of simplifying assumptions about laminar flow, Newton concluded that the viscous force on a surface is proportional to its area A and to the velocity gradient, $\dfrac{dv}{dz}$, which is the rate of change of velocity of flow with distance in a direction perpendicular to the surface. The Newtonian equation for viscous flow is

$$f = \eta A \frac{dv}{dz}$$

The constant of proportionality η is known as the viscosity coefficient (or just viscosity).

In the cgs system of units, the unit of viscosity is the

$$\text{dyne-sec/cm}^2 \left(= \frac{g}{cm\ sec} \right)$$

which is called the poise in honor of Poiseuille who did pioneer work in the measurement of viscosity. Because 1 poise represents a high viscosity, viscosity data for liquids are often given in terms of centipoise units and for gases the micropoise is a useful unit.

By employing the kinetic theory, Maxwell was able to derive an equation for the viscosity coefficient of a gas as a function of molecular properties. We do not present the details of this derivation but merely outline the ideas on which it is based.

Consider two parallel adjacent layers of gas, one of which is moving faster than the other. If no external force is applied, as a consequence of the existence of viscosity the faster layer slows down while the slower layer speeds up. Maxwell accounted for this phenomenon by pointing out that in laminar flow the molecules have velocities which are the sums of the mass or flow velocity of the layer in which they are present and the random velocities the molecules have in the absence of flow velocity. A molecule which leaves the faster layer and enters the slower layer has on the average a larger velocity in the direction of flow than its new neighbors and consequently a larger momentum. Therefore the momentum of the slower layer is increased by arrival of molecules from the faster layer.

Conversely, the momentum of the faster layer is decreased by the arrival of molecules from the slower layer. The viscous force which tends to bring the two layers to the same speed depends on the rate of transfer of momentum from a region of higher momentum to one of lower momentum. The transfer of momentum occurs by means of collisions between the molecules and the rate of transfer should depend on the mean free path, the average molecular speed, and the molecular concentration. The result obtained by Maxwell is

$$\eta = \tfrac{1}{3}mN\bar{v}\lambda = \tfrac{1}{3}\rho\bar{v}\lambda \tag{5-21}$$

in which ρ is the density of the gas.

In Equation 5-21 the gas density is directly proportional to the pressure, the mean free path is inversely proportional to the pressure, and the mean speed is independent of the pressure. This equation thus implies that the viscosity coefficient of a gas should be independent of the gas pressure. Maxwell reached this conclusion before quantitative measurements of gas viscosity had been made. The experimental confirmation of this theoretical prediction, which was in conflict with the prevailing opinion of its time, was one of the major triumphs of the kinetic theory.

❖❖❖❖❖❖

Later and more detailed analyses of the transport of molecular momentum included consideration of the fact that the molecular motion in a gas in a non-uniform state is not isotropic. Consequently, it is not strictly correct to use the average molecular speed obtained from the Maxwell distribution law, which was derived on the assumption of isotropic motion.

When a rigorous (and complicated) treatment of the transport of molecular

Kinetic Theory of Gases

momentum is carried out, an equation of the same form as Equation 5–21 is derived, but the numerical coefficient is $\frac{1}{2}$ instead of $\frac{1}{3}$. Thus the corrected equation is

$$\eta = \tfrac{1}{2}\bar{v}\lambda\rho \qquad (5\text{–}22)$$

in which $\bar{v}$ has the value obtained from Equation 5–15.

The viscosity coefficient of a gas can be measured in several different ways. One of these involves measurements of the rate at which a gas flows through a capillary tube as a result of a constant difference in pressure between the two ends of the tube. Under these conditions, the rate of flow is inversely proportional to the viscosity.

Equation 5–22 can be used to calculate values of the mean free path from measurements of gas viscosity. Solving this equation for λ, we have

$$\lambda = \frac{2\eta}{\bar{v}\rho}$$

Substituting for ρ from the ideal gas law $\left(= \dfrac{PM}{RT}\right)$ and for $\bar{v}\left[= \left(\dfrac{8RT}{\pi M}\right)^{1/2}\right]$ this equation becomes

$$\lambda = 2\eta\frac{RT}{PM}\left(\frac{\pi M}{8RT}\right)^{\tfrac{1}{2}} = \frac{\eta}{\sqrt{2}P}\left(\frac{\pi RT}{M}\right)^{\tfrac{1}{2}}$$

Let us apply this equation to find the mean free path of nitrogen molecules at a pressure of 1 atm ($= 1.01 \times 10^6$ dynes/cm^2) and a temperature of 27°C. At this temperature the viscosity coefficient of nitrogen is found to be 178 micropoise. The mean free path is given by

$$\lambda = \frac{1.78 \times 10^{-4}}{1.41 \times 1.01 \times 10^6}\left(\frac{3.14 \times 8.32 \times 10^7 \times 300}{28.0}\right)^{\tfrac{1}{2}}$$
$$= 6.62 \times 10^{-6} \text{ cm}$$

The mean free path at a pressure of 10^{-6} atm (~ 1 dyne/cm^2) is about 7 cm.

We can now calculate values for collision diameters either from the mean free path or directly from viscosity data. Substituting for λ in Equation 5–22 its equivalent from Equation 5–19, we have

$$\eta = \frac{m\bar{v}}{2\sqrt{2}\pi\sigma^2}$$

On solving this equation for σ^2 and substituting for $\bar{v}$ and $m\left(=\dfrac{M}{\mathfrak{N}}\right)$, we obtain

$$\sigma^2 = \frac{M}{2\sqrt{2}\pi\eta\mathfrak{N}}\left(\frac{8RT}{\pi M}\right)^{\tfrac{1}{2}} = \frac{1}{\eta\mathfrak{N}}\left(\frac{MRT}{\pi^3}\right)^{\tfrac{1}{2}}$$

For nitrogen at 27°C,

$$\sigma^2 = \frac{1}{1.78 \times 10^{-4} \times 6.02 \times 10^{23}} \left(\frac{28.0 \times 300 \times 8.32 \times 10^7}{(3.14)^3}\right)^{\frac{1}{2}}$$
$$= 13.9 \times 10^{-16} \text{ cm}^2$$
$$\sigma = 3.73 \times 10^{-8} \text{ cm} = 3.73 \text{ Å}$$

Here Å is the symbol for the Angstrom unit: $1 \text{ Å} = 10^{-8}$ cm. In general, the collision diameters of simple molecules are of the order of a few Angstrom units.

We have seen how the viscosity of a gas can be understood in terms of the transport of momentum by molecular collisions. In an analogous way the conduction of heat by a gas can be interpreted as a transport of kinetic energy through a gas by molecular collisions as a result of a temperature gradient.

Assume that we have a gas in a container which has two parallel walls held at constant but different temperatures. Measurements show that the rate at which heat is transferred from the hotter wall to the colder is proportional to the area of the walls A and to the temperature gradient, which is the ratio of the temperature difference to the distance between the walls. When a steady state is reached, the temperature in the gas is a linear function of the distance x from the colder wall and the temperature gradient, dT/dx, is constant.

The experimental results are summarized by the following equation:

$$\frac{dq}{dt} = -\kappa A \frac{dT}{dx}$$

in which dq/dt is the rate of heat transfer and κ, the thermal conductivity, is a property of the gas. The negative sign indicates that the flow of heat is in the opposite direction from the temperature gradient.

The last transport process to be discussed here is the diffusion of one gas into another. This process is a transport of matter resulting from a concentration gradient. The rate of transfer of molecules across an area A in the y,z plane is given by an empirical law known as *Fick's law*:

$$\frac{dn_1}{dt} = -DA\frac{dN_1}{dx} \tag{5-23}$$

In this equation dn_1 is the number of molecules of substance 1 transferred across A in the time dt as a result of the concentration gradient dN_1/dx. The constant of proportionality, D, is called the diffusion coefficient.

In general the diffusion of substance 1 through substance 2 is accompanied by the diffusion of substance 2 in the reverse direction. Consequently, the com-

Kinetic Theory of Gases

plete expression of the diffusion coefficient in a binary system consists of two terms, one for each component. Equation 5–23 applies to self-diffusion which is exemplified by the diffusion of isotopically labelled molecules through otherwise identical molecules of the same substance.

By theoretical treatments closely resembling that which led to Equation 5–22, the following expressions for κ and D can be derived:

$$\kappa = 1.25 \, \bar{v}\lambda c_V \rho \qquad D = 0.60 \, \bar{v}\lambda$$

in which c_V is the specific heat of the gas (heat capacity at constant volume, per gram).

Mean free paths and collision diameters can be calculated from measurements of heat conduction and diffusion, as well as from viscosity measurements.

TABLE 5-2 VISCOSITY, THERMAL CONDUCTIVITY, AND DIFFUSION COEFFICIENTS AT 0°C AND 1 ATM, AND MOLECULAR DIAMETERS CALCULATED FROM THESE VALUES*

Gas	η (micropoise)	κ $\left(\dfrac{\text{cal}}{\text{cm sec deg}} \times 10^6\right)$	D (cm^2/sec)	σ (angstroms)	
				From η	From D
Ne	297	110	0.452	2.58	2.42
Ar	210	39.0	0.156	3.64	3.47
N_2	166	—	0.185	3.75	3.48
O_2	192	58.4	0.187	3.61	3.35
CH_4	103	73.4	0.206	4.14	3.79
CO_2	137	34.9	0.0974	4.63	4.28

*J. O. Hirschfelder, C. F. Curtiss, and R. B. Bird, *Molecular Theory of Gases and Liquids*, John Wiley and Sons, New York, 1954.

If the theory for these processes were exact, the values for the mean free path and the collision diameter which are calculated for a given gas from each of the transport properties should all be the same. Actually, as shown in Table 5–2, the calculated values are close to each other, but the differences between them are greater than those which can be attributed to experimental error in the data used.

In addition, if the theory were exact the variation of each of the transport properties with temperature should be the same as that of $\bar{v}$, a variation proportional to $T^{\frac{1}{2}}$. The measured temperature coefficients of viscosity, thermal conduction, and diffusion are all different from each other and significantly larger than the theoretical dependence on $T^{\frac{1}{2}}$.

The source of these discrepancies is not hard to find. All of our treatment has been based on a molecular model which is a rigid sphere, and we assumed that these "molecules" exert no force on each other except during a collision, when very large (or infinite) repulsive forces act for a very short time interval. The behavior of actual molecules is different, however, and the most accurate calculations of the transport properties depend on the magnitude of the intermolecular forces. In fact, it is possible to learn something about the nature of intermolecular forces from a study of the transport properties.

In view of the simple molecular model used, the theoretical results discussed here must be considered to be as good as could be expected. Even if they were exact, at best a collision diameter could provide only a rough estimate of molecular dimensions. Accurate methods for determining interatomic distances are treated in Chapter 8.

Maxwell–Boltzmann Distribution Law

The only kind of molecular energy we have included in our treatment of the kinetic theory so far is the kinetic energy of molecular motion. We can generalize this treatment by assuming that our gas is in some kind of uniform force field. In that case the molecules have potential energy which is a function of their position in the field. The simplest field of this type is the gravitational

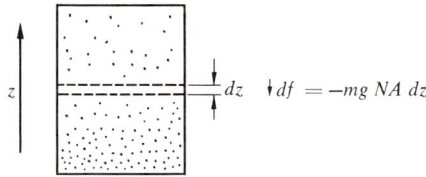

Figure 5–9. Distribution of molecules in a gravitational field.

field which is always with us on the surface of the earth. We wish to obtain the distribution of molecular positions in this field.

Assume that we have a vertical column of gas at equilibrium at a uniform temperature (Fig. 5–9). The gravitational force acting on a molecule of mass m is mg, where g is the acceleration of gravity. The total force acting on a layer of gas whose thickness is dz (this force is the weight of the gas in the volume $A\,dz$) is given by

$$df = -mgA\,N\,dz$$

where N is the molecular concentration and A is the cross-section of the column.

Kinetic Theory of Gases

The negative sign indicates that the gravitational force is directed downwards, the direction of $-z$.

When we divide both sides of this equation by A we obtain the pressure $dP\ (=df/A)$ on the layer:

$$dP = -mg\,N\,dz = -\frac{mgP\,dz}{kT}$$

where we have replaced N by P/kT. Division of both sides by P leads to

$$\frac{dP}{P} = d\ln P = -\frac{mg}{kT}dz$$

Integrating from P_0 at $z = 0$ to P_z at a height z, we have

$$\ln\frac{P_z}{P_0} = -\frac{mgz}{kT};\quad P_z = P_0 e^{-\frac{mgz}{kT}} \tag{5-24}$$

This equation, which is often called the barometric equation, says that the pressure of a pure gas at a constant temperature decreases exponentially with height. With a mixture of gases of different molecular weights each kind has its own distribution; the lower the molecular weight, the greater will be the relative partial pressure of this component at any height. It follows that the composition of the mixture is a function of the height. For columns of gas in the laboratory this variation of composition with height is insignificant. To a limited extent, however, the pressure and distribution of gases in the earth's atmosphere follow the barometric equation although the atmosphere is a system which is neither isothermal nor in equilibrium.

From Equation 5-24 it follows that

$$N = N_0 e^{-\frac{mgz}{kT}} = N_0 e^{-\frac{\epsilon_p}{kT}}$$

where ϵ_p has been used as a symbol for molecular potential energy. In general the potential energy of a molecule may be a function of all three space coordinates. Representing the number of molecules in the element of volume $dx\,dy\,dz$ by $dn_{xyz}\ (=N\,dx\,dy\,dz)$, we can write the following distribution law for molecular positions:

$$dn_{xyz} = N_0 e^{-\frac{\epsilon_p}{kT}}dx\,dy\,dz \tag{5-25}$$

in which N_0 is the molecular concentration in a region of zero potential energy.

This distribution law, which is called the *Boltzmann distribution law*, is independent of molecular velocities. In contrast, the Maxwell distribution law (Eq. 5–13) does not depend on molecular coordinates. In a system in which ϵ_p is a function of the coordinates, how many molecules are in the volume element $dx\,dy\,dz$ and have velocity components v_x, v_y, v_z? The answer to this question is the product of the two independent distribution laws:

$$dn = N_0 e^{-\frac{\epsilon_p}{kT}} dx\,dy\,dz \cdot \left(\frac{m}{2\pi kT}\right)^{\frac{3}{2}} e^{-\frac{m(v_x^2+v_y^2+v_z^2)}{2kT}} dv_x\,dv_y\,dv_z$$

$$= N_0 \left(\frac{m}{2\pi kT}\right)^{\frac{3}{2}} e^{-\frac{1}{kT}\left(\frac{mv^2}{2}+\epsilon_p\right)} dx\,dy\,dz\,dv_x\,dv_y\,dv_z \quad (5\text{--}26)$$

We have thus obtained one form of the Maxwell–Boltzmann distribution law which is one of the most important relationships in physical chemistry. It is not limited to the kinds of energy included so far, namely, kinetic energy of molecular motion and molecular potential energy in an external force field. As we later see, it can be generalized to include other kinds of energy. In addition, in Chapter 9 we return to this equation from a different point of view.

Relationship of Kinetic Theory to Thermodynamics: Equipartition of Energy

Although we have by no means exhausted the scope or the refinements of the kinetic theory, we have treated those portions of the subject which are most relevant to the physical chemistry of ideal gases. At this point it seems desirable to consider in a preliminary way the connections between the molecular theory as developed in this chapter and some of the thermodynamic properties of gases.

By using the simplest possible molecular model, we were able to derive a theoretical equation for the pressure which is consistent with the ideal gas law. We found that we could interpret the absolute temperature as a measure of the average molecular kinetic energy of translation. If we identify the molar energy introduced in thermodynamics with the molar kinetic energy E_k calculated by means of kinetic theory, we can calculate the molar heat capacity at constant volume in the following way:

$$E_k = \mathfrak{N}\bar{\epsilon} = \mathfrak{N} \cdot \tfrac{3}{2}kT = \tfrac{3}{2}RT$$

$$C_V = \left(\frac{\partial E_k}{\partial T}\right)_V = \frac{3}{2}R$$

According to this equation the value of C_V should be independent of the kind of gas and of the temperature. From Equation 2–16 the value of C_P for all gases, considered as ideal, is equal to $C_V + R$, and thus C_P should be equal to $5R/2$.

For all gases whose molecules consist of single atoms this calculated value is strikingly confirmed by measured heat capacities (Table 5–3). We have thus been able to accomplish with a simple molecular theory what cannot be done

Kinetic Theory of Gases

TABLE 5-3   HEAT CAPACITIES AT CONSTANT PRESSURE* FOR MONATOMIC AND DIATOMIC GASES IN CAL/MOLE DEGREE

Gas	C_P at 300°K	C_P at 500°K
He	4.97	4.97
Ar	4.97	4.97
Hg	4.97	4.97
H_2	6.89	6.99
O_2	7.02	7.43
N_2	6.96	7.07
Cl_2	8.11	8.63
HCl	6.96	

*American Institute of Physics Handbook.

on the basis of thermodynamics alone; that is, the calculation from theory of the value of a thermodynamic property.

When we turn our attention from monatomic to diatomic (or polyatomic) molecules, however, we find quite a different situation which we discuss as an introduction to later subject matter. First we note that the only kind of energy we have ascribed to monatomic gas molecules in the calculation of C_V is kinetic energy of translation. A molecule with two or more atoms has, in addition, rotational and vibrational energy.

For a diatomic molecule a more realistic model than the rigid sphere with a collision diameter we have used up to now is the dumbbell model (Fig. 5-10).

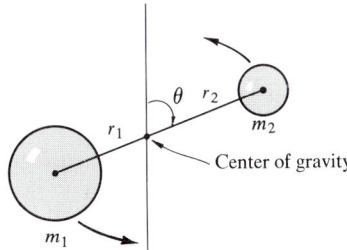

Figure 5-10. Rigid rotor model of a diatomic molecule.

This consists of two masses connected by a rigid rod of length r which represents the chemical bond between the atoms. Such a system has not only kinetic energy of translation as a whole resulting from the motion of the center of gravity, but also kinetic energy of rotation about the center of gravity.

From elementary mechanics the rotational energy is given by

$$\epsilon_r = \tfrac{1}{2} I \, (\dot{\theta}_1^2 + \dot{\theta}_2^2) \qquad (5\text{-}27)$$

in which $\dot{\Theta}_1 \left(= \dfrac{d\Theta_1}{dt}\right)$ is the angular velocity of rotation about any axis 1 perpendicular to the rod and passing through the center of gravity, $\dot{\Theta}_2$ is the angular velocity about axis 2 which is perpendicular both to the rod and axis 1, and I is the moment of inertia of the model about either axis 1 or 2 defined by

$$I = m_1 r_1^2 + m_2 r_2^2$$

m_1 and m_2 are the atomic masses and r_1 and r_2 are the corresponding distances from the center of gravity.

Notice that we do not include any energy of rotation about the rigid connecting rod. This is justified on the basis that practically all of the mass of an atom is concentrated at its center. Consequently the moment of inertia about the molecular axis is essentially zero.

The rigid dumbbell molecule is inadequate for our purpose because it does not permit a change in the interatomic distance resulting from a vibrational motion of the atoms along the molecular axis. A better model consists of two masses held together by a stiff spring (Fig. 5–11). In this oscillator model there is a continuous interchange between vibrational kinetic and potential energy. At the limits of travel the vibrational energy is all potential because the masses stop momentarily to reverse their directions and the spring is stretched or compressed to a maximum extent. At some particular intermediate distance ($r = r_e$) the energy is all kinetic because the spring is neither stretched nor compressed and the masses are moving at maximum speed with respect to the center of gravity. For other values of r, the model has both kinetic and potential energy.

We assume that our model is a harmonic oscillator, that is, it obeys Hooke's law, $f = -k(r - r_e)$, which says that when the masses are displaced from their equilibrium separation, the restoring force is proportional to the displacement and has the opposite direction. The constant k, which is called the force constant, is a measure of the stiffness of the spring. At a later point (p. 294)

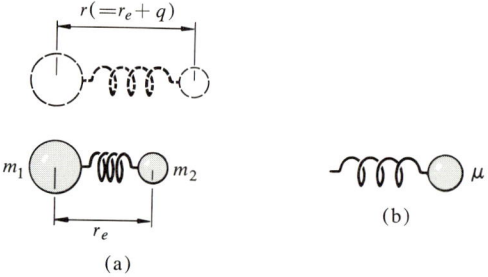

Figure 5–11. (a) Harmonic oscillator model of a diatomic molecule. (b) Equivalent model with mass equal to reduced mass μ.

we will show that the equations for the vibrational motion of our 2-mass model are identical with the equations for a single mass μ [3] attached by a spring to a fixed point. If we represent by q the displacement of the harmonic oscillator from its equilibrium position $(r - r_e)$, the vibrational kinetic energy is given by $\frac{1}{2}\mu\dot{q}^2$ and the vibrational potential energy ϵ_p is given by

$$\tfrac{1}{2}kq^2 \left(= -\int_0^q f\,dq\right)$$

If we now consider our diatomic molecule as a translating, rotating, and vibrating system, we can write for the total molecular energy

$$\epsilon = \tfrac{1}{2}(mv_x^2 + mv_y^2 + mv_z^2 + I\dot{\theta}_1^2 + I\dot{\theta}_2^2 + \mu\dot{q}^2 + kq^2) \qquad (5\text{–}28)$$

This expression for the total molecular energy of a diatomic molecule can be introduced into a generalized form of the Maxwell–Boltzmann distribution law (Eq. 5–26). When this is done, the distribution law can be used to calculate the contributions to the average molecular energy arising from each of the terms in the energy expression. (This use of a distribution law is similar to the previous use of the Maxwell distribution law for molecular velocities.) The result of this calculation, given on page 379, is the general form of the law of equipartition of energy (p. 126), which states that for every term in the expression for the molecular energy which involves the square of a velocity or a coordinate there is a contribution of $kT/2$ to the average energy of the molecule.

We have found it necessary to include seven such square terms in the energy expression of a diatomic molecule. Consequently on the basis of equipartition of energy the average energy of a diatomic molecule should be $7kT/2$, the molar energy should be $7RT/2$, the value of C_V should be $7R/2$ (which is independent of the temperature), and the value of C_P should be $9R/2$.

How does this conclusion compare with experimental results? From Table 5–3 we see that our theoretical value is far from agreement with the data. Most diatomic molecules have a value of C_P near 7, rather than 9, cal/mole deg. Moreover, the experimental values depend to some extent on the temperature. The outstanding example of such a dependence is that of H_2 for which C_P has a value of about 5.5 cal/mole deg at 100°K, and a value of 8 cal/mole deg at 2000°K.

Clearly something is seriously wrong. It required a revolution in physics started by Planck in 1900 to show where the error in this treatment originated. The trouble comes from the use of classical mechanics in calculating the internal energy (particularly vibrational) of a molecule. Although classical mechanics is adequate for calculations involving translational energies, for internal energies of molecules quantum mechanics must be used. The quantum theory underlies most of our treatment of the structure of matter in the chapters ahead.

[3] $\mu = \dfrac{m_1 m_2}{m_1 + m_2}$

Supplementary References

W. Kauzmann, *Kinetic Theory of Gases*, W. A. Benjamin, New York, 1966.
R. D. Present, *Kinetic Theory of Gases*, McGraw-Hill, New York, 1958.

Problems

1. 3.0 moles of helium are contained in a volume of 20 liters at a pressure of 5.0 atm. Calculate (a) the average molecular kinetic energy and (b) the root-mean-square speed of the molecules.
Ans. (a) 8.40×10^{-14} erg/molecule; (b) 1.59×10^5 cm/sec

2. Calculate v_{rms} for mercury atoms whose kinetic energy is 1.00 kcal/mole.
Ans. 2.04×10^4 cm/sec

3. (a) At what temperature will chlorine molecules have the same average kinetic energy that oxygen molecules have at 25°C? (b) At what temperature will the root-mean-square speed of chlorine molecules be the same as the root-mean-square speed of oxygen molecules at 25°C? **Ans.** (a) 25°C; (b) 388°C

4. Derive an equation for the average momentum in the positive x direction for a gas with molecular mass m and temperature T.

5. Derive the Maxwell distribution law for molecular speeds in a 2-dimensional gas.

6. (a) Derive the expression for the most probable speed, v_{mp}, of the molecules of a gas at temperature T. (b) Derive the expression for the most probable kinetic energy of the molecules of a gas at temperature T.

7. (a) Calculate the fraction of nitrogen molecules at 1 atm and 300°K whose speeds are in the range $v_{mp} - 0.005 v_{mp}$ to $v_{mp} + 0.005 v_{mp}$. (b) Calculate the fraction of nitrogen molecules at 1 atm and 300°K whose kinetic energies are in the range $\bar{\epsilon} - 0.005\bar{\epsilon}$ and $\bar{\epsilon} + 0.005\bar{\epsilon}$. **Ans.** (a) 0.083; (b) 0.0046

8. Derive an equation for the distribution of molecular kinetic energy associated with 2 degrees of freedom.

9. Calculate the number of molecules per cubic centimeter in a gas at a temperature of 25°C and a pressure of 10^{-6} torr. **Ans.** 3.23×10^{10}

10. Assuming air to be a mixture of 80% nitrogen and 20% oxygen, calculate the number of molecules which hit a piece of metal foil in 1 sec if the dimensions of the foil are 1 × 2 cm and it is suspended in air at 0.50 atm and 27°C.

11. A mixture of 10 mole % H_2 and 90 mole % Ar leaks through an orifice. What is the composition of the first gas that effuses?

12. 0.87 g of compound X at a pressure of 120 torr at 50°C effuses through a circular hole of radius 0.050 mm in 30.0 min. What is the molecular weight of X? **Ans. 250**

13. A large evacuated vessel is surrounded by nitrogen at a pressure of 1 atm and a temperature of 20°C. Calculate the number of moles of nitrogen which leak into the vessel in 10.0 minutes through a circular hole whose radius is 0.010 cm.

14. Derive the equation for the total number of molecules striking a unit surface per second, from the total number of molecules per second striking a small sphere at a fixed position in the gas, per unit area of the sphere.

15. Calculate the mean free path in CO_2 at 27°C and a pressure of 10^{-6} torr. (Take σ to be 4.60 Å.) **Ans. 3.32×10^3 cm**

16. In a molecular still the distance between the sample surface and a cold receiving surface is 1.50 cm. How low does the pressure have to be for an organic molecule with $\sigma = 12$ Å to distill without undergoing a collision? (Take the temperature of the sample surface to be 30°C.)

17. (a) Calculate the number of bimolecular collisions per second per cubic centimeter in argon at a pressure of 1 atm and a temperature of 100°C if the collision diameter is 3.64 Å. (b) What is the collision rate if the argon pressure is doubled and the temperature is reduced to 50°C?
Ans. (a) 5.03×10^{28} cc^{-1} sec^{-1}; (b) 2.50×10^{29} cc^{-1} sec^{-1}

18. At 280°K the viscosity coefficient of CH_4 is 105 micropoise. Calculate (a) its collision diameter and (b) the number of bimolecular collisions per second per cubic centimeter at atmospheric pressure.

19. Two parallel plates, each with an area of 100 cm², are separated by a distance of 1.5 cm and their temperatures are maintained at 25°C and 20°C. With O_2 at a pressure of 1 atm between the plates, how much heat is conducted in 10 min? **Ans. 11.7 cal**

20. Assuming that the earth's atmosphere is isothermal with a temperature of 250°K, calculate the altitude at which the atmospheric pressure is 0.50 atm.

6

Atomic Structure and Quantum Theory

Introduction

In Chapter 5 a comprehensive theory was discussed in which the molecule was treated as an elementary particle. This theory proved to be extremely successful in accounting for many properties of matter in the gaseous state, but we saw that it fails in those cases in which it is necessary to take into consideration the internal structure and energy of a diatomic or polyatomic molecule.

By the end of the last century it became apparent that classical physics, which is based mainly on Newton's mechanics and Maxwell's theory of electromagnetism, was also facing fundamental difficulties in other areas. For example, although classical physics is capable of treating a wide range of phenomena, it breaks down and gives incorrect results whenever the absorption or emission of light by matter is involved.

The discovery of X-rays by Röntgen (1895), radioactivity by Becquerel (1896), and the electron by J. J. Thomson (1897) initiated a period of rapid development. During the first 30 years of this century drastic changes were made in atomic physics and the foundation was laid for a basic understanding of chemical bonding and molecular structure. One of the essential steps in this development was the recognition that neither the molecule nor the atom is an elementary particle. The atom proved to be a composite structure containing a nucleus and one or more electrons, and nuclei proved to be composite structures composed of protons and neutrons.

Interest in elementary particles and their interactions has continued to the present and a major part of modern physics is concerned with this topic. A variety of short-lived particles has been considered as elementary. It has become evident, however, that the definition of an elementary particle depends on the point of view and the theoretical sophistication of the user of this term.

In this chapter we discuss the structure of atoms and present an introduction to quantum mechanics, which is the basis for our current concepts of the structure and interactions of atoms and molecules.

We start with a review of the elementary particles appropriate to our subject, the electron and the nucleus.

Electrons

An important field of research in physics toward the end of the last century was the study of the conduction of electricity through gases at relatively low pressures (~0.1 torr). It was observed that when two metallic electrodes are introduced into a glass tube (Fig. 6–1) containing a gas at a low pressure and a po-

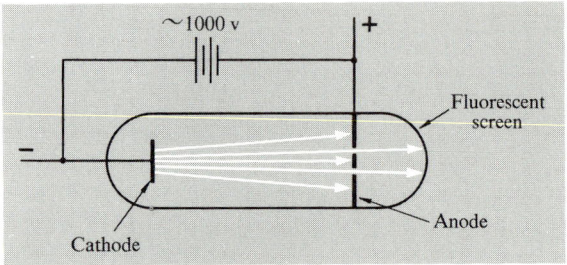

Figure 6–1. Cathode-ray tube.

tential difference of several thousand volts is applied between them, an electric current passes through the gas and rays which are more or less visible, depending on the gas pressure, appear to leave the negative electrode (cathode). If the positive electrode (anode) has a hole in it, some of the cathode rays pass through and strike the walls of the glass tube where they produce a luminous area. In the absence of external electric or magnetic fields the cathode rays travel in straight lines, and if a metallic object partially blocks the beam a sharp shadow is cast.

As part of a study of the nature of cathode rays, J. J. Thomson made measurements on the effects of electric and magnetic fields on these rays, using the apparatus shown in Figure 6–2. When an electric field perpendicular to the

Atomic Structure and Quantum Theory

beam was established by applying a potential difference between the auxiliary electrodes D and E, with E positive, the beam was deflected downward. This direction was evidence that cathode rays have a negative charge. Assuming that the beam is composed of particles, each carrying a charge ϵ, the force on a particle has the magnitude $E\epsilon$ where E is the field strength which is the force acting on a unit charge in the field.

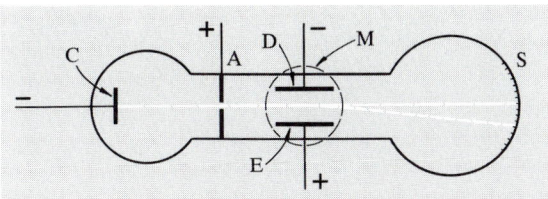

Figure 6–2. Path of the cathode rays in electric and magnetic fields. C is the cathode, A is the anode with a hole in it, D and E are electrodes, M is the pole piece of a magnet whose field is perpendicular to the plane of the paper, and S is a scale for measuring deflection of the cathode rays.

A beam of charged particles is essentially an electric current. In a magnetic field an electrical conductor carrying a current is acted on by a force which is perpendicular to both the direction of the magnetic field and the current. A particle carrying a charge ϵ and having a velocity v in a magnetic field of strength H is acted on by a constant force $\epsilon v H$ perpendicular to both v and H. Because this force is always perpendicular to v, the path of a charged particle in a uniform magnetic field is the arc of a circle. The radius r of the circle is determined by the fact that the magnetic force and the centripetal force are equal. It follows that

$$\epsilon v H = \frac{mv^2}{r}; r = \frac{mv}{\epsilon H}$$

where m is the mass of the charged particle. In a given magnetic field the radius depends on both the ϵ/m ratio and the particle's velocity. When a magnetic field is applied to a beam of cathode rays, the beam is deflected in a direction that is consistent with the negative charge on the particles.

In order to obtain the ratio of the charge of a cathode-ray particle to its mass Thomson used perpendicular electric and magnetic fields applied simultaneously. By adjusting the strengths of these fields he found a combination which left the beam undeflected. Under this condition,

$$\epsilon E = \epsilon v H; v = \frac{E}{H}$$

When the electric field was turned off the resulting deflection of the beam was measured and from this the radius of curvature, r, was calculated. This provided all of the information needed to calculate ϵ/m:

$$\frac{\epsilon}{m} = \frac{E}{rH^2}$$

When Thomson performed this experiment using a variety of gases and cathode materials the values of ϵ/m that he found were all the same within experimental error. This important observation led him to conclude that the cathode-ray particles were a constituent of all matter. These particles were called electrons. Subsequently it was found that electrons could be separated from matter in other ways, such as by heating a metal to a high temperature (thermionic effect) or by irradiation of a metal with ultraviolet rays or X-rays (photoelectric effect).

The presently accepted value of the ϵ/m ratio for the electron at speeds which are relatively small is

$$5.2727 \times 10^{17} \text{ esu/g} = 1.7589 \times 10^8 \text{ coulombs/g}$$

(When electrons are accelerated to speeds approaching the speed of light the value of ϵ/m decreases in agreement with the increase in mass required by Einstein's theory of relativity.)

To determine the electronic charge and mass separately, one or the other of these quantities must be measured. Rough values of the charge were obtained by Thomson and his students but precise data were obtained by Millikan in the well-known oil drop experiment (Fig. 6–3). An electric field directed ver-

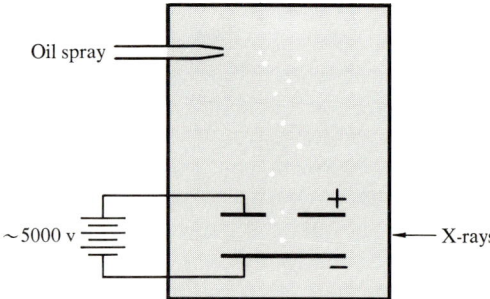

Figure 6–3. Schematic diagram of Millikan's oil drop apparatus.

tically was produced between a pair of horizontal plates which formed an electrical condenser. Very small oil droplets were sprayed into this field and then it was irradiated with X-rays, liberating electrons which became attached to the droplets. The motion of selected droplets was followed through a telescope.

Millikan observed that when the electric field was turned on some of the droplets abruptly acquired a smaller rate of fall and some even reversed their direction. By adjusting the field strength a droplet could be held motionless. This meant that the force of gravity, which depended on the mass of the drop, just balanced the upward force exerted by the field, which depended on the charge on the drop ($\epsilon E = mg$). The mass of the drop was determined indirectly from the terminal rate of fall in air with the field turned off.

When the charges on a number of drops were calculated, it was found that they were all integral multiples of a unit charge, which, of course, is the charge on an electron. No charge smaller than this unit has ever been observed. Today a number of different ways of obtaining the charge on an electron are known and the resulting values are consistent with each other. The presently accepted value is

$$\epsilon = 4.8029 \times 10^{-10} \text{ esu} = 1.6021 \times 10^{-19} \text{ coulomb}$$

This leads to the following value for the rest mass of the electron:

$$m = \frac{4.8029 \times 10^{-10}}{5.2731 \times 10^{17}} = 9.1091 \times 10^{-28} \text{ g}$$

To get some appreciation for the magnitude of the mass of the electron we can compare it to the mass of the hydrogen atom, using the gram atomic weight as the mass of Avogadro's number of hydrogen atoms:

$$m_\text{H} = \frac{1.0079 \text{ g}}{6.0232 \times 10^{23}} = 1.6734 \times 10^{-24} \text{ g}; \quad \frac{m_\text{H}}{m_e} = \frac{1.6734 \times 10^{-24} \text{ g}}{9.1091 \times 10^{-28} \text{ g}} = 1837$$

It is clear that for nearly all purposes the contribution of the electrons to the mass of the atom is negligible.

Nuclei

Because atoms are electrically neutral and electrons carry a negative charge, it could be inferred that some part of the atom must be charged positively. After a period of uncertainty about the distribution of the positive charge in an atom, Rutherford in 1911 reached some conclusions that unambiguously settled this problem. These conclusions were based on observations of the scattering of α rays in passing through thin metallic foils (Fig. 6–4).

Rutherford had shown that α rays consist of positively charged particles whose ϵ/m ratio is about 0.001 that of the electron, and that they were emitted with high kinetic energy during some kinds of radioactive processes. Most of the α particles passed through a foil with essentially no change in direction, but some were deflected at large angles and a few were deflected at angles larger than 90°.

This behavior was completely unexpected. In Rutherford's own words,

"It was almost as incredible as if you fired a 15-inch shell at a piece of tissue paper and it came back and hit you." The only explanation he could find for the experimental results was the idea that most of the mass of the atom is concentrated in a positively charged particle he called the nucleus whose diameter

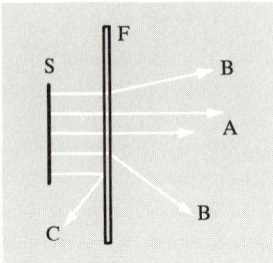

Figure 6–4. Schematic diagram of the scattering of α particles from a radioactive source S by metallic foil F. Most of the α particles (A) go through the foil with no change in direction. A few (B) are deflected at moderately large angles, and a still smaller number (C) are deflected at angles greater than 90° and thus emerge on the same side of the foil as the source.

is only a minute fraction of the atomic diameter. Only when the α particle made a direct hit or a near miss on a nucleus was its path significantly affected. This nuclear model of the atom turned out to be of basic importance for all subsequent work in atomic physics.

From observations on the maximum angle of α-particle scattering, Rutherford could estimate an upper limit to the diameter of the nucleus. This he found to be about 10^{-12} cm. Since the diameter of an atom is of the order of 1 Å ($= 10^{-8}$ cm), the ratio of the diameters is 10^{-4} and the volume occupied by a nucleus is roughly 10^{-12} times the volume of an atom.

From measurements of the angular distribution of α-particle scattering Rutherford calculated the magnitude of the nuclear charge for a number of elements. He found that the charge increased with increasing atomic weight of the element. It was first recognized by van den Broek (1913) that the nuclear charge of an element could be expressed as an integral multiple of the charge on an electron (with opposite sign) and that this integer, called the *atomic number*, gives the position of the element in the sequence of elements in the Periodic Table, starting with hydrogen as 1. It follows that the atomic number must also be the number of electrons in the neutral atom.

◈◈◈◈◈◈

The concept of atomic number soon received powerful support from the work of Moseley (1914) on the characteristic X-rays produced from a series of

Atomic Structure and Quantum Theory

elements. Röntgen had discovered X-rays by allowing cathode rays to strike an anode and observing the emitted penetrating radiation. Moseley studied the X-rays emitted by a number of elements, using a crystal as a diffraction grating to produce an X-ray spectrum. Photographs of these spectra showed a few sharp characteristic lines superimposed on a continuous background. The positions of the characteristic lines shifted in a regular manner with the atomic number of the anode element and this regular shift could be used to determine the atomic number where ambiguities existed. For example, the positions in the Periodic Table of cobalt and nickel, argon and potassium, and iodine and tellurium had to be reversed with respect to the order of their atomic weights to put them into their proper groups on the basis of chemical properties. With respect to their atomic numbers as found from X-ray spectra, no inconsistency was found. The existence of gaps, such as that at atomic number 43 where no element between molybdenum and ruthenium had been found, was also made evident by Moseley's work on X-ray spectra.

The fact that hydrogen has the atomic number 1 suggested that the nucleus of the hydrogen atom, named the *proton* by Rutherford, is an elementary particle with unit positive charge and that protons are contained in all other nuclei. This concept has been confirmed by a vast amount of information on nuclear processes.

For all nuclei other than the proton the ratio of atomic number to mass is one-half or less. This implies that these nuclei must contain at least one other component in addition to the proton. For some years the electron was thought to be this second component. In 1932, however, Chadwick identified the neutron as a particle with nearly the same mass as the proton but with no charge. It was soon realized that the assumption that the neutron was the second component of composite nuclei led to complete consistency with all of the results of nuclear investigations.

The work just described led to the following atomic model: In the center of an atom with atomic number Z there is a nucleus with a diameter of about 10^{-12} cm with a charge equal to that on Z protons and with a mass equal to the sum of the masses of the protons and neutrons in the nucleus. Around the nucleus, Z electrons circulate somewhat like a miniature solar system.

In many ways this is a satisfying picture. The only thing wrong with it is that it is inconsistent with the laws of classical physics! Electronic motion in orbits analogous to planetary orbits is accelerated motion. From Maxwell's electromagnetic theory it follows directly that when charged particles are accelerated they emit energy in the form of electromagnetic radiation. This emitted energy must come from the potential energy of the electrons in the electric field of the nucleus, and a decreasing potential energy means that the distance between the electron and the nucleus must decrease. Consequently the electrons must spiral in and eventually fall into the nucleus.

We reach the conclusion (as we did at the end of the last chapter) that something is seriously wrong with the laws of classical physics when they are applied to the structure of atoms and molecules. Before pursuing this topic any further, we turn our attention to some related developments.

Nature of Light

The nature of light has been a fascinating subject of inquiry since the days of the Greek philosophers. Most of the early speculation was based on one or the other of two assumptions: light consists either of a stream of particles (or "corpuscles") emitted by the source or of some form of wave motion. In the seventeenth century these opposing views were supported by Newton, who favored the corpuscular theory, and by Huygens, who showed that the laws of reflection and refraction of light could be derived on the basis of a wave theory. Because of Newton's great authority the corpuscular theory was generally accepted well into the nineteenth century, when the experimental demonstration of the interference of light by Young in England and Fresnel in France showed that light must be considered to have wavelike properties.

Interference is a characteristic phenomenon shown by any kind of wave — sound waves, water waves, or waves in a stretched string, for example. If two wave trains coincide or cross at a small angle, the effect of the combined wave on a point in the overlap region may be either the sum or the difference of the effects of the separate waves; in other words, either constructive or destructive interference may occur. It is possible for two waves to cancel each other completely.

The demonstration of interference is a crucial test for wave, as opposed to particle, nature because there is no way in which particles can annihilate each other on collision. The wave theory of light gained rapid acceptance after Young showed that when light was allowed to pass through adjacent pinholes in a screen, the transmitted beams interfered with each other.

The characteristic properties of any kind of wave are its *wavelength* (λ), *frequency* (ν), and *speed* (c), which are related by the equation $\lambda\nu = c$. The simplest kind of wave, called a *harmonic* wave, can be represented in one dimension by the equation

$$y = a \begin{Bmatrix} \cos \\ \sin \end{Bmatrix} 2\pi\nu t = a \begin{Bmatrix} \cos \\ \sin \end{Bmatrix} \frac{2\pi c t}{\lambda} \qquad (6\text{-}1)$$

where y, in general, represents the displacement at time t at a point in the medium through which the wave is traveling and a is the maximum displacement, or amplitude, of the wave. The sine and cosine functions differ only in the value of the displacement when $t = 0$.

Another way of expressing the same equation, and one which we will use later, is based on the mathematical identity ($i = \sqrt{-1}$)

$$e^{i\phi} = \cos \phi + i \sin \phi$$

in which ϕ is any angle. In algebraic operations with complex numbers such as this the real and the pure imaginary parts can be treated independently. Representing the real part of $e^{i\phi}$ by the symbol $Re^{i\phi}$, we can write Equation 6-1 in the form

$$y = aRe^{2\pi i \nu t}$$

The symbol R is often omitted. It is then understood that the real part of the complex exponential function is used, unless stated otherwise.

In the latter half of the nineteenth century the supremacy of the wave theory of light was confirmed when Maxwell developed his electromagnetic theory which brought electrical, magnetic, and optical phenomena together in a comprehensive theoretical structure. He showed that light waves could be considered to be a rapidly alternating electric field and a synchronous alternating magnetic field perpendicular to each other and perpendicular to the direction of travel of the waves. The electric field strength at a point in such a wave can be represented as a function of time by

$$E = E_o \cos 2\pi \nu t$$

where E_o is the maximum field strength.

One of the important consequences of Maxwell's theoretical work was the recognition that light is only one kind of electromagnetic radiation. In fact, the only distinguishing characteristic of light is its ability to produce the sensation of sight in human eyes.

In a vacuum all electromagnetic radiation travels with the same speed, the speed of light, which is very close to 3×10^{10} cm/sec. This radiation includes an enormous range of frequencies (or wavelengths) — from a few cycles/sec to more than 10^{20} cycles/sec. In Figure 6-5 the various regions of the electromagnetic spectrum are plotted as a function of the frequency. The discovery and interpretation of this entire spectrum is one of the major accomplishments of science.

Just about the time that the wave theory of light seemed most firmly established, a phenomenon was discovered (Hertz, 1887) which proved to be incompatible with it. This phenomenon, known as the photoelectric effect, consists in the ejection of electrons from a negatively charged metallic surface when light of sufficiently small wavelength is incident on it. As the wavelength of the incident light is decreased below a maximum value, it is observed that the kinetic energy of the ejected electrons increases. As the intensity of the inci-

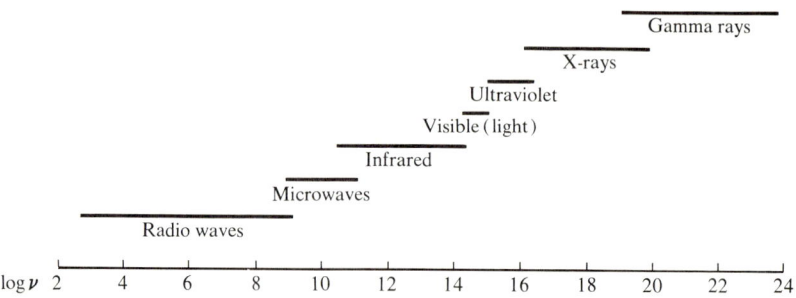

Figure 6–5. Regions of the electromagnetic spectrum.

dent light decreases with the wavelength held constant, the energy of the photoelectrons remains unchanged but the number of electrons ejected per unit time decreases. No matter how weak the incident light beam might be, no time lag is observed between the incidence of the beam on the photoelectric surface and the emission of electrons.

There is no way in which these experimental results can be interpreted on the basis of a transfer of energy from a light wave to an electron in the metal. According to the wave theory, as the intensity of the light beam decreases the amplitude of the light wave decreases and the kinetic energy of the photoelectrons should also decrease. In addition, a measurable time interval should be required for the electrons to acquire the observed amount of kinetic energy from a very weak wave.

Here, again, we have evidence of a basic difficulty in classical physics.

Spectroscopy

The most important source of information about the electron distribution in, and structural details of, atoms and molecules is undoubtedly the field of *spectroscopy*. Spectroscopy is the study of the interaction of matter and electromagnetic radiation as a function of the frequency (or wavelength) of the radiation.[1]

The simplest example of a spectrum is the array of colored light from red to violet observed when sunlight passes through a glass prism. Even though this example is simple, it illustrates the basic requirements for obtaining spectra.

[1] The term spectroscopy has been extended from its original meaning to include other phenomena which have nothing to do with electromagnetic radiation. For example, β-ray spectroscopy is the study of the relative number of electrons emitted in certain radioactive changes as a function of the kinetic energy of the electrons. Mass spectroscopy is the study of the abundance of gaseous ions as a function of their charge-to-mass ratio. If the charge is constant (which is usually the case) then this function becomes a function of the ionic masses.

There must be a source of radiation (in this case, the sun), a device for spreading the radiation into a spectrum (the prism), and a detector for radiation (the human eye).

An instrument for recording spectra on a photographic plate or film is called a spectrograph. With this instrument a range of wavelengths is recorded simultaneously. Another type of instrument, the spectrometer, makes use of a detector which converts radiant energy into an electrical signal which can be measured and recorded. A spectrum is obtained with a spectrometer by allowing successive wavelengths to pass through a slit in front of the detector and plotting the resulting electrical signal against the wavelength.

The spreading of a beam of radiation into a spectrum is known as dispersion. A prism disperses light because the bending of the rays (refraction) is different for different wavelengths. The angle through which a ray of a particular wavelength is bent depends on the index of refraction of the prism material at that wavelength and, in turn, the index of refraction depends on the wavelength. The index of refraction is defined as the ratio of the speed of the wave in a vacuum to the speed in the medium through which the wave is traveling.

A beam may be dispersed not only by refraction but also by diffraction. The dispersing component used in this case is a diffraction grating.

Spectra may be classified in several different ways. One obvious way is in terms of spectral regions, such as the infrared, ultraviolet, or X-ray regions represented in Figure 6-5. Each of these regions requires its own type of apparatus and technique for obtaining spectra. The first region of the spectrum to be studied was, of course, the visible because of the availability of the eye as a detector.

In any spectral region we can distinguish between continuous spectra and discrete spectra. A continuous spectrum contains all of the wavelengths in a wavelength range under consideration, in contrast to a discrete spectrum which has at least some missing wavelengths. A solid heated to incandescence emits a continuous spectrum. Discrete spectra are emitted by gases and vapors that are heated or subjected to an electric discharge in an arc or spark. They are subdivided into line spectra, which are the spectra of atoms, and band spectra, which are the spectra of molecules. We defer consideration of molecular spectra to Chapter 8.

If we place a suitable sample of matter in the beam of radiation between a source which emits a continuous spectrum and the detector of a spectrograph or spectrometer, the sample may absorb selectively some of the wavelengths emitted by the source. The resulting spectrum is called the absorption spectrum of the sample. In general, we distinguish between emission spectra, from which we get information about the source, and absorption spectra, for which the nature of the absorbing sample is of interest.

From the last few paragraphs it can be seen that several descriptive adjectives are needed for accurate reference to a particular type of spectrum. We can, for example, talk about ultraviolet atomic (or line) emission spectra, or infrared band (molecular) absorption spectra, or X-ray continuous emission spectra.

◈◈◈◈◈◈◈

During the nineteenth century the scope of spectroscopy gradually broadened as new sources, dispersers, and detectors were discovered. It was found that the spectrum of each element is unique and that the characteristic pattern of lines could be used to identify an element in qualitative analysis. In fact, Bunsen and Kirchhoff, in an important series of investigations started in 1859, used the newly developed spectroscope as an essential aid in the discovery of new elements, such as cesium and rubidium. The spectroscope also played an important role in the discovery of the noble gases.

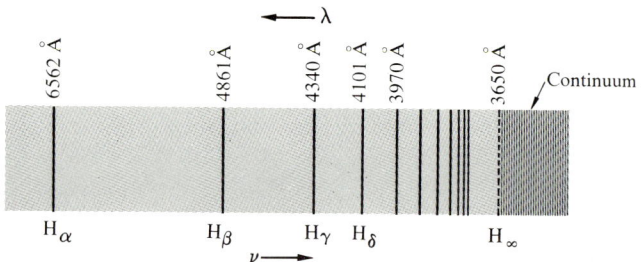

Figure 6-6. Lines in Balmer series of the hydrogen atom.

By the end of the nineteenth century a large number of atomic spectra had been observed and carefully measured. These spectra ranged all the way in complexity from that of hydrogen, which consists of about 10 lines (Fig. 6-6) to that of iron or copper which consists of many hundreds of lines. In 1885 Balmer found an empirical equation from which he could calculate within experimental error the wavelengths of all of the hydrogen lines known at that time. Balmer's equation is

$$\lambda_n = 3645.6 \frac{n^2}{n^2 - 4} \text{ Angstrom units}$$

Atomic Structure and Quantum Theory

The wavelengths of the successive lines in the hydrogen spectrum are obtained by substituting the successive integers for n, starting with $n = 3$ for the line of longest wavelength.

Later Rydberg pointed out that there were advantages in expressing the positions of spectral lines in terms of the *wavenumber* which is the reciprocal of the wavelength ($\omega = 1/\lambda$). The common unit of wavenumber is the reciprocal centimeter (cm^{-1}). The wavenumber in these units is the number of waves per centimeter in the direction in which the wave is traveling.

In terms of wavenumbers the Balmer equation becomes

$$\omega_n = R\left(\frac{1}{2^2} - \frac{1}{n^2}\right) \text{cm}^{-1} \qquad (6\text{-}2)$$

R is known as the *Rydberg constant*. Because of the high accuracy with which spectroscopic measurements can be made, the value of R is known more precisely than most physical constants. The value currently adopted is 1.096782×10^5 cm^{-1}.

In the spectra of some elements other than hydrogen there was a suggestion of regularity in the observed pattern of lines. In most cases, however, the pattern seemed to be completely random. It was strongly suspected that atomic spectra contained a large amount of information about atoms, but the methods of classical physics were incapable of unraveling the mystery of the spectral lines, even in the case of hydrogen.

At about the same time, in another type of spectroscopic investigation quantitative measurements were being made on the distribution of energy as a function of wavelength in the continuous spectra emitted by so-called black bodies. A black body is defined as a solid which absorbs all radiation incident on it (neither reflecting nor transmitting any and hence appearing black) and which is a perfect emitter. Actually, there is no solid which is a perfect emitter although some substances, such as carbon, are very good emitters. It can be shown, however, that when a cavity is formed in an opaque solid and the solid is heated to some suitable temperature, the spectrum of the radiation which escapes through a small hole in the wall of the cavity depends only on the temperature of the solid and not on its composition. This is the characteristic feature of black-body spectra. At constant temperature, the cavity walls emit at the same rate as they absorb and equilibrium is established between the walls and the radiant energy in the cavity.

Some black-body spectra at several temperatures are graphed in Figure 6–7. It is apparent that at each temperature there is a wavelength at which the rate of energy emission is a maximum and this wavelength shifts to smaller values as the temperature is raised. It is for this reason that the color of a solid, such as a piece of iron, which is being heated changes from dull red to cherry red, to yellow, and eventually to white as the temperature is raised. Wien found that the wavelength of maximum emission is inversely proportional to the absolute temperature (Wien's displacement law).

The total amount of energy radiated per unit time and unit area of emitter at all wavelengths, given by the area under the spectral curves, increases rapidly with rising temperature. It was shown experimentally by Stefan and theoretically by Boltzmann that the rate at which energy is radiated by a black body is directly proportional to T^4.

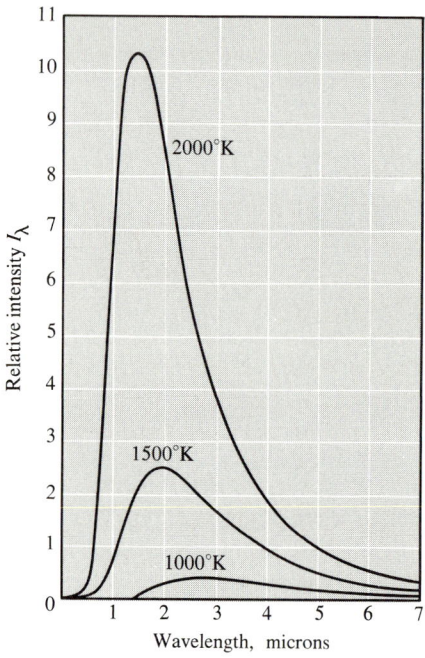

Figure 6-7. Distribution of black-body radiation as a function of wavelength for three temperatures.

At the end of the last century the existence of precise data on black-body radiation was a challenge to physicists and the problem of a theoretical interpretation was attacked in several different ways. Rayleigh and Jeans carried out a treatment based on a generalization of kinetic theory and the equipartition of energy. Their theoretical curves matched the experimental ones at long wavelengths, but with decreasing wavelength the theoretical values increased without a limit and without a trace of a maximum. Wien based his treatment on thermodynamic arguments, considering the cavity and the radiation it contained as a system in equilibrium and introducing his empirical displacement law. His theoretical curve fitted the data well at short wavelengths but deviated significantly at long wavelengths.

Here again was a situation in which classical physics seemed incapable of giving the correct answer. We have seen that failure had occurred in (1) calculating the heat capacity of diatomic and polyatomic molecules; (2) accounting for the stability of the nuclear model for an atom; (3) interpreting the photoelectric effect; (4) explaining the existence of line spectra; and (5) calculating the observed energy distribution in black-body spectra.

Clearly some new ideas were needed. When they arrived on the scene they revolutionized whole branches of physics and affected the development of theoretical chemistry in a fundamental way.

Quantum Theory

The "break-through" came in connection with point (5) above. In 1900 Planck found an empirical equation which reproduced the shape of the black-body radiation curves within experimental error over the entire wavelength range. Soon after, he developed a theoretical derivation for this equation. In order to accomplish this, however, he had to make what seemed like a completely unreasonable assumption; namely, that in the absorption and emission of radiation by the walls of the cavity the energy exchange takes place *only* in discrete amounts he called *energy quanta*, and that the amount of energy in a quantum depends on the frequency of the radiation absorbed or emitted:

$$E = h\nu = \frac{hc}{\lambda} = hc\omega$$

in which h is a constant of nature, now called Planck's constant, with the dimensions of energy $\times$ time.

Planck was a very conservative man. He found his assumption to be distasteful and he spent much of his life in trying to avoid it or, if it was necessary, to limit its scope. Einstein, however, took Planck's assumption seriously and in 1905 pointed out that if the energy exchange between matter and radiation takes place in discrete units one could conclude that radiant energy itself is composed of discrete units. He showed that if a beam of light (or other kind of electromagnetic radiation) is assumed to consist of a flow of particles of energy (now called *photons*), an explanation could be given of the photoelectric effect which accounts quantitatively for all experimental observations.

Einstein's explanation of the photoelectric effect is, in brief, the following: When an electron in a photosensitive surface is struck by a photon it acquires the energy of the photon ($h\nu$) which is absorbed. A certain amount of work (W) must be done on the electron to eject it from the surface. This work, which is called the *work function* of the surface, can be expressed as $h\nu_o$. Only if the photon energy is larger than W, which implies that $\nu > \nu_o$, can the electron leave.

Any excess energy above W goes into electronic kinetic energy which is given by the following equation:

$$\epsilon_{kin} = h\nu - W = h\nu - h\nu_o$$

Decreasing the intensity of a monochromatic light beam with $\nu > \nu_o$ lowers the photocurrent because it lowers the rate at which photons reach the surface and are absorbed. It does not, however, affect the kinetic energy of the ejected electrons which, according to the equation above and in agreement with experimental data, depends on the frequency of the light only.

Einstein's success in explaining the photoelectric effect gave a new look to the wave-particle conflict in the theory of light. Because the earlier evidence for wave properties based on the interference and diffraction of light could not be refuted or ignored, it gradually became apparent that neither the wave model nor the particle model is adequate by itself. In some way electromagnetic radiation has *both* wave and particle properties. This concept, which is referred to as the duality of electromagnetic radiation, is implied by Planck's law: $E_{ph} = h\nu$. The left-hand side of this equation has meaning only in terms of particle properties. The right-hand side has meaning only in terms of waves because frequency is inherently a wave property.

We see that Planck's law is a bridge between the particle and the wave aspects of radiation. In different types of experiments one or the other aspect is evident. In general it is easier to demonstrate wave properties with low-frequency radiation and particle properties in the high-frequency range in which the energy of a photon is relatively large.

The existence of particle-like properties of electromagnetic radiation received strong support from the work of A. H. Compton on the scattering of X-rays by electrons. He directed a beam of nearly monochromatic X-rays at a target and observed the wavelength and intensity of the X-rays scattered by the target over a range of angles. Most of the beam passed through the target without deviation. The X-rays observed at other angles had wavelengths longer than those of the incident radiation and the magnitude of the wavelength shift was a function of the scattering angle. This shift of the X-ray wavelength is called the Compton shift.

Compton explained his results by considering an X-ray photon as a particle which could collide with an electron in the target and give up some energy to it. In such a collision the laws of conservation of energy and momentum should be valid. The energy of a photon is given by Planck's law: $E_{ph} = hc/\lambda$. It also can be expressed in terms of Einstein's relativistic equation relating mass and energy: $E_{ph} = mc^2$. When these two expressions for E_{ph} are equated and the momentum of the photon ($p = mc$) is introduced, it follows that

$$\frac{hc}{\lambda} = mc^2 = pc$$

and

$$p = \frac{h}{\lambda} \qquad (6-3)$$

By substituting this expression for the momentum of a photon into the equations representing the conservation laws, Compton derived an equation relating the angle of scattering of an X-ray photon to its wavelength that quantitatively fits the observed data.

Bohr's Theory of the Hydrogen Atom

Planck's development of the quantum theory of black-body spectra left unsolved the problems of atomic structure and the interpretation of line spectra. In 1913 Bohr took a giant step forward by combining Rutherford's nuclear model of the atom with Planck's quantum theory and Einstein's concept of the photon in order to derive the Balmer equation (Eq. 6-2) for the positions of the hydrogen lines.

Subsequent developments have made much of Bohr's theory obsolete, but some of the basic ideas he introduced are as valid today as when he introduced them. Consequently we present a brief resume of his theory of the hydrogen atom.

Bohr started with Rutherford's model: an electron of mass m revolving in a circular orbit of radius r with a relatively massive nucleus of atomic number Z at its center. The centripetal force, which by elementary mechanics is mv^2/r, is equal to the electrostatic force, $Z\epsilon^2/r^2$, given by Coulomb's law:

$$\frac{mv^2}{r} = \frac{Z\epsilon^2}{r^2} \qquad (6-4)$$

The radius of the orbit is given by

$$r = \frac{Z\epsilon^2}{mv^2} \qquad (6-5)$$

This by itself is not useful because on the basis of classical mechanics the radius can have any value. Bohr now made several assumptions which were arbitrary and inconsistent with classical mechanics, but which led to significant results. He assumed that an atom could exist only in one of a series of states he called stationary states, which were characterized by the requirement that the angular momentum of the electron in its orbit is an integer times $h/2\pi$.

From elementary mechanics the angular momentum of the electron in its orbit is mvr. Bohr's assumption is expressed by the following equations:

$$mvr = \frac{nh}{2\pi} \qquad (n = 1, 2, 3, \ldots)$$

$$v = \frac{nh}{2\pi mr}$$

When this value of v is substituted in Equation 6–5, the values of the radius in the stationary states specified by the values of n are given by

$$r_n = \frac{n^2 h^2}{4\pi^2 m Z \epsilon^2} \tag{6-6}$$

The total energy of the electron in its orbit is the sum of the kinetic and potential energies:

$$E = \frac{1}{2}mv^2 - \frac{Z\epsilon^2}{r} \tag{6-7}$$

The negative sign of the potential energy term results from a choice of the zero of potential energy as the value when the electron and the nucleus are infinitely far apart and so exert no force on each other.[2]

From Equation 6–4

$$\frac{1}{2}mv^2 = \frac{Z\epsilon^2}{2r}$$

When this expression for the kinetic energy is introduced into Equation 6–7 it follows that $E = -\dfrac{Z\epsilon^2}{2r}$. Substituting into this equation the value of r from Equation 6–6, we obtain

$$E_n = -\frac{2\pi^2 m \epsilon^4 Z^2}{n^2 h^2} \tag{6-8}$$

This equation states that the total energy of the electron depends on the square of the integer n, which is called a quantum number.

To account for the existence of atomic spectra Bohr made two other arbitrary assumptions:

1. An atom can absorb or emit radiation only by making a transition from one stationary state to another. (Nothing was said about the details of such a transition.)

2. In a transition an atom absorbs or emits a photon whose energy is equal to the difference of the energies of the initial and final states. This requirement determines the frequency (and therefore the wavelength and wavenumber) of the absorbed or emitted radiation in terms of the values of the quantum num-

[2] $E_p = \displaystyle\int_\infty^r \frac{Z\epsilon^2}{r^2}\,dr = -\frac{Z\epsilon^2}{r}$

bers for the initial and final states, n_1 and n_2, according to the following equations:

$$E_{ph} = h\nu = \frac{hc}{\lambda} = hc\omega$$
$$= E_{n_2} - E_{n_1}$$
$$= \frac{2\pi^2 m\epsilon^4 Z^2}{h^2}\left(\frac{1}{n_1^2} - \frac{1}{n_2^2}\right)$$

The wavenumber of the radiation is given by

$$\omega = \frac{E_{n_2} - E_{n_1}}{hc} = \frac{2\pi^2 m\epsilon^4 Z^2}{h^3 c}\left(\frac{1}{n_1^2} - \frac{1}{n_2^2}\right) \text{cm}^{-1} \quad (6\text{-}9)$$

A comparison of this purely theoretical equation, setting $Z = 1$ and $n_1 = 2$, with the empirical Equation 6-2 shows that the two have the same mathematical form and, even more remarkably, when the combination of five constants was evaluated it equaled the Rydberg constant within the uncertainty with which the constants were known.

The Bohr theory had many other successes. For example, the radius of the smallest orbit ($n = 1$) for hydrogen, usually represented by a_o, as calculated from Equation 6-6 is 0.529 Å. This is of the same order of magnitude as the collision diameters discussed in Chapter 5.

The Bohr theory also correctly implied that there should be other series of lines in the hydrogen spectrum for which n_1 might be 1, 3, or some other integer. Actually another series had been discovered in the infrared region by Paschen in 1908 and the positions of these lines were in complete agreement with Equation 6-9 when n_1 was equal to 3. Subsequently still other series were discovered by Lyman ($n_1 = 1$), Brackett ($n_1 = 4$), and Pfund ($n_1 = 5$) (Fig. 6-8).

With the appropriate value of Z, Equation 6-9 accounted quantitatively for the spectra of other 1-electron systems such as He$^+$ and Li^{++}. With extensions which involved the introduction of more than one kind of quantum number, the Bohr theory supplied a basis for fruitful qualitative discussions of atomic structure and some degree of understanding of the spectra of multi-electron atoms and ions.

In spite of its successes the Bohr theory had fundamental deficiencies. It could not account quantitatively for the spectrum of any element other than hydrogen. Even for helium with only two electrons it gave wrong results. In addition, the arbitrary nature of the assumptions that Bohr had to make about stationary states and transitions made the theory unsatisfactory. Some additional ideas were needed and not much more than a decade after the publication of Bohr's theory they appeared.

Before we turn to these new ideas it is worthwhile to note what part of Bohr's theory has survived and what part has been discarded. The concept that an atom (or molecule) can exist for a finite time only in a series of stationary states, or energy levels, characterized by a set of quantum numbers is still fundamental today. This is also true of the concept that transitions between energy levels are accompanied by the absorption or emission of a photon and, conversely, that each emission or absorption line in a spectrum is associated with a transition between two energy levels.

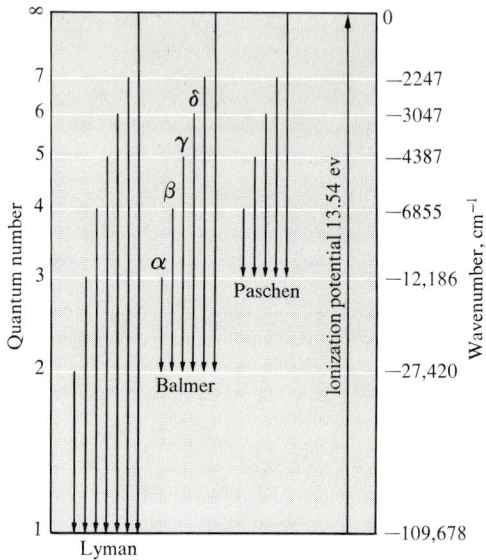

Figure 6-8. Energy level diagram for the hydrogen atom.

What has been discarded is the use of classical mechanics modified by arbitrary assumptions. This has been replaced by quantum mechanics which has proved to be particularly suited for all problems involving atoms and molecules. Also abandoned has been the use of easily visualized models such as the model of an atom in which electrons travel in definite orbits around a nucleus. We return to this point later.

Wave Nature of Matter

One of the new ideas which was to have far-reaching consequences started out in 1924 as a highly speculative thesis by a university student in France. This student, de Broglie, was very much impressed by the wave-particle duality of electromagnetic radiation and the fact that a photon carries momentum which is related to its wavelength (Eq. 6–3). Essentially on the basis of analogy with electromagnetic radiation, he suggested that there might be waves associated

Atomic Structure and Quantum Theory

with a particle of matter, with mass m and velocity v, and that these waves have a length given by the same relation which is valid for electromagnetic radiation:

$$\lambda = \frac{h}{p} = \frac{h}{mv}$$

At the time it was published, de Broglie's proposal seemed "far-out." To give it significance evidence was needed for the interference and diffraction of particles of matter. Within a few years the diffraction of electrons by crystals had been observed in several laboratories, in quantitative confirmation of de Broglie's idea. Later the diffraction of protons, neutrons, and α particles was observed.

Schrödinger was stimulated by de Broglie's idea to develop a radically new theory of mechanics, known as wave mechanics, which has turned out to be remarkably successful and which has had and still is having a profound influence on theoretical chemistry. Wave mechanics is one branch of the general subject of quantum mechanics. Another branch called matrix mechanics was developed by Heisenberg at the same time that Schrödinger was developing wave mechanics. Subsequently it was shown that the two forms of mechanics are essentially the same, but are expressed in different mathematical language. The wave-mechanical formulation has proved to be the more useful in applications to chemistry.

The Classical Wave Equation

Before starting a consideration of wave mechanics it is helpful to summarize some results of classical wave theory. The classical wave equation for any kind of wave motion in the x direction in a medium in which the speed c is independent of the wavelength is

$$\frac{\partial^2 \phi}{\partial x^2} = \frac{1}{c^2} \frac{\partial^2 \phi}{\partial t^2} \qquad (6\text{--}10)$$

The function $\phi(x,t)$ usually (but not always) represents some disturbance of the medium through which the wave is passing. For example, in the case of a wave in a string, ϕ represents the displacement of a point on the string (at a position x and time t) in a direction perpendicular to the direction in which the wave is traveling. In the case of a sound wave ϕ represents a change in the density of the medium. As we have seen, electromagnetic waves are fundamentally different from waves in a string or sound waves because there is no medium to "carry" the waves. For electromagnetic waves ϕ represents either the electric field strength or the associated magnetic field strength.

The wave equation is a partial differential equation. Such an equation has an infinite number of solutions. While the solution of an ordinary differential equation contains arbitrary constants of integration, the general solution of a partial differential equation contains arbitrary functions. If this solution is to have physical significance, the functions must be well-behaved; that is, they must have no discontinuities and they must not become infinite anywhere.

As is easily verified by differentiation, a general solution of the classical wave equation (Eq. 6–10) can be written in the following form:

$$\phi = f(x - ct)$$

where f is any well-behaved function. Now let us see what this solution represents, using a wave in a long string as an example. Consider the function f at time $t = 0$ (Fig. 6–9). It represents a pattern of displacements in the string

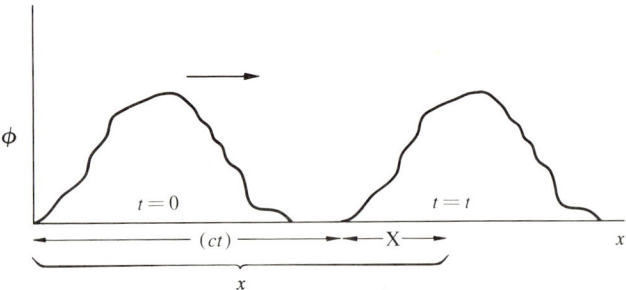

Figure 6–9. Wave form with arbitrary profile, $f(x)$, moving in the x direction with speed c; $f(x) = f(X + ct)$.

which is called the wave shape or the wave profile. At a later time t the pattern of displacements has moved along the string in the x direction. If the origin of coordinates is assumed to move in the x direction with the speed of the wave, the equation for the wave form is $f(X)$ where X is the distance of a point from the moving origin. The relation between X and x, which is always measured from the fixed origin used when $t = 0$, is $x = X + ct$.

It follows that $f(X) = f(x - ct)$ and that this function represents a wave form which is moving in the x direction with speed c. It also follows that $f(x + ct)$ represents the same wave form moving in the $-x$ direction.

Mathematicians have shown that any general wave function such as $f(x - ct)$ can be expressed as a sum of simple functions, known as harmonic functions, each multiplied by an appropriate factor. Such a sum is called a linear combination. A harmonic wave function, a special case of which we have met before (Eq. 6–1), can be expressed either as a real function or as a complex exponential function. The real form can be written as follows:

$$\phi = a \cos m(x - ct)$$

Atomic Structure and Quantum Theory

where a and m are constants for a particular wave. The value of a is the amplitude of the wave. The sine function is equivalent to the cosine except for an additive constant, the phase angle, which is not important for our purpose.

The cosine function is a periodic function of x and t. For any particular value of t it repeats itself whenever $\Delta mx = 2\pi$. The repeat distance Δx is the wavelength λ. Consequently,

$$\Delta x = \frac{2\pi}{m}; \quad m = \frac{2\pi}{\lambda}$$

Substituting this value for m, we can write the harmonic wave function in several equivalent ways:

$$\phi = a \cos \frac{2\pi}{\lambda}(x - ct) = a \cos 2\pi \left(\frac{x}{\lambda} - \frac{ct}{\lambda}\right) = a \cos 2\pi \left(\frac{x}{\lambda} - \nu t\right)$$

We can also express the wave function in the exponential form when this is convenient:

$$\phi = ae^{2\pi i \left(\frac{x}{\lambda} - \nu t\right)}$$

(This does not exhaust the number of ways this function can be written.)

Because Equation 6–10 is a linear partial differential equation, the sum of any two wave functions, each multiplied by an arbitrary constant, is also a solution of the wave equation. For example, if ϕ_1 and ϕ_2 are any wave functions and if a_1 and a_2 are any numbers, then

$$a_1 \frac{\partial^2 \phi_1}{\partial x^2} = \frac{a_1}{c^2} \frac{\partial^2 \phi_1}{\partial t^2}$$

and

$$a_2 \frac{\partial^2 \phi_2}{\partial x^2} = \frac{a_2}{c^2} \frac{\partial^2 \phi_2}{\partial t^2}$$

On adding these equations we obtain

$$a_1 \frac{\partial^2 \phi_1}{\partial x^2} + a_2 \frac{\partial^2 \phi_2}{\partial x^2} = \frac{a_1}{c^2} \frac{\partial^2 \phi_1}{\partial t^2} + \frac{a_2}{c^2} \frac{\partial^2 \phi_2}{\partial t^2} = \frac{\partial^2}{\partial x^2}(a_1 \phi_1 + a_2 \phi_2)$$

$$= \frac{1}{c^2} \frac{\partial^2}{\partial t^2}(a_1 \phi_1 + a_2 \phi_2)$$

This shows that the *linear combination* $(a_1\phi_1 + a_2\phi_2)$ is a solution of the wave equation.

In general, any linear combination of solutions of a wave equation is itself a solution of the wave equation. This general statement, which is often called the principle of superposition, has its physical counterpart in the fact that a number of waves of any type with arbitrary amplitudes can be combined to form a resultant wave.

To summarize: any linear combination of wave functions is a wave function, and any wave function can be expressed as a linear combination of harmonic functions. Consequently there is no loss of generality in limiting our attention at present to harmonic wave functions.

◈◈◈◈◈◈◈

In the most general case, any well-behaved function, whether it is periodic or not, csn be expressed as an infinite series of sine and cosine terms. Such a series is known as a Fourier series for which the general expression is:

$$f(x) = a_0 + \sum_{n=1}^{n=\infty} a_n \cos \frac{n\pi x}{\lambda} + \sum_{n=1}^{n=\infty} b_n \sin \frac{n\pi x}{\lambda}$$

◈◈◈◈◈◈◈

Let us now consider the effect of superimposing two identical harmonic waves traveling in a string in opposite directions. It is convenient to use the sine function in this case, and we write the function representing the combination of the two waves in the following way:

$$\phi = a \sin 2\pi \left(\frac{x}{\lambda} - vt\right) + a \sin 2\pi \left(\frac{x}{\lambda} + vt\right)$$

By use of the trigonometric identities for $\sin(m-n)$ and $\sin(m+n)$, this becomes

$$\phi = 2a \sin \frac{2\pi x}{\lambda} \cos 2\pi vt \tag{6-11}$$

This is the wave function for a standing wave. Because the two variables are in different factors, at all points for which $2x/\lambda$ is an integer $\phi = 0$ at all times. These points, which are known as nodes, occur at $x = n\lambda/2$ with $n = 0, 1, 2 \ldots$. At all other points on the string ϕ is a periodic function of the time (Fig. 6–10).

Now let us assume that the string has the length L and that it is fixed at each end. From a mathematical point of view we impose the boundary conditions $\phi = 0$ at $x = 0$ and at $x = L$. These conditions greatly limit the number of possible values of λ in the solutions of the wave equation. The length of the string must be equal to an integral number of half wavelengths in the standing wave. Therefore the only possible values of the wavelength are given by

$$\lambda = \frac{2L}{n} \quad (n = 1, 2 \ldots) \tag{6-12}$$

where n is the number of nodes after the one at $x = 0$.

Atomic Structure and Quantum Theory 181

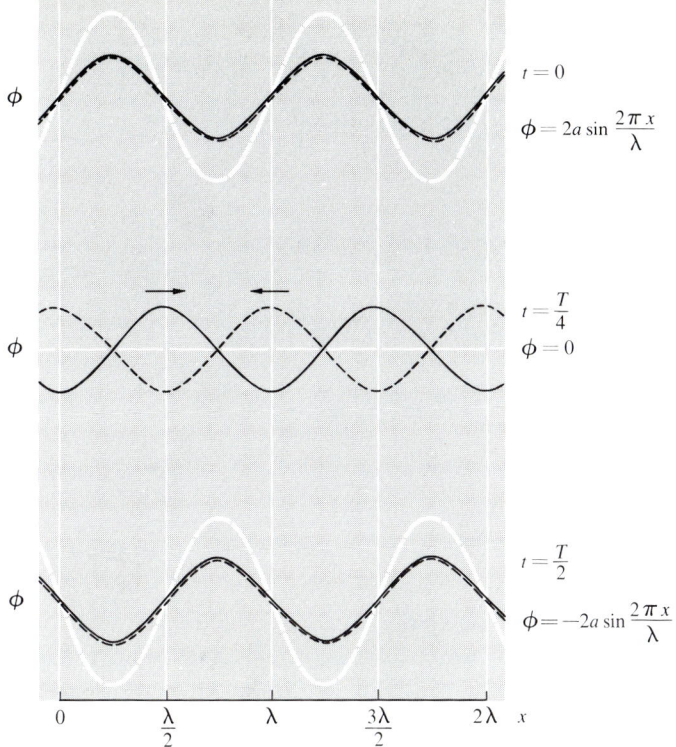

Figure 6–10. Graph of a standing wave formed by two waves traveling in opposite directions. ———— Wave traveling in $+x$ direction. -------- Wave traveling in $-x$ direction. ------- Resultant standing wave. Nodes are located at: $x = 0$, $\frac{\lambda}{2}$, λ, $\frac{3\lambda}{2}$, and 2λ.

◊◊◊◊◊◊◊

It is noteworthy that the case of standing waves is the only part of classical physics in which integers occur in a natural and essential way. This was one of the considerations which led de Broglie to think about matter waves as a way for accounting for the existence of quantum numbers.

◊◊◊◊◊◊◊

It is readily shown that the standing wave function (Eq. 6–11) is a solution of the general wave equation. By direct partial differentiation of this function with respect to x we obtain

$$\frac{\partial^2 \phi}{\partial x^2} = -\frac{4\pi^2}{\lambda^2} \phi \tag{6-13}$$

By partial differentiation with respect to t we obtain

$$\frac{1}{c^2}\frac{\partial^2 \phi}{\partial t^2} = -\frac{4\pi^2 \nu^2}{c^2} \phi = -\frac{4\pi^2}{\lambda^2} \phi = \frac{\partial^2 \phi}{\partial x^2}$$

For simplicity we have limited our discussion so far to the 1-dimensional wave equation. There is no difficulty in a generalization to two dimensions (waves in a drum head, for example), or to three. In Cartesian coordinates Equation 6-13 can be written in the form

$$\frac{\partial^2 \phi}{\partial^2 x} + \frac{\partial^2 \phi}{\partial^2 y} + \frac{\partial^2 \phi}{\partial^2 z} = -\frac{4\pi^2}{\lambda^2} \phi$$

It is convenient to use the symbol $\nabla^2 \phi$, called the Laplacian of ϕ, to represent the sum of the three partial derivatives. The wave equation then is

$$\nabla^2 \phi = -\frac{4\pi^2}{\lambda^2} \phi \tag{6-14}$$

As we have seen before, this partial differential equation is very general. Of the infinite number of solutions it has, the only ones which can have physical significance are functions that are finite and continuous over the whole range of the variables. In addition, the functions must be single-valued; i.e., at any point in space a useful solution must have one and only one value. This still leaves an infinite number of possible solutions. The ones that are wanted for a particular problem are obtained by imposing conditions which are appropriate to the problem.

The Schrödinger Equation

At this point we are ready to apply the background of wave theory to the wave properties of matter. Because we are interested in states of atoms and molecules which do not depend on time, we use Equation 6-14 which contains only partial derivatives with respect to coordinates and in which time does not occur as a variable. Substituting for λ in this equation the value proposed by de Broglie (h/mv, for a particle of mass m and speed v), we obtain

$$\nabla^2 \psi = -\frac{4\pi^2 m^2 v^2}{h^2} \psi$$

We use ψ as the symbol for wave functions representing matter (or de Broglie) waves.

If E is the total energy of the particle and U is its potential energy expressed

Atomic Structure and Quantum Theory

as a function of the coordinates, $E - U = \frac{1}{2}mv^2$ and the equation above can be written in the following form:

$$\nabla^2 \psi + \frac{8\pi^2 m}{h^2}(E - U)\psi = 0 \qquad (6\text{-}15)$$

This is one way of writing the *Schrödinger equation*, with which we will be concerned for several chapters. It should be noticed that we have not derived this equation. In fact, no real derivation has been found so far. We have merely shown that if one assumes a wave equation such as Equation 6-14 and introduces the de Broglie wavelength, the Schrödinger equation is obtained.

To write down the Schrödinger equation corresponding to a particular physical system, the potential energy function U must be known. The statement of this function is an essential part of the specification of a physical problem. In general, there is no difficulty in writing the Schrödinger equation for such a problem. The difficulty arises in finding the desired solutions of this equation.

Like the classical wave equation, the Schrödinger equation is a linear, partial differential equation. The solutions we want are those which are well-behaved functions; that is, they must be finite, continuous, and single-valued over the whole range of the coordinates. A solution that satisfies these requirements is often called an *eigenfunction*. We will see that only for certain values of the total energy E is it possible to obtain solutions of the Schrödinger equation which are eigenfunctions.

The value of an eigenfunction ψ at a point in space specified by particular values of the coordinates may be positive or negative, real or complex. If ψ is complex, the symbol ψ^* represents the complex conjugate of ψ; that is, the function that is obtained by changing the sign of i wherever it occurs in ψ. The product of a function and its complex conjugate is always real and positive. For example, if a value of ψ is represented by $a + ib$ (a and b are real numbers), then the value of ψ^* is $a - ib$ and $\psi\psi^* = a^2 + b^2$. If ψ is real, $\psi\psi^* = \psi^2$. In almost all cases the eigenfunctions we are concerned with in this book are real functions.

In order to make a connection between eigenfunctions and physical quantities, we should be able to state what an eigenfunction represents. It does not represent a displacement in a medium as does a wave in a string or a sound wave. It does not represent the strength of an electric or magnetic field as does an electromagnetic wave. In fact, no interpretation of ψ, itself, has been found.

What does have physical significance, however, is $\psi\psi^*$. One of the basic postulates of wave mechanics can be stated in the following way: If ψ is a solution of the Schrödinger equation for a particle of matter, $\psi\psi^* \, dx \, dy \, dz$ represents the probability that this particle is in the volume element $dx \, dy \, dz$, located at the point x,y,z. The implications of this statement are far-reaching and we return to them later.

For the present, we note that if it is known that a particle exists, the probability of its being somewhere in space must equal 1 — which means certainty. This requirement imposes a restriction on the eigenfunction because it implies that

$$\int_{-\infty}^{+\infty}\int_{-\infty}^{+\infty}\int_{-\infty}^{+\infty} \psi\psi^* \, dx\, dy\, dz = 1 \qquad (6\text{–}16)$$

A function which satisfies this relationship is said to be *normalized*.

It turns out that for only a few problems or, in other words, for only a few forms of the potential energy function can the Schrödinger equation be solved exactly. Each of these problems has application to chemistry.

Before we take up these problems we approach the Schrödinger equation from a different and more basic point of view that emphasizes the word *mechanics* rather than the word *wave* in wave mechanics. First we take another look at classical mechanics which got its start when Newton formulated his laws of motion. These laws were not derived from more fundamental relationships but were postulates which proved to be extremely successful in dealing with a wide range of phenomena from planetary motion to the translational motion of molecules.

Newton's laws of motion are relatively simple from a mathematical point of view as long as the physical system under consideration can be described in terms of Cartesian coordinates. In many problems it is simpler to make use of other types of coordinates, such as polar coordinates, in describing a system and for these coordinates the Newtonian equations can be very complex. For this reason several other formulations of classical mechanics have been developed which avoid the difficulties of Newton's laws.

A particularly powerful version of classical mechanics was developed by Hamilton early in the nineteenth century. While Newton's equations deal with forces, masses, and accelerations, Hamilton used momenta and coordinates as the variables in his equations. An important part of his treatment of mechanics is the Hamiltonian function which is an expression of the total energy of a system, kinetic plus potential, with the kinetic energy written as a function of momenta and the potential energy as a function of the coordinates.

For a simple example, we can write the Hamiltonian function for a harmonic oscillator. As mentioned previously (p. 153), this oscillator can be pictured as a mass on a spring which obeys Hooke's law, $f = -kx$. In this equation x represents the displacement of the mass from its equilibrium position, f is the restoring force, and k is the force constant, which is a measure of the strength of the spring. The potential energy function of the harmonic oscillator is

$$U = -\int_0^x f \, dx = \tfrac{1}{2}kx^2$$

and its kinetic energy in terms of the momentum is $p_x^2/2m$. Its Hamiltonian function is

$$H = \frac{p_x^2}{2m} + \frac{kx^2}{2} \tag{6-17}$$

Now let us leave mechanics for a short time to introduce a topic in mathematics, the algebra of operators. An operator is a symbol which indicates that some specified operation is to be performed on a function whose symbol is written after the operator. Examples of operators are: $a \cdot$ (multiplication by a), $\frac{d}{dx}$ (the differential operator), and the Laplacian operator $\nabla^2 \left(= \frac{\partial^2}{\partial x^2} + \frac{\partial^2}{\partial y^2} + \frac{\partial^2}{\partial z^2} \right)$. If an operator is applied to a function n times, this is indicated by writing the symbol for the operator as though it were raised to the power n; thus,

$$\frac{\partial}{\partial x}\left(\frac{\partial}{\partial x}\right) = \left(\frac{\partial}{\partial x}\right)^2 = \frac{\partial^2}{\partial x^2}$$

Although operators by themselves have no meaning, mathematicians have found that they can be manipulated like algebraic variables.

The operators with which we are concerned are linear operators. An operator, α, is said to be linear if $\alpha(f + g) = \alpha f + \alpha g$, where f and g are functions.

Mathematicians have studied an important type of equation which can be written in the following way:

$$\alpha f = af \tag{6-18}$$

In this equation α is a linear operator such that when it operates on a certain function f the result is the multiplication of the function by a number which is called an eigenvalue. Any function that satisfies this equation is called an eigenfunction of the operator α. This definition of an eigenfunction is more general than the one given on page 183.

Now let us return to the subject of quantum mechanics. Just as classical mechanics can be based on a set of postulates, quantum mechanics can also be based on a set of postulates. One way of formulating these postulates (necessarily rather abstract) is the following:

1. The state of a physical system is described completely by a well-behaved function of the coordinates of the system, $\psi(x_i, y_i, z_i)$, such that the value $\psi\psi^* \, dx_i \, dy_i \, dz_i$ is the probability that the coordinates of the system have values in the ranges x_i to $x_i + dx_i$, y_i to $y_i + dy_i$, z_i to $z_i + dz_i$, and so forth.

2. To every observable quantity, such as a coordinate, velocity, momentum, or energy, there corresponds a certain linear operator.

3. The only values of an observable quantity which can be found by measurement are the eigenvalues of the corresponding operator in the equation

$\alpha\psi_n = a_n\psi_n$, where ψ_n is a member of the set of eigenfunctions of the operator and a_n is the related eigenvalue.

For our purpose in this book the only operators we need are those for a coordinate and a momentum. It has been found that the following operators lead to correct results: For a Cartesian coordinate such as x the corresponding operator is $x \cdot$ meaning a multiplication by x. The operator corresponding to the x component of linear momentum is $\dfrac{h}{2\pi i}\dfrac{\partial}{\partial x}$.

The observable quantity that is of most interest to us is the total energy of an atom or a molecule in a stationary state. We obtain the operator corresponding to the total energy by replacing each coordinate and each momentum in the classical Hamiltonian function by its corresponding operator. This energy operator is called the *Hamiltonian operator*.

To obtain the Hamiltonian operator for the harmonic oscillator, for example, the quantity p_x^2 in Equation 6–17 is replaced by the linear momentum operator operating two times in succession:

$$p_x^2 \rightarrow -\frac{h^2}{4\pi^2}\frac{d^2}{dx^2}$$

and x^2 now represents multiplication by x^2. (We use the total differential operator in this case because there is only one coordinate.) The resulting operator is

$$H_{op} = -\frac{h^2}{8\pi^2 m}\frac{d^2}{dx^2} + \frac{kx^2}{2}$$

If the function ψ is an eigenfunction of the Hamiltonian operator for a system, we can rewrite Equation 6–18 in the following form:

$$H_{op}\psi = E\psi \tag{6-19}$$

in which the eigenvalue E represents the energy of the system when it is in the state represented by the function ψ. This equation is the Schrödinger equation written in its most compact form.

As soon as we know the classical Hamiltonian function for any system, we can express the Hamiltonian operator explicitly and thus write the Schrödinger equation for the system. Again using the harmonic oscillator as an example, we obtain the corresponding Schrödinger equation as follows:

$$\left(-\frac{h^2}{8\pi^2 m}\frac{d^2}{dx^2} + \frac{kx^2}{2}\right)\psi = E\psi$$

We can readily put this equation in the same form as Equation 6–15; it then becomes

$$\frac{d^2\psi}{dx^2} + \frac{8\pi^2 m}{h^2}\left(E - \frac{kx^2}{2}\right)\psi = 0 \tag{6-20}$$

Atomic Structure and Quantum Theory

If we have been able to solve the Schrödinger equation for a system and have thus obtained a particular eigenfunction ψ_m, we can calculate the corresponding energy E_m by multiplying both sides of Equation 6-19 by this function (or its complex conjugate) and integrating over the whole range of the variables in ψ_m:

$$\int \psi_m^* H_{op} \psi_m \, d\tau = \int \psi_m^* E_m \psi_m \, d\tau = E_m \int \psi_m^* \psi_m \, d\tau$$

The differential $d\tau$ represents the element of volume in whatever coordinate system is used; with Cartesian coordinates, $d\tau = dx \, dy \, dz$. The last equality follows from the fact that E_m is a number independent of the coordinates and it therefore can be factored out of the integral. The energy is then given by

$$E_m = \frac{\int \psi_m^* H_{op} \psi_m \, d\tau}{\int \psi_m^* \psi_m \, d\tau} \qquad (6-21)$$

If ψ_m is normalized (Eq. 6-16),

$$E_m = \int \psi_m^* H_{op} \psi_m \, d\tau$$

This equation is not useful, of course, if we do not know the eigenfunction, and to obtain the eigenfunction we must solve the Schrödinger equation.

We have mentioned before that for only a small number of physical problems can the Schrödinger equation be solved exactly. These problems are: (1) the particle in a box; (2) the hydrogen atom; (3) the rigid rotor; and (4) the harmonic oscillator. We defer consideration of the last two problems to Chapter 8.

The Particle in a Box

Before we can write the Schrödinger equation for a particle in a box, we must first describe our particle and say what is meant by a "box." The particle, which may be an electron, a molecule, or even a baseball, is completely described by the statement that it has a mass m and a momentum p; it follows that its kinetic energy is $p^2/2m$. This particle is assumed to be inside a rectangular parallelepiped with edge lengths a, b, and c in the x, y, and z directions, respectively. The particle is also assumed to be force-free inside the box or, in other words, its potential energy is zero. Outside the box, the potential energy is assumed to be infinitely large. We take one corner of the box as our origin of coordinates.

The classical Hamiltonian function for this particle is:

$$H = \frac{1}{2m}(p_x^2 + p_y^2 + p_z^2) + U(xyz)$$

with $U = 0$ for $0 < x < a$, $0 < y < b$, $0 < z < c$
and $U = \infty$ for $0 > x > a$, $0 > y > b$, $0 > z > c$

The corresponding Schrödinger equation is

$$\nabla^2 \psi(xyz) + \frac{8\pi^2 m}{h^2}\left[E - U(xyz)\right]\psi(xyz) = 0$$

In order to solve a differential equation of this type we must always first separate the variables. We do this by looking for solutions which can be written as a product of three factors, each depending on only one variable. Thus the desired solutions must have the form

$$\psi(xyz) = X(x)Y(y)Z(z)$$

where $X(x)$ is a function of x only, and so forth. We assume that E and U can each be written as the sum of contributions associated with each coordinate.

By partial differentiation of ψ two times with respect to each coordinate we find

$$\nabla^2 \psi(xyz) = YZ\frac{d^2 X}{dx^2} + XZ\frac{d^2 Y}{dy^2} + XY\frac{d^2 Z}{dz^2}$$

Introducing this expression into the Schrödinger equation, we obtain

$$YZ\frac{d^2 X}{dx^2} + XZ\frac{d^2 Y}{dy^2} + XY\frac{d^2 Z}{dz^2}$$
$$+ \frac{8\pi^2 m}{h^2}\left[(E_x - U_x) + (E_y - U_y) + (E_z - U_z)\right]XYZ = 0$$

On division by XYZ, this equation becomes

$$\frac{1}{X}\frac{d^2 X}{dx^2} + \frac{1}{Y}\frac{d^2 Y}{dy^2} + \frac{1}{Z}\frac{d^2 Z}{dz^2} + \frac{8\pi^2 m}{h^2}\left[(E_x - U_x) + (E_y - U_y) + (E_z - U_z)\right] = 0$$

Because each of the three variables occurs in separate terms, this equation can be considered to be the sum of three equations, each of exactly the same form. The equation in x is typical:

$$\frac{1}{X}\frac{d^2 X}{dx^2} = -\frac{8\pi^2 m}{h^2}(E_x - U_x)$$

For all values of x less than 0 or greater than a, by hypothesis $U = \infty$ and

$$\frac{d^2 X}{dx^2} = X\infty$$

Atomic Structure and Quantum Theory

If X is to be a factor in an eigenfunction it must not become infinite for any value of x. There is only one solution of this equation which satisfies this requirement and that is $X = 0$. Then $X^2 = 0$, and the probability is zero that the particle has a value of the x coordinate less than 0 or greater than a. Analogous statements can be made about the y and z coordinates and this result is consistent with our assumption that the particle is inside the box.

For values of x between 0 and a, $U = 0$ by hypothesis, and the equation in x becomes

$$\frac{d^2X}{dx^2} = -\frac{8\pi^2 mE_x}{h^2} X \tag{6-22}$$

This equation has the same form as Equation 6-13 with $\dfrac{8\pi^2 mE_x}{h^2} = \dfrac{4\pi^2}{\lambda_x^2}$ where λ_x is the de Broglie wavelength $\dfrac{h}{\sqrt{2mE_x}}$. The solution of Equation 6-22 can be written in the following form:

$$X = N_x \sin\left(\frac{8\pi^2 mE_x}{h^2}\right)^{\frac{1}{2}} x \tag{6-23}$$

in which N_x is a constant of integration. A graph of this function is given in Figure 6-11.

The points located at $x = 0$ and $x = a$ are necessarily nodes (that is, $X = 0$) because the eigenfunction between these limits must be continuous with

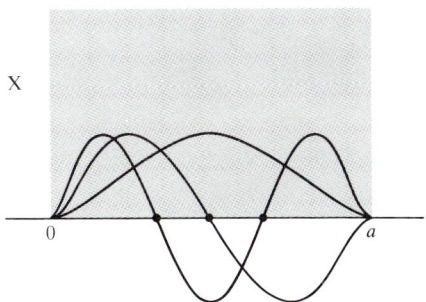

Figure 6-11. Graph of first three wave functions for a particle in a 1-dimensional box. For $\lambda_x = 2a$, a, and $\tfrac{2}{3}a$, the values of n_x are 1, 2, and 3 respectively.

the part outside this range which vanishes. The possible values for λ_x are given by the equation for the values of the wavelength in a standing wave (Eq. 6-12). Consequently $\lambda_x = 2a/n_x$ where the integer n_x is a quantum number associated with the x coordinate, and

$$\frac{8\pi^2 mE_x}{h^2} = \frac{4\pi^2}{\lambda_x^2} = \frac{\pi^2 n_x^2}{a^2} \tag{6-24}$$

Solving for E_x, we obtain

$$E_x = \frac{n_x^2 h^2}{8ma^2} \qquad (n_x = 1, 2, 3, \ldots)$$

The equations for the energies associated with the other two coordinates, E_y and E_z, have the same form as the equation for E_x. On addition of the energy contributions from all three coordinates, we obtain the following expression for the total energy of the particle:

$$E_{xyz} = \frac{h^2}{8m}\left(\frac{n_x^2}{a^2} + \frac{n_y^2}{b^2} + \frac{n_z^2}{c^2}\right) \tag{6-25}$$

We see that for definite values of m, a, b, and c, the energy of the particle, which is all kinetic energy and therefore necessarily positive, is quantized. It can have only the values associated with a particular set of the three quantum numbers n_x, n_y, and n_z. For a given set of quantum numbers and particle mass, the smaller the dimensions of the box, the larger are the associated energy values; also the energy values are inversely proportional to the mass of the particle.

To gain some idea of the magnitude of the energy levels given by Equation 6–25 let us evaluate this expression for two significant physical problems. First we calculate the lowest energy level ($n_x = n_y = n_z = 1$) for a hydrogen molecule in a cube of edge length 1 cm.

$$E_{111} = \frac{3(6.62 \times 10^{-27})^2}{8 \times 3.32 \times 10^{-24} \times 1^2} \sim 10^{-30} \text{ erg}$$

The average kinetic energy of a hydrogen molecule at a temperature of 300°K is approximately $kT \sim 10^{-14}$ erg. The quantized translational energy we calculated is practically infinitesimal in comparison. There is no way of measuring energies this small. Adjacent translational energy levels are so close together that in effect they form a continuous set. Although the direct effects of quantized translational energy levels have never been observed, there are indirect consequences of quantization which agree completely with experimental results.

Now let us calculate the lowest energy level for an electron in a cube of edge length 1 angstrom unit.

$$E_{111} = \frac{3(6.62 \times 10^{-27})^2}{8 \times 9.1 \times 10^{-28}(10^{-8})^2} \sim 10^{-10} \text{ erg}$$

This represents a relatively large amount of energy. If we imagine Avogadro's number of electrons, each confined to its 1 Å cube and in its lowest energy level, the total energy would be about $6 \times 10^{23} \times 10^{-10} = 6 \times 10^{13}$ ergs $= 6 \times 10^6$ joules $\sim 10^6$ calories per "mole" of electrons.

Turning to the eigenfunction for the particle in a box, we see that in view of Equation 6–24 the function X (Eq. 6–23) can be written in the following form:

$$X = N_x \sin \frac{\pi n_x}{a} x \qquad (6\text{–}26)$$

We now determine the value of the constant N_x which is called a normalizing constant. Because by hypothesis the particle is in the box, the probability that its x coordinate is between 0 and a is 1. Consequently,

$$\int_0^a X^2\, dx = N_x^2 \int_0^a \sin^2 \frac{\pi n_x x}{a}\, dx = 1$$

Setting $\theta = \frac{\pi n_x x}{a}$ we note that $dx = \frac{a}{\pi n_x}\, d\theta$, and the value of θ for which $x = a$ is πn_x. When we substitute these quantities in the integral above it becomes a standard form $\int_0^{n_x \pi} \sin^2 \theta\, d\theta$ whose value is $n_x \pi / 2$. Therefore

$$\frac{a}{\pi n_x} \int_0^{n_x \pi} \sin^2 \theta\, d\theta = \frac{a}{n_x \pi} \frac{n_x \pi}{2} = \frac{a}{2}$$

It follows that $\dfrac{N_x^2 a}{2} = 1; \; N_x = \sqrt{\dfrac{2}{a}}; \; X = \sqrt{\dfrac{2}{a}} \sin \dfrac{n_x \pi x}{a}$

The normalizing factors for Y and Z can be found similarly and they have the same form as the factor for X.

Now we can write the complete normalized set of eigenfunctions for the particle in a box:

$$\psi_{n_x n_y n_z} = \sqrt{\frac{8}{abc}} \sin \frac{n_x \pi x}{a} \sin \frac{n_y \pi y}{b} \sin \frac{n_z \pi z}{c}$$

Where is the particle? All that can be said is that the probability of finding the particle in the state specified by the set of quantum numbers n_x, n_y, and n_z and in the volume element $d\tau$ with coordinates x, y, and z is given by

$$\psi^2\, d\tau = \frac{8}{abc} \sin^2 \frac{n_x \pi x}{a} \sin^2 \frac{n_y \pi y}{b} \sin^2 \frac{n_z \pi z}{c}\, d\tau$$

At those points at which the eigenfunction has a node, the probability of finding the particle is zero. Apart from this, however, there is in principle no way of knowing where in the box the particle is.

This conclusion is consistent with the *uncertainty principle* which was stated by Heisenberg in connection with his development of matrix mechanics. He reached the conclusion that for some related pairs of variables, of which coordinate and momentum are an example, there is a limit to the precision with which both of these variables can be determined simultaneously. If Δx represents the uncertainty in the value of the x coordinate of a particle at some instant and Δp_x is the uncertainty in the related momentum, then according to Heisenberg

$$\Delta x \, \Delta p_x \geqslant \frac{h}{2\pi} \tag{6-27}$$

The more closely either x or p_x is known, the greater is the uncertainty in the other. In the case of the particle in a box, when we assumed that the particle was represented by an eigenfunction with a particular set of quantum numbers, we specified the momentum exactly $\left(p_x = \frac{n_x h}{2a}\right)$. Consequently there is complete uncertainty in the value of the coordinate.

Let us now consider two special cases of the particle in a box. First we assume that the box is a cube; then $a = b = c$ and

$$E_{\text{tot}} = \frac{h^2}{8ma^2}(n_x^2 + n_y^2 + n_z^2)$$

Every set of three quantum numbers labels a different eigenfunction but in this case more than one set may lead to the same energy. For example, if one of the quantum numbers has the value 2 while the others have the value 1, we can have three different eigenfunctions ψ_{211}, ψ_{121}, and ψ_{112}, all of which are associated with the energy $6h^2/8ma^2$. Whenever more than one eigenfunction is associated with the same energy level, the set of eigenfunctions is said to be *degenerate* and the *degree of degeneracy* is equal to the number of eigenfunctions in the set.

The other special case is the free particle. As the size of the box increases, the energy levels come closer and closer together and in the limit of an infinitely large box they form a continuous set; that is, the energy levels of a free particle are not quantized. At the same time the value of the normalizing constant in the eigenfunction decreases and approaches zero. Consequently, the probability of finding the particle with a definite momentum in any volume element approaches zero as the box becomes infinitely large.

We have examined the particle in a box not because of its intrinsic importance, but because it is the only problem for which the Schrödinger equation can be solved without some mathematical complications. It has served to introduce some general results which can be summarized as follows:

1. In order to solve a Schrödinger equation which involves more than one coordinate the eigenfunction must be expressed as a product of functions, each depending on only one variable.

2. Each coordinate in the Schrödinger equation is associated with a quantum number.

3. The requirements that the solutions of the Schrödinger equation should be well-behaved and fit the boundary conditions necessarily lead to discrete energy levels.

Atomic Structure and Quantum Theory

4. As the number of nodes in an eigenfunction increases, the associated energy increases.

5. Any eigenfunction can be normalized by multiplication with a normalizing factor given by

$$N = \frac{1}{(\int \psi\psi^* \, d\tau)^{\frac{1}{2}}}$$

6. The smallest amount of energy is associated with the smallest values of the quantum numbers. Since for the particle in a box these smallest values are $n_x = n_y = n_z = 1$, this particle cannot have zero kinetic energy. The energy which the particle cannot lose, even at a temperature of $0°K$, is often called the *zero-point energy*.[3]

It can be shown in general that the set of exact solutions of any Schrödinger equation has the mathematical property of orthogonality. Two functions, ψ_a and ψ_b, are said to be mutually orthogonal if $\int \psi_a\psi_b \, d\tau = 0$. For example, the functions $\sin \theta$ and $\sin 2\theta$ are mutually orthogonal because $\int_0^\pi \sin \theta \sin 2\theta \, d\theta = 0$. Any two members of the set of functions represented by Equation 6-26 are orthogonal with respect to each other. For the first two functions, those with $n_x = 1$ and $n_x = 2$, we obtain

$$\int_0^a N_x \sin\left(\frac{\pi x}{a}\right) \cdot N_x \sin\left(\frac{2\pi x}{a}\right) dx = 0$$

The Hydrogen Atom

The hydrogen atom and 1-electron hydrogen-like ions are the only atomic systems for which the Schrödinger equation can be solved exactly. The importance of this problem can scarcely be exaggerated because, as we see later, its exact solutions provide a foundation on which much of the theoretical study of other atoms has been based and on which our present concepts of chemical bonds have been developed.

Our system consists of a relatively massive nucleus with charge Z and one electron. The most general Hamiltonian function for this system requires six

---

[3] The existence of a zero-point energy can be considered to be the consequence of the Heisenberg uncertainty principle. If a particle, such as a molecule, could be stationary at some point in the box, we could in principle determine its position with high precision and we would also know its momentum exactly because it has a value of zero.

coordinates (three for each particle) and six momentum components. This function includes the translational energy of the system as a whole, which is related to the particle-in-a-box problem. Because this translational energy is of no interest to us here, we simplify our problem by assuming that the nucleus is stationary at the origin of a coordinate system.[4] In this way we eliminate three coordinates and three momentum components. We use as our zero of potential energy the value corresponding to a very large separation of the nucleus and the electron.

The Hamiltonian function for our system is

$$H = \frac{1}{2\mu}(p_x^2 + p_y^2 + p_z^2) - \frac{Z\epsilon^2}{(x_e^2 + y_e^2 + z_e^2)^{\frac{1}{2}}}$$

The related Schrödinger equation is

$$\frac{h^2}{8\pi^2\mu}\nabla^2\psi + \left(E + \frac{Z\epsilon^2}{(x_e^2 + y_e^2 + z_e^2)^{\frac{1}{2}}}\right)\psi = 0$$

It is clear from the form of the potential energy function that a Cartesian coordinate system is a very poor choice for this problem. The potential energy of the system depends only on the distance between the electron and the nucleus and not on the direction. This fact suggests that spherical polar coordinates would be a better choice because of the simple form of the potential energy function in these coordinates ($-Z\epsilon^2/r$). To express the Schrödinger equation in polar coordinates, we need to carry out a transformation of the Laplacian operator from Cartesian coordinates to the set of polar coordinates defined by the equations:

$$z = r \cos \theta$$
$$x = r \sin \theta \cos \phi$$
$$y = r \sin \theta \sin \phi$$

(Compare p. 135.)

The derivation of the form of the Laplacian operator in polar coordinates is given in many mathematics textbooks. The use of this result leads to the Schrödinger equation in the following form:

$$\frac{1}{r^2}\frac{\partial}{\partial r}\left(r^2\frac{\partial \psi}{\partial r}\right) + \frac{1}{r^2 \sin \theta}\frac{\partial}{\partial \theta}\left(\sin \theta \frac{\partial \psi}{\partial \theta}\right) + \frac{1}{r^2 \sin^2 \theta}\frac{\partial^2 \psi}{\partial \phi^2} + \frac{8\pi^2\mu}{h^2}\left(E + \frac{Z\epsilon^2}{r}\right)\psi = 0 \quad (6\text{--}28)$$

The variables in this equation are rather badly mixed up and we need to separate them in order to solve the equation. We want our solution to have the

[4] We make a correction for this simplification by replacing the mass of the electron by its reduced mass defined by $\mu = \dfrac{m_e m_n}{m_e + m_n}$, whose value is very close to m_e.

Atomic Structure and Quantum Theory

form of a product of factors, each depending on only one variable, which we write as follows:

$$\psi = R(r)\Theta(\theta)\Phi(\phi)$$

When we carry out on this function the operations indicated in Equation 6-28 and multiply the result by $\dfrac{r^2 \sin^2 \theta}{R\Theta\Phi}$ we arrive at the following equation in which the right-hand side depends only on the variable ϕ:

$$\frac{\sin^2 \theta}{R} \frac{d}{dr}\left(r^2 \frac{dR}{dr}\right) + \frac{\sin \theta}{\Theta} \frac{d}{d\theta}\left(\sin \theta \frac{d\Theta}{d\theta}\right) + r^2 \sin^2 \theta \frac{8\pi^2 \mu}{h^2}\left(E + \frac{Z\epsilon^2}{r}\right) = -\frac{1}{\Phi}\frac{d^2\Phi}{d\phi^2} \quad (6\text{-}29)$$

This equation must be satisfied for all values of r, θ, and ϕ. There is only one condition for which this is possible and that is to have each side equal to the same constant. It is customary to represent this constant by m^2. Then we have

$$-\frac{1}{\Phi}\frac{d^2\Phi}{d\phi^2} = m^2$$

When this differential equation for Φ is written in the following way:

$$\frac{d^2\Phi}{d\phi^2} + m^2\Phi = 0 \quad (6\text{-}30)$$

we see that it has the same form as Equation 6-22. We can write its solutions in either the trigonometric or the equivalent exponential form:

$$= N' \begin{Bmatrix}\sin \\ \cos\end{Bmatrix} m\phi = Ne^{\pm im\phi}$$

The normalizing constants N' and N are different for the two forms.

Now let us focus on the exponential form. So far all that we have said about m is that it is a constant, either positive or negative. To obtain more information about m we make use of the requirement that if Φ is to be a factor in an eigenfunction it must be single-valued. If the angle ϕ is increased by 2π we return to the same point in space and Φ must be unchanged. Consequently,

$$e^{\pm im\phi} = e^{\pm im(\phi + 2\pi)} = e^{\pm im\phi}e^{2\pi im}$$

Because $e^{2\pi im}$ is equal to 1 only when m is an integer or zero it follows that m is restricted to the values $0, \pm 1, \pm 2, \ldots$ We have thus obtained a quantum number which is usually called the magnetic quantum number because the energy of a hydrogen atom in a magnetic field depends on its value.

We can readily find the normalizing constant N for the function Φ in the following way:

$$\int_0^{2\pi} \Phi^*\Phi \, d\phi = N^2 \int_0^{2\pi} e^{im\phi} e^{-im\phi} \, d\phi = N^2 \int_0^{2\pi} d\phi = 1$$

$$N = \frac{1}{\sqrt{2\pi}}$$

For some purposes the real trigonometric form of Φ is more convenient than the exponential form. Making use of the fact that any linear combination of solutions of Equation 6–30 is a solution and the fact that

$$e^{im\phi} \pm e^{-im\phi} = \begin{cases} 2\cos m\phi \\ 2i\sin m\phi \end{cases}$$

we can express Φ in the following normalized form:

$$\Phi_{m\,\cos} = \frac{1}{\sqrt{\pi}}\cos m\phi; \quad \Phi_{m\,\sin} = \frac{1}{\sqrt{\pi}}\sin m\phi$$

We still have the left side of Equation 6–29 to deal with. When we set this equal to m^2 and then divide by $\sin^2\theta$, the result can be written as follows:

$$\frac{1}{R}\frac{d}{dr}\left(r^2 \frac{dR}{dr}\right) + \frac{8\pi^2 \mu r^2}{h^2}\left(E + \frac{Z\epsilon^2}{r}\right) = \frac{m^2}{\sin^2\theta} - \frac{1}{\Theta \sin\theta}\frac{d}{d\theta}\left(\sin\theta \frac{d\Theta}{d\theta}\right)$$

This equation must be valid for all values of r and θ. As we have seen before, it follows that each side must be equal to the same constant which we express as $l(l+1)$. Setting each side in turn equal to this constant we obtain the following two differential equations:

$$\frac{1}{\sin\theta}\frac{d}{d\theta}\left(\sin\theta \frac{d\Theta}{d\theta}\right) + \left[l(l+1) - \frac{m^2}{\sin^2\theta}\right]\Theta = 0 \qquad (6\text{–}31)$$

and

$$\frac{d}{dr}\left(r^2 \frac{dR}{dr}\right) - l(l+1)R + \frac{8\pi^2 \mu r^2}{h^2}\left(E + \frac{Z\epsilon^2}{r}\right)R = 0 \qquad (6\text{–}32)$$

These differential equations and their solutions were well-known to mathematicians in the nineteenth century. The equation in θ is the Legendre differential equation and the one in r is named for Laguerre. Because the process of solving these equations is somewhat lengthy we omit the details of this process. We might expect that l would turn out to be an integer and so it does. The function Θ then depends on the values of two integers, l and m. The function R also depends on the values of two integers, n and l. Consequently the complete eigenfunction depends on three integers in the following way:

$$\psi_{nlm} = R_{nl}(r)\Theta_{lm}(\theta)\Phi_m(\phi)$$

The integer n is called the principal quantum number; it is analogous to the quantum number n in Bohr's theory. The integer l is called the angular momen-

Atomic Structure and Quantum Theory

tum (or azimuthal) quantum number because the angular momentum associated with a particular value of l is given by $\sqrt{l(l+1)}\,h/2\pi$. For a given value of n, l can have any of the values $n-1, n-2, \ldots 0$. For a given value of l, the possible values of m are $\pm l, \pm(l-1), \ldots 0$.

An eigenfunction which, like ψ_{nlm}, is a function of the coordinates of one electron is usually called an *orbital*. Although a particular orbital can be designated by giving the values of the three quantum numbers, it is more common to use the value of n and one of the letters[5] $s, p, d,$ or f to represent a value of l of 0, 1, 2, or 3, respectively. Thus a $3d$ orbital has $n = 3$ and $l = 2$.

The normalized forms of the two s orbitals ($l = 0$) of lowest energy are:

$$\psi_{1s} = \psi_{100} = \frac{1}{\sqrt{\pi}}\left(\frac{Z}{a_0}\right)^{\frac{3}{2}} e^{-\frac{Zr}{a_0}} \tag{6-33}$$

$$\psi_{2s} = \psi_{200} = \frac{1}{4\sqrt{2\pi}}\left(\frac{Z}{a_0}\right)^{\frac{3}{2}}\left(2 - \frac{Zr}{a_0}\right) e^{-\frac{Zr}{2a_0}}$$

There are three $2p$ orbitals ($l = 1$) which, when written as complex functions, differ in the value of the quantum number m ($= 0, \pm 1$). In real and normalized form, these p orbitals are:

$$\psi_{210} = \psi_{2p_z} = \frac{1}{4\sqrt{2\pi}}\left(\frac{Z}{a_0}\right)^{\frac{3}{2}} e^{-\frac{Zr}{2a_0}} \frac{Zr}{a_0} \cos\theta$$

$$\psi_{211} \left\{ \begin{array}{l} \psi_{2p_x} = \dfrac{1}{4\sqrt{2\pi}}\left(\dfrac{Z}{a_0}\right)^{\frac{3}{2}} e^{-\frac{Zr}{2a_0}} \dfrac{Zr}{a_0} \sin\theta \cos\phi \\[1em] \psi_{2p_y} = \dfrac{1}{4\sqrt{2\pi}}\left(\dfrac{Z}{a_0}\right)^{\frac{3}{2}} e^{-\frac{Zr}{2a_0}} \dfrac{Zr}{a_0} \sin\theta \sin\phi \end{array} \right.$$
$$\psi_{21-1}$$

These orbitals are labeled by Cartesian coordinates because they contain as factors $r\cos\theta$ ($= z$), $r\sin\theta\cos\phi$ ($= x$), and $r\sin\theta\sin\phi$ ($= y$), respectively.

Wherever r appears in these functions, it is divided by a_0 ($= h^2/4\pi^2\mu\epsilon^2$) which is the value calculated by Bohr for the radius of the smallest electron orbit for the hydrogen atom (Eq. 6-6). Although nothing is said in wave mechanics about orbits, the quantity a_0 appears as a fundamental atomic unit of length equal to 0.529 Å.

Each of these orbitals contains an exponential factor, $e^{-\frac{Zr}{na_0}}$, which guarantees an asymptotic approach to a value of zero as r increases. In principle, an orbital extends indefinitely far from the nucleus, but its value beyond a distance of a few Angstrom units is very small. One way of graphically representing an orbital is to choose, more or less arbitrarily, a fixed value of r beyond

---

[5]In the early days of atomic spectroscopy these letters were used respectively as abbreviations for the words sharp, principal, diffuse, and fundamental applied to series of spectral lines. They have completely lost their original meanings.

which the orbital is considered to have a negligible value. Then a contour surface is drawn for this value of r. Another type of graphical representation is obtained by plotting the value of the orbital against r/a_0 for fixed values of the angles θ and ϕ.

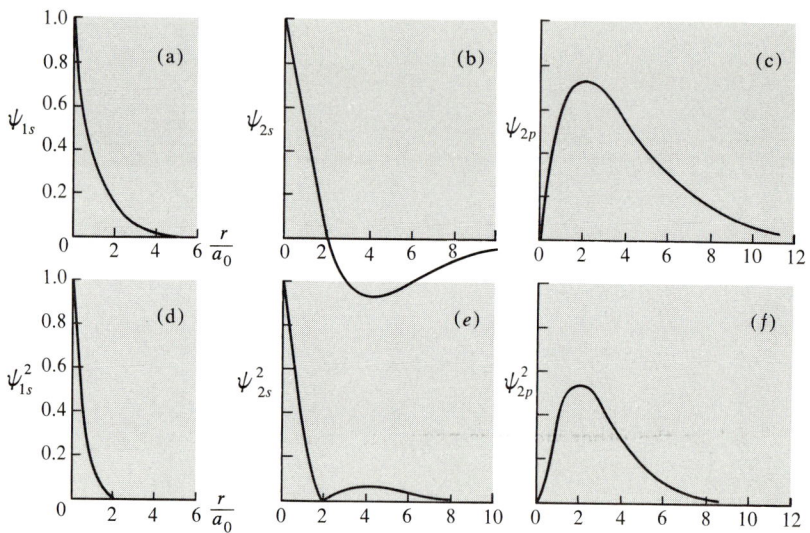

Figure 6–12. (a), (b), (c) Relative values of ψ_{1s}, ψ_{2s} and the radial factor of ψ_{2p} as a function of $\dfrac{r}{a_0}$. (d), (e), (f) Relative values of squares of these functions.

Because the s orbitals are functions of r only, and thus do not involve the angles, their contour surfaces are spheres. The $1s$ and $2s$ orbitals and the squares of these orbitals are plotted as a function of r/a_0 in Figure 6–12. Notice that ψ_{2s} has a value of zero when $\dfrac{r}{a_0} = \dfrac{2}{Z}$, which means that this orbital has a spherical nodal surface. In the two regions of space separated by this surface, the orbital has different signs. It is meaningless, however, to ask in which region the sign is positive because only the square of the orbital has physical significance.

Turning now to the p orbitals, we see that it is more difficult to represent them graphically than s orbitals because they include both a radial factor (a function of r) and an angular factor. The radial factor for a $2p$ orbital is graphed in Figure 6–12c.

The angular factors for the three $2p$ orbitals are all different. Let us consider the $2p_z$ orbital first. For a fixed value of r this orbital is simply $C \cos \theta$. The graph of this function in any plane passing through the z axis consists of two circles tangent to each other at the origin. Because the function is independent of ϕ, it has cylindrical symmetry about the z axis. Consequently the graph of the angular factor consists of two spheres tangent to each other and to the x,y plane (Fig. 6-13). On one side of this plane the orbital has positive

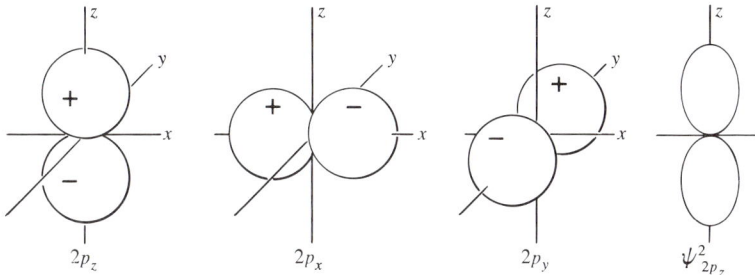

Figure 6-13. Angular factors of $2p$ orbitals and of $\psi_{2p_z}^2$ for constant value of r (contour surfaces).

values, and on the other side it is negative. Clearly, the x,y plane is a nodal surface for a p_z orbital.

It is somewhat tedious but not difficult to show that the $2p_x$ and $2p_y$ orbitals have exactly the same shape as $2p_z$, but they are oriented differently. For example, the x axis is the symmetry axis for $2p_x$, and its nodal surface is the y,z plane.

◇◇◇◇◇◇

The shapes of the orbital contour surfaces, as given by the angular factors in the orbitals, are important in valence theory (Chapter 7). For future reference we list the normalized factors $\Theta\Phi$ of the $2s$ and the three $2p$ orbitals as follows:

$$
\begin{array}{ll}
2s & \dfrac{1}{2\sqrt{\pi}} \\[6pt]
2p_z & \dfrac{1}{2\sqrt{\pi}} \cos\theta \\[6pt]
2p_x & \sqrt{3}\,\dfrac{1}{2\sqrt{\pi}} \sin\theta \cos\phi \\[6pt]
2p_y & \sqrt{3}\,\dfrac{1}{2\sqrt{\pi}} \sin\theta \sin\phi
\end{array}
\qquad (6\text{-}34)
$$

◇◇◇◇◇◇

Once we have an expression for a normalized orbital it is a simple task to find the probability of finding the electron in a volume element located at the point r,θ,ϕ. The square of the orbital (if it is real) gives us the probability per unit volume, sometimes called the probability density, which is multiplied by the volume differential. For an electron in the ground state of hydrogen this probability is

$$\psi_{1s}^2 \, d\tau = \frac{1}{\pi a_0^3} e^{-\frac{2r}{a_0}} r^2 \, dr \sin\theta \, d\theta \, d\phi$$

Often more interesting than this is the probability of finding the electron in an element of length at a certain distance from the nucleus, independent of direction. This is the same as the probability that the electron is in a spherical shell of radius r and thickness dr. We can calculate this probability per unit length by integrating over the angle coordinates (compare p. 135) to obtain:

$$\psi_{1s}^2 r^2 \, dr \int_0^\pi \sin\theta \, d\theta \int_0^{2\pi} d\phi = 4\pi r^2 \psi_{1s}^2 \, dr = \frac{4r^2}{a_0^3} e^{-\frac{2r}{a_0}} \, dr$$

This function, which is often called the radial distribution function, is graphed in Figure 6–14.

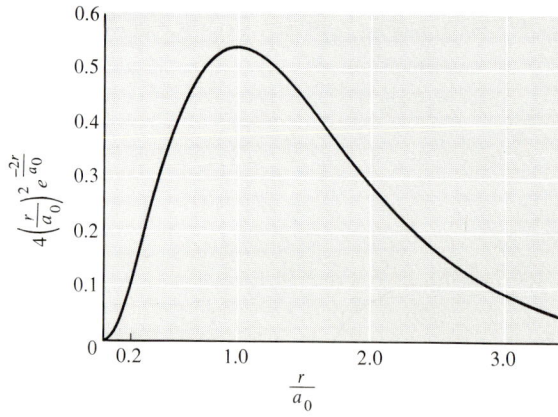

Figure 6–14. Radial distribution function for the 1s orbital of the hydrogen atom.

We can make use of the radial distribution function in several different ways. For example, we can find the most probable distance of the electron from the nucleus by determining the position of the maximum of the radial distribution function. For the hydrogen 1s orbital we have:

$$\frac{d}{dr} 4\pi r^2 \psi_{1s}^2 = \frac{4}{a_0^3} \frac{d}{dr} r^2 e^{-\frac{2r}{a_0}} = \frac{4}{a_0^3} e^{-\frac{2r}{a_0}} \left(2r - \frac{2r^2}{a_0}\right) = 0$$

It follows that
$$r = a_o$$

This remarkable result shows that the most probable separation of electron and proton turns out to have just the value that Bohr calculated for the radius of the first orbit.

In another application of the radial distribution function we can calculate the probability that the electron is at a distance from the proton less than some specified value. Let this specified distance in units of a_o be b and let $x = r/a_o$. Then
$$4\pi r^2 \psi_{1s}^2\, dr = 4x^2 e^{-2x}\, dx$$
and
$$4\int_0^b x^2 e^{-2x}\, dx = 1 - e^{-2b}(2b^2 + 2b + 1)$$

With $b = 3$ (~ 1.59 Å), the desired probability is $1 - 25e^{-6} = 0.937$.

Now let us turn to the energy levels associated with the hydrogen-like (1-electron) orbitals. Notice that E appears only in the differential equation involving r (Eq. 6–32). In solving this equation it is found that in order for R to be a well-behaved function the negative[6] values of E must be given by

$$E_n = -\frac{2\pi^2 \mu \epsilon^4 Z^2}{n^2 h^2} = -\frac{Z^2 \epsilon^2}{2a_o n^2}$$

This is exactly the equation for the energy levels found by Bohr (Eq. 6–8).

It is apparent that the energy associated with a hydrogen-like orbital depends only on the value of the principal quantum number. (This is not true of many-electron systems.) For $n > 1$ several orbitals with the same value of n and different values of l and m are obtained. It follows that all of the orbitals except ψ_{1s} exist in degenerate sets. To obtain the degree of degeneracy we can tabulate the various orbitals as follows:

n	1	2		3		
l	0	1	0	2	1	0
m	0	$1, 0, -1$	0	$2, 1, 0, -1, -2$	$1, 0, -1$	0
Number of orbitals	1	4		9		

For each value of l there are $2l + 1$ orbitals with different values of m. Consequently the degree of degeneracy for a given value of n is the sum of the first n odd integers, which is equal to n^2:

$$\sum_{l=0}^{l=n-1} (2l + 1) = n^2$$

---

[6] Positive values of E are not quantized but form a continuous distribution. They correspond to the situation in which the electron is relatively far from the nucleus and its kinetic energy is greater in absolute magnitude than its potential energy. In other words, for positive values of E the electron is not bound to the nucleus.

Spin Quantum Number

We have found that the solutions of the Schrödinger equation with three coordinates as variables depend on three quantum numbers. Even before the development of quantum mechanics it had become clear that three quantum numbers were not enough to account for certain observations in atomic spectroscopy. In 1925 Uhlenbeck and Goudsmit, on a purely empirical basis, proposed a fourth quantum number which they called the *spin* quantum number. The name spin came from their model of an electron as a charged particle which could rotate in either of two directions about its axis. While the model was subsequently discarded, the name has been retained.

Shortly after Schrödinger and Heisenberg published their fundamental work on quantum mechanics, Dirac showed that when the Schrödinger equation was written in a way which was consistent with Einstein's theory of relativity four coordinates were required, the fourth being time. Consequently four quantum numbers were obtained from the solution of this equation, the fourth being the spin quantum number introduced by Uhlenbeck and Goudsmit. A common symbol for this quantum number is s. (This use of s should not be confused with its use to designate an orbital with $l = 0$; fortunately, the context always indicates which usage is intended.)

The spin quantum number is quite different from the other quantum numbers we met previously. It can have only two values: $+1/2$ and $-1/2$. Associated with these two values are two spin functions represented by α and β, whose mathematical form is of no interest here. The complete expression for an atomic wave function is a product of an orbital and a spin function:

$$\Psi_{nlms} = \psi_{nlm}\alpha \text{ or } \psi_{nlm}\beta \qquad (6\text{-}35)$$

When there is only one electron in the system the distinction between α and β is unimportant, but for the treatment of more complex systems the spin quantum number is essential.

Wave Functions for Many-Electron Atoms

When there is more than one electron in the atom under consideration complications arise in both experimental observation and theoretical treatment. Even for helium the atomic spectrum is considerably more complex than for a hydrogen atom. On the theoretical side, although the Schrödinger equation for the helium atom can easily be written no way has been found to solve this equation exactly. The potential energy function is:

$$U = -\frac{2\epsilon^2}{r_1} - \frac{2\epsilon^2}{r_2} + \frac{\epsilon^2}{r_{12}}$$

Atomic Structure and Quantum Theory

in which r_1 and r_2 are the distances of electrons 1 and 2 from the nucleus, and r_{12} is the distance between the two electrons. The last term represents the interelectronic Coulombic repulsion. With the introduction of this function, the Schrödinger equation for the helium atom becomes

$$(\nabla_1^2 + \nabla_2^2)\psi + \frac{8\pi^2 m}{h^2}\left(E + \frac{2\epsilon^2}{r_1} + \frac{2\epsilon^2}{r_2} - \frac{\epsilon^2}{r_{12}}\right)\psi = 0 \qquad (6\text{-}36)$$

The difficulty in the solution of this equation is the fact that the interelectronic repulsive forces cause the motion of either electron to be affected by the position of the other; mathematically, the variables in this equation cannot be separated. It turns out to be a good approximation for complex atoms to assume that the potential energy of each electron depends on the combined field of the nucleus and the averaged field due to all of the other electrons. In order to calculate this average field, however, the orbitals of all of these electrons must be known and we seem to be back where we started. A numerical method of solving the Schrödinger equation by successive approximations was developed by Hartree. This method, known as the self-consistent field method, starts with crude wave functions of single electrons (obtained by "educated guessing") and then improves these for one electron after another by iterative computations. These are tedious but the use of fast electronic computers has made it possible to carry out calculations of this type for many atoms. The resulting orbitals are in the form of tables of numbers. Because these self-consistent field orbitals are awkward to use and have limited chemical application we say no more about them.

Even relatively crude wave functions are useful for understanding the structure and chemical properties of atoms. For this purpose an atomic wave function is approximated by a product of orbitals, each depending on the coordinates of one electron. Each orbital is designated by the same quantum numbers used previously.

The angular factors in these orbitals are the same as for the hydrogen orbitals. Consequently the shapes of the s, p, d . . . orbitals are the same as for the exact orbitals of hydrogen. The radial factors, however, are significantly different, one of the main differences being the use of an "effective" nuclear charge, rather than the actual atomic number. The reason for this change is the fact that the electrostatic force of the nucleus on an electron is partially screened by the other electrons.

As a result of the interelectronic repulsions, the energies associated with the orbitals depend not only on the value of the quantum number n, as for hydrogen, but also on the value of l. For a given value of l, the $2l + 1$ orbitals form a degenerate set. Thus there are three p orbitals with the same energy, five d orbitals ($l = 2$), and so forth. A simple expression for these orbital energies cannot be obtained. Up to about $Z = 20$, the sequence of the orbitals in the order of increasing energy is

$$1s < 2s < 2p < 3s < 4s \sim 3d$$

The approximate energy levels for these orbitals are graphed in Figure 6–15.

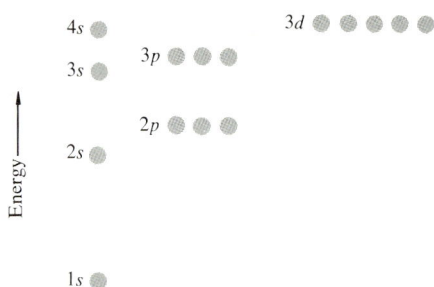

Figure 6–15. Energy levels of atomic orbitals (schematic). The actual position of each energy level in an atom depends on the atomic number of the atom.

Electron Configurations and the Pauli Exclusion Principle

The electron configuration of an atom in a certain energy state is given by indicating the number of electrons occupying each of the atomic orbitals. The electron configuration in the ground state of an atom is the result of (1) the tendency of each electron to occupy an available orbital with the lowest energy and (2) the *Pauli exclusion principle*. This principle is an empirical generalization which preceded the development of quantum mechanics. It has not been derived from any theory. In its simplest form, the Pauli exclusion principle states that in an atom or a molecule two electrons cannot have the same values of the four quantum numbers. It follows that at most two electrons can occupy a single orbital, and they must have different values of the spin quantum number.

In some situations another generalization, known as *Hund's rule*, is applicable. This rule states that when a degenerate set of orbitals is partially occupied by electrons, the energy of the atom is minimized when the electrons occupy (to the greatest extent possible) different orbitals and have the same value of the spin quantum number. In other words, the electrons in the partially filled degenerate set of orbitals tend to have parallel spins.

With these physical principles it is possible to set up what might be called a physicist's Periodic Table in terms of the electron configurations in the ground states of atoms. One form of this table is presented in Table 6–1. It is a re-

TABLE 6-1 ELECTRON CONFIGURATIONS OF THE ELEMENTS AS GASEOUS ATOMS

Shell:	K	L		M			N			
Element	1s	2s	2p	3s	3p	3d	4s	4p	4d	4f
1. H	1									
2. He	2									
3. Li	2	1								
4. Be	2	2								
5. B	2	2	1							
6. C	2	2	2							
7. N	2	2	3							
8. O	2	2	4							
9. F	2	2	5							
10. Ne	2	2	6							
11. Na	2	2	6	1						
12. Mg	2	2	6	2						
13. Al	2	2	6	2	1					
14. Si	2	2	6	2	2					
15. P	2	2	6	2	3					
16. S	2	2	6	2	4					
17. Cl	2	2	6	2	5					
18. Ar	2	2	6	2	6					
19. K	2	2	6	2	6		1			
20. Ca	2	2	6	2	6		2			
21. Sc	2	2	6	2	6	1	2			
22. Ti	2	2	6	2	6	2	2			
23. V	2	2	6	2	6	3	2			
24. Cr	2	2	6	2	6	5	1			
25. Mn	2	2	6	2	6	5	2			
26. Fe	2	2	6	2	6	6	2			
27. Co	2	2	6	2	6	7	2			
28. Ni	2	2	6	2	6	8	2			
29. Cu	2	2	6	2	6	10	1			
30. Zn	2	2	6	2	6	10	2			
31. Ga	2	2	6	2	6	10	2	1		
32. Ge	2	2	6	2	6	10	2	2		
33. As	2	2	6	2	6	10	2	3		
34. Se	2	2	6	2	6	10	2	4		
35. Br	2	2	6	2	6	10	2	5		
36. Kr	2	2	6	2	6	10	2	6		

Shell:	K	L	M	N				Q				P		Q
Element				4s	4p	4d	4f	5s	5p	5d	5f	5g	6s	6p	6d	7s
37. Rb	2	8	18	2	6			1								
38. Sr	2	8	18	2	6			2								
39. Y	2	8	18	2	6	1		2								
40. Zr	2	8	18	2	6	2		2								
41. Nb	2	8	18	2	6	4		1								
42. Mo	2	8	18	2	6	5		1								
43. Tc	2	8	18	2	6	(5)		(2)								

TABLE 6-1 (continued)

Shell:	K	L	M	N				O					P			Q
Element				4s	4p	4d	4f	5s	5p	5d	5f	5g	6s	6p	6d	7s
44. Ru	2	8	18	2	6	7		1								
45. Rh	2	8	18	2	6	8		1								
46. Pd	2	8	18	2	6	10										
47. Ag	2	8	18	2	6	10		1								
48. Cd	2	8	18	2	6	10		2								
49. In	2	8	18	2	6	10		2	1							
50. Sn	2	8	18	2	6	10		2	2							
51. Sb	2	8	18	2	6	10		2	3							
52. Te	2	8	18	2	6	10		2	4							
53. I	2	8	18	2	6	10		2	5							
54. Xe	2	8	18	2	6	10		2	6							
55. Cs	2	8	18	2	6	10		2	6				1			
56. Ba	2	8	18	2	6	10		2	6				2			
57. La	2	8	18	2	6	10		2	6	1			2			
58. Ce	2	8	18	2	6	10	2	2	6				2			
59. Pr	2	8	18	2	6	10	3	2	6				2			
60. Nd	2	8	18	2	6	10	4	2	6				2			
61. Pm	2	8	18	2	6	10	5	2	6				2			
62. Sm	2	8	18	2	6	10	6	2	6				2			
63. Eu	2	8	18	2	6	10	7	2	6				2			
64. Gd	2	8	18	2	6	10	7	2	6	1			2			
65. Tb	2	8	18	2	6	10	9	2	6				2			
66. Dy	2	8	18	2	6	10	10	2	6				2			
67. Ho	2	8	18	2	6	10	11	2	6				2			
68. Er	2	8	18	2	6	10	12	2	6				2			
69. Tm	2	8	18	2	6	10	13	2	6				2			
70. Yb	2	8	18	2	6	10	14	2	6				2			
71. Lu	2	8	18	2	6	10	14	2	6	1			2			
72. Hf	2	8	18	2	6	10	14	2	6	2			2			
73. Ta	2	8	18	2	6	10	14	2	6	3			2			
74. W	2	8	18	2	6	10	14	2	6	4			2			
75. Re	2	8	18	2	6	10	14	2	6	5			2			
76. Os	2	8	18	2	6	10	14	2	6	6			2			
77. Ir	2	8	18	2	6	10	14	2	6	7			2			
78. Pt	2	8	18	2	6	10	14	2	6	9			1			
79. Au	2	8	18	2	6	10	14	2	6	10			1			
80. Hg	2	8	18	2	6	10	14	2	6	10			2			
81. Tl	2	8	18	2	6	10	14	2	6	10			2	1		
82. Pb	2	8	18	2	6	10	14	2	6	10			2	2		
83. Bi	2	8	18	2	6	10	14	2	6	10			2	3		
84. Po	2	8	18	2	6	10	14	2	6	10			2	4		
85. At	2	8	18	2	6	10	14	2	6	10			2	5		
86. Rn	2	8	18	2	6	10	14	2	6	10			2	6		
87. Fr	2	8	18	2	6	10	14	2	6	10			2	6		1
88. Ra	2	8	18	2	6	10	14	2	6	10			2	6		2
89. Ac	2	8	18	2	6	10	14	2	6	10			2	6	1	2
90. Th	2	8	18	2	6	10	14	2	6	10			2	6	2	2
91. Pa	2	8	18	2	6	10	14	2	6	10	2		2	6	1	2
92. U	2	8	18	2	6	10	14	2	6	10	3		2	6	1	2
93. Np	2	8	18	2	6	10	14	2	6	10	5		2	6		2
94. Pu	2	8	18	2	6	10	14	2	6	10	6		2	6		2
95. Am	2	8	18	2	6	10	14	2	6	10	7		2	6		2
96. Cm	2	8	18	2	6	10	14	2	6	10	7		2	6	1	2

TABLE 6-1 (*continued*)

Shell:	K	L	M	N				O					P		Q	
Element				4s	4p	4d	4f	5s	5p	5d	5f	5g	6s	6p	6d	7s
97. Bk	2	8	18	2	6	10	14	2	6	10	8		2	6	1	2
98. Cf	2	8	18	2	6	10	14	2	6	10	9		2	6	1	2
99. E	2	8	18	2	6	10	14	2	6	10	10		(2	6	1	2)
100. Fm	2	8	18	2	6	10	14	2	6	10	11		(2	6	1	2)
101. Mv	2	8	18	2	6	10	14	2	6	10	12		(2	6	1	2)
102. No	2	8	18	2	6	10	14	2	6	10	13		(2	6	1	2)
103. Lw	2	8	18	2	6	10	14	2	6	10	14		(2	6	1	2)

markable and fortunate fact that this table coincides with the chemist's Periodic Table, originally developed mainly from a consideration of chemical properties. This is the basis for relating the chemical properties of an atom to its electron configuration.

It can be seen from Table 6-1 that with increasing atomic number up to 18 (argon) the order in which the successive orbitals are occupied by electrons is quite regular. The orbitals with a certain value of n are occupied by the maximum number of electrons before an orbital with a higher value of n is occupied. In view of this fact, one might expect that the 19th electron in potassium would go into a $3d$ orbital. It turns out, however, that although the energies of the $3d$ and $4s$ orbitals are nearly the same, the $4s$ orbital has the lower energy and it is the one that is occupied in potassium.

With calcium the 20th electron also enters the $4s$ orbital. With scandium, however, the 21st electron enters the $3d$ orbital rather than the $4p$ because of the lower energy of the $3d$ orbital. In the succeeding 7 elements the $3d$ orbital is gradually filled. The characteristic physical and chemical properties of these eight elements, which form the first transition series, are related to the fact that they have a partially filled d subshell. Similar considerations account for the second transition series (elements 39 to 45 with a partially filled $4d$ subshell) and the rare earth elements with partially filled $4f$ or $5d$ subshells.

One of the physical properties of atoms which clearly shows a periodicity with increasing atomic number is the ionization potential, which is defined as the energy (usually expressed in electron-volts) required to remove one of the electrons of highest energy from the atom.[7] From the graph of ionization potentials in Figure 6-16 it can be seen that the noble gases, which from a chemical point of view are relatively inert as a group, have relatively high ionization potentials. Within this group the ionization potential decreases with increasing atomic number. This fact may be related to the fact that of the noble gases

---

[7]This is actually the first ionization potential. The second, third, and higher ionization potentials are the energies required to remove in succession two, three, or more electrons from the atom.

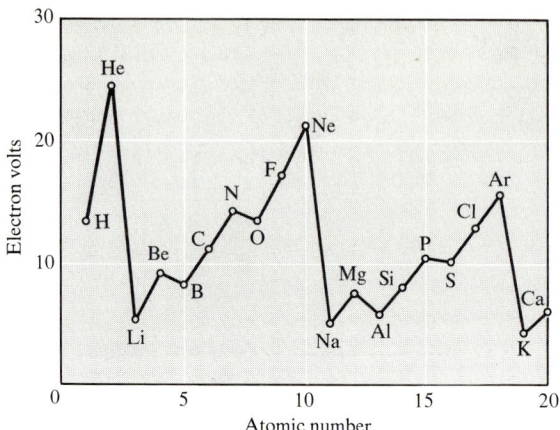

Figure 6-16. Ionization potentials of the first 20 elements in the Periodic Table.

only xenon and krypton have been shown to form compounds and krypton compounds are much less stable than xenon compounds.

Methods of Approximation

Because all molecules and nearly all atoms contain more than one electron, it is clear that methods for finding approximate solutions of the Schrödinger equation and the energies associated with them are of great importance in chemistry. In addition to the Hartree self-consistent field method mentioned previously, there are two general methods which have been extensively applied: (1) the perturbation method and (2) the variation method.

1. The *perturbation method* is based on the idea that in some cases a physical problem for which an exact solution cannot be obtained is only slightly different from one of the problems which can be solved exactly. A case in point is an anharmonic oscillator for which the potential energy function is $U = kx^2/2 + k'x^4$. The second term is called a perturbation. If k' is small enough, the known harmonic oscillator wave functions can be used to calculate approximate corrections to these functions and the associated energies to adapt them to the anharmonic oscillator. Although this method is useful for many physical problems, it has not found much application in problems of chemical interest and no more will be said about it.

2. The *variation method* is particularly well suited for approximating the wave function and the energy of the ground state of a molecule. This method has played an important role in the development of theoretical chemistry, and we treat it in some detail. Let us begin by assuming that we are interested in a

system for which the Schrödinger equation cannot be solved. Because we can always write the expression for the Hamiltonian operator, we can write the Schrödinger equation $H_{op}\psi_n = E_n\psi_n$, but we know neither the functions ψ_n nor the energy values E_n. In a purely formal way, however, we can write the expression for the exact value of the lowest energy level of this system (Eq. 6-21):

$$E_o = \frac{\int \psi_o^* H_{op} \psi_o \, d\tau}{\int \psi_o^* \psi_o \, d\tau}$$

Now let us take any well-behaved function F which depends on the same variables as ψ_o and write the analogous expression for the approximate energy associated with this function:

$$\mathcal{E} = \frac{\int F^* H_{op} F \, d\tau}{\int F^* F \, d\tau} \tag{6-37}$$

We might expect that no general statement could be made about the relation between the exact energy E_o and the approximate energy $\mathcal{E}$. There is, however, a useful mathematical theorem which states that for any choice of the function F, $E_o < \mathcal{E}$. Because the approximate energy is always higher than the exact energy, it follows that any change in the function F which lowers the approximate energy can only be an improvement in the approximation. If a mathematical form for F is chosen which includes an adjustable parameter, then the value of this parameter for which the approximate energy is a minimum yields the best possible function of this form.

Let us outline the application of the variation method to a particular problem, the ground state of the helium atom. It was pointed out previously that the Schrödinger equation for the helium atom (Eq. 6-36) cannot be solved exactly because the variables in this equation cannot be separated. If the term ϵ^2/r_{12} in the potential function representing the interelectronic repulsion were not present, the variables for the two electrons could be separated. Then the Schrödinger equation would become the sum of two Schrödinger equations, one for each electron separately, for the hydrogen-like orbitals of the He$^+$ ion.

Because the electron repulsion term is present, it is necessary to approximate the solutions to the Schrödinger equation. A reasonable choice for an approximate wave function F for the ground state of the helium atom is a product of normalized orbitals, one for each electron, of the same mathematical form as the 1s hydrogen-like orbitals (Eq. 6-33) for the He$^+$ ion, but with an unknown effective nuclear charge represented by Z'. Thus our trial function is written in the following way:

$$F = \frac{1}{\pi}\left(\frac{Z'}{a_o}\right)^3 e^{-\frac{Z'r_1}{a_o}} e^{-\frac{Z'r_2}{a_o}} \tag{6-38}$$

By substitution of this function in Eq. 6-37 the approximate energy is found to be

$$\varepsilon = \left[\frac{1}{\pi}\left(\frac{Z'}{a_o}\right)^3\right]^2 \int \int e^{-\frac{Z'r_1}{a_o}} e^{-\frac{Z'r_2}{a_o}} H_{op}\left(e^{-\frac{Z'r_1}{a_o}} e^{-\frac{Z'r_2}{a_o}}\right) d\tau_1\, d\tau_2$$

When these integrals are evaluated (we omit the details), there results

$$\varepsilon = \frac{\epsilon^2}{a_o}\left(Z'^2 - \frac{27}{8}Z'\right)$$

In this equation we have the approximate energy expressed as a function of the parameter Z'. The best value of Z' is that value which minimizes ε and we obtain this in the usual way, namely, by setting the derivative of ε with respect to Z' equal to zero and then solving the resulting equation for Z'.[8] Thus we have

$$\frac{d\varepsilon}{dZ'} = \frac{\epsilon^2}{a_o}\left(2Z' - \frac{27}{8}\right) = 0$$

$$Z' = \frac{27}{16}$$

In this way we have found that for the approximate wave function in Equation 6-38 the effective nuclear charge Z' is lower than the value of 2 for the actual nuclear charge. This lowering of the effective nuclear charge from 2 to 27/16 can be interpreted as a result of the partial shielding (or screening) of each electron by the other. The energy corresponding to this value of Z' is about 2% higher than the value determined from spectroscopic data.

If a better value for the energy is desired it can be obtained by using a more complicated form for the approximate wave function with a larger number of adjustable parameters. The approximate energy then must be minimized with respect to each of these parameters. The computations, however, rapidly become more involved and time-consuming as the number of parameters is increased, and these computations can strain the capability of even a large electronic computer. The most complicated wave function used for helium so far contains 50 adjustable parameters. It yields an energy which is equal to the experimental value within the uncertainty of this value.

For our purpose it is better to use a simpler, though cruder, form for atomic orbitals rather than a complicated one because in applying wave mechanics to chemical problems we are not primarily interested in extremely accurate energy values.

In Chapter 7 we apply the variation method to obtain approximate wave functions for molecules.

[8]In general this procedure leads to either a maximum or a minimum. It turns out that in this application a minimum is obtained.

Supplementary Material

Symmetry of Wave Functions

The complete wave function for every atomic or molecular system has a fundamental property called its symmetry. For definiteness in the discussion of this concept, let us take as an example the wave function for the first excited state (1s2s) of helium. As mentioned previously (p. 202), a complete wave function consists of a product of a space factor, which is a solution of a Schrödinger equation, and a spin factor (Eq. 6-35).

For the space factor, ϕ, we take a product of helium atomic orbitals; thus

$$\phi(1,2) = \psi_{1s}(1)\psi_{2s}(2)$$

This function implies that we know that electron 1 is in the 1s orbital and that electron 2 is in the 2s orbital. The electrons are indistinguishable, however, and we might just as well have written

$$\phi(2,1) = \psi_{1s}(2)\psi_{2s}(1)$$

Which of these two expressions for the space factor is correct? The answer is, neither. Since there is no way of knowing which electron is in either orbital, it is incorrect to write a function which implies that we have this knowledge. Consequently the space factor, ϕ, must be expressed as a linear combination of $\phi(1,2)$ and $\phi(2,1)$; thus we have

$$\phi = \phi(1,2) \pm \phi(2,1)$$

Now let us note that the probability density, ϕ^2, for the electron distribution cannot depend on the way we happen to designate the electrons. Therefore ϕ^2 must be unchanged if we exchange the labels of the two electrons. This requirement allows only the following two possibilities: (1) This exchange of labels leaves ϕ unaltered, in which case ϕ is said to be a symmetric function. (2) The exchange transforms ϕ into its negative, in which case ϕ is an antisymmetric function. In the first case we have the symmetric linear combination $\phi_S = \phi(1,2) + \phi(2,1)$. In the second case we have the antisymmetric linear combination $\phi_{AS} = \phi(1,2) - \phi(2,1)$.

Turning now to the spin factor, we find the following possibilities: (1) Both electrons have the same spin quantum number, $s = +1/2$, and the corresponding spin factor is $\alpha(1)\alpha(2)$. (2) Both electrons have the spin quantum number, $s = -1/2$ with spin factor $\beta(1)\beta(2)$. If one electron has $s = +1/2$ and the other has $s = -1/2$, since we cannot distinguish the electrons the only possibilities for the spin factor are (3) the symmetric combination, $\alpha(1)\beta(2) + \alpha(2)\beta(1)$, and (4) the antisymmetric combination, $\alpha(1)\beta(2) - \alpha(2)\beta(1)$.

We can now multiply each of the two space factors by each of the four spin factors to obtain eight different products. Because the product of two functions of the same symmetry type is symmetric, and the product of two functions of opposite symmetry type is antisymmetric, we see that of the eight possible products, four are symmetric and four are antisymmetric.

The wave mechanical form of the Pauli exclusion principle states that the only atomic or molecular systems that exist are those for which the complete wave function is antisymmetric for the exchange of two electrons. Accordingly, only the following four products of space factors and spin factors are allowed:

$$[\psi_{1s}(1)\psi_{2s}(2) + \psi_{1s}(2)\psi_{2s}(1)][\alpha(1)\beta(2) - \alpha(2)\beta(1)]$$

and

$$[\psi_{1s}(1)\psi_{2s}(2) - \psi_{1s}(2)\psi_{2s}(1)] \begin{cases} \alpha(1)\alpha(2) \\ \beta(1)\beta(2) \\ [\alpha(1)\beta(2) + \alpha(2)\beta(1)] \end{cases}$$

Each of these wave functions is associated with a state of the helium atom. Thus we see that the electron configuration we started with, $1s2s$, turns out to be associated with four states. The last three wave functions above are degenerate in the absence of an external electric or magnetic field. Collectively, they form what is known as a triplet state, which is characterized by parallel electron spins. The first wave function above is associated with a singlet state in which the spins are opposed; that is, for one electron $s = +1/2$, and for the other $s = -1/2$.

In the ground state of helium, both electrons are in a $1s$ orbital. Consequently the antisymmetric space factor vanishes and the complete wave function is the product of a symmetric space factor and an antisymmetric spin factor. Thus the ground state of helium is a singlet, with opposed electron spins.

It is not difficult to show that the wave mechanical form of the Pauli exclusion principle implies the more elementary statement given on page 204. If for two electrons the values of the quantum numbers n, l, and m are the same, the antisymmetric space factor in the complete wave function vanishes. If in addition the values of s are the same, the antisymmetric spin factor vanishes. Consequently at least one quantum number must be different for the two electrons if the complete wave function is to have a value other than zero.

Supplementary References

M. W. Hanna, *Quantum Mechanics in Chemistry*, 2nd edition, W. A. Benjamin, New York, 1969.

G. Herzberg, *Atomic Spectra and Atomic Structure*, Dover, New York, 1944.
W. Kauzmann, *Quantum Chemistry*, Academic Press, New York, 1957.

I. Cohen and T. Bustard, "Atomic Orbitals." *J. Chem. Ed.*, **43**, 187 (1966).

Problems

1. (a) Calculate the kinetic energy in ergs of a photoelectron emitted by a sodium surface when light of wavelength 4000 Å is incident on it. The work function of sodium is 2.28 ev. (b) Calculate the value of the longest wavelength which can result in the emission of a photoelectron from a sodium surface.
Ans. (a) 1.32×10^{-12} erg; (b) 5450 Å

2. Calculate from Equation 6–6 the value of r when $n = 1$.

3. Calculate the wavelength and the wavenumber for the hydrogen atomic lines for the following transitions: (a) $n = 3 \rightarrow n = 2$; (b) $n = 4 \rightarrow n = 2$; (c) $n = 2 \rightarrow n = 1$; (d) $n = 4 \rightarrow n = 3$. **Ans.** (a) 6560 Å; (b) 4860 Å

4. Calculate the wavenumber and the wavelength of the spectral line associated with the transition $n = 2 \rightarrow n = 1$ of the Li^{++} ion.

5. What is the speed and de Broglie wavelength of an electron that has been accelerated by a potential difference of 500 v?
Ans. 1.33×10^9 cm/sec, 0.55 Å

6. Calculate the de Broglie wavelength associated with: (a) a 10 g mass moving with a speed of 10 meters per second; (b) a hydrogen molecule moving with a speed equal to its root-mean-square speed at 300°K; (c) an electron whose kinetic energy is 10^4 ev; (d) an α particle whose kinetic energy is 10^4 ev.

7. (a) What is the speed of an electron whose de Broglie wavelength is 1 Å? (b) By what potential difference must such an electron have been accelerated from an initial speed of zero? (c) If all of the energy of such an electron is transferred to a single photon, what is the wavelength of this photon?
Ans. (a) 7.29×10^8 cm/sec; (b) 150 v; (c) 82 Å

8. Show that $\psi = ke^{-ar}$ is an eigenfunction of the operator $\dfrac{d^2}{dr^2}$ for positive values of r. What is the eigenvalue? What is the normalization constant?

9. Calculate the value of n_x for a hydrogen molecule in a box whose x dimension is 10 cm, if its energy is equal to the average kinetic energy of hydrogen molecules at 300°K.

10. (a) Calculate the energy absorbed by an electron in a box of 5Å edge length when one of its quantum numbers is changed from 1 to 2. (b) If this energy comes from electromagnetic radiation, calculate the wavelength of the absorbed radiation. **Ans.** (a) 7.24×10^{-12} ergs; (b) 2750 Å

11. Calculate the probability of finding a particle in a 1-dimensional box of length a in an element of length dx located at $x = a/3$ if the particle is in its lowest energy state.

12. What is the degree of degeneracy of the eigenfunctions for the particle in a cubic box associated with the energy $14h^2/8ma^2$? **Ans.** 6

13. Show that the $1s$ and the $2s$ orbitals of the hydrogen atom are mutually orthogonal.

14. Graph the function $f = r \cos \theta$. (It is convenient to use polar coordinate paper.)

15. Calculate the probability that the electron in the ground state of He^+ is at a distance greater than 1.06 Å from the He nucleus. **Ans.** 0.014

16. Find the values of r for which the radial distribution function for the $2s$ state of the hydrogen atom has a maximum or a minimum.

17. Show that the electron distribution in the ground state of the neon atom has spherical symmetry (that is, is independent of the angles). (Hint: express p_x^2, p_y^2, and p_z^2 in terms of trigonometric functions.)

18. Calculate the ionization potential in electron-volts of (a) the hydrogen atom, and (b) the Li^{++} ion. **Ans.** (a) 13.6 ev; (b) 122 ev

19. With the function e^{-cr} as a trial function for the ground state of the hydrogen atom, use the variation method to find the best value of the parameter c. Compare the resulting function with the exact $1s$ orbital of the hydrogen atom.

7

Chemical Bonding and Valence

Introduction

Ever since the early days of chemistry as a science, the nature of the forces holding atoms together in a molecule has been an important problem in chemical theory. The discovery that some compounds could be decomposed by an electric current suggested that chemical bonding is due to electrical forces. Early in the nineteenth century Berzelius (the outstanding chemist of his day) developed a theory of chemical "affinity," known as the dualistic theory, which attributed the existence of molecules to the electrostatic attraction between positively and negatively charged atoms or radicals.

Because chemistry at that time dealt mostly with the behavior of acids, bases, and salts, the dualistic theory could account for a wide range of observations. With the development of organic chemistry, however, this theory soon met fatal difficulties. It could not explain the fact that an electronegative atom, such as chlorine, could replace an electropositive atom, such as hydrogen, in many compounds with relatively little change in the properties of the compound. It could not explain the fact that the molecules of some elements consisted of several identical atoms. (Incidentally, the reluctance of many chemists to accept this fact, which follows directly from Avogadro's "hypothesis," was the main factor which delayed the general acceptance of this hypothesis.)

With the demise of the dualistic theory chemists abandoned for many years any attempt to account for the force which holds a molecule together. The concept of a specific bond (of unknown origin) between certain atoms in a molecule was introduced by Kekulé, Couper, and Butlerov, and proved to be very fruitful. This concept, together with the related concept of a definite arrangement in space of the atoms introduced by van't Hoff and Le Bel, was the basis for the remarkable development of organic chemistry in the latter part of the nineteenth century.

The underlying problem of the nature of the chemical bond still remained in the background. Why are some elements so inert that no compounds of these elements have been found? Why can a chlorine atom, for example, combine with one, and only one, hydrogen atom? Why do the four hydrogen atoms in the methane molecule have a tetrahedral arrangement about the carbon atom? The answers to these and many similar questions could not even be approached until developments in physics provided the essential foundation.

When J. J. Thomson recognized that the electron was a universal constituent of all matter, he pointed out that the electrons in a molecule must play an important role in chemical combination. After the development of the Bohr theory it became clear that the distribution of electrons in a molecule is intimately related to its chemical properties, including its valence.

In 1916 Kossel showed that when elements in Groups I and II of the Periodic Table react with elements in Groups VI and VII they do so by transferring electrons in such a way that the resulting ions have the same electron configuration as the noble gas which is closest in atomic number. In the resulting compounds the force holding the ions together is essentially the coulombic attraction between ions of opposite charge. This interpretation of ionic bonding is clearly related to Berzelius's dualistic theory. Because most compounds which contain ionic bonds are solids at ordinary temperatures, we leave further consideration of ionic bonding to Chapter 12, which deals with solids.

Also in 1916, G. N. Lewis called attention to the fact that nearly all molecules contain an even number of electrons. He suggested that chemical bonds between identical or similar atoms, called covalent bonds, could be ascribed to the sharing of a pair of electrons between the bonded atoms, thus giving each atom the noble gas configuration. This concept of the electron-pair bond led to the correlation and clarification of a large amount of chemical information. It was unsatisfactory from a physical point of view, however, because a configuration in which a pair of electrons remained in the region between the bonded atoms seemed inherently unstable.

It was only with the development of quantum mechanics and its application to chemical bonding that a fundamental understanding of the covalent bond became possible. On this basis a successful attack has been made on the central problem in valence theory, namely, the determination of the distribution of nuclei and electrons associated with minimum molecular energy. One of the consequences of the success of this application of quantum mechanics is the fact that concepts and vocabulary of quantum mechanical origin dominate much of theoretical chemistry today.

Approximation Methods

In principle, all that we have to do to obtain the energy and structure of a molecule in which we are interested is to write the Schrödinger equation for

Chemical Bonding and Valence

this molecule and then solve this equation. There is no real difficulty in writing Schrödinger equations, but solving them is another matter. We saw in the last chapter that for atoms containing two or more electrons approximation methods are essential. Even as simple a molecule as H_2, which consists of only four particles, is in some ways a more complicated problem than an atom which consists of many more particles because the molecule does not have the spherical symmetry of the atom.

In addition to translational energy and electronic energy, which are the only forms of energy an atom can have, a molecule has rotational energy which depends on the rate of change of the direction of the molecular axis, and vibrational energy resulting from a periodic change in the internuclear separations. In this chapter we focus on the electronic energy of molecules and defer consideration of rotational and vibrational energies to the next chapter.

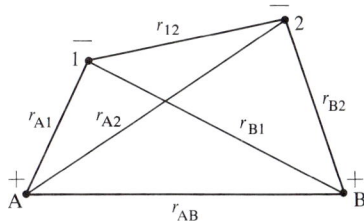

Figure 7–1. Model of the H_2 molecule showing the six internal coordinates.

When we write the Schrödinger equation for H_2 in terms of the Cartesian coordinates of each particle, we have an equation involving 12 variables. Because the electronic energy does not depend on the translational or rotational motion of the molecule as a whole, we can simplify the problem without any approximation by assuming that the center of gravity of the molecule is fixed in space (no translational energy) and that the direction of the molecular axis is fixed (no rotational energy). Then we can write the Schrödinger equation in terms of the six internal coordinates shown in Figure 7–1 in which the subscripts 1 and 2 designate the electrons, and A and B designate the nuclei.

Now we can carry out a further simplification which makes use of a very good approximation introduced by Born and Oppenheimer. They concluded that the nuclei in a molecule, because of their much greater mass, move much less rapidly than the electrons. As a consequence, the internuclear distance may be considered as constant in the calculation of the electronic energy. Derivatives of the wave function with respect to nuclear coordinates then drop out of the Schrödinger equation, although the electronic energy still depends on the internuclear separation.

Even with the simplifications mentioned above, the Schrödinger equation for H_2 is still rather complicated; it assumes the following form:

$$H_{op}\Psi = \left[-\frac{h^2}{8\pi^2 m}(\nabla_1^2 + \nabla_2^2) + \epsilon^2\left(-\frac{1}{r_{A1}} - \frac{1}{r_{A2}} - \frac{1}{r_{B1}} - \frac{1}{r_{B2}} + \frac{1}{r_{12}} + \frac{1}{r_{AB}}\right)\right]\Psi$$
$$= E\Psi \tag{7-1}$$

This equation has not been solved exactly even when r_{AB} is assumed to be constant because the presence in the Hamiltonian operator of the term which involves r_{12} makes it impossible to separate the variables. (Compare the discussion of the helium atom in Chapter 6.) Consequently for H_2 and all other molecules we must be content with approximate solutions. There are various degrees of approximation, however, and an approximate result is not necessarily an inaccurate result. As we deal with molecules of greater complexity we must be satisfied with cruder results because of computational difficulties. The best way to obtain approximate solutions of the Schrödinger equation is still a subject of much research.

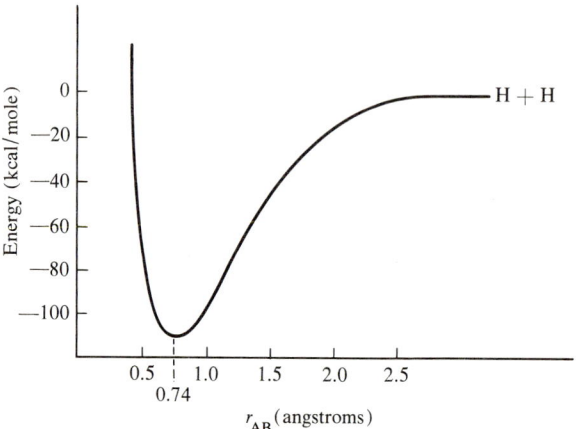

Figure 7–2. Potential energy curve of the H_2 molecule.

If an approximate solution of a molecular Schrödinger equation has been obtained for a particular value of the internuclear separation r_{AB}, the corresponding molecular energy can be calculated. When this energy has been calculated for a number of values of r_{AB} the results are often represented by a graph of energy versus r_{AB}, Figure 7–2. A graph of this type is called a potential energy curve. (Potential energy curves can also be obtained from spectroscopic data. We come back to this point later.) As the internuclear separation becomes large, the energy approaches the value for the completely separated atoms. At very small internuclear separations repulsive forces become strong and the energy becomes large and positive. At some intermediate separation there is a minimum in the potential energy curve; this distance is the equilibrium separation of the nuclei in the ground state of the molecule. If the potential energy curve does not have a minimum, a stable molecule cannot be formed.

Each electronic state of a molecule is associated with a potential energy curve. From the point of view of valence theory, we are concerned with molecules in their ground electronic states because this is the state in which molecules exist under ordinary conditions. Consequently little is said in this chapter about excited states and their potential energy curves, but we return to this subject in Chapter 8.

Two approaches have been widely used in arriving at approximate solutions of the Schrödinger equation for molecules in their lowest electronic energy states. These approaches differ in the choice of trial wave functions whose parameters are then optimized by the variation method. One of these approaches, the *valence bond* method developed by Heitler and London, Slater, and Pauling, is related to Lewis' concept of the electron-pair bond. The other, the *molecular orbital* method developed by Hund, Mulliken, Lennard-Jones, and Hückel, is related to the treatment of complex atoms. In their elementary forms, both of these methods give results that are crude from a quantitative point of view. When both are refined by the use of more complicated trial functions, the difference between them disappears.

Although the valence bond method was the first to be applied to molecules (Heitler and London, 1927), the molecular orbital method has turned out to be more useful, and we devote more attention to it here.

Molecular Orbital Method: Application to H_2^+ and H_2

The basic assumption of the molecular orbital method is that the wave function of a molecule can be approximated by a product of *molecular orbitals* (MO's) in the same way that the wave function of a many-electron atom is approximated by a product of atomic orbitals (AO's) (p. 203). MO's, like AO's, are defined as functions of the coordinates of single electrons. Each electron in a molecule is associated with an MO that is specified by certain quantum numbers. The energy of such an MO is roughly the energy required to remove the electron from the molecule.

In order to apply the MO method we need some basis for choosing a mathematical form for an MO. Let us use the H_2^+ molecular ion as the simplest example. This, of course, is not a kind of molecule that can be kept on the shelf in a bottle. It does exist, however, in electric discharges in H_2, and its properties are quite well known from spectroscopic measurements. Because H_2^+ has only one electron, its Schrödinger equation can be solved exactly for a fixed internuclear distance. The exact solutions, however, are quite complicated and we do not discuss them here.

With the distances of the electron from the two protons represented by r_A and r_B, the Hamiltonian operator for H_2^+ is

$$H_{op} = \frac{-h^2}{8\pi^2 m}\nabla^2 + \epsilon^2\left(-\frac{1}{r_A} - \frac{1}{r_B} + \frac{1}{r_{AB}}\right)$$

It is reasonable to expect that when the electron is in the neighborhood of proton A the influence of proton B is small. Then the form of the ground-state MO in this region should be similar to that of a $1s$ hydrogen AO, represented by ψ_A. Likewise in the region near proton B, the MO should resemble the $1s$ hydrogen AO, ψ_B. To take other regions of space into consideration, the simplest assumption is that the MO can be expressed as a linear combination of ψ_A and ψ_B which we write as follows:

$$\psi_{MO} = c_A\psi_A + c_B\psi_B \qquad (7\text{-}2)$$

in which c_A and c_B are constants. This use of linear combinations (LC) of atomic orbitals for expressing approximate MO's is usually referred to as the LCAO–MO approximation.

Now we can use the variation method to find the best values of c_A and c_B by minimizing the approximate electronic energy ($\mathcal{E}$) calculated with the use of the above MO as a trial function. Along the way we will obtain an expression for the corresponding value of $\mathcal{E}$. We first write an equation for $\mathcal{E}$ as a function of the coefficients c_A and c_B in the following way:

$$\mathcal{E} = \frac{\int \psi_{MO} H_{op} \psi_{MO}\, d\tau}{\int \psi_{MO}^2\, d\tau} = \frac{\int (c_A\psi_A + c_B\psi_B)H_{op}(c_A\psi_A + c_B\psi_B)\, d\tau}{\int (c_A\psi_A + c_B\psi_B)^2\, d\tau}$$

Expanding the parentheses and making use of the fact that $\int \psi_A H_{op}\psi_B\, d\tau = \int \psi_B H_{op}\psi_A\, d\tau$,[1] we obtain

$$\mathcal{E} = \frac{c_A^2 \int \psi_A H_{op}\psi_A\, d\tau + c_B^2 \int \psi_B H_{op}\psi_B\, d\tau + 2c_A c_B \int \psi_A H_{op}\psi_B\, d\tau}{c_A^2 \int \psi_A^2\, d\tau + c_B^2 \int \psi_B^2\, d\tau + 2c_A c_B \int \psi_A\psi_B\, d\tau}$$

[1] It can be shown that any Hamiltonian operator is a member of the class of operators for which $\int \psi_A^* \alpha \psi_B\, d\tau = \int \psi_B^* \alpha \psi_A\, d\tau$. These operators are known as Hermitian operators in honor of the French mathematician Hermite who first studied their properties. In the case under consideration here, the wave functions are taken to be real functions and consequently $\psi_A^* = \psi_A$.

Chemical Bonding and Valence

We can write this equation in more convenient form by using the following symbols for the integrals:

$$H_{AA} = \int \psi_A H_{op} \psi_A \, d\tau \qquad H_{AB} = \int \psi_A H_{op} \psi_B \, d\tau$$
$$S_{AA} = \int \psi_A^2 \, d\tau \qquad S_{AB} = \int \psi_A \psi_B \, d\tau \qquad (7\text{-}3)$$

The integral H_{AA} is usually called a *coulomb integral*. Its value is close to E_A, the energy of an electron in the AO ψ_A in the isolated atom A. An integral of the type H_{AB} is called a *resonance integral*. The integrals S_{AA} and S_{BB} are simply normalization integrals whose value is 1 if ψ_A and ψ_B are normalized. An integral of type S_{AB} is called an *overlap integral*. The use of the word overlap comes from the fact that only those regions of space in which both ψ_A and ψ_B have significant values (that is, in which they overlap) contribute to the value of this integral. For two atoms more than a few angstrom units apart, the value of the overlap integral is essentially zero.

In terms of these symbols the expression for the approximate energy becomes

$$\mathcal{E} = \frac{c_A^2 H_{AA} + c_B^2 H_{BB} + 2c_A c_B H_{AB}}{c_A^2 S_{AA} + c_B^2 S_{BB} + 2c_A c_B S_{AB}}$$

We wish to find the values of c_A and c_B which make the approximate energy a minimum, and also the value of this energy. We do this in the usual way by differentiating $\mathcal{E}$ partially, first with respect to c_A and then with respect to c_B, and then setting the resulting derivatives equal to zero. This step yields the following equations:

$$\frac{\partial \mathcal{E}}{\partial c_A} = c_A H_{AA} + c_B H_{AB} - \mathcal{E}(c_A S_{AA} + c_B S_{AB}) = 0$$

$$\frac{\partial \mathcal{E}}{\partial c_B} = c_B H_{BB} + c_A H_{AB} - \mathcal{E}(c_B S_{BB} + c_A S_{AB}) = 0$$

We have here a pair of linear simultaneous equations in c_A and c_B which can be rewritten in the following form:

$$c_A(H_{AA} - \mathcal{E}S_{AA}) + c_B(H_{AB} - \mathcal{E}S_{AB}) = 0$$
$$c_A(H_{AB} - \mathcal{E}S_{AB}) + c_B(H_{BB} - \mathcal{E}S_{BB}) = 0 \qquad (7\text{-}4)$$

Now we make use of a theorem in algebra which states that a system of linear homogeneous simultaneous equations can have nontrivial solutions only if the determinant of the coefficients vanishes. It follows that

$$\begin{vmatrix} H_{AA} - \mathcal{E}S_{AA} & H_{AB} - \mathcal{E}S_{AB} \\ H_{AB} - \mathcal{E}S_{AB} & H_{BB} - \mathcal{E}S_{BB} \end{vmatrix} = 0 \qquad (7\text{-}5)$$

A determinantal equation of this type is obtained in every application of the variation method. Such an equation is often called a *secular equation*.

When this determinant is expanded we obtain the following quadratic equation with the approximate energy ε as unknown:

$$(H_{AA} - \varepsilon S_{AA})(H_{BB} - \varepsilon S_{BB}) - (H_{AB} - \varepsilon S_{AB})^2 = 0$$

We can simplify this equation slightly by assuming that our AO's ψ_A and ψ_B are normalized. Then $S_{AA} = S_{BB} = 1$, and we can drop the subscript on S_{AB}. The quadratic equation then becomes

$$(H_{AA} - \varepsilon)(H_{BB} - \varepsilon) - (H_{AB} - \varepsilon S)^2 = 0 \qquad (7\text{--}6)$$

Because the molecule under consideration is homonuclear ψ_A has the same mathematical form as ψ_B. It follows that $H_{AA} = H_{BB}$ and consequently

$$H_{AA} - \varepsilon = \pm(H_{AB} - \varepsilon S)$$

When this equation is solved for ε, we obtain the following values corresponding to the two choices of sign:

$$\varepsilon_+ = \frac{H_{AA} + H_{AB}}{1 + S} \quad \text{and} \quad \varepsilon_- = \frac{H_{AA} - H_{AB}}{1 - S} \qquad (7\text{--}7)$$

To interpret these equations we recall that H_{AA} is a negative number because it is approximately equal to the energy of an electron in the AO ψ_A. Resonance integrals such as H_{AB} also have negative values. The value of the overlap integral S turns out to be small (~ 0.1) for normal internuclear separations and it is sometimes neglected in rough calculations. The equations above tell us that in place of the two identical energy levels E_A and E_B we would have if an electron were in either ψ_A or ψ_B, we have obtained two new energy levels,

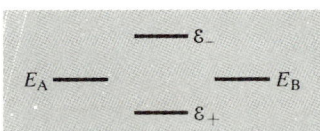

Figure 7–3. Energy levels associated with bonding and antibonding molecular orbitals for homonuclear diatomic molecule ($E_A = E_B$).

one of which (ε_+) is lower than E_A (more negative) and the other (ε_-) is higher (Fig. 7–3).

The lowering of ε_+ relative to E_A is responsible for the formation of a chemical bond, and the MO associated with ε_+ is called a *bonding MO*. Conversely, because ε_- is higher than E_A, the associated orbital is called an *antibonding MO*.

To calculate the coefficients in these MO's we substitute the expressions for ε_+ and ε_- from Equations 7–7 into Equations 7–4 and we find that $c_A = c_B$ and $c_A = -c_B$, respectively. Then we can write the MO's in the following way:

$$\psi_+ = c_+(\psi_A + \psi_B)$$
$$\psi_- = c_-(\psi_A - \psi_B) \qquad (7\text{–}8)$$

When the bonding MO ψ_+ is normalized, the value of c_+ is $(2 + 2S)^{-\frac{1}{2}}$; likewise, for ψ_- the value of c_- is $(2 - 2S)^{-\frac{1}{2}}$.

◈◈◈◈◈◈◈

We see that although we started this derivation with the intention of obtaining the "best" values of the coefficients c_A and c_B, that is, those associated with a minimum value of the energy $\mathcal{E}$, we have actually obtained two sets of values of the energy and the related coefficients. We interpret the lower energy value as the upper limit of the energy of the ground state of the hydrogen molecular ion. The other energy value is the upper limit of the energy of the lowest excited state of this ion.

We can obtain the values $c_A/c_B = \pm 1$ in the special case of a homonuclear diatomic molecule without carrying out the variation calculation. The two AO's are identical by hypothesis and they both must contribute the same amount to the probability density ψ_{MO}^2. Consequently $c_A^2 = c_B^2$ and $c_A = \pm c_B$. For heteronuclear molecules, however, symmetry is not sufficient to determine the ratio of the coefficients and the use of the variation method is essential.

◈◈◈◈◈◈◈

By squaring the molecular orbitals in Equation 7–8 we can obtain the associated electron probability distributions (Fig. 7–4). This operation yields the following equations:

$$\psi_+^2 = \frac{\psi_A^2 + \psi_B^2 + 2\psi_A\psi_B}{2 + 2S} \qquad (7\text{–}9)$$

and

$$\psi_-^2 = \frac{\psi_A^2 + \psi_B^2 - 2\psi_A\psi_B}{2 - 2S}$$

The significant feature of Equation 7–9 is the presence of the term $+2\psi_A\psi_B$ in the numerator. In the region between the two nuclei both ψ_A and ψ_B have appreciable values, and so does their product. It follows that when an electron is in a bonding MO such as ψ_+, the probability of finding this electron between the nuclei is relatively high. Since the electron in this region interacts with both nuclei, its potential energy is lower, that is, has a larger negative value, than it would be if it interacted with only one nucleus. Because the lowering of the electronic energy is greater than the increase in potential energy resulting from internuclear repulsion, a stable bond is formed between a hydrogen atom and a proton.

On considering the antibonding MO ψ_-, we see that its value is zero wherever ψ_A and ψ_B have the same absolute value; in other words, wherever $r_A = r_B$. Consequently this MO has a value of zero in a plane (the nodal surface) which is perpendicular to and bisects the axis between the nuclei. From the square of this orbital it is evident that the probability of finding an electron is zero in

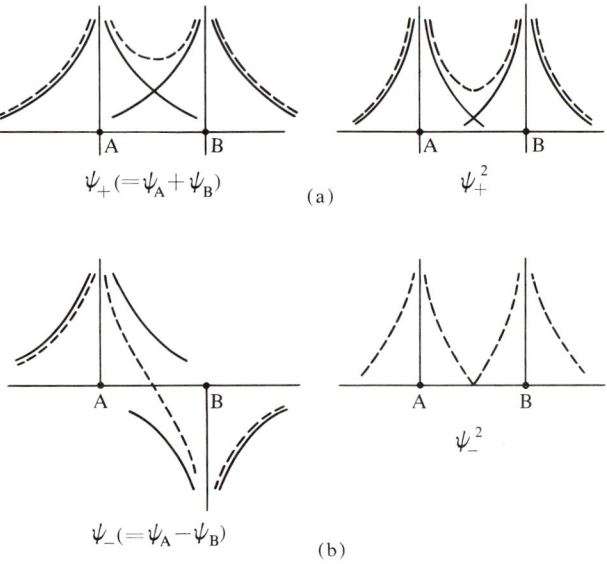

Figure 7–4. (a) Graphs of bonding MO, ψ_+, and ψ_+^2 as a function of distance along the molecular axis. (b) Corresponding graphs for antibonding MO, ψ_-, and ψ_-^2. Notice the significant value of ψ_+^2 near the mid-point between the nuclei and the zero value of ψ_-^2 in the same region.

the nodal surface and is relatively low in the region between the nuclei. This means that the electron tends to avoid the region in which it interacts with both nuclei, and thus a bond cannot be formed. This behavior is typical for the presence of an electron in an antibonding MO.

For the H_2^+ molecular ion the integrals H_{AA}, H_{AB}, and S can be evaluated at a series of values of r_{AB}. Then the total energy in the ground state can be calculated as a sum of the electronic energy, $\mathcal{E}_+$, and the nuclear repulsion energy, ϵ^2/r_{AB}, which is always positive. The potential energy curves in Figure 7–5 were obtained in this way. The existence of a minimum in the curve for the ground state is consistent with the stability of H_2^+ with respect to a dissociation into a hydrogen atom and a proton.

As might be anticipated, the calculated values of the energy and the internuclear separation at the minimum of the potential energy curve are rather rough approximations. With a more complicated trial function values are obtained that are in quantitative agreement with experimental results.

Now let us imagine that an electron is added to H_2^+ to form H_2 in its ground state. Qualitatively, we see that this second electron can enter the bonding MO ψ_+ provided its spin is opposed to that of the electron already present to

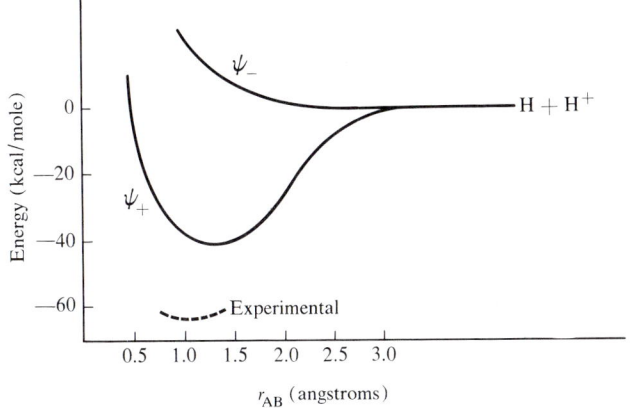

Figure 7-5. Potential curves for H_2^+ molecular ion in the ground state (ψ_+) and the excited state (ψ_-). The experimental values are obtained from spectroscopic measurements.

satisfy the requirement of the Pauli exclusion principle. With 2 electrons in a bonding MO, the binding energy of H_2 should be considerably larger than that of H_2^+, and this conclusion is confirmed by spectroscopic data. Because the electronic energy is larger for H_2 than for H_2^+, it can be expected that the equilibrium internuclear distance is smaller for H_2 than for H_2^+, and this is also consistent with spectroscopic data.

From a quantitative point of view, the presence of two electrons in H_2 produces mathematical complications because of the term in the Hamiltonian operator that depends on r_{12}. The simple LCAO–MO treatment presented above leads to the following form for the approximate molecular wave function of the ground state:

$$\Psi \text{ molecular} = \psi_+(1)\psi_+(2) = \frac{1}{2+2S}[s_A(1) + s_B(1)][s_A(2) + s_B(2)] \qquad (7\text{-}10)$$

in which $s_A(1)$ represents a $1s$ AO for electron 1 on hydrogen atom A, and so on. The value of the binding energy calculated with this wave function is only about

60% of the experimental value, and the calculated equilibrium internuclear distance is 0.850 Å compared with the experimental value of 0.742 Å. The calculated values can be improved to any desired extent by the use of more complicated trial functions, as was done with the helium atom and H_2^+. The most ambitious effort so far involved the use of a linear combination of 50 functions and this led to calculated values that agree with experimental ones to within the experimental error. This extremely difficult calculation required the use of a large electronic computer for as simple a molecule as H_2. It is clear that this approach is not practical for more complicated molecules. We restrict our attention to the simplest form of the theory.

Application of the Molecular Orbital Method to Other Diatomic Molecules

In order to apply the LCAO–MO method to other molecules we need to extend our treatment in several directions. We have attributed the bond in H_2 to the presence of two electrons in a bonding MO which is formed from the overlap of two AO's. Clearly, the overlap of AO's is of basic importance. Now we ask if there are any limitations in the AO's which can be combined to form MO's. The answer is that there are limitations and not all AO's can be combined to form MO's.

To obtain one of these limitations let us return to Equation 7–6. We solved this equation for the special case of two identical AO's for which $c_A = \pm c_B$ and $H_{AA} = H_{BB}$. Suppose we assume that the AO's are different and that $H_{BB} > H_{AA}$. When we expand Equation 7–6 and write the exact solutions for the quadratic, we obtain fairly complicated expressions. By neglecting S^2 with respect to 1 and using an approximation for the square root, we can write the following approximate solutions which are sufficient for our purpose:

$$\mathcal{E}_+ = H_{AA} - \frac{(H_{AB} - H_{AA}S)^2}{H_{BB} - H_{AA}}; \quad \mathcal{E}_- = H_{BB} + \frac{(H_{AB} - H_{BB}S)^2}{H_{BB} - H_{AA}}$$

H_{AA} is approximately the energy of an electron in ψ_A, E_A, and likewise $H_{BB} \sim E_B$. Introducing these further approximations, we write the expressions for the energies of the bonding and antibonding MO's as follows:

$$\mathcal{E}_+ = E_A - \frac{(H_{AB} - E_AS)^2}{E_B - E_A}; \quad \mathcal{E}_- = E_B + \frac{(H_{AB} - E_BS)^2}{E_B - E_A}$$

The squared quantities in the above expressions have positive values, and by hypothesis $E_B - E_A$ is also positive. Therefore the second term in each expression is necessarily positive. Thus in place of the atomic energy levels,

E_A and E_B, we obtain two MO energy levels, ε_+ and ε_- with ε_+ lower (more negative) than E_A, while ε_- is higher than E_B (Fig. 7-6). If E_B is much higher than E_A the second term in each expression is small; then ε_+ is approximately equal to E_A and ε_- is approximately equal to E_B. This implies that there is little overlap between ψ_A and ψ_B and consequently little binding of the atoms.

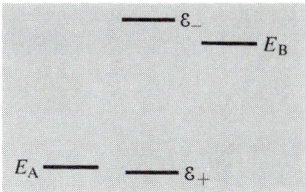

Figure 7-6. Energy levels associated with bonding and antibonding molecular orbitals when $E_B > E_A$.

This conclusion is confirmed when ε_+ and ε_- are introduced in succession into Equation 7-4 in order to obtain the ratios of the coefficients c_A and c_B in the MO's ψ_+ and ψ_-. For ψ_+, c_A/c_B is very large, which means that ψ_+ is nearly the same as ψ_A. Conversely, for ψ_- c_A/c_B is very small and thus ψ_- is nearly the same as ψ_B.

We conclude that for a bond to be formed between two atoms as a result of the overlap of AO's, the AO's that can be combined in an MO are limited to those whose energies are not greatly different.

The second limitation on the AO's that can be combined to form MO's comes from the shape of the AO's and the fact that except for a $1s$ orbital, all AO's have both positive and negative values in different regions. Let us suppose the nuclei of atoms A and B are located on the x axis. (We always take the molecular axis as the x axis.) When we try to form an MO by combining a $1s$ orbital of atom A with a $2p_y$ orbital of atom B, we find (Fig. 7-7) that for every volume element in which $\psi_{1s}\psi_{2p_y}$ has a certain positive value there is another volume element which has the same absolute value but is negative. It follows that the overlap integral S is equal to zero and it can be shown that then $H_{AB} = 0$. This means that an MO cannot be formed from the combination of an s orbital and a p_y orbital. Because in this case p_y and p_z are indistinguishable, the same conclusion applies to p_z orbitals.

The situation is different for a p_x orbital of atom B which has a lobe extending toward atom A. When this lobe has the same sign as the s orbital on atom A, we see that an effective combination can result because the value of the s orbital in the region where p_x has the opposite sign is very small. A similar situation exists for the combination of two p_x orbitals whose overlapping lobes have the same sign.

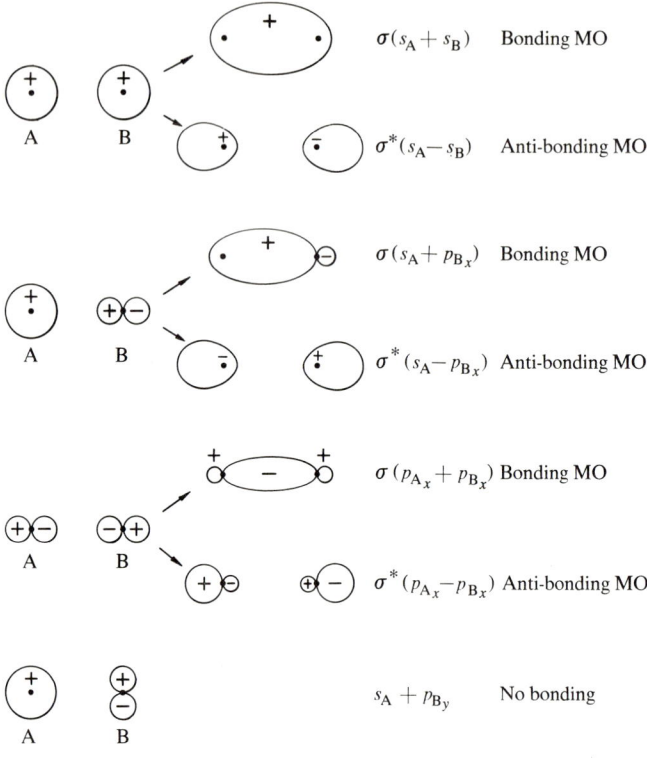

Figure 7-7. Bonding and antibonding σ molecular orbitals formed from combination of two s, two p, and one s and one p atomic orbitals.

When we consider possible combinations of p orbitals whose axes are perpendicular to each other, we find that $S = 0$ and no MO can be formed. In contrast p_y can combine with p_y and p_z can combine with p_z because for both of these a "sidewise" overlap is possible (Fig. 7-8).

To distinguish the types of LCAO–MO's mentioned above we need ways of classifying them and designating them by appropriate symbols. In order to do this we make use of the fact that MO's are specified by molecular quantum numbers in much the same way that AO's are specified by atomic quantum numbers. One useful way to classify MO's is on the basis of the quantum number λ which gives the electronic angular momentum about the molecular axis in units of $\dfrac{h}{2\pi}$. When this quantum number has the value 0 or 1, the corresponding MO is said to be a σ or a π MO, in analogy with the use of the letters s and p for AO's with the quantum number l equal to 0 and 1. A σ MO is symmetrical about the molecular axis. In contrast, a π orbital has a

Chemical Bonding and Valence

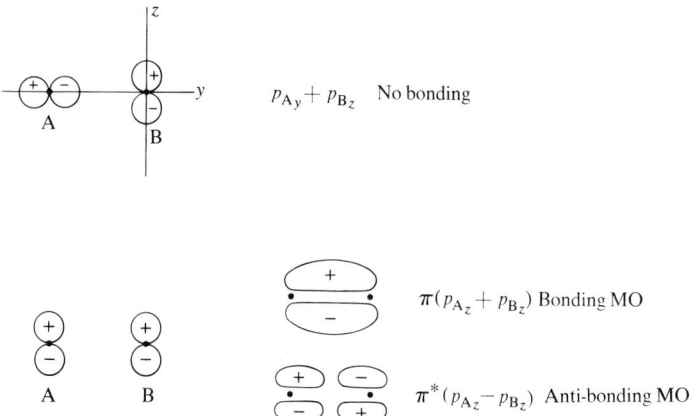

Figure 7-8. Bonding and antibonding π molecular orbitals formed from combination of two p atomic orbitals with parallel axes.

nodal plane which separates two regions in which the value of the orbital has opposite signs.

The combination of two s, two p_x, or one s and one p_x AO's leads to σ orbitals. The combination of $2p_y$ orbitals leads to a π_y MO which has the x,z plane as the nodal plane. Similarly, the combination of $2p_z$ orbitals leads to a π_z MO. The only difference between the π_y and π_z MO's is the different orientation of the nodal planes.

Another way of classifying MO's is based on their bonding character. In general when two AO's are combined two MO's are formed, one of which is bonding and the other is antibonding. It is customary to distinguish these types by placing an asterisk as a superscript on the symbol of the antibonding MO. Thus σ^* represents an antibonding σ MO.

For a homonuclear diatomic molecule the AO's that are combined to form an MO are always of the same type. For such molecules the AO's used to form the MO's are often indicated by writing the symbol for the AO after the symbol for the MO. For example, the bonding and antibonding MO's for H_2 are written in the following way: $\sigma 1s$ and $\sigma^* 1s$. The notation for heteronuclear molecules is mentioned later.

◊◊◊◊◊◊

One further characteristic of a MO of a homonuclear diatomic molecule which is of importance in spectroscopy but not for our purpose here is the behavior of the MO on "inversion" through the center of the molecule. With this center at the origin of our coordinate system, inversion means that x, y, and z are replaced by $-x$, $-y$, and $-z$. On inversion, the probability density represented by ψ_{MO}^2, which cannot depend on the direction of the coordinate axes, must be unchanged. Consequently ψ_{MO} must either be unchanged by inversion or it must go into $-\psi_{MO}$. If the sign of the MO is unchanged, the subscript g is written after the orbital symbol. (The letter g is an abbreviation for *gerade*, the German word for even.) If the sign is changed, the subscript u (for *ungerade*, odd) is employed. For example, since the bonding MO of H_2 is unchanged on inversion it is represented by $\sigma_g 1s$. The corresponding antibonding MO changes sign and thus is represented by $\sigma_u^* 1s$. For the bonding and antibonding MO's formed from p_y orbitals, the complete symbols are $\pi_u 2p_y$ and $\pi_g^* 2p_y$, respectively.

Electron Configurations of Diatomic Molecules

Before we can consider the electron configurations of molecules, we need information about the relative energies of the MO's. To obtain these energies by direct calculation is, as we have seen, a formidable undertaking which has been completed for only a small number of simple molecules. Consequently, indirect methods must be relied on. One such method depends on the detailed interpretation of molecular spectra, which itself can be rather complicated.

Another indirect method starts with knowledge of the sequence of the energy levels of the isolated atoms. For homonuclear molecules, the two atoms are imagined to be pushed together until their nuclei coincide, thus forming a hypothetical "atom" of twice the atomic number of the original atoms. The energy levels of the "united atom" can be estimated fairly reliably. Next the two nuclei are imagined to be gradually separated. As the internuclear separation increases from zero to infinity, the energy levels of the united atom must change smoothly to their values in the separated atoms (Fig. 7–9). At some intermediate distance which is the equilibrium internuclear distance in the diatomic molecule formed from these atoms, the sequence of the interpolated MO energies can be found.

For homonuclear molecules of the elements in the first row of the Periodic Table, the sequence of MO's in the order of increasing energy has been found by these methods to be the following:

$$\sigma 1s < \sigma^* 1s < \sigma 2s < \sigma^* 2s < \sigma 2p_x < \pi 2p_y = \pi 2p_z < \pi^* 2p_y = \pi^* 2p_z < \sigma^* 2p_x$$

To use this sequence for obtaining the electron configuration of a diatomic molecule, we first imagine that the bare nuclei are fixed at their equilibrium separation. Then we add electrons, one at a time, to build up the configuration by employing the same rules we applied to atoms: (1) An electron will go into

Chemical Bonding and Valence

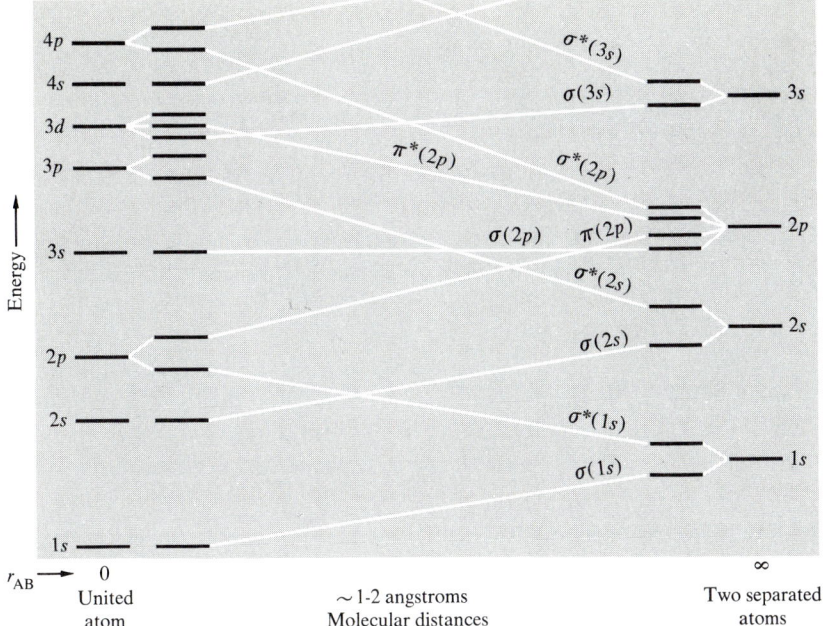

Figure 7-9. Schematic correlation diagram for homonuclear diatomic molecules. The energy levels of the united atom are lower than the corresponding ones of the separated atoms because the nuclear charge of the united atom is higher. The correlation lines connect atomic states of the same symmetry. Correlation lines do not cross if both are associated with states of σ, σ^*, π, or π^* symmetry.

the available MO of lowest energy. (2) Because of the Pauli exclusion principle, one MO can hold two electrons at most. (3) If there is a degenerate set of MO's, one electron will go into each of these until each orbital in the set is singly occupied, and all of these electrons have parallel spins (Hund's rule).

Let us apply these rules to the ground states of the diatomic molecules of a few typical elements. We have already seen that in the ground state of H_2 the two electrons are in the lowest bonding MO. This electron configuration is represented by the symbol $(\sigma 1s)^2$. When we consider the possible formation of He_2, we see that if it existed its configuration would be $(\sigma 1s)^2(\sigma^*1s)^2$. This configuration, however, implies that no bonding occurs because the bonding effect of the two electrons in the bonding MO is cancelled by the antibonding effect of the two electrons in the antibonding MO. It follows that when two He atoms collide, they repel each other.

In contrast, we would expect He_2^+ to have some stability because it has only one electron in an antibonding MO. This conclusion is confirmed by spectroscopic evidence for the existence of He_2^+ in electric discharges in helium.

In the lithium atom, the two $1s$ electrons are in a closed shell and the corresponding energy level is much lower than that of the $2s$ electron. Also, because of the $3+$ charge on the Li nucleus the average distance of a $1s$ electron from the nucleus is smaller than the average distance of the hydrogen $1s$ electron from the proton. It follows that when two Li atoms are close together there is no significant interaction between the $1s$ electrons of one atom and any of the electrons of the other atom. The $2s$ orbitals of the two Li atoms, however, can combine to form MO's, and two electrons can occupy the resulting $\sigma 2s$ MO. Consequently we expect the Li_2 molecule to be stable, in agreement with the fact that it is observed in lithium vapor. Because the bond between the Li atoms is attributed to the presence of two electrons in a σ bonding MO it is often called a σ bond. Since the $1s$ electrons in the Li_2 molecule remain essentially unaffected by the formation of the bond, and thus retain their atomic character, the electron configuration of this molecule is usually written as $KK(\sigma 2s)^2$ where each K represents the two electrons of one atom in the K shell ($n = 1$).

These conclusions about the formation of Li_2 molecules can be extended to the other alkali metals. Since the atom of each of these elements has an s electron outside closed shells, and the electrons in closed shells do not participate significantly in bonding, all of the alkali metals would be expected to form diatomic molecules whose bonds are σ bonds, and it is found that this is indeed the case. The configuration of Na_2, for example, is $KKLL(\sigma 3s)^2$, where L represents the eight electrons in the closed shell with $n = 2$.

When, however, we consider the possible existence of Be_2, we see that the four valence electrons would fill both the bonding and the antibonding $\sigma 2s$ MO's. Therefore we conclude that Be_2 should not exist. It has never been observed.

Turning to the other side of the Periodic Table, we come to fluorine. Since the F atom has 7 electrons outside the closed shell, we must accommodate 14 electrons in the F_2 molecule (Fig. 7–10). Accordingly, we write its configuration as

$$KK(\sigma 2s)^2(\sigma^* 2s)^2(\sigma 2p_x)^2(\pi 2p_y)^2(\pi 2p_z)^2(\pi^* 2p_y)^2(\pi^* 2p_z)^2$$

Both the bonding and the antibonding $\sigma 2s$, $\pi 2p_y$, and $\pi 2p_z$ orbitals are filled and the corresponding pairs of orbitals effectively cancel each other. This leaves the pair of electrons in the $\sigma 2p_x$ MO as the source of the single bond between the F atoms. Since the other halogen atoms differ from the F atom only in the number of closed shells, the electron configurations of all of the halogen molecules are similar.

If we now imagine that one electron is removed from each of the orbitals of highest energy (the π^* orbitals) in the F_2 configuration, we obtain the configuration of the electrons in the O_2 molecule. In this configuration there are four more electrons in bonding orbitals than in antibonding orbitals. By associating two electrons with a single bond, we can say that the atoms in O_2 are

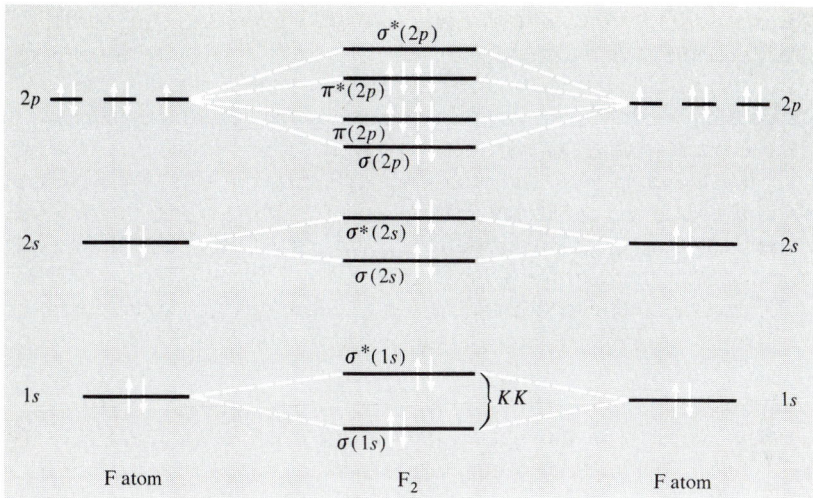

Figure 7-10. Energy level diagram of the F_2 molecule showing occupied molecular orbitals.

bonded by a double bond. Since two of the bonding electrons are in a σ orbital, ($\sigma 2p_x$), and two are in a pair of π orbitals (πp_y and πp_z), the double bond is said to consist of a σ bond and a π bond.

◈◈◈◈◈◈

According to Hund's rule, in O_2 there is one electron in each of the $\pi^* p_y$ and $\pi^* p_z$ orbitals and these electrons have parallel spins. An important consequence of the presence of these unpaired electrons is the fact that O_2 is paramagnetic; that is, it is attracted to a magnetic field. The S_2 molecule, found in sulfur vapor, is also paramagnetic for the same reason.

◈◈◈◈◈◈

By removing the two electrons in the π^* orbitals of O_2 we arrive at the configuration of N_2. In this case there is an excess of six electrons in bonding MO's, with two in each of the π orbitals. Hence we can say that N_2 has a triple bond which consists of one σ bond and two π bonds. In the formation of a N_2 molecule from two N atoms, all of the electrons which originally were in $2p$ AO's go into bonding MO's. Consequently, we would expect the bonding energy of N_2 to be high relative to that of O_2 or F_2. This conclusion is supported by the data in Table 7–1 from which the effect of electrons in antibonding MO's on bonding energy is evident.

For the consideration of heteronuclear diatomic molecules, the concepts developed above are still applicable but some of the details are different. The MO's for the molecule AB can be expressed in terms of the AO's of atoms A and B in the same way that we used before (Eq. 7–2), but because the AO's in this case are associated with different kinds of atoms the ratio of the coefficients, c_A/c_B, is different from 1. The best value of this ratio and the corresponding energy for the ground state MO are found from the solution of the appropriate secular equation (Eq. 7–5). Then an approximate expression

TABLE 7-1 BONDING ENERGIES OF SOME DIATOMIC MOLECULES AND MOLECULAR IONS*
(in electron-volts)

N_2	9.76
$N_2{}^+$	6.34
O_2	5.08
$O_2{}^+$	6.48
NO	6.49
F_2	2.2

*G. Herzberg, *Spectra of Diatomic Molecules*, 2nd ed., Van Nostrand, Princeton, 1950.

for the distribution of electrons in this MO can be found by evaluating ψ_{MO}^2. The larger the value of c_A/c_B, the closer to atom A is the most probable position of an electron in this MO. Thus the magnitude of the ratio of the coefficients is related to the polarity of the bond in AB. Unfortunately, accurate calculations of this ratio are very difficult.

The relative energies of the MO's can be found from a consideration of the "united atom" and interpreted spectra, as in the case of homonuclear molecules. In describing the electron configuration of a heteronuclear molecule, however, we cannot associate an MO with a particular kind of AO as we did for homonuclear molecules. For this purpose the following notation, introduced by Mulliken, is often used: The MO's are simply labeled in the order of increasing energy by the letters of the alphabet in reverse order, omitting the MO's involving the 1s orbitals of atoms other than hydrogen. Thus the sequence of MO's corresponding to the set on page 230 is

$$z\sigma < y\sigma < x\sigma < w\pi < v\pi < u\sigma$$

The MO's $z\sigma$, $x\sigma$, and $w\pi$ are bonding; $y\sigma$, $v\pi$, and $u\sigma$ are antibonding. We can illustrate the use of this notation with nitric oxide, NO, as an example. Its electron configuration is

$$KK(z\sigma)^2(y\sigma)^2(x\sigma)^2(w\pi)^4 v\pi$$

Because NO is one of the rare molecules with an odd number of electrons, it has an unpaired electron spin and is therefore paramagnetic. Since there is one electron in the antibonding $v\pi$ orbital, the binding energy of NO would be expected to be less than that of N_2, in agreement with experimental data (Table 7–1).

Another example of a heteronuclear molecule is hydrogen fluoride, HF. This molecule has ten electrons, four of which are in low-energy $1s$ and $2s$ AO's concentrated around the fluorine nucleus. We have already seen that an s orbital on one atom cannot combine with a p_y or p_z orbital on another atom. Consequently the bond in HF can be ascribed to the presence of two electrons in the bonding MO formed from the $1s$ AO of hydrogen and the $2p_x$ AO of fluorine, and it is thus a σ bond. The remaining four electrons are essentially unchanged $2p_y$ and $2p_z$ electrons of the fluorine atom.

In recent years quite accurate wave functions for simple molecules have been calculated by the use of large electronic computers. It has been found possible to combine such a computer with a curve plotter to draw automatically a set of quantitative contour lines of electron density calculated from the individual molecular orbitals or calculated for the molecule as a whole. Figure 7–11 includes a set of such plots for the oxygen molecule.

Valence Bond Method for Diatomic Molecules

We now turn to the valence bond (VB) method which was mentioned on page 219. The basic goal of this method is the same as that of the MO method, namely the interpretation of chemical bonding on the basis of an approximate solution for the Schrödinger equation for a molecule. The VB method starts from a consideration of the separate atoms between which a bond is formed and the pairing of electrons of opposite spin as the atoms approach each other.

Since the VB method is not used at present as widely as the MO method, a fairly brief discussion of it is sufficient here. Again let us use the H_2 molecule as an example. By regrouping the terms in the Hamiltonian operator for this molecule (Eq. 7–1), we can write it in the following form:

$$H_{op} = \left(-\frac{h^2}{8\pi^2 m}\nabla_1^2 - \frac{\epsilon^2}{r_{A1}}\right) + \left(-\frac{h^2}{8\pi^2 m}\nabla_2^2 - \frac{\epsilon^2}{r_{B2}}\right) + \left(-\frac{\epsilon^2}{r_{A2}} - \frac{\epsilon^2}{r_{B1}} + \frac{\epsilon^2}{r_{12}} + \frac{\epsilon^2}{r_{AB}}\right)$$

When the three groups of terms in parentheses are represented by $H_A(1)$, $H_B(2)$, and H', respectively, this equation becomes

$$H_{op} = H_A(1) + H_B(2) + H' \qquad (7-11)$$

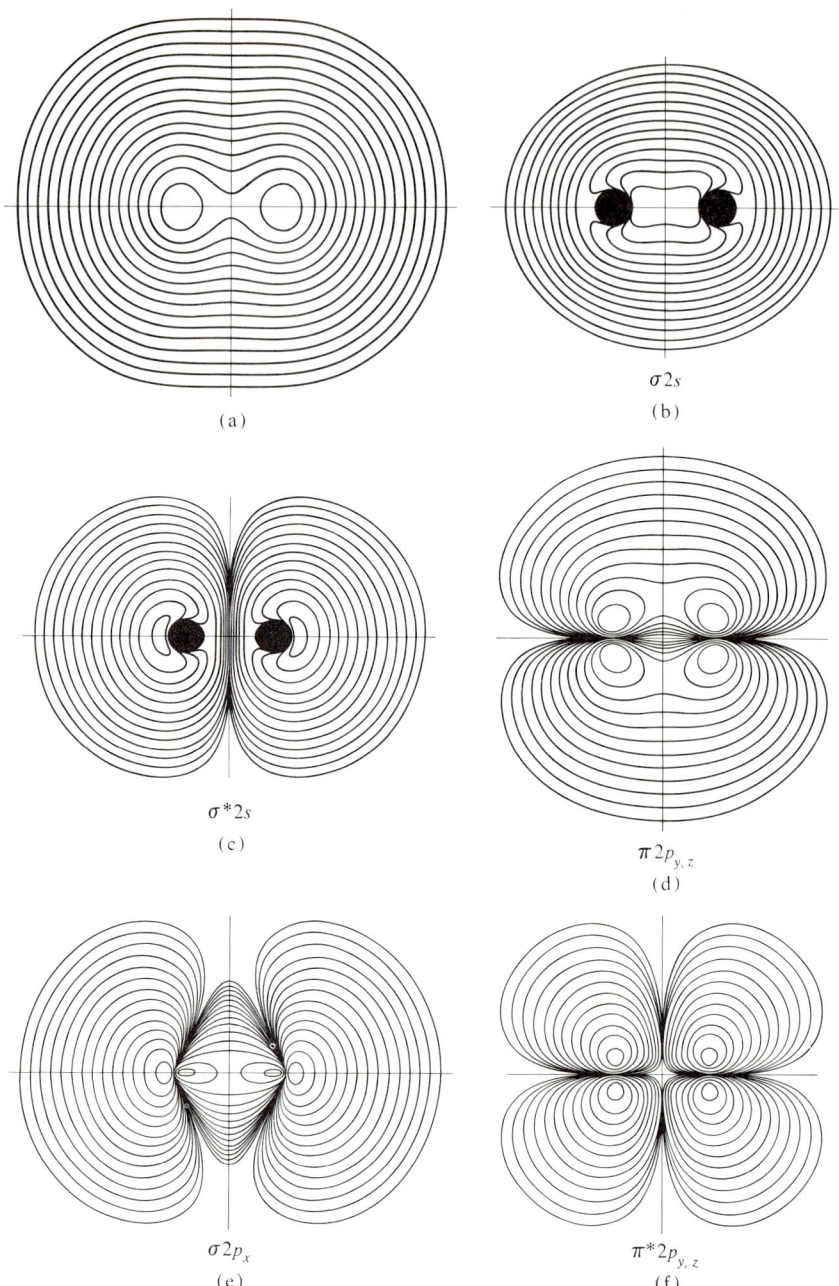

Figure 7–11. Plots of contour lines of electron density for the oxygen molecule and its molecular orbitals. (a) Total electron density; (b) $\sigma 2s$; (c) $\sigma^* 2s$; (d) $\pi 2p_{yz}$; (e) $\sigma 2p_x$; (f) $\pi^* 2p_{yz}$. (Reproduced by permission from Arnold C. Wahl, *Science*, **151**, 961 1966). For more information, see *Atomic and Molecular Structure: A Pictorial Approach*, Arnold C. Wahl and Maria T. Wahl, McGraw-Hill, New York, 1969.)

$H_A(1)$ is identical with the Hamiltonian operator for an isolated hydrogen atom consisting of proton A and electron 1, and $H_B(2)$ is the Hamiltonian operator for a hydrogen atom consisting of proton B associated with electron 2. If the two hydrogen atoms are assumed to be far apart, all of the terms grouped in H' vanish. Consequently the Hamiltonian operator for the system of two isolated hydrogen atoms is $H_A(1) + H_B(2)$. The total energy is then the sum of the energies of the separate atoms, and the wave function for this system is the product of the atomic orbitals of the separate atoms. Representing the $1s$ AO's of atoms A and B by $s_A(1)$ and $s_B(2)$, respectively, we have

$$\psi = s_A(1)s_B(2)$$

Now let us imagine that the two hydrogen atoms approach each other until their separation is the same as in a normal H_2 molecule. The term H' in Equation 7–11 is then different from zero. What is the wave function for this system? We might be tempted to say that it is still approximately $s_A(1)s_B(2)$, but when the two hydrogen atoms are close together we cannot distinguish electron 1 from electron 2. Consequently the function $s_A(2)s_B(1)$ is just as appropriate as the original one, with which it forms a degenerate pair, and we express the approximate molecular wave function as a linear combination of the two degenerate functions $\psi_I [= s_A(1)s_B(2)]$ and $\psi_{II} [= s_A(2)s_B(1)]$ which we write as follows:

$$\psi_{mol} = c_I \psi_I + c_{II} \psi_{II}$$

Next we can use the variation method to obtain the best values for the ratio c_I/c_{II}. The resulting secular equation looks very much like Equation 7–5, but the integrals represented in it are different. Actually, in this special case we do not need to solve the secular equation. Since ψ_I and ψ_{II} have the same mathematical form, it follows that $c_I^2 = c_{II}^2$ and $c_I/c_{II} = \pm 1$. Corresponding to these two values of the coefficient ratio, we obtain the following expressions for the molecular wave function:

$$\psi_+ = \psi_I + \psi_{II} \qquad (7\text{–}12)$$

and

$$\psi_- = \psi_I - \psi_{II}$$

When the expressions for the energies, $\mathcal{E}_+$ and $\mathcal{E}_-$, associated with these wave functions are obtained, it can be shown that $\mathcal{E}_+$ is always lower than $\mathcal{E}_-$. Thus ψ_+ represents the simplest VB wave function for the ground state of H_2. For this simple molecule a numerical value for $\mathcal{E}_+$ can be calculated. This value is somewhat closer to the experimental value than the one calculated by the use of the LCAO–MO method, but it is still only a fair approximation.

In order to obtain a better energy value, a wave function is used which contains terms in addition to ψ_+. For example, one of the functions used is the following:

$$\psi = \psi_+ + c[s_A(1)s_A(2) + s_B(1)s_B(2)] \quad (7\text{--}13)$$

The value of the coefficient c determined by the variation method is 0.25. Although the energy value associated with this wave function is only slightly better than ε_+, even better values have been obtained by the use of still more complicated functions, but we do not pursue this point any further.

In the context of the VB method, a wave function in which the overlap of AO's of particular electrons on particular atoms is implied is said to be the wave function of a "*structure*." A wave function like ψ_+, which implies that one electron is associated with each of the bonded atoms, is called the wave function for a covalent structure. A function like $s_A(1)s_A(2)$, which implies that both electrons are associated with one atom, is called the wave function for an ionic structure which for hydrogen can be represented by $H^- - H^+$. The complete VB wave function for a molecule is formed from a linear combination of wave functions of structures. Pauling interpreted the type of molecular wave function in Equation 7–13 as representing *resonance* between covalent and ionic structures.

For diatomic molecules in general, we can say that the bond is formed as a result of the overlap of AO's for electrons which are unpaired in the separate atoms. In the case of fluorine, for example, each atom has one unpaired electron in an orbital that we can take to be the $2p_x$ orbital. Then by analogy with H_2 the simplest molecular wave function can be written in the following way:

$$\psi = p_x(1)p_x(2) + p_x(2)p_x(1)$$

For homonuclear molecules, the inclusion of terms for ionic structures has little effect on the associated energy. For heteronuclear molecules, however, the contributions of the ionic structures are important and must be included in the molecular wave function. Thus for hydrogen fluoride, in which the bond is attributed to the overlap of the $1s$ AO of hydrogen with the $2p_x$ orbital of F, the VB wave function is

$$\psi = [s(1)p_x(2) + s(2)p_x(1)] + c[p_x(1)p_x(2)]$$

The conceivable ionic structure $s(1)s(2)$ is omitted because, in view of the large electronegativity of F relative to H, a structure such as H^-F^+ would make a negligible contribution to the molecular wave function.

Valence Bond Wave Functions Including Electron Spin

Since the VB wave functions involve the AO's of two electrons, there is an analogy with the treatment of the helium atom (p. 211), for which the two

AO's are associated with the same atom. Let us illustrate this point by returning to the treatment of H_2. The ground state wave function represented by ψ_+ (Eq. 7–12) is symmetric for the exchange of the two electrons. The Pauli exclusion principle (p. 211) requires that the complete wave function, including the spin factor, be antisymmetric for the exchange of two electrons. It follows that the spin factor must be antisymmetric, which implies that the two electrons which are "paired" to form the bond must have opposite spins. If as before we represent by α and β the spin functions corresponding to the values of $+\frac{1}{2}$ and $-\frac{1}{2}$, respectively, for the spin quantum numbers, we can write the complete VB wave function for the ground state of H_2 in the following form:

$$\psi_{mol} = [s_A(1)s_B(2) + s_A(2)s_B(1)][\alpha(1)\beta(2) - \alpha(2)\beta(1)]$$

Similarly, the excited state wave function which we represented by ψ_- is antisymmetric. Consequently the spin factor must be symmetric. There are three different ways in which a symmetric spin factor can be written, and these lead to the three following complete wave functions:

$$\psi_{mol} = [s_A(1)s_B(2) - s_A(2)s_B(1)] \begin{cases} \alpha(1)\alpha(2) \\ \beta(1)\beta(2) \\ \alpha(1)\beta(2) + \alpha(2)\beta(1) \end{cases}$$

These three functions form a degenerate set in the absence of external fields and thus the excited state of the H_2 molecule associated with ψ_- is a triplet state.

Polyatomic Molecules

In our treatment of polyatomic molecules we must be content with cruder approximations than we used for diatomic molecules. We are not concerned with detailed calculations of molecular energy, but rather with the more qualitative aspects of chemical bonding, including consideration of the arrangement in space of the atoms (directed valence).

Using the LCAO–MO method we assume that the nuclei are in their equilibrium positions and that they are surrounded by those of their electrons that are in filled atomic shells. Having formed MO's for the molecule as a whole by linear combinations of all of the AO's for the valence electrons, we proceed to add valence electrons which occupy these MO's in the order of increasing energy. In this procedure, we have "lost" the concept of a chemical bond between particular atoms in the molecule, and this is one of the basic concepts in chemistry. Fortunately, it is possible to show that for many polyatomic molecules a mathematical transformation can be carried out which replaces the set of MO's which extend over a whole molecule by an equivalent set of *bond* (or *localized*) MO's formed by linear combinations of the AO's of the bonded atoms. These bond MO's are essentially like those of diatomic molecules. We restrict our attention in this section to those molecules for which such a transformation can be carried out. The others are treated in the next section.

The bond MO's introduced here are similar to VB wave functions. Since for both methods the requirement for bonding between atoms is the overlapping (or combination) of suitable AO's, it makes no difference in the remainder of this section which method we use. We extend our previously developed theory by the following postulate due to Pauling: A bond will be formed in the direction in which maximum overlapping of orbitals occurs. This statement implies that the further the contour surface of an AO extends in a certain direction, the better will be the bond formed in that direction because of the more favorable overlap with the orbital of another atom.

Let us take the water molecule as a simple example of a polyatomic molecule. The oxygen atom has four electrons in its three p orbitals. One of these orbitals is filled by two electrons, while the other two orbitals hold 1 electron each. These two partly filled p orbitals can be overlapped by the s orbitals of two hydrogen atoms located on the axes of the p orbitals. Thus we can account for the fact that an oxygen atom forms bonds with just two hydrogen atoms. Because the axes of the p orbitals are perpendicular to each other, we would expect the two identical O–H bonds to form a right angle. Qualitatively, this expectation is confirmed by experimental methods described in the next chapter. The water molecule does have the shape of an isosceles triangle, but the valence angle is 104.5° instead of the predicted 90°.

This discrepancy in the magnitude of the valence angle is somewhat disconcerting. When we consider H_2S, however, the bonds of which involve the same type of AO's as H_2O, the experimental valence angle turns out to be 92°. We conclude that although our theory leads to reasonably good results it needs some further attention from a quantitative point of view.

Turning to the nitrogen atom, we see that it has three partly filled $2p$ orbitals with their axes at right angles to each other. When these orbitals are overlapped by the s orbitals of three hydrogen atoms, we have the three N–H bonds of ammonia. We would predict that NH_3 has the shape of a regular pyramid with the N atom at the vertex, and that the angle between any two N–H bonds is 90°. Several different types of measurements, however, lead to a bond angle close to 107°, a larger discrepancy than before. For PH_3, which involves the same type of AO's as NH_3, the experimental value of the bond angle is 93°. Again we have a mixture of qualitative agreement and quantitative disagreement.

When we come to the carbon atom, we must conclude that something is wrong or incomplete in our theory. The electron configuration of an isolated carbon atom is $1s^2 2s^2 2p^2$. Since this configuration has two partly filled p orbitals, on the same basis as before we would predict the existence of a triatomic CH_2 molecule with a bond angle of 90°, in glaring disagreement with the facts about compounds of hydrogen and carbon.

The difficulty with our theoretical treatment lies in the assumption that the orbitals to be used in forming bond orbitals are the AO's of isolated atoms. This is the simplest approach, but now we see that it is so crude that for carbon compounds it yields a result that is qualitatively incorrect. Pauling improved the theory by showing that it is possible to obtain more effective bonds and, in some cases, a larger number of bonds by forming suitable linear combinations of the AO's of a single atom. Such linear combinations of AO's are called *hybrid orbitals*. A hybrid orbital can be overlapped by an orbital of another atom to form a better bond than could be formed from the AO's themselves. The number of hybrid orbitals formed from a set of AO's is equal to the number of AO's in the set.

Pauling applied the concept of orbital hybridization to the methane molecule in the following way: By adding a moderate amount of energy to a carbon atom in its ground state, one of the $2s$ electrons can be "promoted" to the $2p$ energy level. In this state the atom has four unpaired electrons and thus it

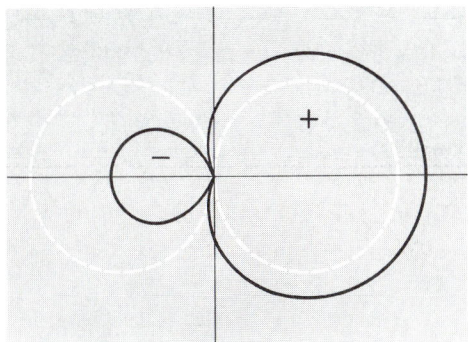

Figure 7–12. Contour of sp^3 hybrid orbital in a plane passing through its axis. The dotted line is a graph of the angular factor of $2p$ atomic orbital on the same scale.

could form four bonds. These bonds, however, would not be equivalent because three of them would involve electrons in p orbitals and one would involve the remaining electron in an s orbital. In contrast, by forming linear combinations of one s orbital and three p orbitals four hybrid orbitals, known as sp^3 hybrid orbitals, can be obtained which are all equivalent, have their axes pointing toward the vertices of a regular tetrahedron, and have shapes (contour surfaces) that are well suited for overlap with orbitals of other atoms, such as hydrogen atoms (Fig. 7–12). Pauling showed that the energy lowering as a result of the formation of four strong bonds, as in methane, more than compensates for the promotion energy of the $2s$ electron, mentioned above, in agreement with the known stability of the methane molecule.

The simplest way of representing the four sp^3 hybrid orbitals is the following:

$$\begin{aligned}
\psi_1 &= \tfrac{1}{2}(s + p_x + p_y + p_z) \\
\psi_2 &= \tfrac{1}{2}(s + p_x - p_y - p_z) \\
\psi_3 &= \tfrac{1}{2}(s - p_x + p_y - p_z) \\
\psi_4 &= \tfrac{1}{2}(s - p_x - p_y + p_z)
\end{aligned} \qquad (7\text{--}14)$$

(The factor of $\tfrac{1}{2}$ is a normalization constant.) It can be shown that these four hybrids all have the same spatial form, and that the angles between their axes are the angles formed by lines from the center of a regular tetrahedron to any two of its vertexes (109°28′).

In order to derive a set of four hybrid orbitals by forming linear combinations of one s and three p orbitals (each assumed to be normalized) we begin by representing the hybrid orbitals in the following general way:

$$\psi_i = a_i s + b_i p_x + c_i p_y + d_i p_z$$

in which a, b, c, and d are coefficients to be determined, and the subscript i is given in succession the values of the integers from 1 through 4. We want these hybrid orbitals to be normalized and orthogonal (p. 193) to each other. (Orthogonality is the mathematical requirement for the mutual independence of these hybrids.) In the present case, orthogonality means that $\int \psi_i \psi_k \, d\tau = 0$, where i and k refer to any two different hybrids. Thus,

$$\int \psi_i \psi_k \, d\tau = 0$$

$$= a_i a_k \int s^2 \, d\tau + b_i b_k \int p_x^2 \, d\tau + c_i c_k \int p_y^2 \, d\tau + d_i d_k \int p_z^2 \, d\tau$$

$$+ a_i b_k \int s p_x \, d\tau + a_i c_k \int s p_y \, d\tau + a_i d_k \int s p_z \, d\tau$$

$$+ b_i c_k \int p_x p_y \, d\tau + b_i d_k \int p_x p_z \, d\tau + c_i d_k \int p_y p_z \, d\tau$$

The integrals involving a product of different AO's are each equal to zero because the AO's themselves are individually orthogonal to each other. The integrals involving the square of an orbital are each equal to 1 because the AO's are normalized. It follows that the condition for orthogonality of the two hybrids is

$$a_i a_k + b_i b_k + c_i c_k + d_i d_k = 0 \qquad (7\text{--}15)$$

Similarly, the condition for a hybrid to be normalized is

$$\int \psi_i^2 \, d\tau = 1$$

$$= a_i^2 \int s^2 \, d\tau + b_i^2 \int p_x^2 \, d\tau + c_i^2 \int p_y^2 \, d\tau + d_i^2 \int p_z^2 \, d\tau \qquad (7\text{-}16)$$

Since the AO's are normalized, we have

$$a_i^2 + b_i^2 + c_i^2 + d_i^2 = 1 \qquad (7\text{-}17)$$

Because the orientation of the x, y, and z axes is arbitrary, we are free to choose any convenient direction for the axis of the first hybrid to be derived. Let us assume that it is directed along the x axis. Then p_y and p_z do not make a contribution to ψ_1, and thus $c_1 = d_1 = 0$. The equation for ψ_1 reduces to

$$\psi_1 = a_1 s + b_1 p_x$$

From Equation 7–17, $a_1^2 + b_1^2 = 1$, and

$$b_1 = \sqrt{1 - a_1^2} \qquad (7\text{-}18)$$

To find the value of a_1, let us note that if we sum Equation 7–16 over all four hybrids we obtain

$$\Sigma \int \psi_i^2 \, d\tau = 4$$

$$= \Sigma a_i^2 \int s^2 \, d\tau + \Sigma b_i^2 \int p_x^2 \, d\tau + \Sigma c_i^2 \int p_y^2 \, d\tau + \Sigma d_i^2 \int p_z^2 \, d\tau$$

This equation represents the total probability of finding an electron in all four hybrids. This probability must be equal to the sum of the probabilities of finding an electron in the one s and the three p orbitals from which the hybrid orbital is formed. Since the AO's are normalized, these probabilities are all equal and thus

$$\Sigma a_i^2 = \Sigma b_i^2 = \Sigma c_i^2 = \Sigma d_i^2 = 1$$

Because the four hybrids are all equivalent and the s orbital is independent of direction, this orbital must contribute equally to each of the hybrids. Therefore $a_1^2 = a_2^2 = a_3^2 = a_4^2$, $4a_1^2 = 1$, and $a_1 = \frac{1}{2}$. When this value is substituted in Equation 7–18, it follows that $b_1 = \sqrt{3}/2$ and the equation for the first hybrid is

$$\psi_1 = \frac{1}{2} s + \frac{\sqrt{3}}{2} p_x \qquad (7\text{-}19)$$

We now derive the equation for ψ_2. The axes of ψ_1 and ψ_2 lie in a plane which we take to be the x,z plane; then $c_2 = 0$ and the general equation for ψ_2 is

$$\psi_2 = \tfrac{1}{2} s + b_2 p_x + d_2 p_z$$

Because ψ_1 and ψ_2 are orthogonal, it follows from Equation 7–15 that

$$\frac{1}{4} + \frac{\sqrt{3}}{2} b_2 = 0$$

and thus

$$b_2 = -\frac{1}{2\sqrt{3}}$$

From Equation 7–17

$$d_2^2 = 1 - a_2^2 - b_2^2$$
$$= \tfrac{2}{3}$$

and

$$d_2 = \sqrt{\tfrac{2}{3}}$$

We have thus obtained the following equation for ψ_2:

$$\psi_2 = \frac{1}{2} s - \frac{1}{2\sqrt{3}} p_x + \sqrt{\frac{2}{3}} p_z \tag{7-20}$$

In the same way we can obtain the following equations for ψ_3 and ψ_4:

$$\psi_3 = \frac{1}{2} s - \frac{1}{2\sqrt{3}} p_x + \frac{1}{\sqrt{2}} p_y - \frac{1}{\sqrt{6}} p_z$$

$$\psi_4 = \frac{1}{2} s - \frac{1}{2\sqrt{3}} p_x - \frac{1}{\sqrt{2}} p_y - \frac{1}{\sqrt{6}} p_z$$

The set of equations we have just derived for the four sp^3 hybrid orbitals is mathematically equivalent to the set in Equation 7–14. The only difference is the fact that the two sets are oriented differently with respect to the chosen coordinate axes. Because the orientation of our coordinate axes is arbitrary and has no physical significance, it follows that the hybrid orbitals can be expressed in an infinitely large number of different but mathematically equivalent ways.

To learn more about these hybrid orbitals, let us introduce into their equations approximate expressions for the AO's represented by s, p_x, p_y, and p_z. The radial factors of the s and p orbitals are not greatly different from each other. Following Pauling, we neglect the difference between these radial factors and focus on the normalized angular factors as given in Equation 6–34, because these factors determine the shapes of the hybrid orbitals. Since the s orbital is independent of angles, its contour surfaces are spheres. For the purpose of comparison, we take the radius of a spherical contour surface, representing the magnitude of the s orbital, to be equal to 1. Relative to this unit, the magnitudes of the p orbitals are given by the following equations:

$$\begin{aligned} p_z &= \sqrt{3} \cos \theta \\ p_x &= \sqrt{3} \sin \theta \cos \phi \\ p_y &= \sqrt{3} \sin \theta \sin \phi \end{aligned} \tag{7-21}$$

When the unit magnitude of the s orbital and the expression for p_x given above are introduced into Equation 7-19, this equation becomes

$$\psi_1 = \frac{1}{2} \cdot 1 + \frac{\sqrt{3}}{2} \sqrt{3} \sin \theta \cos \phi$$

The axis of ψ_1 is the x axis, for which $\theta = 90°$ and $\phi = 0°$. On substituting these values for the angles in the above equation for ψ_1, we obtain the following relative magnitude of ψ_1 in the x direction:

$$\psi_1 = \tfrac{1}{2} + \tfrac{3}{2} = 2$$

Since the relative magnitude of the p_x orbital in the x direction is $\sqrt{3}$ ($= 1.732$), we see that ψ_1 has a larger value than p_x. We interpret this result to mean that ψ_1 is better suited than p_x for overlapping the orbital of another atom located on the x axis, and thus ψ_1 can form a better bond with this atom than p_x itself (Fig. 7-12).

Turning now to ψ_2, we wish to find the direction in which it has a maximum value because, according to the postulate stated on page 240, that is the direction in which a bond resulting from the overlap of ψ_2 is formed. We assumed that the axis of ψ_2 lies in the x,z plane. With the value of 180° for ϕ in this plane, $\cos \phi = -1$, and thus $p_x = -\sqrt{3} \sin \theta$. Substituting in Equation 7-20 this expression for p_x, the expression in Equation 7-21 for p_z, and the value of 1 for s, we obtain the following equation for ψ_2 as a function of the angle θ:

$$\psi_2 = \tfrac{1}{2} + \tfrac{1}{2} \sin \theta + \sqrt{2} \cos \theta \tag{7-22}$$

To find the value of θ for which ψ_2 has a maximum magnitude, we proceed in the usual way by differentiating ψ_2 with respect to θ and then finding the value of θ for which the derivative is equal to zero. Thus

$$\frac{d\psi_2}{d\theta} = \frac{1}{2} \cos \theta - \sqrt{2} \sin \theta = 0$$

The solution of this equation is

$$\theta = \sin^{-1} \tfrac{1}{3} = 19°28'$$

Since θ is the angle between the axis of ψ_2 and the z axis, the angle between the axes of ψ_1 and ψ_2 is $90° + 19°28' = 109°28'$. This is just the tetrahedral angle mentioned before. When the procedure we used for ψ_2 is used for ψ_3 and ψ_4, it turns out that the angle between the axes of any two of the sp^3 hybrid orbitals is the tetrahedral angle.

When we substitute the value of 19°28' for θ in Equation 7-22 for ψ_2 to find the relative magnitude of this orbital in the direction of its axis, we obtain a value of 2 — the same as the relative magnitude of ψ_1. Similarly, it can be shown that ψ_3 and ψ_4 have the relative magnitude of 2 in the direction of their axes. Thus it is clear that all four of the sp^3 hybrid orbitals are equivalent.

◈◈◈◈◈◈

Knowing the shape and directions of the axes of the sp^3 hybrid orbitals, we can readily account for the observed tetrahedral structure of CH_4 in terms of the overlap of each of these orbitals with the s orbital of an H atom to form four identical C–H bonds. The carbon-to-carbon bond in ethane is attributed to the overlap of two sp^3 hybrid orbitals, one on each carbon atom, with their axes along the same line.

The NH_4^+ ion, which has the same number of electrons as CH_4, is known also to have the symmetry of a regular tetrahedron. When a proton is removed from NH_4^+ to form NH_3, the resulting molecule as noted above has bond angles of about 107° — only slightly less than the tetrahedral angle. This fact strongly suggests that in NH_3 we also have hybridization of one s and three p orbitals. In this case, three of the nitrogen hybrid orbitals overlap the s orbitals of three hydrogen atoms to form three N–H bonds, and the remaining hybrid orbital is occupied by a nonbonding pair of electrons. Because the four hybrid orbitals are not equivalent, however, a different "mixture" of s and p orbitals is involved than in the case of a tetrahedral molecule. In this way we can account for the deviation of the bond angles in ammonia from the tetrahedral value. We can interpret the valence angle in H_2O in an analogous fashion by considering hybridization of the oxygen AO's. For H_2O there are two pairs of nonbonding electrons ("lone pairs") in hybrid orbitals.

Hybridization is not limited to the combination of one s and three p AO's. For some molecules the number of bonds and the bond angles can be accounted for in terms of sp^2, or trigonal hybridization. With the same ideas and technique that we used to obtain the tetrahedral hybrids, it is not difficult to show that the three sp^2 hybrids are equivalent and have their axes in a plane with angles of 120° between them (Fig. 7–13).

Let us apply these results to a simple compound of boron in which it is trivalent,[2] such as BF_3. The electron configuration of the isolated boron atom in its ground state is $1s^2 2s^2 2p$. We can imagine that this atom is "prepared" for bond formation by assuming that one of the $2s$ electrons is promoted to a $2p$ orbital. Then sp^2 hybridization leads to three trigonal hybrids, each of which is overlapped by a partly filled $2p$ orbital of a fluorine atom. The resulting molecule of BF_3 would be expected to be planar, with bonds directed to the vertices of an equilateral triangle. This expectation is completely consistent with the known structure of BF_3.

The formation of many carbon compounds can also be attributed to the overlapping of trigonal hybrids of carbon. The simplest compound in this class is ethylene, C_2H_4. We can say that in this molecule a trigonal hybrid of one carbon atom overlaps a trigonal hybrid of another carbon atom to form a bond. The other four hybrids on the two carbon atoms at angles of 120° to the C–C bond are overlapped by the s orbitals of four hydrogen atoms, and

---

[2] The simplest compound of boron and hydrogen is not used as an example in this case because BH_3 dimerizes to B_2H_6.

thus form four C–H bonds. One p orbital remains on each of the carbon atoms with the axis of this orbital perpendicular to the plane of the corresponding trigonal hybrids. When the trigonal hybrids of both carbon atoms have their

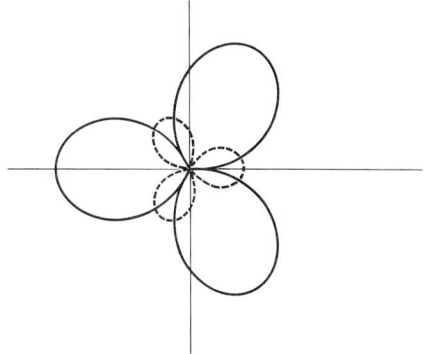

Figure 7–13. Contours of sp^2 hybrid orbitals.

axes in the same plane, the two p orbitals overlap each other, thus forming a bond MO.

We can now interpret the existence of a double bond in ethylene and in olefinic compounds in general. The bond MO that is formed as a result of the overlapping of the trigonal hybrids of the two carbon atoms is symmetrical about the bond axis, and thus the corresponding bond is a σ bond. The bond that is formed as a result of the overlapping of the p orbitals of the two carbon atoms is a π bond. Just as in the case of the O_2 molecule (p. 232), the carbon-to-carbon double bond consists of a σ bond and a π bond. On this basis we would expect the ethylene molecule to be planar, with the H–C–H angles close to 120°. Because of the additional bonding due to the π bond we would expect the distance between the carbon atoms in ethylene to be less than the corresponding distance in ethane. All of these expectations are confirmed by a variety of observations on ethylene and its simple derivatives.

Another type of hybridization is the combination of an s orbital with one p orbital, say the p_x orbital. The two resulting sp, or digonal, hybrids are given by the following equations:

$$\psi_1 = \frac{1}{\sqrt{2}}(s + p_x)$$

and

$$\psi_2 = \frac{1}{\sqrt{2}}(s - p_x)$$

These orbitals extend in opposite directions along the x axis, corresponding to $+p_x$ and $-p_x$, respectively.

The bonds in acetylene, C_2H_2, may be interpreted in terms of these digonal hybrids. The two carbon atoms form a σ bond by overlap of two digonal hybrids, and each of the other two hybrids forms a C–H σ bond by overlapping an s orbital of a hydrogen atom. The two unhybridized p orbitals on each carbon atom overlap to form two π bonds whose nodal planes are mutually perpendicular. This interpretation implies that the four nuclei in acetylene are located on a line, and this is confirmed by experimental data.

The bonding of atoms other than carbon can be interpreted in terms of digonal hybrids. Beryllium, for example, whose electron configuration is $1s^2 2s^2$ forms a chloride, $BeCl_2$, which has a linear structure in the vapor state. We can imagine that one of the $2s$ electrons is promoted to the $2p$ level and that the resulting sp hybrids overlap the partially filled p orbitals of two chlorine atoms. Likewise mercury forms linear dihalides, and in $Hg(CH_3)_2$ the three heavy nuclei have a linear arrangement.

Electron Delocalization and Conjugation

In the last section we considered the bonding in those molecules for which a mathematical transformation can be carried out from MO's for the molecule as a whole to localized MO's for the individual bonds. Now we take up the problem of those molecules for which this transformation cannot be carried out. Molecules, such as butadiene, $H_2C=C-C=CH_2$, that contain conju-
$\qquad\qquad\qquad\qquad\qquad\qquad\quad\; |\; |$
$\qquad\qquad\qquad\qquad\qquad\qquad\;\; H\; H$
gated double bonds are in this class. It was known for a long time that the properties of such molecules cannot be described simply in terms of single, double, and triple bonds. In fact, the existence of conjugation was a troublesome point in chemical theory before quantum mechanics was applied to valence problems.

We can best approach the problem of butadiene and related molecules by taking a closer look at the ethylene molecule. We saw that we could account for its structure in terms of σ bond orbitals formed from the combination of trigonal hybrids of carbon and the $1s$ AO's of hydrogen. In addition, there are two electrons in a π bonding orbital formed from the overlap of two carbon p_y orbitals (with the y axis perpendicular to the plane of the molecule). These π electrons are mainly responsible for the differences in physical and chemical properties that distinguish ethylene from ethane, and unsaturated compounds from saturated compounds in general.

It is a good approximation to assume that we can focus our attention on the

π electrons without explicit consideration of the σ electrons. This assumption greatly simplifies our problem for it means that we can treat the two π electrons in ethylene in a way that is similar to our treatment of the two electrons in H_2 by the LCAO–MO method. Accordingly, we express the π MO's of ethylene as a linear combination of carbon p_y AO's in the following way (compare Eq. 7–2):

$$\psi_\pi = c_A p_{y_A} + c_B p_{y_B}$$

On applying the variation method to this function, we obtain the following secular equation which is the analog of Equation 7–5:

$$\begin{vmatrix} \alpha - \varepsilon & \beta \\ \beta & \alpha - \varepsilon \end{vmatrix} = 0$$

In this equation the quantities α and β[3] are the analogs of the coulomb integral H_{AA} (Eq. 7–3) and the resonance integral H_{AB}, respectively, and the overlap integral S_{AB} has been assumed to be equal to zero.[4] In the case of H_2 all of these integrals can be evaluated, but here the computational difficulties are so great that no attempt is made to carry out this evaluation. Instead the integrals are treated like semiempirical parameters. We return to this point later.

The solutions of the secular equation above are $\varepsilon_+ = \alpha + \beta$ and $\varepsilon_- = \alpha - \beta$. The quantities α and β, like H_{AA} and H_{AB} for the hydrogen molecule, are both negative numbers, and thus ε_+ represents a lower energy level than ε_-. The lower energy level is associated with a value of the coefficient ratio c_A/c_B of $+1$ and ε_- is associated with a value of -1. As before, we call the corresponding wave functions bonding and antibonding MO's, respectively. To represent these MO's in normalized form, the sum of the squares of c_A and c_B must equal 1. It follows that the two π MO's in ethylene are given by

$$\psi_\pi = \frac{1}{\sqrt{2}} p_{y_A} \pm \frac{1}{\sqrt{2}} p_{y_B} \qquad (7\text{–}23)$$

in which the $+$ sign is used for the bonding MO. In the ground state of ethylene, the two π electrons are in the bonding π MO, and their total energy is represented by $2\alpha + 2\beta$.

Now let us turn to the butadiene molecule. We can describe its structure in terms of sp^2 hybridization for each carbon atom and a resulting σ bond

[3] α and β are defined by $\int p_{y_A} H_{\text{eff}} p_{y_A} \, d\tau$ and $\int p_{y_A} H_{\text{eff}} p_{y_B} \, d\tau$, respectively, where H_{eff} is an effective Hamiltonian operator which includes terms related to the carbon nuclei and the averaged effects of all of the σ electrons.

[4] In a basic sense, this assumption is illogical for if S_{AB} actually had a value of zero, β would also be zero, the AO's would not overlap, and there would be no bond. What is meant here is that S_{AB} is so small that it can be neglected relative to 1 in this approximate treatment.

framework that contains 18 valence electrons. This leaves one p orbital on each carbon atom and these four p AO's overlap to form four π MO's. For effective overlap to be possible the axes of the p orbitals must all be parallel, and this implies a planar structure for the butadiene molecule (Fig. 7-14). There is experimental evidence for this planar structure.

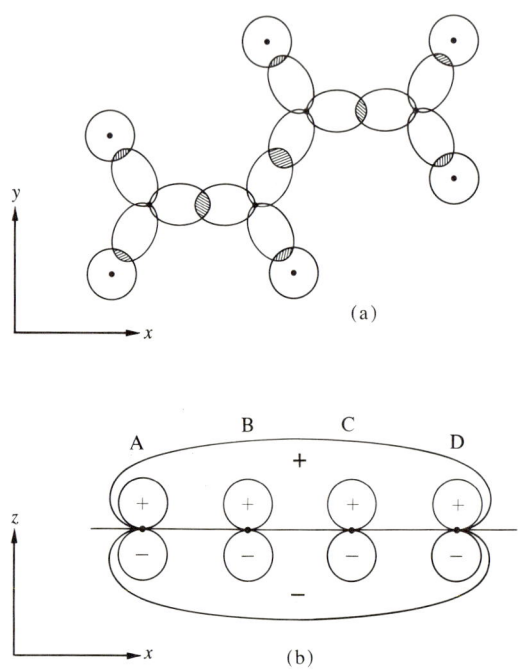

Figure 7-14. (a) Contours in the x,y plane for σ bond framework for butadiene. (b) Contour in the x,z plane for π molecular orbital formed from the combination of four p_z atomic orbitals.

Next, we obtain the four π MO's as linear combinations of the p orbitals associated with each of the four carbon atoms which we designate by the letters A, B, C, and D. We write the general expression for the π MO's in the following way:

$$\psi_i = c_{iA}p_A + c_{iB}p_B + c_{iC}p_C + c_{iD}p_D$$

in which $i = 1, 2, 3$, or 4. Again we employ the variation method to obtain the best values of the coefficients. In this case, however, before writing the resulting secular equation, we introduce the following simplifying assumptions which, collectively, are known as the *Hückel approximation:*

1. All coulomb integrals are assumed to have the same value, α.
2. All resonance integrals are set equal to zero, except those involving adjacent atoms which have the value β.
3. All overlap integrals are set equal to zero.

These are rather drastic approximations, but it turns out that they yield useful results. When they are applied to butadiene, the secular equation has the following form:

$$\begin{vmatrix} \alpha - \varepsilon & \beta & 0 & 0 \\ \beta & \alpha - \varepsilon & \beta & 0 \\ 0 & \beta & \alpha - \varepsilon & \beta \\ 0 & 0 & \beta & \alpha - \varepsilon \end{vmatrix} = 0$$

For convenience in working with such secular equations, each element in the determinant can be divided by β. With $x = (\alpha - \varepsilon)/\beta$, the equation becomes

$$\begin{vmatrix} x & 1 & 0 & 0 \\ 1 & x & 1 & 0 \\ 0 & 1 & x & 1 \\ 0 & 0 & 1 & x \end{vmatrix} = 0$$

When the determinant is expanded, we obtain the following quartic equation:

$$x^4 - 3x^2 + 1 = 0$$

Because this equation happens to be a biquadratic, it can be solved by the elementary method for quadratic equations. The four roots are: $x = \pm 1.618$, ± 0.618. The corresponding four values of ε are: $\alpha \pm 1.618\beta$, $\alpha \pm 0.618\beta$. These are the energy levels of the four π MO's which are represented in Figure 7-15. The two MO's with lowest energy are bonding MO's, and the other two are antibonding. Since there are four π electrons in butadiene and two electrons can occupy one MO, we can conclude that the four π electrons occupy the two bonding MO's, whose energies are $\alpha + 1.618\beta$ and $\alpha + 0.618\beta$, respectively. Then the total energy of the π electrons is

$$2(\alpha + 1.618\beta) + 2(\alpha + 0.618\beta) = 4\alpha + 4.472\beta$$

Although we do not know the numerical value of either α or β, we can compare our calculated π energy with the value we would obtain if the p orbitals on carbon atoms B and C did not overlap each other. In this case we would have the equivalent of two separate ethylene molecules whose combined π energy is $2(2\alpha + 2\beta)$. The difference between the π energy of butadiene and that of two ethylene molecules is 0.472β, which is a negative number. This energy lowering of the two conjugated double bonds in butadiene relative to two isolated double bonds is attributed to the fact that the MO's in butadiene

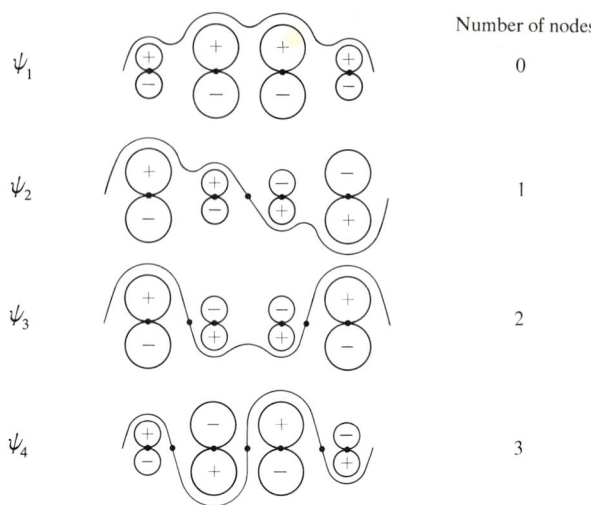

Figure 7–15. Schematic graph of four π molecular orbitals of butadiene.

extend over the whole molecule, and thus the π electrons are not associated with any particular bond. The energy lowering due to this delocalization of the π electrons is called the *delocalization energy*.

The coefficients in the equations for the normalized π MO's can be calculated from the 4 roots of the secular equation. (This calculation is outlined on p. 262.) The following equations, written in the order of increasing energy, are obtained:

$$
\begin{array}{ll}
\varepsilon & \\
\alpha + 1.618\beta & \psi_1 = 0.372 p_A + 0.602 p_B + 0.602 p_C + 0.372 p_D \\
\alpha + 0.618\beta & \psi_2 = 0.602 p_A + 0.372 p_B - 0.372 p_C - 0.602 p_D \\
\alpha - 0.618\beta & \psi_3 = 0.602 p_A - 0.372 p_B - 0.372 p_C + 0.602 p_D \\
\alpha - 1.618\beta & \psi_4 = 0.372 p_A - 0.602 p_B + 0.602 p_C - 0.372 p_D
\end{array}
\quad (7\text{-}24)
$$

These MO's are represented schematically in Figure 7–15. Observe that the number of nodes increases from 0 to 3 with increasing energy.

Because the equations for the π MO's give us information, even though approximately, of the probability density and thus of the distribution of the electrons occupying these orbitals, these equations can be used to calculate several quantities of chemical interest. One of these quantities is the π *bond order* which is defined by Coulson in the following way: The order of a π bond between atoms j and k (P_{jk}) is the product of the coefficients of the AO's p_j and p_k in a particular π MO times the number of electrons in the MO (either 1 or 2), summed over all the occupied π MO's. Expressed as a mathematical formula, this definition is

Chemical Bonding and Valence

$$P_{jk} = \sum_i n_i c_{ij} c_{ik}$$

in which n_i is the number of electrons in the ith π MO.

To illustrate the use of this formula, let us first apply it to the π bond in ethylene. In this case there is only one bonding π MO and it is occupied by two electrons. Using the coefficients of the normalized MO in Equation 7-23, we find

$$P_{AB} = 2\,\frac{1}{\sqrt{2}}\,\frac{1}{\sqrt{2}} = 1$$

Thus we see that the above definition of π bond order leads to a value of 1 for the carbon-to-carbon bond in ethylene. We have seen that in addition to the π bond there is a σ single bond whose order we can take as 1. Thus the total bond order in ethylene is 2, which is consistent with the fact that this bond is a typical double bond.

In butadiene, the π bond order for the bond between carbon atoms A and B (or C and D) is $2 \cdot 0.372 \cdot 0.602 + 2 \cdot 0.602 \cdot 0.372 = 0.89$. Likewise, for the order of the π bond between the central carbon atoms we have $2 \cdot 0.602 \cdot 0.602 - 2 \cdot 0.372 \cdot 0.372 = 0.45$. Since each of the carbon-to-carbon bonds in butadiene also includes a σ single bond of order 1, we see that the total bond order of the end and central bonds is 1.89 and 1.45, respectively. This conclusion is in agreement with much information which indicates that the end bonds, represented as double bonds in the formula on page 248, are somewhat different from the typical double bond as found in ethylene. The central bond, formulated as a single bond on page 248, has some double bond character; that is, it behaves to some extent as a double bond.

We now apply the Hückel MO method to benzene which, ever since the work of Kekulé, has occupied an important place in theoretical chemistry. The benzene molecule is known to be planar with the symmetry of a regular hexagon. This shape is readily accounted for in terms of sp^2 hybridization for each carbon atom. The trigonal hybrids overlap similar hybrids of each adjacent carbon atom and also the $1s$ orbital of a hydrogen atom. In this way we obtain the σ bond framework for the benzene molecule with H–C–C and C–C–C angles of 120°.

The remaining six p AO's are oriented with their axes all parallel to each other and perpendicular to the plane of the molecule, and this plane is their common nodal plane (Fig. 7-16). Overlap of these six p orbitals results in six π MO's which extend all around the ring of carbon atoms on both sides of the nodal plane. As before, we can obtain equations for these MO's by forming linear combinations of the p AO's. The general equation can be written in the following form:

$$\psi_i = c_{iA} p_A + c_{iB} p_B + c_{iC} p_C + c_{iD} p_D + c_{iE} p_E + c_{iF} p_F$$

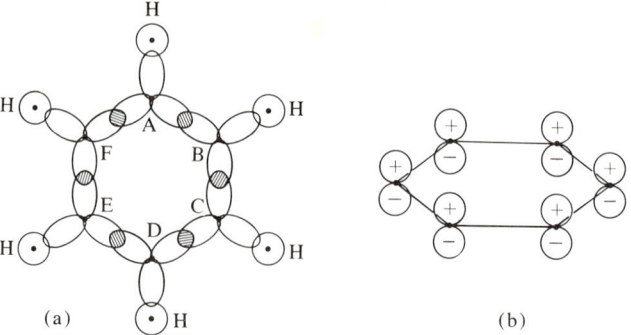

Figure 7-16. (a) σ bond framework for benzene. (b) Six p atomic orbitals which overlap to form π molecular orbitals.

We now apply the variation method to this trial function, again using the Hückel approximation and the abbreviation $x = (\alpha - \varepsilon)/\beta$. In this way the following secular equation is obtained:

$$\begin{vmatrix} x & 1 & 0 & 0 & 0 & 1 \\ 1 & x & 1 & 0 & 0 & 0 \\ 0 & 1 & x & 1 & 0 & 0 \\ 0 & 0 & 1 & x & 1 & 0 \\ 0 & 0 & 0 & 1 & x & 1 \\ 1 & 0 & 0 & 0 & 1 & x \end{vmatrix} = 0$$

The expansion of determinants such as this by an elementary method is a tedious process. Fortunately, there are shortcut methods for accomplishing this expansion in many cases. Here we merely write the result, which is

$$x^6 - 6x^4 + 9x^2 - 4 = (x^2 - 4)(x^2 - 1)^2 = 0$$

Since the six roots of this equation are ± 2, ± 1, and ± 1, it follows that the values of the energies of the six MO's, in the order of increasing energy, are: $\alpha + 2\beta$, $\alpha + \beta$, $\alpha + \beta$, $\alpha - \beta$, $\alpha - \beta$, and $\alpha - 2\beta$. The first three energy values are lower than α, and the corresponding MO's are bonding MO's. The other three MO's are antibonding MO's. It should be noted that there are two degenerate pairs of orbitals; that is, the orbitals in each pair have the same energy.

The six π electrons fill the three bonding orbitals and thus the energy of these electrons is given by $2(\alpha + 2\beta) + 4(\alpha + \beta) = 6\alpha + 8\beta$. Because the energy of the six π electrons in three isolated double bonds is $6\alpha + 6\beta$, we conclude that the delocalization energy in benzene is equal to 2β.

When the secular equation for benzene has been solved, the coefficients of the carbon p AO's in the equations for the six LCAO–MO's can be calculated. The equations for the three bonding MO's in the ground state of benzene are as follows:

$$\varepsilon$$

$$\alpha + 2\beta \qquad \psi_1 = \frac{1}{\sqrt{6}}(p_A + p_B + p_C + p_D + p_E + p_F)$$

$$\psi_2 = \tfrac{1}{2}(p_B + p_C - p_E - p_F) \tag{7-25}$$

$$\alpha + \beta$$

$$\psi_3 = \frac{1}{\sqrt{3}}\left(p_A + \frac{p_B}{2} - \frac{p_C}{2} - p_D - \frac{p_E}{2} + \frac{p_F}{2}\right)$$

These orbitals are represented schematically in Figure 7–17. It can be seen that ψ_1 has no nodal plane perpendicular to the plane of the molecule while ψ_2 and ψ_3, which form a degenerate pair of orbitals, each has one such nodal plane. The mathematical form of the equations given above implies a particular orientation of the nodal planes with respect to the molecule. Because the orientation of the nodal planes is arbitrary, provided that the planes for ψ_2 and ψ_3 are mutually perpendicular, the equations for these two orbitals can be written in an infinite number of equivalent ways.

The π bond order for the bond between carbon atoms A and B in benzene is $2 \cdot \frac{1}{\sqrt{6}} \cdot \frac{1}{\sqrt{6}} + 2 \cdot \frac{1}{\sqrt{3}} \cdot \frac{1}{2\sqrt{3}} = \frac{2}{3}$. It is easily verified that the same value is found for each of the other bonds in the ring. When the σ carbon-to-carbon single bond is included, the total bond order is 1.67. This value, which is con-

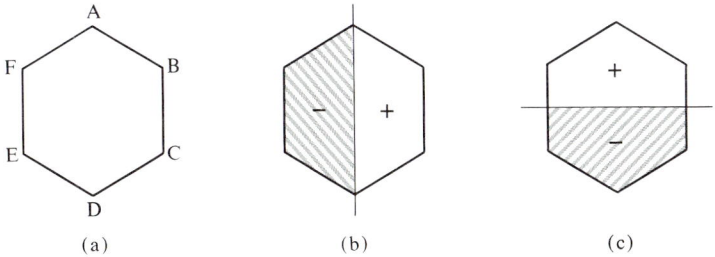

Figure 7–17. (a) ψ_1 of benzene — no nodal plane perpendicular to the plane of the molecule. (b) ψ_2 — one nodal plane passing through carbon atoms A and D. (c) ψ_3 — one nodal plane passing between carbon atoms B and C and between E and F. ψ_2 and ψ_3, each with one nodal plane, form a degenerate pair of molecular orbitals.

siderably less than the bond order of an ethylenic double bond, is consistent with the extent of conjugation and the related π electron delocalization in benzene.

The use of the Hückel MO approximation described here can be extended without serious difficulty to larger aromatic molecules. If atoms other than carbon are included in the conjugated system, then different values of α and β are needed in those terms of the secular equation which involve such atoms. The appropriate changes in the values of α and β are related to the electronegativity of the atoms involved, but there is no general and reliable way of obtaining these changes. The introduction of more than one value of α and of β complicates the treatment considerably, and we do not pursue this topic any further.

It was mentioned above that in this semiempirical MO treatment no attempt is made to calculate theoretically the value of the resonance integral β, and thus the delocalization energy. Instead some measured molecular properties which are believed to be related to β are used for this evaluation.

One such property is the absorption spectrum of a conjugated compound in the visible or ultraviolet spectral regions. From the energy levels we have calculated, it can be seen that the energy difference between levels is a function of β. It follows that if a π electron in the highest level occupied in the ground state is excited to the next higher level, the amount of energy absorbed depends on β. If this excitation is produced by the absorption of radiation and if the absorption band due to this transition can be identified in the absorption spectrum of a compound, a value of β can be calculated from the observed wavelength.

Another property used for obtaining approximate values of β for conjugated compounds is the heat of hydrogenation. For example, the heat of the reaction of benzene with hydrogen to form cyclohexane is 49.8 kcal/mole, while the heat of hydrogenation of cyclohexene to form cyclohexane is 28.6 kcal/mole. If we take the double bond in cyclohexene to be a "normal" isolated double bond, we can compare the heat of hydrogenation of 3 moles of cyclohexene, containing 3 "moles" of double bonds (85.8 kcal) with the value for benzene. The energy difference of 36 kcal is attributed to the effect of the delocalization of the π electrons in benzene, and thus is assumed to be approximately equal to 2β.

The values of β obtained from one property for a series of similar compounds are found to be quite consistent with each other.[5] The values obtained

[5]A. Streitweiser, *Molecular Orbital Theory for Organic Chemists*, John Wiley and Sons, New York, 1961.

for a single compound by use of several properties are in only rough agreement. This fact is not surprising in view of the oversimplified nature of the theoretical treatment. We can conclude that although the Hückel MO method has proved useful in correlating much chemical information, it has obvious deficiencies from a quantitative point of view. More refined methods of calculation of varying degrees of rigor have been, and are being developed, but these are beyond the scope of this book.

Valence Bond Treatment of Benzene

In this section we consider briefly the treatment of benzene and related molecules by the valence bond method. We start from the same σ bond framework we used in the MO method. Then we note that there are six $2p$ electrons on the six carbon atoms available for bond formation by overlap of the $2p$ AO's for pairs of electrons with opposed spins. For benzene, and conjugated molecules in general, there is more than one way of pairing the electrons. Each particular arrangement of spin-paired electrons with each electron on a separate atom is a covalent structure. It is possible to write a wave function for each structure, and by forming a linear combination of these structure wave functions we can obtain an approximate molecular wave function.

The number of different covalent structures for a molecule is the number of different ways the electrons can be paired. This number (N) is given by an equation which we merely quote:

$$N = \frac{(2n)!}{n!(n+1)!}$$

in which $2n$ is the number of electrons to be paired, and $n!$ (n factorial) is $n(n-1)(n-2) \cdots 1$. For benzene, $2n = 6$ and $N = \frac{6!}{3!4!} = 5$. (The number of covalent structures goes up very rapidly as the number of electrons to be paired increases. For naphthalene with ten p electrons, $N = 42$.) The five covalent structures for benzene can be represented in the following way:

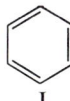 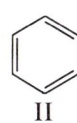   
I   II   III   IV   V

The first two are called the Kekulé structures because they resemble the structural formulas for benzene proposed by Kekulé a hundred years ago; for a similar reason, the others are called the Dewar structures.

Each of these structures is described by a wave function that is formed from linear combinations of products of the six carbon p AO's, including the spin. Then the wave function for the benzene molecule takes the following form:

$$\psi = c_I\psi_I + c_{II}\psi_{II} + c_{III}\psi_{III} + c_{IV}\psi_{IV} + c_V\psi_V$$

In this expression, $c_I = c_{II}$ because the two Kekulé structures are equivalent and make equal contributions to the molecular wave function; similarly, $c_{III} = c_{IV} = c_V$.

Next, the best values of the coefficients in this approximate wave function are obtained by applying the variation method. This leads to a secular determinant with five rows and five columns, whose elements contain a coulomb integral, Q, and a resonance integral, K, both defined in terms of the structure wave functions.

When the secular equation is solved, from the root that corresponds to the lowest energy level of the molecule the value of the energy is given by $Q + 2.61\,K$. (Like α and β in the MO method, Q and K are both negative quantities.) The energy calculated for just one of the Kekulé structures is $Q + 1.50\,K$. Consequently the calculated energy of benzene has been lowered relative to that of a single Kekulé structure by an amount

$$1.11\,K\,[=\,Q + 2.61\,K - (Q + 1.50\,K)]$$

as a result of the inclusion in the molecular wave function of contributions from all five covalent structures. The value of the integral K is obtained semiempirically from a measured property, such as the heat of hydrogenation or the heat of combustion, by comparison with a value calculated for a hypothetical compound, such as one represented by a single Kekulé structure. Pauling described the use of a linear combination of structure wave functions for benzene by saying that the five structures resonated among themselves, and he called the energy lowering of $1.11\,K$ the *resonance stabilization energy*.

In general, for structures to resonate they must differ only in the position and pairing of electrons, and the arrangement of the atomic nuclei must be the same in all structures. It should be understood that the individual structures introduced in the VB method have no physical significance. They are employed because of a desire to preserve the concept of bond formation by coupled pairs of electrons in those cases where a unique arrangement of paired electrons does not exist.

One consequence of the concept of resonance is that we can picture the electrons, which are associated with particular atoms in a particular structure, as moving through the whole resonating system. In this way the VB method accomplishes the delocalization of electrons, which in the MO method results from π MO's extending over the atoms in a conjugated system. The resonance stabilization energy of the VB method, expressed as a function of the integral K, is roughly analogous to the delocalization energy of the MO method, expressed as a function of the integral β.

The problem of atoms other than carbon in a conjugated system is treated in the VB method by the addition of ionic structures to the linear combination of covalent structures in order to take account of the effect of electronegativity differences. The simplest example is pyridine which, like benzene, has six p electrons to be paired. Because the nitrogen atom is more electronegative than the carbon atom, there is a greater probability of finding a π electron near N than near C. Thus in addition to the five covalent structures like the ones described for benzene, the following ionic structures are included in the set of resonating structures:

The Hydrogen Bond

On the basis of wave mechanics, we have interpreted the formation of covalent chemical bonds as a result of the overlapping of orbitals of the bonded atoms. From this point of view a single hydrogen atom should form only one bond because it has only one electron in a $1s$ orbital. The physical and chemical properties of many compounds, however, indicate that a hydrogen atom can form two bonds to other atoms, one of these bonds being a covalent bond while the other is known as a *hydrogen bond*. In contrast to covalent bonds, hydrogen bonds are formed only by hydrogen atoms already bonded to electronegative atoms, particularly to oxygen or to nitrogen.

Hydrogen bonding is mainly the result of an electrostatic attraction between the proton of a hydrogen atom covalently bonded to an electronegative atom and the lone pair electrons of another electronegative atom. This type of bond is limited to the hydrogen atom because this atom is the only one that does not have a filled shell of electrons surrounding, and thus shielding, its nucleus. When a hydrogen atom is covalently bonded to an electronegative atom, such as oxygen, there is a relatively high probability that the bonding electrons are near the oxygen atom, thus leaving the proton almost "bare." This proton can therefore exert an attractive force on the lone pair electrons of an oxygen atom in an O–H group, for example, on a neighboring molecule, and in this way form a hydrogen bond. For two water molecules we may picture the situation in the following way:

$$\begin{matrix} H & & H \\ \cdot\cdot & & \cdot\cdot \\ H:\ddot{O}:&---H:\ddot{O}: \end{matrix}$$

Hydrogen bonds in alcohols and phenols can be similarly pictured.

Hydrogen bonds are considerably weaker than typical covalent bonds. It is estimated that hydrogen bonds have a dissociation energy of about 5 kcal/mole, in contrast to the range of dissociation energies of 50–100 kcal/mole for covalent bonds. Since hydrogen bonds are weak, it is believed that in liquids these bonds are continually being formed and broken as the molecules move about.

Although hydrogen bonds are relatively weak, they have a significant effect on the properties of compounds of oxygen and nitrogen, particularly in the liquid or solid state. For example, water, which is extensively hydrogen-bonded, has a boiling point of 100°C and a heat of vaporization of 9700 cal/mole while the corresponding values for hydrogen sulfide, which is slightly, if at all, hydrogen-bonded, are −61.8°C and 4500 cal/mole. In general, hydrogen-bonded compounds have higher boiling points and heats of vaporization than related compounds that are not hydrogen-bonded. Analogous "abnormal" values of other physical properties (viscosity, dielectric constant, absorption spectrum, and others) are used to infer the presence of hydrogen bonds.

Carboxylic acids, such as acetic acid, are known from molecular weight determinations to exist as dimers even in the vapor state. Hydrogen bonding in this case is particularly strong because a pair of molecules can be bonded by two hydrogen bonds, as shown by the following formula for the dimer of acetic acid:

$$H_3CC\begin{matrix}O-H-----O\\ \\O-----H-O\end{matrix}CCH_3$$

For some compounds internal hydrogen bonds can be formed between different parts of the same molecule. A simple example is salicylaldehyde:

For molecules of this type the most stable configuration is determined by the presence of the hydrogen bond.

Hydrogen bonds are of great importance in the behavior of many compounds of interest in biochemistry and molecular biology, such as enzymes and nucleic acids. These bonds are responsible for the configurations of polypeptide chains in proteins and for the double-stranded helical configuration of the deoxyribonucleic acid (DNA) molecule, which is intimately related to its role as the carrier of genetic information.

Supplementary Material

Use of d Orbitals in Bonding

For the theoretical treatment of compounds formed of elements in the first row of the Periodic Table, it is sufficient to include the use of s and p AO's only. With the remaining elements, however, for which electrons may have a principal quantum number n whose value is larger than 2, d orbitals ($l = 2$) are often involved in bonding.

Like the three p orbitals, the five d orbitals, which correspond to the five values of the quantum number m (± 2, ± 1, 0), form a degenerate set in the absence of external fields. For our purpose we can ignore the radial factors of these orbitals and concentrate on the angular factors. There is no unique way of representing these angular factors as real functions. With our previously defined polar coordinate system, the conventional way is the following:[6]

d_{z^2}	$3\cos^2\theta - 1$
d_{xz}	$\sin\theta \cos\phi \cos\theta$
d_{yz}	$\sin\theta \sin\phi \cos\theta$
d_{xy}	$\sin\theta \cos\phi \sin\theta \sin\phi = \sin^2\theta \sin 2\phi$
$d_{x^2-y^2}$	$\sin^2\theta \cos^2\phi - \sin^2\theta \sin^2\phi = \sin^2\theta \cos 2\phi$

Each of these d orbitals has two angular nodal surfaces (Fig. 7–18). For d_{z^2}, these surfaces are represented by the equation $\theta = \pm\cos^{-1}\sqrt{\frac{1}{3}}$ and thus θ has the approximate values of 55° and 125°. This equation for θ represents two cones whose common axis is the z axis and whose apexes are located at the origin. For d_{xz}, the nodal surfaces are given by $\theta = 90°$ (the x,y plane) and $\phi = 90°$ (the z,y plane). Similarly for d_{yz} and d_{xy} the nodal surfaces are the coordinate planes x,y and x,z, and y,z and x,z, respectively. For $d_{x^2-y^2}$, the nodal planes ($\phi = 45°$ and 135°) bisect the angles between the x,z and y,z planes.

Hybrid orbitals can be constructed from s, p, and d AO's in the same way they were constructed from s and p AO's. The most common of these hybrids are formed from one s, three p, and two d orbitals. These six sp^3d^2 hybrids are all equivalent and their axes point toward the vertexes of a regular octahedron.

An example of a molecule whose structure can be interpreted on the basis of sp^3d^2 hybridization is SF_6, which is known to have the shape of a regular octahedron. Although the outer shell of an isolated S atom has the configuration $3s^2 3p^4$, we can imagine that two electrons are promoted to give $3s 3p^3 3d^2$

---

[6]The normalization constant is omitted in these expressions, which are an extension of the list of angular factors for p orbitals given in Equation 6–34.

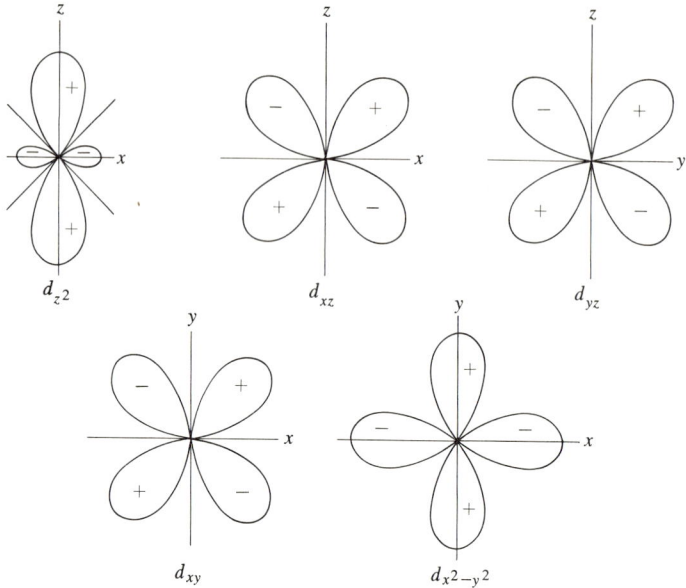

Figure 7–18. Polar diagrams of the angular factors of the five *d* orbitals.

and then these six orbitals hybridize in the way just described. Overlap of each of these hybrids by a *p* orbital of a F atom leads to the observed structure.

Another type of hybrid is sp^3d in which the axes of three of the hybrids lie in a plane with the other two axes perpendicular to this plane. This type has the symmetry of a trigonal bipyramid. A molecule with this symmetry is PF_5 resulting from the promotion of a 3s electron of phosphorus to a 3d orbital. Still another type of hybridization involving *d* orbitals is sp^2d which results in square planar symmetry.

The transition elements all have *d* orbitals available for bond formation. The nature of these orbitals is particularly important for the study of the covalent compounds and complex ions formed by these elements.

Calculation of Coefficients for Normalized Butadiene MO's

Having found the four roots of the secular equation for butadiene (p. 251), we can use these values to calculate the coefficients of the AO's in the four MO's of this compound. Only the ratios of these coefficients can be obtained directly, but then we normalize the MO's and this step yields definite values of the coefficients.

The easiest way to obtain the ratios of the coefficients is to make use of the fact that these ratios are equal to the ratios of the corresponding minors[7] of the elements of any row or column of the secular determinant, with due regard to sign. Let us use the elements of the top row of the secular determinant. The first element in this row corresponds to the AO p_A, and its minor is

$$\begin{vmatrix} x & 1 & 0 \\ 1 & x & 1 \\ 0 & 1 & x \end{vmatrix}$$

The second element corresponds to p_B and its minor is

$$-\begin{vmatrix} 1 & 1 & 0 \\ 0 & x & 1 \\ 0 & 1 & x \end{vmatrix}$$

The ratio c_{iB}/c_{iA} is then

$$\frac{c_{iB}}{c_{iA}} = -\frac{\begin{vmatrix} 1 & 1 & 0 \\ 0 & x & 1 \\ 0 & 1 & x \end{vmatrix}}{\begin{vmatrix} x & 1 & 0 \\ 1 & x & 1 \\ 0 & 1 & x \end{vmatrix}}$$

When these determinants are expanded the result is

$$\frac{c_{iB}}{c_{iA}} = -\frac{x^2 - 1}{x^3 - 2x}$$

Similarly, using the minors of the third and fourth elements of the top row we obtain

$$\frac{c_{iC}}{c_{iA}} = \frac{\begin{vmatrix} 1 & x & 0 \\ 0 & 1 & 1 \\ 0 & 0 & x \end{vmatrix}}{x^3 - 2x} = \frac{1}{x^2 - 2}$$

and

$$\frac{c_{iD}}{c_{iA}} = \frac{\begin{vmatrix} 1 & x & 1 \\ 0 & 1 & x \\ 0 & 0 & 1 \end{vmatrix}}{x^3 - 2x} = -\frac{1}{x^3 - 2x}$$

Next we substitute into these equations in succession each of the four values of x we have found. For example, for $x = -1.618$, which we associate with ψ_1, we obtain

[7] The minor of any element in a determinant is the determinant that is obtained by striking out the row and column in which the element is located. The sign of a minor is given by $(-1)^{m+n}$ where m is the number of the row and n is the number of the column in which the element is located.

$$\frac{c_{1B}}{c_{1A}} = 1.618; \quad \frac{c_{1C}}{c_{1A}} = 1.618; \quad \frac{c_{1D}}{c_{1A}} = 1$$

Thus the unnormalized form of ψ_1 is

$$\psi_1 = p_A + 1.618 p_B + 1.618 p_C + p_D$$

For a normalized orbital the sum of the squares of the coefficients must equal 1 as in Equation 7-17. We can achieve this result by (1) adding the squares of the above coefficients; (2) taking the square root of this sum; and (3) dividing each coefficient by this square root. For these steps we find

$$1 + (1.618)^2 + (1.618)^2 + 1 = 7.236$$
$$\sqrt{7.236} = 2.690$$

and

$$\frac{1}{2.690} = 0.372; \quad \frac{1.618}{2.690} = 0.602$$

It follows that the normalized form of ψ_1 as given on page 252 is

$$\psi_1 = 0.372 p_A + 0.602 p_B + 0.602 p_C + 0.372 p_D$$

The calculation of the coefficients for the other three MO's follows the same pattern.

Supplementary References

C. A. Coulson, *Valence*, 2nd edition, Oxford University Press, 1961.
M. W. Hanna, *Quantum Mechanics in Chemistry*, 2nd edition, W. A. Benjamin, New York, 1969.
J. W. Linnett, *Wave Mechanics and Valency*, Methuen, London, 1960.
A. Streitweiser, *Molecular Orbital Theory for Organic Chemists*, Wiley, New York, 1961.

Problems

1. Show that the substitution of ε_+ and ε_- from Equation 7-7 into Equation 7-4 leads to the values $c_A = \pm c_B$.

2. Account for the facts (Table 7-1) that (a) the binding energy of N_2^+ is smaller than that of N_2 and (b) the binding energy of O_2^+ is larger than that of O_2.

3. Describe the electron configuration of the ground states of the following molecules: (a) Cl_2; (b) S_2; (c) K_2; (d) HCl.

4. Show that the sp^3 hybrid orbitals as represented by Equation 7–14 are mutually orthogonal.

5. Show that the axes of the three sp^2 hybrid orbitals lie in a plane with an angle of 120° between any two of the axes.

6. Describe the bonds in the following molecules in terms of the VB or the localized MO approximation: (a) propylene ($CH_3CH\!\!=\!\!CH_2$); (b) methyl acetylene ($CH_3C\!\!\equiv\!\!C$); (c) allene ($CH_2\!\!=\!\!C\!\!=\!\!CH_2$); (d) trimethyl boron ($B(CH_3)_3$).

7. Show that the molecular orbitals in Equations 7–24 are normalized and mutually orthogonal.

8. Using Equations 7–25, find the order of the bond between carbon atoms B and C and between carbon atoms A and F in the benzene molecule as represented in Figure 7–17.

8

Determination of Molecular Structure

Introduction

The meaning of the words "molecular structure" depends to some extent on the context in which they are used. Here we take them to mean the shape (or symmetry) of a molecule, the interatomic distances, and the angles between the bonds. In the last chapter we had a number of occasions to mention these quantities, and now we take up the principal physical methods used for their determination.

In a real sense the subject of molecular structure started with Kekulé's concept of chains and rings of bonded carbon atoms. By the end of the nineteenth century organic chemists had obtained a large amount of qualitative structural information by employing an extremely effective method of inference. Their main weapon was the isomer number; that is, the number of compounds with the same molecular formula. For example, from the existence of *cis* and *trans* isomers the "rigidity" of the double bond was inferred, in contrast to the "free rotation" about a single bond. From the existence of only one isomer of doubly substituted methane the concept of a molecule existing in 3-dimensional space was inferred. This concept was strongly supported by the interpretation of optical isomerism by van't Hoff and LeBel on the basis of the asymmetric carbon atom.

Although the organic chemists' method was highly successful, it could not supply any information about molecular dimensions. The earliest source of such data — crude as they were — came from the physicists who developed the kinetic theory of gases, principally Maxwell. This approach, presented in Chapter 5, was limited to the rigid sphere model of a molecule. From the collision diameter found for such a model of a simple molecule the order of magnitude of interatomic distances could be estimated.

Further progress depended on the use of more powerful quantitative methods. In the last 50 years a number of such methods have been developed.

They involve the use of low frequency (essentially static) electric fields and magnetic fields, electromagnetic radiation over a very wide range of frequencies, and electron and neutron beams.

It is a remarkable fact that when quantitative physical methods were applied to a variety of molecules, the qualitative conclusions reached by nineteenth century chemists were nearly always confirmed.

The shape of a molecule (or of any object) can be most clearly described in terms of its symmetry. When we come to a consideration of the structure of crystals in Chapter 12 we will find that crystal symmetry is an important part of this topic. Many physical properties depend on the symmetry of molecules and crystals. For these reasons we begin our treatment of molecular structure with an introduction to the subject of molecular symmetry.

Molecular Symmetry

In order to be quite definite in our use of the word symmetry as applied to molecules, let us first consider an object with the shape of a prism, whose end faces are equilateral triangles (Fig. 8–1) with their vertices designated by letters. Now let us imagine that this object is rotated through an angle of 120° about an axis passing through the centers of the triangular end faces. Vertices a, b, and c then occupy the positions initially occupied by vertices c, a, and b, respectively. If the vertices were not marked and thus were identical, there would

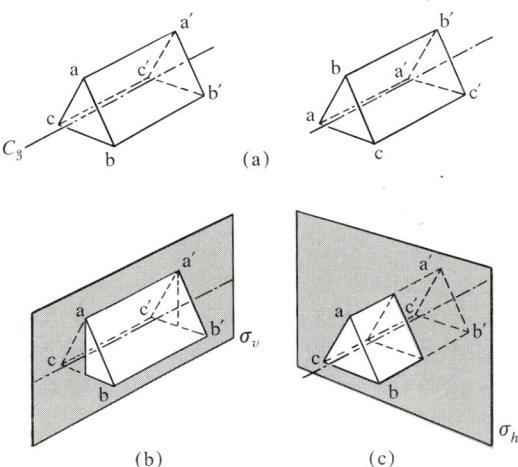

Figure 8–1. (a) Rotation of a prism by 120° about its axis (C_3). (b) Mirror plane (σ_v) passing through edge aa′ and the axis. (c) Mirror plane (σ_h) perpendicular to the axis and bisecting edges aa′, bb′, and cc′.

be no way to tell, by observing the before and after positions, whether the prism had been rotated or not. In other words, the prism would have been moved into an equivalent or indistinguishable position.

The operation of moving an object, a molecule, or a set of points into an indistinguishable arrangement is called a *symmetry operation*. The rotation of the prism through an angle of 120°, just described, is an example of a symmetry operation. A repetition of this operation leads again to an indistinguishable arrangement, and another repetition brings the prism back to its original position. Because three identical rotations are required to return to the initial position, this symmetry operation is called a 3-fold rotation.

In general, this kind of symmetry operation is called an n-fold rotation, where n is an integer. The corresponding angle of rotation is $2\pi/n$ radians or $360°/n$.

Now let us imagine that we have passed a plane mirror through the edge a–a' and the axis of the prism (Fig. 8-1b). When viewed from either side, half of the prism is reflected in the mirror and the prism appears to be unchanged. Thus this reflection is a symmetry operation of the prism. The two other reflections, with the mirror plane passing in turn through the edge b–b' and through the edge c–c', are also symmetry operations. In general, if the mirror plane is considered to be the x,y plane of a Cartesian coordinate system, the operation of reflection changes the value of the z coordinate of each point into its negative.

One other type of reflection is also a symmetry operation of the prism. In this case we imagine the mirror to be oriented perpendicular to the prism edges and passing through their centers. One half of the prism is then reflected into the other half, and again we have an indistinguishable arrangement (Fig. 8-1c).

The symmetry operations mentioned so far include a 3-fold rotation, 3 reflections in planes containing the prism axis, and one reflection in a plane perpendicular to this axis. This set of symmetry operations completely specifies the symmetry of our prism.

In order to describe the symmetry of any object two additional kinds of symmetry operations are needed, namely, *inversion* and *rotation-reflection*. If the position of each point in our object is referred to a coordinate system (Fig. 8-2), the operation of inversion with respect to this origin consists in moving each point with coordinates x,y,z into the position $-x,-y,-z$. When the inverted arrangement is indistinguishable from the initial one, inversion is a symmetry operation.

A rotation-reflection consists in the combination of an n-fold rotation and a subsequent reflection through a plane perpendicular to the rotation axis (Fig. 8-3). Relatively few molecules have this as a symmetry operation.

Each n-fold rotation is associated with an n-fold rotation axis about which the rotation takes place. Likewise, each reflection is associated with a mirror

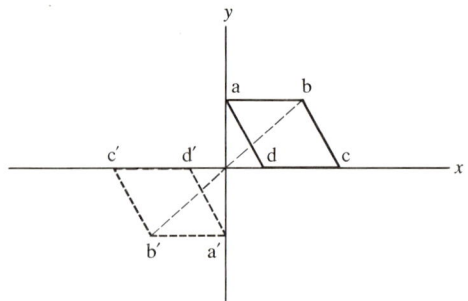

Figure 8–2. Operation of inversion in two dimensions.

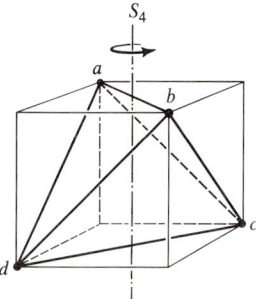

Figure 8–3. Example of rotation-reflection operation. A rotation of 90° about axis S_4 followed by a reflection in a plane perpendicular to this axis replaces point a by c, b by d, c by b, and d by a.

plane. A mirror plane that contains the principal rotation axis is called a *vertical plane* and is represented by σ_v. A mirror plane that is perpendicular to the principal rotation axis is said to be a horizontal plane and is represented by σ_h. The operation of inversion is associated with the point, called the center of symmetry, with respect to which the inversion is carried out.

Rotation axes, mirror planes, and the center of symmetry are called collectively the *symmetry elements*. The symmetry of an object or a molecule can be described in terms of either the set of symmetry operations which can be carried out or the set of symmetry elements which the object possesses. The symmetry operations and the associated symmetry elements are listed in Table 8–1, together with the symbols used to designate them.

In many cases the presence of certain symmetry elements implies the presence of others. For example, in the case of the prism discussed above, the presence of the 3-fold rotation axis and the mirror plane perpendicular to it as symmetry elements means that this prism must also have three 2-fold rotation axes perpendicular to the 3-fold axis, and conversely. In this case not all

Determination of Molecular Structure

TABLE 8-1 POINT SYMMETRY OPERATIONS AND ELEMENTS

Symmetry Operation	Symmetry Element	Symbol
n-fold rotation	n-fold rotation axis	C_n
reflection	mirror plane	σ or C_s
inversion	center of symmetry	i or C_i
n-fold rotation-reflection	n-fold rotation-reflection axis	S_n

of the symmetry elements are needed for an unambiguous specification of the symmetry of the prism.

A set of symmetry elements which is just sufficient to specify the symmetry of an object is known as a *point symmetry group*. The set is called a group because it satisfies the requirements for a mathematical group (p. 325). (The mathematical treatment of symmetry is part of group theory.) It is called a point group because in each of the symmetry operations at least one point is assumed to be unmoved. It is not necessary that there should be a particle or an atom at this point. Each point group is given its own descriptive symbol. The more common point groups with their symbols and symmetry elements are listed in Table 8-2.

TABLE 8-2 SOME COMMON POINT SYMMETRY GROUPS

Group Symbol	Symmetry Elements					Examples
C_{2v}	C_2	σ_v	σ_v'			H_2O, SO_2
C_{3v}	C_3	$3\sigma_v$				NH_3, PCl_3, $CHCl_3$
D_{2h}	$3C_2$	3σ	i			C_2H_4
D_{3h}	C_3	$3\sigma_v$	σ_h			BF_3
D_{6h}	C_6	$6C_2$	$6\sigma_v$	σ_h	i	C_6H_6
T_d (regular tetrahedron)	$4C_3$	$3C_2$	6σ			CH_4
O_h (regular octahedron)	$3C_4$	$4C_3$	$6C_2$	9σ	i	SF_6
$C_{\infty v}$	C_∞					HCl, HCN
$D_{\infty h}$	C_∞	$\infty \sigma_h$	σ_h	i		N_2, C_2H_2, CO_2

There is another type of symmetry group known as a *space group* which includes a translation in 3-dimensional space as a symmetry element. Space groups are important in the study of crystal structure and they are referred to in Chapter 12. Unfortunately, there are two sets of symbols for designating symmetry elements and groups. The set of symbols used in this chapter, called the Schönfliess symbols after their originator, is the one used by those who are interested in the structure of individual molecules.

Electric Dipole Moments

In this section we consider the type of information about molecular structure that can be obtained by the use of an electrostatic field. Imagine that we introduce a hydrogen molecule into the evacuated space between two plane parallel plates which form an electrical condenser. The two halves of the molecule are identical, and we can say that the "center of gravity" of positive charge coincides with the "center of gravity" of negative charge. When a constant electric field is applied to the molecule by charging the condenser plates, the electrons in the molecule are attracted by the positive plate and the protons are attracted by the negative plate. This distortion of the molecule results in the separation of the centers of positive and negative charge which is described by the statement that the field has induced an electric dipole moment m in the molecule.

The *electric dipole moment* of two equal charges, $+q$ and $-q$, separated by a distance r is defined as qr. This is a vector quantity whose direction is conventionally taken to be the direction from the positive charge to the negative one. In this section we use the electrostatic unit (esu) as the unit of electric charge, and then the unit of electric dipole moment is the esu-cm. Since the only dipole moments we discuss here are electric moments, we omit the word electric for brevity.

If the strength of the applied field, E_o, is not too large, the induced dipole moment is directly proportional to it; thus

$$m = \alpha E_o \qquad (8\text{--}1)$$

in which the constant of proportionality, α, is a molecular property known as the *molecular polarizability*. Because the dipole moment has the dimensions of charge times length and electric field strength has the dimensions of charge divided by the square of a length, it can be seen that the molecular polarizability has the dimensions of volume. Its order of magnitude is the same as that of the volume of a molecule, namely, 10^{-24} cm³.

Now let us imagine that many molecules are introduced into the space between the condenser plates. This matter, which is assumed to be an electrical insulator, is called the *dielectric*. It may be a gas, liquid, or solid. The electric

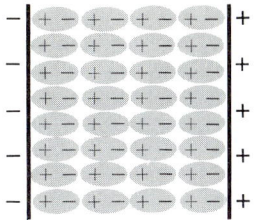

Figure 8-4. Polarized dielectric in an electric field.

field strength in the dielectric is reduced to a value E from its value in a vacuum E_0 as a consequence of the partial compensation of charges on the condenser plates by induced charges on the boundary of the dielectric (Fig. 8-4).

It can be shown that

$$E_0 = E + 4\pi \mathcal{P}$$

in which $\mathcal{P}$, the *dielectric polarization*, is the total induced electric moment per unit volume of dielectric. Thus $\mathcal{P}$ is given by Nm, where N is the number of molecules per unit volume of dielectric. Dividing both sides of the above equation by E leads to

$$\frac{E_0}{E} = 1 + 4\pi \frac{\mathcal{P}}{E}$$

The ratio of E_0 to E is a property of the dielectric known as the *dielectric constant, D*. Since $\mathcal{P}/E = Nm/E$ and $m/E = \alpha$, it follows that

$$D = 1 + 4\pi N\alpha \tag{8-2}$$

The dielectric constant is a quantity that can be measured relatively easily. It is equal to C/C_o, where C and C_o represent the capacitances of a condenser with a dielectric and a vacuum, respectively, between the plates. The capacitance is defined as the ratio of the charge on either one of the plates to the potential difference between the plates.

Let us say that we have 1 mole of substance, with volume V_m, between the condenser plates; then $N = \mathfrak{N}/V_m = \mathfrak{N}\rho/M$, where ρ and M are, respectively, the density and molecular weight of the substance. We can now rewrite Equation 8-2 in the following form:

$$(D - 1)\frac{M}{\rho} = 4\pi \mathfrak{N}\alpha \tag{8-3}$$

This relation states that the ratio of $(D - 1)$ to the density is independent of the density because the polarizability is a constant for a particular substance and thus the right-hand side of Equation 8-3 is independent of the density.

Equation 8-3 is consistent with measurements on some gases at pressures low enough that a dipole moment in any one of them has a negligible interaction with the moments of its neighbors because the molecules are relatively far apart on the average. For gases at high pressures and for condensed phases, however, the interaction between induced moments is important and must be taken into consideration. It has proved to be very difficult to do this exactly. We use a relation, known as the *Clausius–Mosotti equation*, which approximately includes a correction for the effect of the interaction between induced moments. This equation is the following:

$$\frac{D-1}{D+2}\frac{M}{\rho} = \frac{4}{3}\pi \mathfrak{N}\alpha \tag{8-4}$$

The quantity $\dfrac{D-1}{D+2}\dfrac{M}{\rho}$, represented by P_m, is called the *molar polarization*.[1]

Notice that as the pressure of a gas is lowered, D approaches 1, the value for a vacuum, and the Clausius–Mosotti equation reduces to Equation 8-3 when $D + 2 \sim 3$.

The Clausius–Mosotti equation provides a connection between a measurable property, D, of a substance and an important property, α, of its molecules. Because α is a constant, according to this equation the molar polarization of a substance is a constant and thus should be independent of density and temperature. When dielectric constants were measured for a number of gases and vapors over a range of temperatures, it was found that the Clausius–Mosotti equation was obeyed very well by some substances (such as H_2, CH_4, CO_2, and CCl_4), and not at all by others (such as HCl, H_2O, CH_3Cl, and NH_3).

These results were puzzling until Debye in 1912 called attention to the fact that the substances that showed the anomalous behavior were just those whose molecules could be expected to have a permanent dipole moment, that is, a dipole moment even in the absence of an applied electric field, because of an unsymmetrical distribution of electric charge in the molecules. For these polar molecules the effect of an applied electric field is a partial alinement of the molecular dipoles, resulting in an orientation polarization. This alinement is disturbed by molecular collisions which tend to produce random orientations of the molecules. It follows that the orientation polarization, which depends on the average value of the component of the permanent dipole moment in the direction of the applied electric field, should depend on the collision frequency and therefore on the temperature at which the measurement is made.

To evaluate the orientation polarization we calculate the average component of the permanent dipole moment in the field direction in the following way: Consider a molecule with a permanent dipole moment μ oriented at an

---

[1] It is unfortunate that this word is used for P_m. Do not confuse it with the dielectric polarization, $\mathcal{P}$, because they are different concepts.

Determination of Molecular Structure

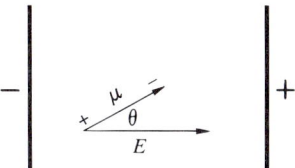

Figure 8–5. Orientation of permanent dipole moment μ in an electric field with strength E.

angle θ to the direction of the applied electric field (Fig. 8–5). The potential energy, U, of this molecule in the field is given by the product of the field strength E and the component of the dipole moment in the direction of the field, which is $-\mu \cos \theta$. (Although two angles are needed to specify the direction of a vector in space, the potential energy depends only on the angle θ.) Thus $U = -E\mu \cos \theta$.

The number of molecules, dn_Ω, whose dipole moment vectors lie within an element of solid angle $d\Omega$, centered at the angle θ, is given by the following special form of the Maxwell–Boltzmann distribution law (Eq. 5–26):

$$dn_\Omega = Ae^{-\frac{U}{kT}} d\Omega$$

in which A is proportional to the total number of molecules under consideration. When U is replaced by the expression in the preceding paragraph, the distribution law becomes

$$dn_\Omega = Ae^{\frac{E\mu \cos \theta}{kT}} d\Omega$$

Now we use this distribution law to find an average value as we have done before (p. 132). The average component of μ in the direction of the field, $\overline{m}_o$, is the average value of $\mu \cos \theta$ which is

$$\overline{m}_o = \overline{\mu \cos \theta} = \frac{\int_0^{4\pi} Ae^{\frac{E\mu \cos \theta}{kT}} \mu \cos \theta \, d\Omega}{\int_0^{4\pi} Ae^{\frac{E\mu \cos \theta}{kT}} d\Omega}$$

We can simplify this equation by introducing the abbreviation $a = E\mu/kT$ and dividing numerator and denominator by A. We then have

$$\overline{m}_o = \mu \frac{\int_0^{4\pi} e^{a \cos \theta} \cos \theta \, d\Omega}{\int_0^{4\pi} e^{a \cos \theta} \, d\Omega} \tag{8–5}$$

To carry out these integrations we need the relation between θ and Ω. In order to obtain that relation, consider a segment of a sphere of radius r, as shown in Figure 8-6. The area, S, of this segment is given by $S = 2\pi rh = 2\pi r(r-y) = 2\pi r(r - r\cos\theta) = r^2(2\pi - 2\pi\cos\theta)$. By definition, the solid angle subtended

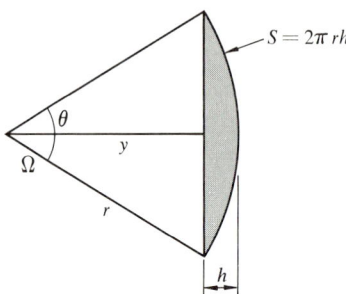

Figure 8-6. Relation between solid angle $\Omega \left(= \dfrac{S}{r^2} \right)$ and the angle θ. S is the area of a segment of a sphere of radius r.

by a spherical segment is the area of the segment divided by the square of the radius of the sphere. Consequently,

$$\Omega = \frac{S}{r^2} = 2\pi - 2\pi \cos\theta$$

and

$$d\Omega = -2\pi \, d(\cos\theta)$$

On dividing Equation 8-5 by μ and letting $\cos\theta = x$, we have (notice that when $\Omega = 0$ or 4π, $x = 1$ or -1)

$$\frac{\overline{m}_0}{\mu} = \frac{\displaystyle\int_{-1}^{+1} e^{ax} x \, dx}{\displaystyle\int_{-1}^{+1} e^{ax} \, dx} = \frac{\left.\dfrac{e^{ax}}{a^2}(ax-1)\right|_{-1}^{+1}}{\left.\dfrac{e^{ax}}{a}\right|_{-1}^{+1}}$$

$$= \frac{e^a + e^{-a}}{e^a - e^{-a}} - \frac{1}{a}$$

When the function [2] of a on the right-hand side of this equation is expanded as a power series in a, it can be shown that $a/3$ is a good approximation for this

[2] This function is often called the Langevin function, represented by $L(a)$, because Langevin derived it first in a theoretical treatment of magnetic dipoles. It is equal to $\coth a - 1/a$, where $\coth a$ is the hyperbolic cotangent of a.

function for small values of a. In those cases of interest to us here, $E\mu$ is small relative to kT and thus $a \ll 1$. It follows that

$$\frac{\overline{m}_o}{\mu} = \frac{E\mu}{3kT}$$

Introducing the orientation polarizability, α_o, defined by $\frac{\overline{m}_o}{E}$, we obtain

$$\alpha_o = \frac{\mu^2}{3kT} \tag{8-6}$$

We have thus obtained the orientation polarizability as a function of the permanent dipole moment and the temperature. The result is consistent with our expectation that $\overline{m}_o$ should decrease with rising temperature.

In addition to the orientation polarizability, α_o, polar molecules, like the nonpolar ones, have a distortion polarizability, α_d. For nonpolar molecules this is the only kind of polarizability, and thus $\alpha = \alpha_d$. In contrast, for polar molecules the molar polarization must be expressed as a sum of two contributions, as follows:

$$P_m = \frac{D-1}{D+2}\frac{M}{\rho} = \frac{4}{3}\pi\mathfrak{N}(\alpha_d + \alpha_o)$$

On substitution of the value of α_o from Equation 8–6, the result is

$$P_m = \frac{4}{3}\pi\mathfrak{N}\left(\alpha_d + \frac{\mu^2}{3kT}\right) \tag{8-7}$$

This equation is called the *Debye equation*.

From the Debye equation it can be seen that the molar polarization is a linear function of the reciprocal of the absolute temperature of the form $P_m = a + b/T$, where $b = 4\pi\mathfrak{N}\mu^2/9k$. When measurements of the dielectric constant and density have been made for a substance over a range of temperatures, the graph of P_m versus $1/T$ should be a straight line whose slope is b, as shown in Figure 8–7. From the value of b, the permanent dipole moment can be calculated from the relation $\mu = \frac{3}{2}\sqrt{kb/\pi\mathfrak{N}}$. When the numerical values of the constants are introduced, this becomes $\mu = 1.28 \times 10^{-20}\sqrt{b}$ esu-cm. The value of the distortion polarizability can be readily calculated from the intercept, a, on the P_m axis ($1/T = 0$), which is equal to $\frac{4}{3}\pi\mathfrak{N}\alpha_d$.

In order to see what the order of magnitude of permanent dipole moments is, let us consider a hypothetical diatomic molecule with a bond length of 1 Å in which one electron has been transferred from one atom to the other. In this case the dipole moment would be 4.80×10^{-10} esu $\times 10^{-8}$ cm $\sim 10^{-18}$ esu-cm. The convenient unit of 10^{-18} esu-cm is usually called the Debye, and its symbol is D. It always follows a number and thus can be distinguished from the symbol for the dielectric constant.

The application of the Debye equation to the determination of permanent dipole moments[3] is most reliable when the substance under consideration is in the form of a gas or vapor. There are, however, relatively few vapors which can be studied over a temperature range that is wide enough for good values of

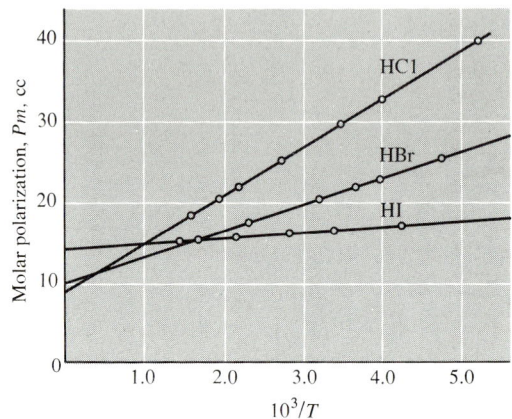

Figure 8–7. Molar polarization as a function of $1/T$.

the dipole moment to be obtained. Consequently, it is desirable to utilize measurements made on substances in the liquid state. The Debye equation is not applicable to polar liquids because the interaction between the dipole moments of closely packed polar molecules is too strong. In contrast, for dilute solutions of polar molecules in nonpolar solvents the polar molecules on the average are relatively far apart, and the Debye equation can be applied. Even in this case, as we see later, interactions between the polar molecules and between polar and solvent molecules are significant and must be taken into consideration.

Let us say that measurements of the dielectric constant, D_{12}, and the density, ρ_{12}, have been made over a range of temperatures for a dilute solution of a polar solute in a nonpolar solvent. At each temperature the molar polarization, P_{12}, of the solution is calculated from the following equation:

$$P_{12} = \frac{D_{12} - 1}{D_{12} + 2} \frac{M_{av}}{\rho_{12}}$$

in which M_{av} is given by $M_1 X_1 + M_2 X_2$ with M and X representing the molecular weight and the mole fraction of the components.

The molar polarization of the solution must now be divided into contribu-

[3]In the remainder of this section we refer to permanent dipole moments simply as dipole moments.

tions from the solvent and the solute. The simplest basis for this division is the assumption that the contributions to P_{12} are proportional to the respective mole fractions. This assumption leads to the following equation:

$$P_{12} = P_1(1 - X_2) + P_2 X_2$$

P_1 is calculated from values of the dielectric constant and density of the pure solvent. Since X_2 is known from the composition of the solution, the apparent molar polarization of the solute, P_2, can be obtained.

The values of P_2 obtained in this way are found to vary when X_2 is varied at constant temperature. This effect, which is attributed to interactions of the polar molecules, can be minimized by obtaining P_2 for a series of solutions with small values of X_2, and then finding the value ($P_{2\infty}$) extrapolated to infinite dilution. (This extrapolation, however, does not eliminate the effect of interactions between the solute and the solvent molecules, and other semiempirical and more complicated methods have been proposed.) From the values of $P_{2\infty}$ at a series of temperatures the value of μ is calculated from the Debye equation.

This method is limited by the temperature range over which solutions can be studied. Fortunately, there is another method which is more generally useful, even though it involves some additional approximations, because it requires measurements at only a single temperature. This method depends on the behavior of polar molecules in alternating electric fields of very high frequency.

Consider a set of polar molecules in an electric field and imagine that the field is suddenly turned off. A finite time is required for the molecules to lose their partial alinement and to reach the random distribution of orientations that is characteristic of equilibrium in the absence of a field. This time is called the relaxation time. When the field is suddenly applied in the reverse direction, a finite time is required for the orientation polarization to build up to its equilibrium value.

Now imagine that the reversal of field direction occurs at an increasing frequency. As long as the frequency is not too high, the molecules can follow the field and both the dielectric constant and the molar polarization remain constant. When, however, the period of the alternations ($= 1/\nu$) becomes comparable with the relaxation time, this is no longer the case. The molecules cannot stay in phase with the field and the polarization drops, as indicated in Figure 8–8, because the contribution from orientation decreases. The frequency range at which this change occurs is usually 10^{10}–10^{12} cycles/sec.

At still higher frequencies only the distortion contribution remains. The distortion polarization results from the movement of both electrons and nuclei in the alternating field. As the frequency continues to rise through the range corresponding to the infrared region of the electromagnetic spectrum to the visible region ($\sim 10^{15}$ cycles/sec), the nuclei can no longer follow the alternating field because of their relatively large mass. Then the polarization drops again until only the contribution from electron displacements remains.

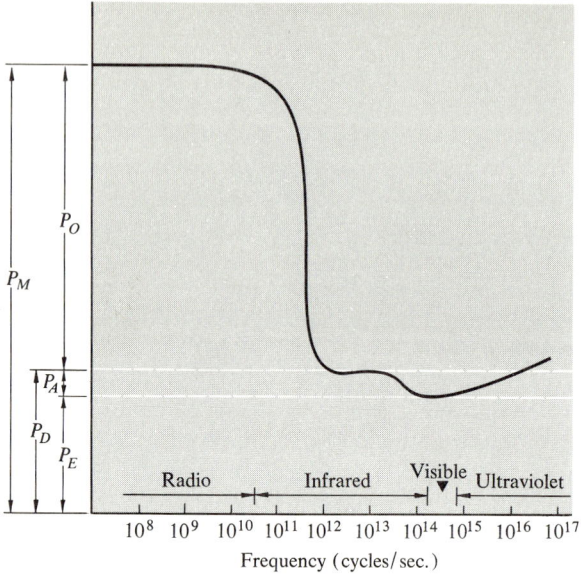

Figure 8–8. Molar polarization as a function of the frequency of the electric field.

We started this section by imagining that the substance under investigation was placed between the plates of a condenser, from whose capacitance the dielectric constant could be obtained. At frequencies higher than about 10^{10} cycles/sec the direct measurement of capacitance cannot be carried out. It is possible, however, to make use of one of the important consequences of Maxwell's electromagnetic theory of light, the relationship $D = n^2$, in which n is the index of refraction.

◈◈◈◈◈◈

Maxwell's relationship requires that the dielectric constant and the refractive index be measured at the same frequency. Normally the dielectric constant is measured by electrical methods at relatively low frequencies ($< 10^6$ cycles/sec) while the refractive index is measured by optical techniques at much higher frequencies ($> 10^{14}$ cycles/sec). For nonpolar molecules, however, $D \sim n^2$ even though these quantities are measured at frequencies that differ by many orders of magnitude. For example, the value of D for benzene is 2.27 and the value of n is 1.50. In contrast, for water $D = 80$ and $n = 1.33$. The

Determination of Molecular Structure

difference between D and n^2 is a consequence of the fact that the water molecule has a fairly large dipole moment.

When the Maxwell relationship is introduced into the Clausius–Mosotti equation, the following equation, which is known as the *Lorentz–Lorenz* equation, is obtained:

$$\frac{n^2 - 1}{n^2 + 2} \frac{M}{\rho} = \frac{4}{3} \pi \mathfrak{N} \alpha_e \tag{8-8}$$

The quantity $\dfrac{n^2 - 1}{n^2 + 2} \dfrac{M}{\rho}$ is called the *molar refraction*, represented by R_m, and α_e is that part of the total molecular polarizability which is due to electron displacements.

The distortion polarizability, α_d, as mentioned above depends on the displacements of both electrons and nuclei. It is the sum of α_e and α_a, the contribution to the molecular polarizability due to displacements of nuclei. Thus the total molar polarization of a polar substance is represented by

$$P_m = \frac{4}{3} \pi \mathfrak{N} (\alpha_o + \alpha_e + \alpha_a)$$

and the Debye equation (Eq. 8–7) can be written in the following form:

$$P_m = \frac{4}{9} \frac{\pi \mathfrak{N}}{kT} \mu^2 + \frac{4}{3} \pi \mathfrak{N} \alpha_e + \frac{4}{3} \pi \mathfrak{N} \alpha_a \tag{8-9}$$

The quantity $\frac{4}{3} \pi \mathfrak{N} \alpha_a$ is often called the atom polarization, a term which is somewhat misleading.

When we subtract Equation 8–8 from Equation 8–9, we obtain

$$P_m - R_m = \frac{D - 1}{D + 2} \frac{M}{\rho} - \frac{n^2 - 1}{n^2 + 2} \frac{M}{\rho}$$

$$= \frac{4}{9} \frac{\pi \mathfrak{N}}{kT} \mu^2 + \frac{4}{3} \pi \mathfrak{N} \alpha_a$$

The atom polarization is difficult to determine experimentally. Fortunately, it is relatively small (often about 10% of R_m) and it is usually neglected. Then we can obtain an approximate value for the dipole moment by solving the equation above for μ; thus

$$\mu = \left(\frac{9}{4} \frac{k}{\pi \mathfrak{N}} (P_m - R_m) T\right)^{\frac{1}{2}}$$

When the values of the constants are introduced the following equation for the value of μ in Debye units is obtained:

$$\mu = 0.0128 \left((P_m - R_m) T\right)^{\frac{1}{2}}$$

When measurements of the dielectric constant have been made on a series of solutions of a polar molecule in a nonpolar solvent, the value of $P_{2\infty}$ is used in the equation above in place of P_m. The value of R_m is calculated from the refractive index and density of the pure polar substance.

We have thus arrived at a method for obtaining dipole moments from measurements at one temperature of the dielectric constant at a relatively low frequency and the refractive index for visible light. Most values of the dipole moment have been determined in this way. The data are not so reliable as those obtained by the temperature variation method because of the uncertainty in the value of the atom polarization. In particular, dipole moments whose magnitude is a few tenths of a Debye unit are probably not significant.

TABLE 8-3   DIPOLE MOMENTS*
(in debyes)

Cl_2	0	HI	0.44	H_2O	1.85
CO_2	0	HBr	0.82	SO_2	1.63
CCl_4	0	HCl	1.08	NH_3	1.47
SiF_6	0	HF	1.82	PCl_3	0.78
C_6H_6	0			CH_3Cl	1.87
C_6Cl_6	0				
BCl_3	0				
		cis-dichloroethylene		1.90	
		trans-dichloroethylene		0	

*R. D. Nelson, Jr., D. R. Lide, Jr., and A. A. Maryott, "Selected Values of Electric Dipole Moments for Molecules in the Gas Phase." NSRDS-NBS 10, U.S. Government Printing Office, Washington, D.C., 1967.

A few typical values of dipole moments are listed in Table 8-3. The values for diatomic molecules are clearly consistent with our expectations. The values for all homonuclear molecules are zero and, as illustrated by the hydrogen halides, the values for heteronuclear molecules increase with increasing electronegativity difference between the atoms.

The dipole moment data for polyatomic molecules provide information about the structure of the molecules. To interpret the data we assume that the dipole moment of a molecule is the vector sum of moments associated with individual bonds in the molecule. On this basis the value of zero for the dipole moment of CO_2 is direct evidence that this molecule is a linear molecule (symmetry group $D_{\infty h}$) because the bond moments cancel each other. The substantial dipole moment of H_2O is evidence for the nonlinearity of this molecule (symmetry group C_{2v}).

If a molecule has a center of symmetry it must have a zero dipole moment because each bond moment is canceled by an equal and oppositely directed bond moment. Likewise, complete cancellation of bond moments occurs if the molecule has more than one rotation axis or a rotation-reflection axis. Only the C_{nv} symmetry groups remain, and a molecule with a permanent dipole moment must belong to one of these groups.

Thus we conclude from dipole moment data that BCl_3 must be planar (D_{3h}) and that PCl_3 (C_{3v}) cannot be planar. Because one isomer of dichloroethylene has a moment of 0 while the other has a moment of $1.89 D$, we conclude that the nonpolar isomer is the trans isomer (D_{2h}), that it is planar, and that the polar isomer is cis (C_{2v}). These conclusions are consistent with the identification of these isomers by organic chemists.

The strength and weakness of the assumption of the additivity of bond moments can be illustrated by the following data for a series of substituted benzenes (benzene itself has a zero moment):

	Chloro-	1,2-dichloro-	1,3-dichloro-	1,4-dichloro-	1,3,5-trichloro-
Observed	1.70 D	2.53 D	1.67 D	0	0
Calculated		2.94	1.70	0	0

The calculated values are based on a planar hexagonal model for the benzene ring with the C–Cl bond lying in the plane of the hexagon. The C–Cl bond moment is assumed to have the same value as the total moment of chlorobenzene. For the resultant of two bond moments, μ_1 and μ_2, forming an angle θ with each other, the following equation for the addition of vectors is useful:

$$\mu = (\mu_1^2 + \mu_2^2 + 2\mu_1\mu_2 \cos \theta)^{\frac{1}{2}}$$

It can be seen that the calculated values for the 1,3-, and 1,4-dichlorobenzenes, and 1,3,5-trichlorobenzene are in excellent agreement with the experimental values, but the discrepancy for 1,2-dichlorobenzene is far beyond experimental error. This discrepancy may be due to either or both of the following reasons: (1) The angle between the C–Cl bond vectors may be larger than 60° because of mutual repulsion of the relatively negative chlorine atoms, or (2) the magnitudes of the bond moments may be modified by their proximity. This ambiguity of interpretation is not uncommon and it introduces uncertainties in the calculation of valence angles or of electron distributions in bonds from quantitative dipole moment data.

The main usefulness of dipole moments in the determination of molecular structure is based on the fact that the presence or absence of a dipole moment provides reliable evidence that certain symmetry elements are or are not present in the molecule. For example, the zero moment of 1,4-dichlorobenzene and 1,3,5-trichlorobenzene is good evidence for the coplanarity of the C–Cl bonds and the benzene ring. In contrast, the value of 1.64 D for the moment of hydroquinone (1,4-dihydroxybenzene) leads to the conclusion that the valence angle of the oxygen atom is different from 180°.

The two cases which have been discussed in this section are: (1) P_m is independent of temperature, which implies that $\mu = 0$; and (2) P_m varies inversely as the temperature, which implies that μ is independent of the temperature. For certain types of molecules μ is found to increase with temperature. One example of such a molecule is 1,2-dichloroethane whose dipole moment as a function of temperature is shown in Figure 8–9. We can interpret this behavior by noting

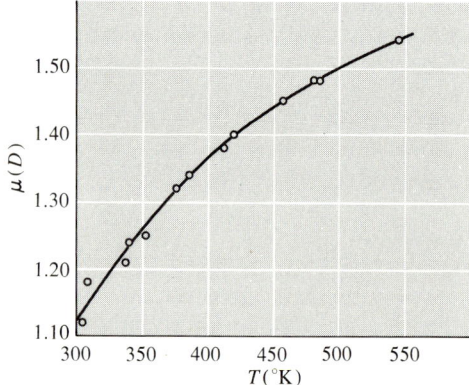

Figure 8–9. Dipole moment of 1,2-dichloroethane as a function of the temperature.

that an internal rotation about the C–C bond can occur. In the conformation in which the Cl atoms are *trans* to each other, a center of symmetry is present and $\mu = 0$. As one half of the molecule is rotated with respect to the other half, the moment increases. The *trans* conformation has the lowest potential energy because the Cl atoms have a maximum separation. If the equilibrium distribution of conformations follows the Boltzmann distribution law, the fraction of molecules with $\mu \neq 0$ increases with rising temperature. Consequently the average dipole moment will increase to a limiting value at high temperatures, in agreement with the observed variation.

In this section we have used only concepts and relations from nineteenth-century classical physics. That might seem peculiar in view of the emphasis in

Determination of Molecular Structure

earlier chapters on the statement that quantum mechanics must be used for the theory of all phenomena involving electrons, atoms, and molecules. There is actually no inconsistency in this case. When the quantum-mechanical calculation of the orientation polarizability is carried out, the result for low-frequency electric fields turns out to be the same as that derived on the basis of classical theory.

The dipole moment and the polarizability of a molecule depend on the arrangement of its nuclei and the distribution of its electrons. If the Schrödinger equation for a molecule could be solved exactly, accurate values for its dipole moment and polarizability could be calculated from the molecular wave function. At present, moderately accurate values for these properties have been calculated for only a few of the simplest molecules.

If a molecule is in the alternating electric field of electromagnetic radiation whose frequency increases to a value for which $h\nu$ is equal to the difference in energy of two molecular energy levels, a phenomenon may occur which has been avoided in this section—namely, the absorption by the molecule of energy from the field. This phenomenon is the basis for molecular absorption spectroscopy. We saw in Chapter 6 how important the role of atomic spectroscopy has been in arriving at our present ideas of atomic energy levels and atomic structure. Molecular spectroscopy is equally important in the study of molecular structure, and the rest of this chapter is devoted to this topic.

Rotation Spectra of Diatomic Molecules and Interatomic Distances

Let us start our consideration of molecular spectroscopy by imagining a highly idealized experiment. Assume that we have a source of electromagnetic radiation which emits a single frequency that is continuously variable over the range 10^{10}–10^{13} cycles/sec (corresponding to wavenumbers between 1 and 1000 cm^{-1}). Assume also that this radiation passes through a container holding gaseous HCl at a pressure of a few torr to a detector that indicates the rate at which the radiant energy is transmitted through the HCl.

As the wavenumber of the radiation rises to a value near 20 cm^{-1} the detector shows a sharp drop in the transmission of the radiant energy. With increasing wavenumber the detector reading returns to its former values but it falls and rises again near 40 cm^{-1}. Continuing in this way, we would observe a series of narrow absorption regions or "lines" which are nearly uniformly spaced on a wavenumber scale (Fig. 8–10). The wavenumbers of these lines have been found to be represented by the following empirical equation: $\omega_j = 20.79j - 0.0016j^3$, where j is an integer which gives the order of the line in the series.

The absorption spectrum under consideration is called a pure rotation spectrum because it is interpreted in terms of transitions between molecular rotational energy levels. Our objective is the use of rotational spectra to obtain information about interatomic distances in molecules.

The rotational energy of diatomic molecules was referred to briefly at the end of Chapter 5. We use the same molecular model introduced at that point,

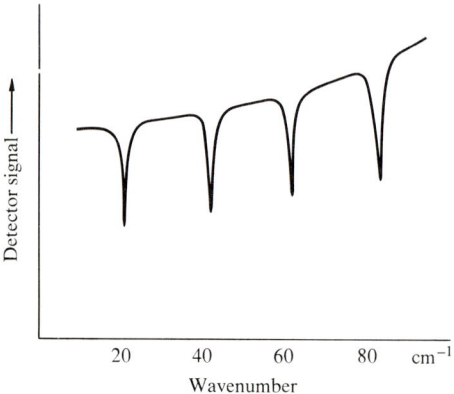

Figure 8-10. Idealized representation of pure rotation-absorption spectrum of HCl.

but now we need to examine the problem in greater detail. Our model consists of two masses, m_1 and m_2, with a constant distance, r, between their centers, rotating with angular velocity, $\dot{\theta}$, about their center of gravity which we assume is stationary (Fig. 5-10). The distance r is the sum of the distances r_1 and r_2 from the centers of the masses to the center of gravity.

Since the rigid rotor model has no potential energy, we need consider only its kinetic energy. If, for simplicity, we assume that the motion of our rigid rotor takes place in a plane, the classical expression for its rotational kinetic energy is:

$$\epsilon_{\text{rot}} = \frac{1}{2} I \dot{\theta}^2 = \frac{p_\theta^2}{2I}$$

in which I is the moment of inertia ($= m_1 r_1^2 + m_2 r_2^2$) and p_θ is the angular momentum ($= I\dot{\theta}$). (For rotational motion I, $\dot{\theta}$, and p_θ play the same role as the mass, velocity, and linear momentum in linear motion.)

It is convenient to express the energy in terms of the single distance r, rather than the two distances r_1 and r_2. We can do this by using the fact that, from the definition of the center of gravity, $m_1 r_1 = m_2 r_2$. It follows that

$$r = r_1 + r_2 = r_1 + \frac{m_1}{m_2} r_1 = \frac{m_2 + m_1}{m_2} r_1$$

and $r_1 = \dfrac{m_2}{m_1 + m_2} r$; likewise, $r_2 = \dfrac{m_1}{m_1 + m_2} r$. When we substitute these expressions for r_1 and r_2 in the equation for I this equation becomes

$$I = m_1 \left(\dfrac{m_2 r}{m_1 + m_2}\right)^2 + m_2 \left(\dfrac{m_1 r}{m_1 + m_2}\right)^2 = \left[\dfrac{m_1 m_2^2 + m_2 m_1^2}{(m_1 + m_2)^2}\right] r^2 = \dfrac{m_1 m_2}{m_1 + m_2} r^2 \tag{8-10}$$

The factor $\dfrac{m_2 m_1}{m_1 + m_2}$ is called the *reduced mass* and its usual symbol is μ.[4] By expressing the moment of inertia in terms of the reduced mass, $I = \mu r^2$, we can treat our 2-mass model as though it consisted of a single mass μ which moves in a circle of radius r with kinetic energy of $p_\theta^2/2\mu r^2$.

According to classical mechanics, the angular momentum and kinetic energy of this model can have any value. We know, however, that classical mechanics is not applicable to our molecular model and we must turn to quantum mechanics to obtain the values of the quantized rotational energy levels.

◈◈◈◈◈◈

We have oversimplified our model by the assumption that the motion of the model takes place in a plane. When we remove this restriction we have the problem of the motion of a particle of mass μ on the surface of a sphere of radius r. The Schrödinger equation for this model has a form which is quite similar to that of the Schrödinger equation for the hydrogen atom. In fact it is Equation 6–28 without the first term (because r is assumed constant) and with $U = 0$. As mentioned before, the rigid rotor is one of the very few problems for which the Schrödinger equation is soluble exactly.

◈◈◈◈◈◈

From the requirement that the solutions of the Schrödinger equation be eigenfunctions, the following equations for the angular momentum and the energy of the rigid rotor are obtained:

$$p_J = \dfrac{h}{2\pi} \sqrt{J(J+1)} \qquad J = 0, 1, 2, \ldots \tag{8-11}$$

and

$$\epsilon_J = \dfrac{h^2}{4\pi^2} \dfrac{J(J+1)}{2I} = \dfrac{h^2}{8\pi^2 \mu r^2} J(J+1) \tag{8-12}$$

[4] It is unfortunate that this is the same symbol used to represent the dipole moment, but no confusion should result.

in which J is the rotational quantum number. If μ is expressed in grams and r in centimeters, the energy values in this equation are expressed in ergs.

It was pointed out previously that in the field of spectroscopy it is convenient to express the magnitude of an energy value not in ergs but in terms of the wavenumber which is directly proportional to the energy ($\omega = \nu/c = \epsilon/hc$ cm^{-1}). The wavenumbers corresponding to the rotational energy values are known as the rotational term values F_J. These are given by the following expressions:

$$F_J = \frac{\epsilon_J}{hc} = \frac{h}{8\pi^2 \mu r^2 c} J(J+1) = BJ(J+1)$$

The coefficient B, which is characteristic for a particular molecule, is called the rotational constant.

The term values corresponding to successive values of the rotational quan-

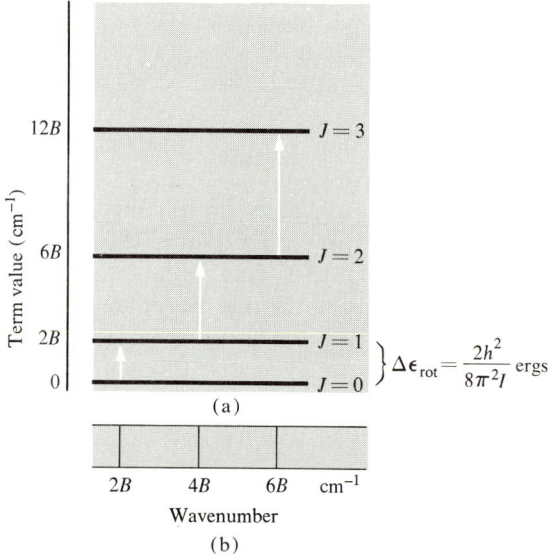

Figure 8-11. (a) Diagram of rotational energy levels, showing allowed transitions. (b) Graph of positions of corresponding rotational absorption lines.

tum number are: $F_J = 0, 2B, 6B, 12B, \ldots$ These values are diagrammed in Figure 8-11.

Now we make use of the concept, introduced by Bohr (p. 174), that a line in a spectrum results from a transition between two energy levels in which a photon is absorbed or emitted. The wavenumber of the radiation involved is always given by the difference between two term values. It does not follow, however, that every difference between term values represents the wavenumber of a spectral line because only certain transitions can occur (or are "allowed"). A statement of which transitions are allowed is called a selection rule. In general, selection rules can be derived when the wave functions are known for the system under consideration.

The selection rule for a transition between rotational energy levels as a result of the absorption or emission of electromagnetic radiation is $\Delta J = \pm 1$, provided that the molecule has a permanent dipole moment. Thus for polar molecules transitions can occur only between neighboring energy levels. Nonpolar molecules do not absorb (or emit) radiation in the spectral region we are considering here.

◊◊◊◊◊◊◊

Although the transition between energy levels is a process which can be treated accurately only by quantum mechanics, we can make the restriction to polar molecules plausible by the following classical argument: As a polar molecule rotates, its dipole moment vector rotates at the same frequency as the molecule. If the molecule is in the alternating electric field of a beam of electromagnetic radiation whose frequency is the same as the rotational frequency of the molecule, the field is in phase with the rotating dipole moment vector, and energy can be transferred from the field to the molecule. If the molecule is nonpolar the field cannot interact with the rotating molecule as a whole. Of course the nonpolar molecule has its set of rotational energy levels, given by Equation 8–12, but transitions between them do not occur by absorption of electromagnetic radiation.

◊◊◊◊◊◊◊

When we know the rotational selection rule we can obtain a general expression for the wavenumber of the radiation absorbed in any allowed transition in the following way:

$$\omega_J = F_J - F_{J-1} = BJ(J+1) - B(J-1)J = 2BJ \text{ cm}^{-1} \qquad (J = 1, 2, 3, \ldots)$$

where J is the quantum number of the upper level involved in the transition. This result tells us that the rotational spectrum should consist of a set of absorption lines which are uniformly spaced on a wavenumber basis with a constant separation between lines of $2B$ cm^{-1}.

Now we can compare our theoretical prediction with the experimental results mentioned above. If we identify the integer j in the empirical equation for the rotational lines of HCl with the quantum number J, the theoretical equation is the same as the empirical equation except for the small term in j^3

which we ignore for the moment. Consequently we obtain the value of the rotational constant B for HCl as follows:

$$\omega_J = 2BJ = 20.79J \text{ cm}^{-1}$$

$$B = 10.40 \text{ cm}^{-1}$$

Thus,

$$\frac{h}{8\pi^2 \mu r^2 c} = 10.40 \text{ cm}^{-1}$$

In this equation the only unknown is r, the interatomic distance, and we can find its value without difficulty. First, we need the reduced mass of HCl^{35}. (We assume that the experimental data refer to molecules containing the more abundant Cl isotope, Cl^{35}.) Recalling that atomic mass numbers are expressed in terms of the atomic mass unit which, to a sufficient degree of approximation, is numerically equal to the reciprocal of Avogadro's number, we have

$$\mu = \frac{1 \times 35}{1 + 35} \times 1.66 \times 10^{-24} = 1.61 \times 10^{-24} \text{ g}$$

It follows that

$$r = \left(\frac{6.62 \times 10^{-27}}{8 \times 9.87 \times 1.61 \times 10^{-24} \times 3.00 \times 10^{10} \times 10.40} \right)^{\frac{1}{2}}$$
$$= 1.28 \times 10^{-8} \text{ cm} = 1.28 \text{ Å}$$

In this section we have seen how rotational spectra are used to calculate moments of inertia and interatomic distances of diatomic molecules.

There are several additional points which should be mentioned here:

1. The discrepancy between theory and experiment involving the term in j^3 is due to the fact that our theoretical model was oversimplified. We assumed that a molecule is a rigid rotor, which means that the interatomic distance is independent of the rotational state of the molecule. Because the force holding the atoms together is not infinite, as a molecule acquires more rotational energy and thus from a classical point of view rotates more rapidly, increasing centrifugal force results in a larger interatomic distance. Consequently, the moment of inertia increases. This increase leads to a closer separation between rotational energy levels and to a closer spacing of the absorption lines. The theoretical treatment of this effect results in a small correction term which depends on J^3.

2. We have seen that the rotational constant of a diatomic molecule depends on the value of its moment of inertia and therefore on the mass of its atoms. It follows that two molecules of the same substance which differ only in the isotopes they contain (such as HCl^{35} and HCl^{37}) each has its own set of energy levels, and the lines in the rotational spectrum occur in closely spaced pairs.

Determination of Molecular Structure

3. The Schrödinger equation for the rigid rotor involves two variables, the angular coordinates θ and ϕ. Consequently the wave functions for the rigid rotor depend on *two* quantum numbers, J and K, but the energy, as we have seen, depends on J only. For each value of J, the quantum number K can have the values $J, J-1, \cdots 0, \cdots -J$. Thus for each value of J there are $2J+1$ wave functions which are associated with the same energy level. We make use of this fact at a later point. The relation between the molecular quantum numbers J and K is analogous to the relation between the atomic quantum numbers l and m. (Cf. p. 197.)

4. The rotational spectra of only a few diatomic molecules with relatively small moments of inertia can be observed in the far infrared region of the spectrum by use of spectrometers described in a general way on p. 167. For other molecules the rotational constants are so small that the rotational spectra are in the microwave region. In this region the radiation sources are electronic oscillators which produce monochromatic waves and the waves are detected by electrical methods.

Rotational Spectra of Polyatomic Molecules

In general, a macroscopic object has three moments of inertia about three perpendicular axes in the object that are known as principal axes. The rotational energy is the sum of three terms, each depending on the value of one of the moments of inertia. Likewise, the rotational energy of a molecule depends on the values of three moments of inertia, designated as I_a, I_b, and I_c, about the three principal axes a, b, and c.

On the basis of the relative values of the three moments of inertia, molecules can be classified as linear, symmetric tops,[5] spherical tops, or asymmetric tops. Each of these classes has distinguishing characteristics.

1. Linear molecules ($I_a = 0; I_b = I_c$). The symmetry groups for these molecules are $D_{\infty h}$ and $C_{\infty v}$, which includes polar diatomic molecules. The a axis is the molecular axis; since the nuclei on this axis have practically all of the mass of the molecule, the value of I_a is zero. The equation for the energy levels is the same as for diatomic molecules (Eq. 8–12), and the spectrum has the same general appearance (lines spaced almost equally). The only problem in determining interatomic distances arises because the line spacing yields only one moment of inertia. From one moment of inertia only one interatomic distance can be calculated, and a linear polyatomic molecule has at least two distances to be determined. The way out of this difficulty is the use of the spectra of isotopic molecules to supply additional moments of inertia. As many moments of inertia are needed as the number of distances to be calculated. This use of isotopic molecules involves the assumption that interatomic distances are inde-

[5]In this usage, a top is anything which rotates about an axis.

pendent of the particular isotope in the molecule. The validity of this assumption has been demonstrated in many cases.

2. Symmetric tops ($I_a \neq 0$; $I_b = I_c$). The symmetry group is C_{nv} with $n > 2$. The rotational energy for molecules in this class depends on two quantum numbers, J and K. (The total angular momentum of the symmetric top depends on J, while the angular momentum of rotation about the molecular axis depends on K.) The rotational term values are given by the following equations:

$$F_{J,K} = J(J+1)\frac{h}{8\pi^2 I_b c} + K^2\left(\frac{h}{8\pi^2 I_a c} - \frac{h}{8\pi^2 I_b c}\right)$$
$$= BJ(J+1) + (A-B)K^2$$

The selection rules for the rotational transitions of symmetric top molecules are $\Delta J = \pm 1$ and $\Delta K = 0$. The physical basis for the second selection rule can be seen from the following argument: Because of the symmetry of the molecule, its dipole moment is necessarily directed along the molecular axis. Rotation of the molecule about this axis does not involve a rotation of the dipole moment vector, and consequently the energy associated with this type of motion cannot be changed by interaction with the electric field of electromagnetic radiation.

It follows that ΔF does not depend on K and only one moment of inertia can be calculated from the rotational absorption spectrum. As in the case of linear molecules, data for isotopic molecules are needed for the calculation of interatomic distances.

3. Spherical tops ($I_a = I_b = I_c$). The symmetry groups are T_d and O_h. Molecules in this class cannot have permanent dipole moments because of their symmetry and therefore they do not absorb radiation in the microwave region of the spectrum.

4. Asymmetric tops ($I_a \neq I_b \neq I_c$). The rotational energy levels of molecules in this class cannot be represented by a simple equation and their rotational spectra, even for triatomic molecules such as H_2O, are extremely complicated. Nothing further is said about them in this book.

As a conclusion to this section it can be said that rotational spectra provide the most direct approach to the evaluation of interatomic distances. Values obtained from the analysis of far infrared or microwave spectra are among the most reliable of all such data. The main disadvantage of rotational spectroscopy is that the method is limited to those molecules with a permanent dipole moment which can be studied in the gas phase.

Some representative data are presented in Table 8–4. The difference of 0.002 Å between the C–N distances in HCN and in CH_3CN is at the borderline of significance. The difference of 0.007 Å in the C–N distances in HCN and ClCN is significant and shows a real difference between these bonds which is a consequence of the differing electron distributions in these molecules.

TABLE 8-4 INTERNUCLEAR DISTANCES AND BOND ANGLES FROM ROTATIONAL SPECTRA

Diatomic Molecules			
Molecule	Distance in Å	Molecule	Distance in Å
HF	0.917	O_2	1.208*
HCl	1.275	N_2	1.095*
HBr	1.414	Li_2	2.672*
CO	1.128	Cl_2	1.989*

Linear Polyatomic Molecules					
Molecule	Distances				
N_2O	NN	1.126		NO	1.191
HCN	CH	1.064		CN	1.156
ClCN	CCl	1.629		CN	1.163

Nonlinear Polyatomic Molecules						
Molecule	Distances				Angle	
CH_3CN	CH	1.092	CN	1.158	HCH	109°
	CC	1.460				
CH_3Cl	CH	1.103	CCl	1.782	HCH	110.5°
$CHCl_3$	CH	1.073	CCl	1.767	ClCCl	110.5°
NH_3	NH	1.106			HNH	107°
SO_2	SO	1.433			OSO	119.5°

*Obtained from electronic spectra.

Vibrational Spectra of Diatomic Molecules

Again let us use HCl as an example of a diatomic molecule, but this time let us shift our attention to the spectral range from about 2000 to 12,000 cm^{-1}. If we were to obtain the absorption spectrum of HCl in this range with a crude spectrometer, this spectrum would consist of a series of nearly equally spaced "lines" whose intensity decreases greatly with increasing wavenumber (Fig. 8–12). We now interpret this spectrum in terms of transitions between vibrational energy levels.

To start our interpretation of vibrational spectra we consider the harmonic oscillator model for a vibrating diatomic molecule which was discussed on page 152. This model (Fig. 5–11) consists of two masses held together by a massless spring obeying Hooke's law, $f = -k(r - r_e)$, where r is the distance between the centers of the masses, r_e is the equilibrium value of this distance, and k, the force constant, is a measure of the strength of the spring. We assume that the center of gravity of this model is fixed in space.

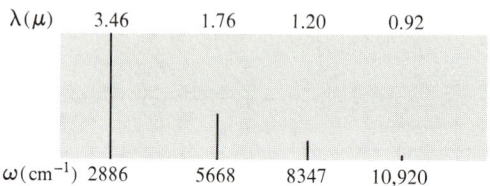

Figure 8–12. Schematic diagram of the vibrational spectrum of HCl.

We now show that our 2-mass model is equivalent in mechanical behavior to a harmonic oscillator which consists of a single mass attached to a fixed point by a spring. Let us call the axis of the 2-mass model the x axis and represent the displacements of atom 1 and atom 2 from their respective equilibrium positions by x_1 and x_2 with opposite signs. Then $r - r_e = x_2 - x_1$, and from Hooke's law the force acting on atom 2 is $f = -k(x_2 - x_1)$. The acceleration of atom 2 at any instant is (d^2x_2/dt^2) which is equal by Newton's second law to the force acting on this atom divided by its mass. In symbols this equality is

$$\frac{d^2x_2}{dt^2} = \frac{f}{m_2} = -\frac{k}{m_2}(x_2 - x_1)$$

From Newton's third law the force acting on atom 1 is equal and opposite to the force acting on atom 2. Consequently, the acceleration of atom 1 is given by

$$-\frac{d^2x_1}{dt^2} = -\frac{f}{m_1} = -\frac{k}{m_1}(x_2 - x_1)$$

Then for the total acceleration of the two atoms we have

$$\frac{d^2x_2}{dt^2} - \frac{d^2x_1}{dt^2} = \frac{d^2(x_2 - x_1)}{dt^2} = \frac{f}{m_2} - \frac{f}{m_1}$$

and

$$\frac{d^2(x_2 - x_1)}{dt^2} = -k\left(\frac{1}{m_2} + \frac{1}{m_1}\right)(x_2 - x_1)$$

When we represent $x_2 - x_1$ by x, this equation becomes

$$\frac{d^2x}{dt^2} = -k\left(\frac{1}{m_2} + \frac{1}{m_1}\right)x$$

$$= -\frac{k}{\mu}x \tag{8-13}$$

Determination of Molecular Structure

The reduced mass, μ, defined in this way is exactly the same as the reduced mass defined by Equation 8–10.

Equation 8–13 is the differential equation for a 1-dimensional harmonic oscillator with force constant k and mass μ. We can write its solution in the following form:

$$x = A \cos \sqrt{\frac{k}{\mu}}\, t$$

where A is the maximum displacement of the mass. We can also write the solution as $x = A \cos 2\pi\nu t$ in which ν is the vibrational frequency of the harmonic oscillator given by[6]

$$\nu = \frac{1}{2\pi} \sqrt{\frac{k}{\mu}} \tag{8-14}$$

It can be seen that the stiffer the spring (larger value of k) the higher the frequency, and the larger the mass the lower the frequency.

Solving this equation for k and substituting the result into the equation for the potential energy of the harmonic oscillator, we find that $U = 2\pi^2\nu^2\mu x^2$. When the displacement has its maximum value, A, all of the energy of the oscillator is in the form of potential energy; consequently

$$\epsilon_{\text{vib}} = 2\pi^2\nu^2\mu A^2$$

According to this classical mechanical treatment the magnitude of the displacement, and thus the energy, can have any value. Because classical mechanics is not applicable to molecular vibration we must use the quantum mechanical expression for the vibrational energy levels of our molecular model. The Schrödinger equation for the harmonic oscillator is (Eq. 6–20):

$$\frac{d^2\psi}{dx^2} + \frac{8\pi^2\mu}{h^2}\left(E - \frac{kx^2}{2}\right)\psi = 0 \tag{8-15}$$

When this equation is solved, the energy values are found to be given by the following equation:

$$\boxed{\begin{aligned} \epsilon_n &= \frac{h}{2\pi}\sqrt{\frac{k}{\mu}}(n + \tfrac{1}{2}) \quad (n = 0, 1, 2, \ldots) \\ &= h\nu(n + \tfrac{1}{2}) \end{aligned}} \tag{8-16}$$

in which ν is the frequency of a classical harmonic oscillator with force constant k and mass μ. The corresponding vibrational term values (G_n) are:

$$G_n = \frac{1}{2\pi c}\sqrt{\frac{k}{\mu}}(n + \tfrac{1}{2}) = \omega(n + \tfrac{1}{2}) \text{ cm}^{-1}$$

---

[6]The mass returns to its original position every time the argument of the cosine changes by 2π. The time required for one complete cycle is the period, T; when $t = 0$, $x = A$ and when $t = T$, $x = A$, again. Consequently $\sqrt{\frac{k}{\mu}}\, T$ must equal 2π and $T = 2\pi\sqrt{\frac{\mu}{k}}$. The frequency $\nu = 1/T$.

These equations represent a set of equally spaced energy levels (Fig. 8–13). The selection rule for vibrational transitions as a result of the absorption or emission of electromagnetic radiation is $\Delta n = \pm 1$, provided that the dipole

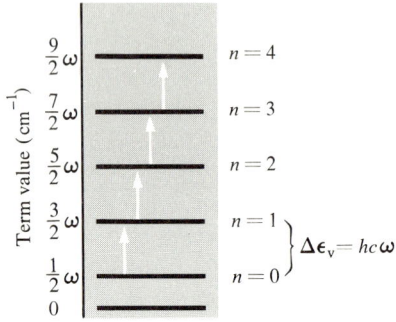

Figure 8–13. Vibrational energy levels of an harmonic oscillator. Levels are equally spaced; only allowed transitions are those for which $\Delta n = \pm 1$.

moment of the molecule changes during the vibration. It follows that symmetric diatomic molecules, which have a zero dipole moment at all times, cannot undergo transitions between vibrational energy levels by absorption of radiation. In fact, such molecules do not absorb anywhere in the infrared or microwave regions of the spectrum because they can undergo neither vibrational nor rotational transitions.

◇◇◇◇◇◇◇

According to Equation 8–16 even in the lowest energy level, for which $n = 0$, the harmonic oscillator still has an amount of vibrational energy equal to $\tfrac{1}{2}h\nu$. This energy which the harmonic oscillator cannot lose is called the zero-point energy. It has no effect on the vibrational spectrum but, as is shown later, it has important indirect effects on other properties. The existence of zero-point energy is required by the Heisenberg uncertainty principle, for if it did not exist the momentum of the harmonic oscillator would be exactly zero when $n = 0$, and both the momentum and coordinate could be known in principle to any degree of accuracy.

◇◇◇◇◇◇

The harmonic oscillator model leads to a prediction that the vibrational spectrum of a heteronuclear diatomic molecule should consist of a single line whose wavenumber is given by $\Delta G_{n \to n+1} = \omega$ cm^{-1}. For HCl, however, as mentioned above, more than one vibrational transition is observed. In addition to the transition corresponding to $\Delta n = 1$, which is often called the fundamental transition, there are transitions corresponding to $\Delta n = 2, 3$, and 4. The fact that more than one vibrational transition is observed is evidence that HCl is not actually an harmonic oscillator; however, the fact that only the fundamental transition results in "strong" absorption of infrared radiation is evidence that this molecule behaves approximately as a harmonic oscillator, particularly for small values of the vibrational quantum number.

The energy levels of the nonharmonic, or anharmonic, oscillator are not equally spaced but the intervals between them decrease with increasing values of n. Consequently, the wavenumber of the $\Delta n = 2$ transition is slightly less than twice the wavenumber of the $\Delta n = 1$ transition, and so on. At sufficiently large values of n the vibrational energy levels converge to a limiting energy above which they have a continuous distribution. This limiting value corresponds to the energy required to dissociate the molecule into atoms. (Notice that a harmonic oscillator could never dissociate, no matter how large the vibrational energy became because the restoring force increases without limit as the displacement increases.) For those molecules for which enough information is available to determine the limiting energy this value is found to be in good agreement with heats of dissociation determined calorimetrically.

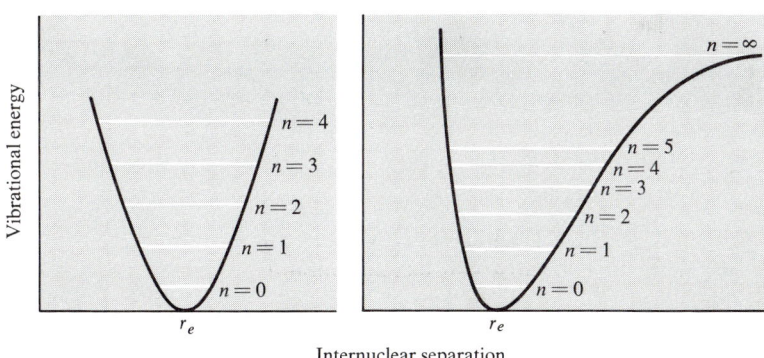

Figure 8–14. (a) Energy levels for a harmonic oscillator on a graph of vibrational potential energy *vs* internuclear separation. (b) Energy levels for an anharmonic oscillator.

The energy levels for both harmonic and anharmonic oscillators are shown schematically in Figure 8–14 in which the potential energy is plotted against the internuclear separation. The curve for the harmonic oscillator is a parabola ($U = \frac{1}{2}kx^2$). The curve for the anharmonic oscillator, which is typical of the potential energy curves for actual molecules (Fig. 7–2), is steeper than the parabola at small internuclear separations because of repulsion between the atoms. It is less steep at internuclear separations larger than the equilibrium value and eventually it becomes horizontal, which means that the atoms are independent of each other and the molecule has dissociated. The shape of the curve is almost parabolic near the minimum. It is for this reason that the harmonic oscillator can be used as a simple model of a diatomic molecule for small values of the vibrational quantum number.

◆◇◆◇◆◇◆

An empirical equation which represents to a good approximation the actual potential energy curve for many molecules with the use of only two parameters was found by Morse. The Morse function is $U = D(1 - e^{-ax})^2$ in which x, as before, is equal to $r - r_e$, D is the dissociation energy of the molecule in ergs, and

$$a = \omega_e \left(\frac{2\pi^2 c^2 \mu}{D} \right)^{\frac{1}{2}}$$

ω_e is the theoretical wavenumber for a vibration of infinitesimal amplitude.

◆◇◆◇◆◇◆

When the vibrational spectrum of a diatomic molecule has been observed, an approximate measure of the force holding the atoms together in the molecule can be readily obtained. From Equation 8–14 we have the following equation for the force constant:

$$\boxed{k = 4\pi^2 \nu^2 \mu = 4\pi^2 \omega^2 c^2 \mu}$$

Substituting the value of ω obtained from the spectrum and the value of μ for HCl^{35}, we obtain

$$k = 4 \times 9.87 \times (2886)^2 \times 9 \times 10^{20} \times 1.61 \times 10^{-24} = 4.80 \times 10^5 \text{ dynes/cm}$$

Instead of the cgs unit for the force constant, the dyne/cm, a related unit, the millidyne/Å, which is equal to 10^5 dynes/cm, is convenient. The value of 4.80 mdyne/Å for HCl^{35} means that a force of 4.80 millidynes would be required to stretch the H–Cl bond by 1 angstrom unit if the molecule were a harmonic oscillator and thus obeyed Hooke's law.

Force constants depend only on the kinds of atoms in a molecule. Consequently, molecules containing different isotopes have different vibrational frequencies and spectra. The values of the force constants for some diatomic molecules are included in Table 8–5.

Determination of Molecular Structure

TABLE 8-5 VIBRATIONAL WAVENUMBERS AND FORCE CONSTANTS

	Diatomic Molecules	
Molecule	Wavenumber (cm^{-1})	Force Constant $(dynes/cm)$
HF	4141	9.6×10^5
HCl^{35}	2990	4.8×10^5
HBr	2650	4.1×10^5
CO	2170	19.0×10^5
Li_2^7	351*	
Cl_2^{35}	565*	

	Triatomic Molecules		
Molecule		Wavenumber (cm^{-1})	
	Symmetric stretch	Asymmetric stretch	Bending
CO_2	1340	2349	667
CS_2	657	1523	397
HCN	2089	3312	712
N_2O	1285	2224	589
H_2O	3652	3756	1595

*Obtained from electronic spectra.

Rotation-Vibration Spectra of Diatomic Molecules

At the beginning of the last section we imagined that we observed the infrared absorption spectrum of HCl with a crude spectrometer. When a "high quality" spectrometer is used, the "line" at 2886 cm^{-1} is found to be a band which consists of a number of lines (Fig. 8-15). To account for this observation we must recognize that the model of a diatomic molecule as a harmonic oscillator which vibrates along a certain direction in space is oversimplified, and that molecules rotate and vibrate at the same time. This behavior, of course, appears to be inconsistent with the assumption of a rigid rotor we used in the treatment of rotational spectra. From a classical point of view, however, the molecular vibrational frequency is much higher than the rotational frequency. Consequently the molecule carries out many cycles of vibration while it makes one complete rotation and the rotational energy levels are determined by the average, or effective, value of the moment of inertia.

For most molecules it is a good approximation to assume that the vibra-

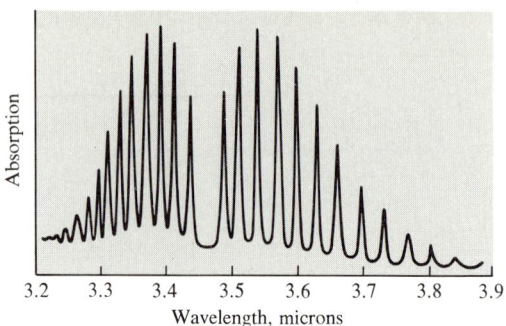

Figure 8-15. Fundamental absorption band of HCl.

tional and rotational energies can be simply added, and we limit our consideration to this case. Then the energy of the rotating harmonic oscillator is given by the following equation:

$$\epsilon_{n,J} = h\nu(n + \tfrac{1}{2}) + \frac{h^2}{8\pi^2 I_{av}} J(J+1)$$

and the related term values are

$$G_{n,J} = \omega(n + \tfrac{1}{2}) + B_{av} J(J+1)$$

Each vibrational energy level has associated with it a whole set of rotational energy levels (Fig. 8-16). The selection rules for transitions between these energy levels are the same as before: $\Delta n = \pm 1$ and $\Delta J = \pm 1$, provided the molecule has a permanent dipole moment. Consequently the wavenumbers of the possible transitions in a fundamental, rotation-vibration absorption band $(n = 0 \rightarrow n = 1)$ are given by

$$\Delta G = G_{1,J} - G_{0,J'} = \omega \pm 2BJ \text{ cm}^{-1} \qquad J = 1, 2, \ldots$$

According to this equation, a rotation-vibration absorption band of a diatomic molecule consists of two sets of uniformly spaced lines separated by a gap of twice the spacing of the lines in either set. This gap results from the absence of the line corresponding to the forbidden transition $\Delta n = 1$, $\Delta J = 0$. The set of lines corresponding to $\omega + 2BJ$ is known as the R branch of the rotation-vibration band and the set corresponding to $\omega - 2BJ$ is the P branch.

A comparison of the predicted spectrum with the observed spectrum of HCl (Fig. 8-15) shows quite good agreement. The line spacing is almost constant. From this spacing, values of B, I_{av}, and r_{av} can be calculated that are in satisfactory agreement with those obtained from the far infrared spectrum.

Determination of Molecular Structure

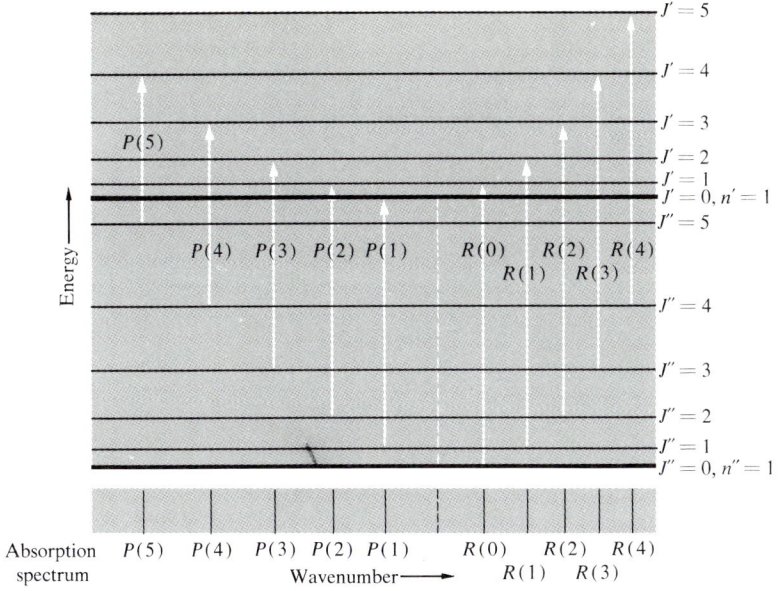

Figure 8-16. Rotational-vibrational energy levels and the corresponding absorption spectrum.

Vibrations of Polyatomic Molecules

There is only one way in which a diatomic molecule can vibrate and, of course, there is at most one fundamental vibrational band in the infrared region. The situation is considerably more complicated for polyatomic molecules. Our model in this case consists of a number of masses held together by springs which obey Hooke's law. If such a model is constructed of balls and springs and if the system is set into vibration, as by striking it, the resulting motion appears complex and chaotic. A particular ball might be observed to be vibrating in one direction for a while, then stop and start vibrating in a different direction with a different frequency, without apparent regularity.

Both theoretical analysis and observation, however, show that the motion is in fact regular and predictable. This is a consequence of the fact that any general vibrational motion can be expressed as a sum of basic types of motion, known as *normal modes of vibration*, which are characterized by the following properties:

1. Each normal mode of vibration has a definite frequency.

2. In a normal mode all of the masses are in phase; that is, all of the masses pass through their equilibrium positions at the same instant.

3. The normal modes are independent of each other. If the system is set

into vibration in one of its normal modes, it will continue to vibrate in that mode alone.

It is not difficult to calculate the number of normal modes of vibration for a system consisting of N masses. To specify the position of each mass at some instant requires $3N$ coordinates (3 for each mass). The system, however, can be specified in another, more useful, way. Three coordinates can be used to locate the center of gravity of the entire system, and changes in these coordinates are associated with translational motion of the system. Three more coordinates (angles, in this case) are needed to give the orientation of the system as a whole with respect to a set of three perpendicular axes passing through the center of gravity. (If the system is linear, only two angles are needed.) Changes in these angular coordinates are associated with rotational motion of the system.

The remaining $3N-6$ (or $3N-5$, for linear systems) coordinates give the relative distances between the masses in the system, and changes in these "internal" coordinates are associated with vibrational motions of the masses. Thus there are $3N-6(5)$ normal modes of vibrations for the N mass system, whether these masses are steel balls or atoms.

The vibrational kinetic and potential energies of a molecule can be expressed in Hamiltonian form as a function of $3N-6$ momenta and $3N-6$ internal coordinates. When the corresponding Hamiltonian operator is used to obtain the Schrödinger equation for molecular vibrations, this equation can be transformed into a sum of $3N-6$ Schrödinger equations (one for each normal mode of vibration), each of which has the same mathematical form as the Schrödinger equation for the simple harmonic oscillator (Eq. 8-15). The coordinates and reduced masses in these equations are complicated functions of the internal coordinates and masses of the atoms in the molecule.

It follows that each normal mode of vibration has a set of energy levels given by an equation similar to Equation 8-16. Because the normal modes are independent of each other, the vibrational energy of a molecule is the sum of the energies associated with each of the normal modes. In symbols,

$$\epsilon_{\text{vib}} = \sum_{i=1}^{i=3N-6} h\nu_i(n_i + \tfrac{1}{2})$$

The determination of the normal modes of vibration for a molecule is a problem in classical mechanics. For even small polyatomic molecules it involves a degree of mathematical manipulation which is beyond the scope of this book. We merely illustrate the type of results obtained for two simple examples, the linear and the bent symmetrical triatomic molecules, BA_2, whose normal models of vibration are represented schematically in Figure 8-17.

For the linear molecule, four $(3 \times 3 - 5)$ normal modes are shown. In the first mode, designated by its frequency ν_1, atom B remains stationary while atoms A move toward and away from it in phase. The dipole moment is zero at all times during this symmetrical stretching vibration because the two equal and opposite bond moments cancel each other. Consequently transitions between the energy levels associated with the symmetrical stretching vibration

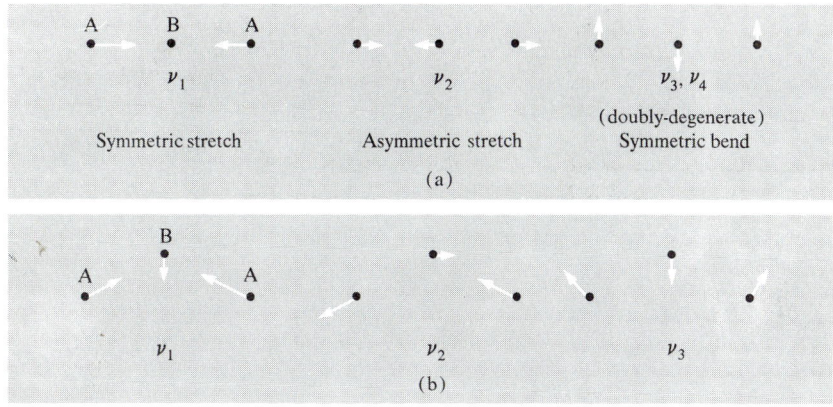

Figure 8-17. (a) Diagram of normal modes of vibration of linear, symmetric triatomic molecules, BA_2. (b) Corresponding diagram for bent BA_2 molecules.

cannot occur by absorption of electromagnetic radiation, and this mode is said to be "inactive" in the infrared spectrum.

In the second normal mode, ν_2, while the two outer atoms move in one direction, the central atom moves in the opposite direction. The bond moments do not cancel each other, and as a result of the changing dipole moment this asymmetrical stretching mode is active in the infrared.

In ν_3, the atoms A and B move in opposite directions approximately at right angles to the molecular axis and the valence angle at the central atom changes. A changing dipole moment perpendicular to the molecular axis is produced in this bending vibration and consequently it is active in the infrared. An exactly similar type of bending motion can occur in a direction which is perpendicular to the motions in ν_3. Because the only difference between these two bending modes is one of direction, the frequency ν_4 is the same as the frequency ν_3. Here we have an example of two normal modes of vibration with the same frequency. Such modes are said to be degenerate. In this case we have 2-fold degeneracy because two normal modes are involved. In some other models it is possible for 3-fold degeneracy to occur, and this is the highest degree of vibrational degeneracy.

In summary, four normal modes of vibration are predicted for the linear symmetrical triatomic model and by application of classical mechanics four are found. There are, however, only three different frequencies because of the degeneracy of two of the normal modes. Of these three frequencies, only two are expected to be observed in the infrared spectrum of a molecule in this category because the appearance of the symmetrical stretching mode is forbidden by the selection rule.

These theoretical predictions are completely consistent with the observed infrared absorption spectra of CO_2 and CS_2. Each of these spectra contains two strong absorption bands whose centers are at 2249 and 667 cm^{-1} for CO_2 and at 1523 and 397 cm^{-1} for CS_2. In each case the band with the higher wavenumber is associated with the asymmetrical stretching mode of vibration. In general, stretching modes have higher wavenumbers than bending modes (Table 8–5).

When we turn to the bent symmetric triatomic molecule we expect to find three (3N–6) normal modes of vibration and the forms of these motions are shown schematically in Figure 8–17. In this case the dipole moment changes in all three modes and each has a different frequency. Consequently we expect to find three infrared absorption bands for a molecule in this category. This expectation is realized in the infrared spectra of, for example, H_2O and SO_2.

With an increasing number of atoms in the molecule, the number of normal modes increases rather rapidly. In general, the number of distinct frequencies is less than the number of normal modes because of vibrational degeneracy, and the number of frequencies which can be found in infrared spectra is even less because some modes may be inactive. In order to obtain more information about molecular vibrations some other types of experimental observation are needed. Another branch of spectroscopy which supplies some of this information is Raman spectroscopy.

Raman Spectroscopy

Imagine that we have a glass vessel containing a substance which does *not* absorb in the visible region of the electromagnetic spectrum (Fig. 8–18). If a beam of monochromatic light of frequency ν_I is sent into this vessel, nearly all

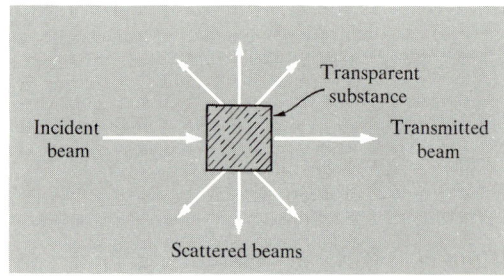

Figure 8–18. Diagram illustrating the scattering of light by a nonabsorbing substance.

of this light will be transmitted; that is, it will pass through the vessel. A small fraction of the incident light, however, will be found to emerge from the vessel in all directions as a result of the process of scattering.

In 1928 Raman and Krishnan allowed some of the light scattered from a substance at right angles to the incident beam to enter a spectrograph. The resulting spectrogram showed, in addition to the line expected at a position corresponding to the frequency of the incident (or "exciting") light, a pattern of relatively weak lines on the low-frequency side of the exciting line and a similar, but still weaker, pattern on the high-frequency side (Fig. 8–19).

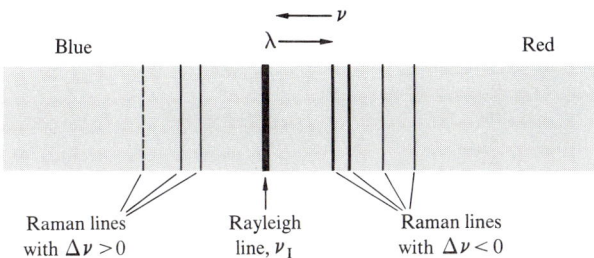

Figure 8–19. Schematic representation of a Raman spectrum. The lines on the high-frequency side of the Rayleigh line are very weak.

The line at the incident frequency ν_I is called the Rayleigh line after Lord Rayleigh who did much of the pioneer study of light scattering. The pattern of weak lines of modified frequency is called the *Raman spectrum* of the scattering substance. The frequency (or wavenumber) difference between the Rayleigh line and a Raman line is related to a vibrational or rotational transition of the scattering substance. This difference, often called the Raman shift, can be represented in the following ways:

$$\Delta\nu = |\nu_I - \nu_R| = \nu_{mol} \text{ sec}^{-1}$$

$$\Delta\omega = \left|\frac{1}{\lambda_I} - \frac{1}{\lambda_R}\right| = \omega_{mol} \text{ cm}^{-1}$$

in which the subscripts I, R, and mol refer to the incident light, the modified scattered light, and the molecule, respectively. The vertical bars signify the absolute value of the corresponding difference.

Some understanding of modified scattering may be obtained from consideration of this process from a classical point of view. Assume that the scattering sample consists of diatomic molecules whose vibrational frequency is $c\omega$ and that the incident light has the frequency ν_I. The electric field of the incident light wave ($E = E_o \cos 2\pi\nu_I t$) induces an electric dipole moment in a molecule whose magnitude depends on the molecular polarizability (Eq. 8–1) due to electron displacements. In general, this polarizability, α_e, depends on the internuclear distance. For small displacements from the equilibrium distance we can express the polarizability in the following way:

$$\alpha_e = \alpha_o + \left(\frac{\partial \alpha}{\partial x}\right) x_o \cos 2\pi c\omega t$$

in which x_o is the maximum value of $r - r_e$. Consequently the induced moment is given by

$$m = \alpha_e E$$
$$= \left[\alpha_o + \left(\frac{\partial \alpha}{\partial x}\right) x_o \cos 2\pi c\omega t\right] E_o \cos 2\pi \nu_I t$$

Making use of the trigonometric identity

$$2 \cos \theta \cos \phi = \cos (\theta + \phi) + \cos (\theta - \phi)$$

we find

$$m = \alpha_o E_o \cos 2\pi \nu_I t + \frac{x_o E_o}{2}\left(\frac{\partial \alpha}{\partial x}\right)[\cos 2\pi t(\nu_I + c\omega) + \cos 2\pi t(\nu_I - c\omega)] \quad (8\text{-}17)$$

What this equation says is that the induced dipole moment contains three components. One of these components changes with the same frequency as the incident light wave, while the other two change with frequencies which are the sum and difference of the frequency of the light wave and the vibrational frequency of the molecule. According to the electromagnetic theory, a vibrating electric dipole emits radiation whose frequency is the same as its vibrational frequency. Consequently from Equation 8–17 we would expect to find three frequencies in the scattered light: ν_I, $\nu_I + c\omega$, and $\nu_I - c\omega$. The existence of the modified frequencies depends on the change of polarizability during the vibration of the molecule.

The dipole moment induced in a molecule by an electric field depends on the orientation of the molecule with respect to the field, as well as on the internuclear distance. (These dependences were ignored in the treatment of dipole moments because only the average polarizability was of interest at that point.) Therefore the change in polarizability as a result of molecular rotation would be expected to lead to a rotational Raman spectrum with shifts from the Rayleigh line which are related to rotational frequencies.

To a considerable extent these conclusions are consistent with observations of Raman spectra. The first term in Equation 8–17 represents the unmodified Rayleigh radiation, while the other two terms represent the Raman radiation. Equation 8–17 predicts, however, that the pair of Raman lines with the same absolute value of the frequency displacement should appear in the spectrum of the scattered light with the same intensity. This prediction is in marked disagreement with experimental observations, and we must go beyond this classical interpretation of modified scattering.

From a quantum point of view we can picture the incident light as a stream of photons which, by hypothesis, cannot be absorbed by the molecules of the scattering substance. These photons, however, can "collide" with the molecules. Most of the collisions are elastic with no exchange of energy between the molecule and the photon, which merely has its direction of travel altered. A small fraction of the collisions are inelastic with exchange of energy between molecule and photon.

These processes are illustrated schematically in Figure 8–20. For the extremely short time ($\sim 10^{-15}$ sec) that a photon is near a molecule, the two can

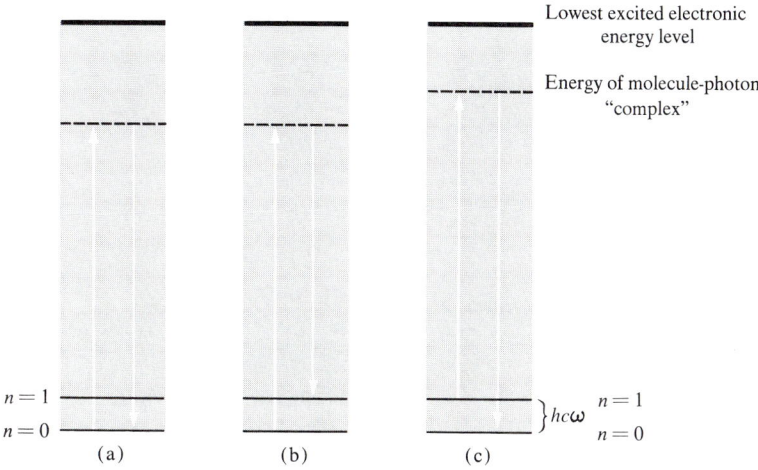

Figure 8–20. (a) Diagram for Rayleigh scattering with $\Delta v = 0$. (b) Raman scattering with $\Delta v < 0$. (c) Raman scattering with $\Delta v > 0$.

be considered to form a "complex" whose energy does not correspond to a molecular energy level. After the photon has moved away there is a high probability that the molecule will be in the same energy level that it occupied before the photon was scattered. In this case the photon energy is unchanged, and Rayleigh scattering has occurred.

There is a small probability that if the molecule was initially in its ground vibrational energy level ($n = 0$), it will be in its first excited energy level ($n = 1$) after the photon has moved away. The energy gained by the molecule is lost by the photon, which consequently has a lower frequency.

A small fraction of the molecules are initially in the excited energy level with $n = 1$. There is a very small probability that when a photon interacts with a vibrationally excited molecule and then moves away, the molecule will be in its ground energy level and the photon will have gained the energy lost by the molecule.

The modified scattering described in the last two paragraphs is Raman scattering. For a pair of Raman lines with the same value of the frequency displacement, the line with frequency higher than that of the incident light is weaker than the line with lowered frequency because of the relatively small number of molecules in an excited energy level.

For a complete treatment of light scattering, quantum mechanical perturbation theory must be applied. The vibrational selection rule which is derived from this theory is the following: the quantum number for a normal mode of vibration can change by ± 1 by Raman scattering provided that the molecular polarizability changes during this mode of vibration. The rotational selection rule is $\Delta J = \pm 2$. (This selection rule is related to the fact that, in contrast to the permanent dipole moment, the molecular polarizability is not a vector quantity. After a rotation of the molecule in the inducing electric field by an angle of 180°, the polarizability returns to its original value. Consequently when a molecule rotates with a certain frequency the polarizability changes with a frequency which is twice the molecular frequency.)

We can say that for Raman scattering the change in polarizability plays a role analogous to the change in dipole moment for the absorption and emission of radiation. It follows that homonuclear diatomic molecules and symmetrical linear molecules have both rotational and vibrational Raman spectra. In fact, some of the most reliable structural data for these molecules have been provided by their Raman spectra.

In general, because of the difference in selection rules Raman and infrared spectra supplement each other in the determination of molecular structure. Both types of spectra are usually needed to determine the symmetry group of a molecule of unknown symmetry. With the molecular formula known, it is possible to predict for each symmetry group to which the molecule might belong the number of normal modes of vibration which are active in the Raman and in the infrared spectrum. When one of these predictions is consistent with the observed spectra, the corresponding symmetry group is likely to be the correct one.

Let us take the BF_3 molecule as an example. Since this molecule contains four atoms, it has six normal modes of vibration. The two simplest possibilities for its molecular shape are the regular pyramid, symmetry group C_{3v}, or the plane equilateral triangle with the boron atom at its center, symmetry group D_{3h}.

If BF_3 belongs to symmetry group C_{3v}, it can be predicted that there are two pairs of degenerate normal modes and that all normal modes are active in both infrared and Raman spectra. Thus the infrared absorption spectrum should contain four absorption bands that can be considered as fundamentals, and the Raman spectrum should contain four lines whose Raman shifts have the same values as the wavenumbers of the infrared absorption bands.

If BF_3 belongs to symmetry group D_{3h}, it can be predicted that again there are two 2-fold degenerate normal modes but of the four observable wavenumbers two should appear only in the infrared absorption spectrum, one only in the Raman spectrum, and one in both infrared and Raman spectra. That the observed spectra are completely consistent with D_{3h} symmetry is shown by the following data:

Symmetric stretch	ν_1	888 cm^{-1} Raman active
Out-of-plane bending	ν_2	691 cm^{-1} Infrared active
Asymmetric stretch	ν_3	1446 cm^{-1} Infrared active
In-plane bending	ν_4	480 cm^{-1} Raman and infrared active

The correctness of the conclusion that BF$_3$ has D_{3h} symmetry is confirmed by the fact that it has a zero dipole moment.

The next step in the determination of molecular structure from vibrational frequencies is the expression of the vibrational potential energy in terms of an appropriate set of coordinates. The unknown force constants in this potential energy function cannot be calculated directly, but if a set of force constant values is assumed the frequencies of the normal modes of vibration can be calculated and compared with the spectroscopic data. By a trial and error process the force constants are varied until the calculated and observed vibrational frequency values are in agreement. When this has been accomplished the symmetry group of the molecule is known with a high degree of reliability, and much information about interatomic forces has been obtained. In addition, the degree of degeneracy associated with each vibrational frequency is known, a point to which we return in the next chapter.

Electronic Spectra

In the preceding section we considered the molecular scattering of visible radiation, assuming that none of this radiation was absorbed by the molecules. Now we briefly consider the case in which radiation in the visible or ultraviolet spectral regions is absorbed. Such absorption always produces a transition between electronic energy levels. From the fact that the absorption occurs in a spectral region with a shorter wavelength and higher frequency than the infrared region, it can be concluded that the differences in energy between electronic levels are much larger than those between rotational or vibrational energy levels, as is illustrated in Figure 8–21.

Electronic spectra provide one of the important sources of information about the state of electrons in molecules. Although the emission spectra of many simple molecules have been studied, we confine our discussion to absorption spectra.

An electronic transition is accompanied by changes in vibrational and rotational quantum numbers. The resulting change in the vibrational quantum

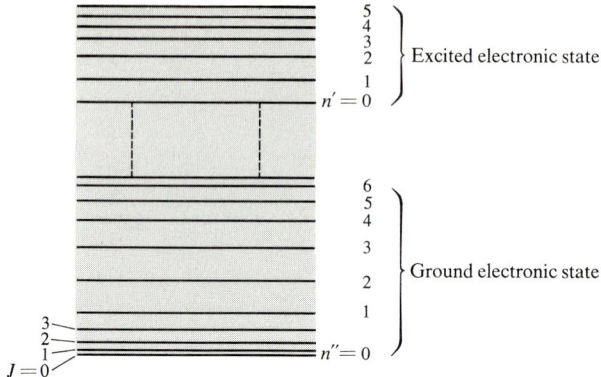

Figure 8-21. Diagram of two electronic energy levels with associated vibrational energy levels. Each vibrational level has its own set of rotational levels as indicated for the lowest vibrational level of the ground electronic state. The spacing between the electronic levels is not in correct proportion to the other spacings.

number is not limited by the selection rule mentioned before in connection with rotation-vibration transitions (p. 300). Consequently in the simple case represented by a few diatomic molecules the spectrum resulting from a single electronic transition consists of a number of absorption bands, each resulting from a particular change in the vibrational quantum number, and each resolvable into a set of closely spaced rotational lines (Fig. 8-22). When such a band

Figure 8-22. Schematic representation of part of a band system showing four bands with resolved rotational lines.

system has been analyzed, a great deal of detailed information has been obtained about the structure of the molecule in both the ground and the excited electronic states.

The resolution of a band system into individual bands with rotational fine structure is possible only for some substances and then only when they are in the gas phase. For these substances in the liquid or dissolved states, the perturbing effect of neighboring molecules smears out all or nearly all of the band structure observed in the gas phase.

An electronic transition for a diatomic molecule can be represented conveniently by a graph of the two potential energy curves for the electronic states

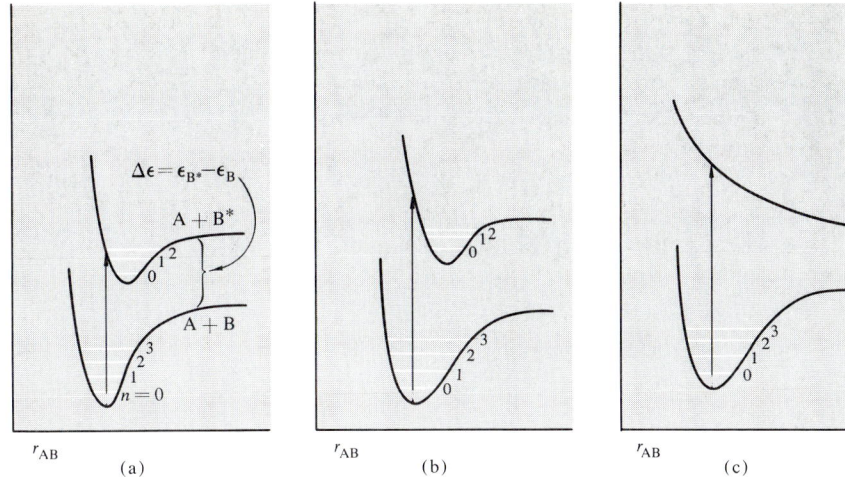

Figure 8–23. Potential energy curves for ground and excited electronic states. (a) Indicated transition results in band of maximum intensity. (b) Indicated transition results in continuous absorption. (c) All transitions lead to continuous absorption and dissociation in the upper state.

involved in the transition. (In principle, every electronic state has such a curve and these curves should be calculable by quantum mechanical methods. Because of computational difficulties, however, essentially all potential energy curves are inferred from observed spectra.) Out of a large number of possibilities, the three pairs of curves shown in Figure 8–23 have been selected as illustrative. For simplicity, the associated rotational energy levels have been omitted.

In the first case, the properties of the excited state are not greatly different from those of the ground state. It can be seen that the equilibrium internuclear distance r_e is slightly larger and the force constant, which determines the curvature near the minimum of the curve, is slightly smaller. Dissociation in the ground state leads to the normal atoms A and B, while dissociation in the excited state leads to normal A and excited B. The vertical separation of the curves at large values of r represents the energy of excitation of atom B.

Although it is possible for a transition to occur between any of the vibrational levels in the lower curve to any vibrational level in the upper one, thus producing a large number of bands, in general only a small fraction of the possible bands are observed. The relative probability of a particular change in the vibrational quantum number, which is related to the relative intensity of the corresponding band, is determined by the *Franck–Condon principle*. In essence, this principle asserts that the time required for an electronic transition to occur is so short relative to the time required for one cycle of vibration

($\sim 10^{-13}$ sec) that the transition is over and the molecule is in the excited state before the nuclei have had an opportunity to move a significant distance from their positions in the initial state. The lines representing the transitions in Figure 8-23 are vertical because the values of r are the same in both the initial and final states.

Now let us imagine that most of the molecules in a sample are in the ground vibrational state ($n = 0$). The internuclear distance for a molecule in this state changes between the limits shown in the graph, and the most probable distance is close to the one that corresponds to the minimum of the curve. It follows that the strongest band in the band system belonging to the electronic transition of interest is the one that results from the transition to the upper vibrational level ($n \sim 2$) for which the nuclei have most nearly the same separation that they had in the initial state.

Another application of the Franck–Condon principle is shown in Figure 8-23b. In this case the bond in the excited state is much weaker than that in the ground state as evidenced by the smaller curvature and depth of the upper curve and the displacement of its minimum to larger internuclear separations. If a transition occurs, such as that indicated in the figure, to a level whose energy is higher than that corresponding to the asymptote, dissociation of the molecule into A + B* will occur with the energy excess above that needed for dissociation going into kinetic energy of the dissociation products. Because this kinetic energy is not quantized, any amount of energy above a certain minimum can be absorbed. The spectrum in this case consists of a series of bands which, in the direction of shorter wavelength or higher frequency, suddenly terminates in a region of continuous absorption.

In the third case, Figure 8-23c, the curve for the upper state has no minimum, which means that the excited state has no stability with respect to dissociation products. Only continuous absorption can be observed and every transition leads to dissociation.

Every molecule has a number of excited electronic states and the detailed study of these states for even diatomic molecules is complex. For polyatomic molecules the problem is even more difficult, and one must be content with results that are cruder, both in experimental data and theoretical interpretation, than those obtained for diatomic molecules.

In favorable cases it is possible to correlate absorption bands with the various types of electronic transition which can occur in polyatomic molecules. When all valence electrons are in σ bonds, as in saturated hydrocarbons, the absorption band with the longest wavelength lies in the far ultraviolet region ($\lambda < 1400$ Å for CH_4 and C_2H_6) for which the energy per photon is high. (This region of the spectrum is often called the vacuum ultraviolet because the oxygen of the atmosphere absorbs so strongly that the use of evacuated spectrographs is a necessity.)

When the molecule under consideration contains a double bond, such as the C=C bond in olefins or the C=O bond in ketones, an absorption band is found at somewhat longer wavelengths ($\sim$1800 Å), which is attributed to the excitation of an electron in a π bonding molecular orbital to an antibonding π^* orbital. A transition of this type is called a $\pi \rightarrow \pi^*$ transition. In some molecules, such as ketones or alkyl halides, absorption at still longer wavelengths is attributed to the excitation of a nonbonding electron (a lone pair electron on the oxygen atom of the carbonyl group or the halogen atom) to an antibonding π orbital ($n \rightarrow \pi^*$ transition).

Magnetism

So far in this chapter we have been concerned with the interaction of molecules with alternating electric fields whose frequencies range from nearly zero to roughly 10^{15} cycles per second. Now we turn our attention to the type of information that can be obtained from the interaction of molecules with magnetic fields.

In some ways the behavior of matter in a magnetic field is analogous to its behavior in an electric field, but there are important differences. We can define a magnetic dipole moment in a way which parallels the definition of an electric dipole moment, although there is no isolated magnetic pole. The total magnetic moment per unit volume of matter produced by a magnetic field H is called the intensity of magnetization, M. For most substances the ratio of M to H is a constant which is a property of the substance called the *magnetic susceptibility*, χ. (We exclude from consideration here the case of ferromagnetism for which this statement is not valid.)

The magnetic susceptibility, in contrast to its electrical counterpart ($\mathcal{P}/E$), can have either positive or negative values. If the value for a substance is negative, the magnetization has a direction which is opposed to that of the applied magnetic field and the substance is said to be diamagnetic. When a diamagnetic substance is in a nonuniform magnetic field, a force acts on the substance which tends to push it out of the field. A paramagnetic substance has a positive value of the magnetic susceptibility and such a substance is attracted by a nonuniform magnetic field. This fact is the basis for the most frequently used methods for the measurement of magnetic susceptibilities.

Magnetic susceptibilities are usually measured by the Gouy method, in which the sample is in the form of a cylinder; a glass tube is used for liquids and powdered solids. The sample is suspended from one arm of an analytical balance and it is positioned with its lower end in the center of the field between the pole pieces of an electromagnet. When the field is turned on, the magnetic force acting on the sample is proportional to the susceptibility of the sample.

This force, which is equivalent to a change in the weight of the sample (Δmg) as measured by the balance, is given by the equation

$$f = \frac{\chi}{2} H^2 A$$

in which H is the magnetic field strength at the lower end of the sample and A is the area of the cross-section of the cylinder. This equation assumes that the field strength at the upper end of the sample and the susceptibility of the air displaced by the sample are both negligible. The apparatus is usually calibrated with a sample of known susceptibility.

The magnetic susceptibility measured in this way is usually referred to as the volume susceptibility, although its value is independent of the volume of the sample. It is often convenient to multiply this value by the molar volume of the substance comprising the sample and thus obtain the molar susceptibility.

Every substance has a diamagnetic contribution to its susceptibility which is nearly independent of the temperature. For diamagnetic substances this is the only contribution. In general, those substances that are paramagnetic have total susceptibilities that are larger than the diamagnetic contributions by a factor of about 100. In many cases paramagnetic susceptibilities vary inversely with the absolute temperature according to the following equation:

$$\chi_m = \frac{\mathfrak{N}\mu^2}{3kT}$$

in which μ is the paramagnetic moment of a molecule. This equation, known as *Curie's law*, has the same mathematical form as the Debye equation (Eq. 8–6) for the orientation polarizability of a polar molecule. It was derived by Langevin some years before Debye applied it to dielectric polarization.

Paramagnetic moments result from the presence of one or more unpaired electrons in an atom, ion, or molecule. In the case of an atom, the paramagnetic moment includes contributions from both the orbital motion of the electron, which depends on the quantum number m, and its spin with quantum number $+\frac{1}{2}$ or $-\frac{1}{2}$.

In molecules it is found that the orbital contribution is effectively zero, and only the spin contribution is observed. Thus from measurements of magnetic susceptibility the number of unpaired electrons can be calculated. This information has been valuable in studying the electron distribution and bonding in odd-electron molecules such as NO and ClO_2, free radicals, and complex ions containing one of the transition elements.

Nuclear Magnetic Resonance

The magnetic effects discussed in the last section are all connected with the properties of electrons. In this section we consider the important consequences

Determination of Molecular Structure

of the fact that many kinds of nuclei have a magnetic moment and behave to some extent like little bar magnets. The development of this subject leads us into the field of radio-frequency spectroscopy.

Atomic nuclei may be characterized by the following four properties: charge Z, mass M, spin quantum number I, and magnetic moment μ. The spin angular momentum of a nucleus is determined by the value of the quantum number I in the same way that the orbital angular momentum of an electron in an atom is determined by the quantum number l (p. 197) and the rotational angular momentum of a molecule is determined by the rotational quantum number J (Eq. 8–11). The spin angular momentum of a nucleus (p_n) is given by

$$p_n = \frac{h}{2\pi} \sqrt{I(I+1)} \qquad (8\text{–}18)$$

This represents a vector quantity directed along the axis of rotation.

The theory of nuclear structure has not yet advanced to such a point that values of I can be calculated from values of Z and M. Some empirical rules, however, are valid. For example, for those nuclei with even numbers for the values of both Z and M, $I = 0$. If M is even while Z is odd, I is an integer, and if M is odd, I is an integer times $\frac{1}{2}$.

The magnetic moment of a nucleus is proportional to its spin angular momentum and has the same direction. If we take as a classical model of the simplest nucleus, the proton, a rotating spherical shell of charge ϵ and mass M_p with angular momentum p, the value of the magnetic moment is given by

$$\mu = \frac{\epsilon}{2M_p c} p \qquad (8\text{–}19)$$

The speed of light, c, enters this expression as the ratio of the electromagnetic to the electrostatic system of units. Although this simple rotating shell model is not an adequate model for a proton, Equation 8–19 correctly implies that the nuclear magnetic moment is proportional to the spin angular momentum. It is necessary, however, to introduce an additional factor which cannot be calculated theoretically up to now but must be determined for each kind of nucleus from experimental data. This factor, which is represented by g, is called the nuclear splitting factor, or simply the nuclear g value. When p in Equation 8–19 is replaced by its value from Equation 8–18 and the factor g is included, the expression for the magnetic moment becomes

$$\mu = g \frac{\epsilon h}{4\pi M_p c} \sqrt{I(I+1)}$$

If we substitute for p in Equation 8–19 the value of the unit of angular momentum $\left(\frac{h}{2\pi}\right)$ introduced by Bohr, we obtain the unit of nuclear magnetic moment called the nuclear magneton, μ_o:

$$\mu_o = \frac{\epsilon h}{4\pi M_p c}$$

It is customary to express nuclear magnetic moments in terms of the nuclear magneton. When this is done we have

$$\frac{\mu}{\mu_o} = g\sqrt{I(I+1)}$$

Because nuclei with even values of Z and M have $I = 0$ and therefore zero magnetic moment, they are of no interest to us here. Although this fact excludes some of the most important and abundant nuclei (He^4, C^{12}, O^{16}, S^{32}), fortunately for many useful nuclei $I \neq 0$ and $\mu \neq 0$. Among these nuclei, the proton, with $I = \frac{1}{2}$, is outstanding.

Let us now assume that we have a nucleus of magnetic moment μ in a magnetic field of strength H. The potential energy of the nucleus in the field is given by an equation that is completely analogous to the equation for the energy of an electric dipole moment in an electric field:

$$U = -H\mu \cos \theta$$

where θ is the angle between the magnetic moment vector and the direction of the magnetic field.

At this point we must turn to the quantum-mechanical treatment of this system. According to quantum mechanics, θ is not a continuous variable and it can have only a few discrete values. This is the result of a phenomenon known as space quantization which is shown by any type of angular momentum when there is some unique direction in space, such as that provided by an external electric or magnetic field. Not only is the total angular momentum quantized in units of $h/2\pi$ but also the component of the angular momentum in the special direction is quantized.

In the particular case with which we are now concerned, the unique direction in space is the direction of the applied magnetic field, H. The possible values of the component of the nuclear spin angular momentum in the direction of the magnetic field are given by $m_I \frac{h}{2\pi}$, where the quantum number m_I can have any of the values $I, I - 1, \cdots -I + 1, -I$. For each of these $2I + 1$ different values of m_I there is a definite value of θ and a corresponding value of the component of the nuclear magnetic moment in the direction of the applied field (Fig. 8–24). Thus there are $2I + 1$ values of $\mu \cos \theta$ which we can represent by $\mu_{m_I} = \mu_o g m_I$, and $2I + 1$ different energy levels which are given by

$$U = -H\mu_o g m_I$$

The selection rule for transitions between these energy levels is found to be $\Delta m_I = \pm 1$. If such a transition can be carried out we expect the frequency of the radiation absorbed or emitted to be given by the equation

$$\nu = \frac{\Delta U}{h} = \frac{\mu_o g}{h} H \qquad (8\text{–}20)$$

Determination of Molecular Structure

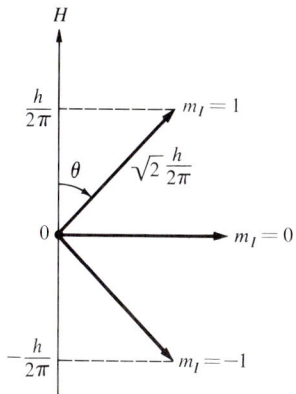

Figure 8-24. Orientation of spin angular momentum vectors for a nucleus with $I = 1$ in a magnetic field H. The possible values of the component in the direction of the magnetic field are $\frac{h}{2\pi}$, 0, and $\frac{-h}{2\pi}$.

(In terms of the angular frequency in radians per second ($\omega = 2\pi\nu$) this equation becomes

$$\omega = \left(\frac{2\pi\mu_0 g}{h}\right) H = \gamma H$$

The characteristic nuclear constant γ is called the *gyromagnetic ratio*.)

We see that, in contrast to transitions between the energy levels of atoms and molecules which occur in the absence of external fields, transitions between nuclear magnetic energy levels occur only in the presence of an applied magnetic field. The frequency associated with such a transition depends on the strength of the magnetic field at the nucleus undergoing the transition.

The study of transitions between nuclear magnetic energy levels is the subject of *nuclear magnetic resonance (NMR) spectroscopy*. It is probable that more NMR spectra are obtained for protons than for all other nuclei taken together. In the remainder of this section, for simplicity we confine our attention to NMR spectra of protons as representative nuclei.

In order to make use of Equation 8-20 we must know the value of g for protons. Careful measurements by physicists have yielded the value 5.5854. The magnitude of the nuclear magneton μ_0 is 5.049×10^{-24} ergs/gauss. Because I for the proton is $\frac{1}{2}$, the only values for m_I are $+\frac{1}{2}$ and $-\frac{1}{2}$ (Fig. 8-25). The frequency for the transition from the level with $m_I = +\frac{1}{2}$ to $m_I = -\frac{1}{2}$ is given by

$$\nu = \frac{5.049 \times 10^{-24} \times 5.5854}{6.62 \times 10^{-27}} H = 4.26 \times 10^3 \, H \text{ sec}^{-1}$$

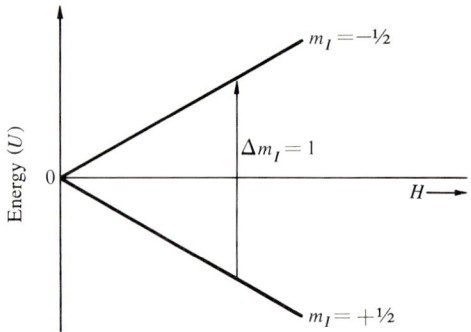

Figure 8-25. Transition between magnetic energy levels of the proton. $\Delta U = h\nu = \mu_0 g H$

The order of magnitude of the magnetic fields normally used for NMR spectroscopy is 10,000 gauss. For this field strength the frequency for protons is 42.6 megacycles per second, which is in the radiofrequency range of the electromagnetic spectrum.

The energy difference between the levels of the proton in this magnetic field is extremely small, even on a molecular energy scale. $\Delta U = 2.822 \times 10^{-19}$ ergs in a field of 10,000 gauss, as compared with a value of kT of 4.14×10^{-14} ergs at 300°K. Assuming that the ratio of the numbers of protons in these energy levels at equilibrium is given by the Boltzmann distribution law (the analog of Eq. 5-25), we have for this ratio

$$\frac{n_{-1/2}}{n_{+1/2}} = e^{-\frac{\Delta U}{kT}}$$

Because $\Delta U/kT$ is so small relative to 1, we may approximate the exponential by $1 - \Delta U/kT$. Thus for the case under consideration,

$$\frac{n_{-1/2}}{n_{+1/2}} = 1 - \frac{2.82 \times 10^{-19}}{4.14 \times 10^{-14}} = 1 - 6.82 \times 10^{-6} = 0.999993$$

This means that an excess of only four protons out of about a million points in the direction of the applied magnetic field.

The possibility of observing transitions between these energy levels depends on the difference in their populations. If upward transitions take place at a rate which is greater than the rate of the reverse transitions, the populations

Determination of Molecular Structure

in the two states will become equal and no net absorption can occur. In this condition the system of protons is said to be saturated. Clearly, the time required for equilibrium to be reestablished after the field is turned off, called the relaxation time, is of great importance in the observation of NMR spectra.

There still remains the question of how to carry out the transitions between proton energy levels. These transitions, which are magnetic dipole transitions, cannot be observed by simply sending electromagnetic radiation of the frequency given by Equation 8-20 through a sample containing hydrogen atoms. It is helpful to look at this problem from a classical point of view, recognizing the limitations of such an approach.

From a classical point of view a proton, which may be pictured as a bar magnet rotating on its axis, is a gyroscope. If a force (torque) is applied to the axis of a spinning gyroscope, the gyroscope will precess about the direction of

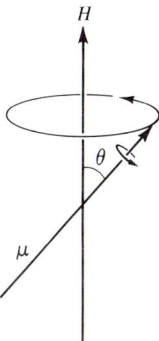

Figure 8–26. Precession of magnetic gyroscope about the direction of the applied magnetic field, with constant angle θ.

the force. When a magnetic gyroscope is placed in a magnetic field, its axis will precess about the direction of the field (Fig. 8–26), keeping the angle the axis makes with the field (θ) constant. As long as this angle is constant, the potential energy in the field is constant and no energy is absorbed from the field.

In order to change the angle θ it is necessary to apply a second magnetic field perpendicular to the main field. If the secondary field is stationary, however, whatever effect it has in one half cycle of the precessional motion is canceled by the effect in the other half cycle. To have a net effect the secondary field must rotate about the direction of the main field in synchronism with the precessional motion of the proton; that is, the secondary field must be in resonance with the precessional motion produced by the main field. Then the magnetic gyroscope will gradually aline itself in the direction of the rotating secondary field and in this process it absorbs energy from the secondary field.

Early in this century the British physicist Larmor calculated the precessional frequency for an electron in a circular orbit which is in a magnetic field, and the expression for the precessional frequency of a spinning proton has the same form. This expression for the Larmor frequency can be written in the following way:

$$\nu = \frac{g_\epsilon H}{4\pi M_p c}$$

This equation is identical with Equation 8–20.

On the basis of this discussion it can be concluded that in order for magnetic dipole transitions to occur between proton energy levels in a magnetic field, it is necessary to have a secondary magnetic field whose direction rotates in a plane perpendicular to the main field with a frequency given by Equation 8–20.

In NMR spectrometers the rotating secondary magnetic field is produced in a simple way. The output of a radiofrequency oscillator is sent through a helical coil (solenoid) of wire whose axis is perpendicular to the direction of the main field produced by an electromagnet. The sample under investigation is placed in a glass tube positioned along the axis of the coil. An electric current passing through such a coil produces a magnetic field in its center and directed along its axis. This magnetic field reverses its direction with the same frequency as the current from the oscillator.

An alternating magnetic field is entirely equivalent to two rotating magnetic fields which rotate in opposite directions with the same frequency. One of these directions is the same as the direction of the precessional motion of the protons and thus the field rotating in this direction is the secondary magnetic field mentioned above. The effect of the field which rotates in the opposite direction averages out to zero and thus it can be ignored.

The equivalence between any alternating motion represented by a vector whose magnitude changes periodically and two circular motions (Fig. 8–27) is a general property of vectors. For example, a beam of plane-polarized light (electric vector vibrating in a plane) is equivalent to two superimposed beams of

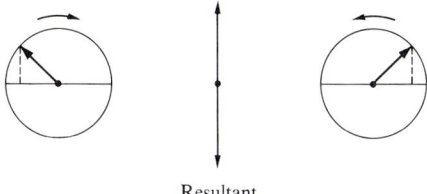

Resultant

Figure 8-27. Equivalence between the resultant of two identical vectors rotating with the same frequency in opposite directions, and a vector with constant direction and periodic change in magnitude.

circularly polarized light (electric vectors rotating in opposite directions) of the same frequency.

When the frequency of the alternating current supplied to the coil and the magnetic field at the proton have values that satisfy Equation 8–20, a condition of resonance exists and protons can absorb energy from the radiofrequency field if the nuclear system is not saturated. This resonance condition can be obtained by varying either the frequency of the oscillator or the field strength of the magnet. In practice, it is preferable for most instruments to use a fixed frequency (which at present is usually 60 megacycles per second) supplied by a crystal-controlled oscillator and to vary the magnetic field applied to the sample by the electromagnet. When resonance is reached the absorption of energy from the radiofrequency circuit is measured electrically and can be displayed on the chart of a recorder.

Now we must take into consideration the fact that the magnetic field strength represented in Equation 8–20 is the field strength at the protons in the sample. This is *not* the same as the strength of the applied magnetic field because the applied field may be modified in several ways by the sample.

1. The sample always contains paired electrons whose motion results in a diamagnetic susceptibility, as mentioned previously. The effect of this bulk diamagnetic susceptibility is to partially shield the protons from the applied magnetic field.

2. The sample usually contains many protons or other magnetic nuclei which produce magnetic fields in their neighborhoods. If the sample is a solid, these nuclei have a fixed relationship to each other in space. The resulting magnetic dipole interaction has the effect of "smearing-out" the desired spectrum. This effect is eliminated by melting or dissolving the solid because the molecular motion in the liquid phase averages the dipole interaction to zero. (For most chemical applications of NMR spectroscopy a sample in the liquid phase is used.)

3. The sample often contains protons in different chemical environments. In this case the average electron concentrations in these environments are different and so is the shielding effect of the electrons on these protons. Consequently as the applied magnetic field is increased, those protons that are relatively "bare" because they are bonded to an electronegative atom or group, will come into resonance before protons which have a relatively large electron concentration around them. The separation between the resonances for the same nuclei in different environments is called the chemical shift. The magnitude of the chemical shift is proportional to the strength of the applied magnetic field.

To illustrate the chemical shift we use the NMR spectrum of ethanol taken with a relatively crude instrument (Fig. 8–28a). This spectrum shows three

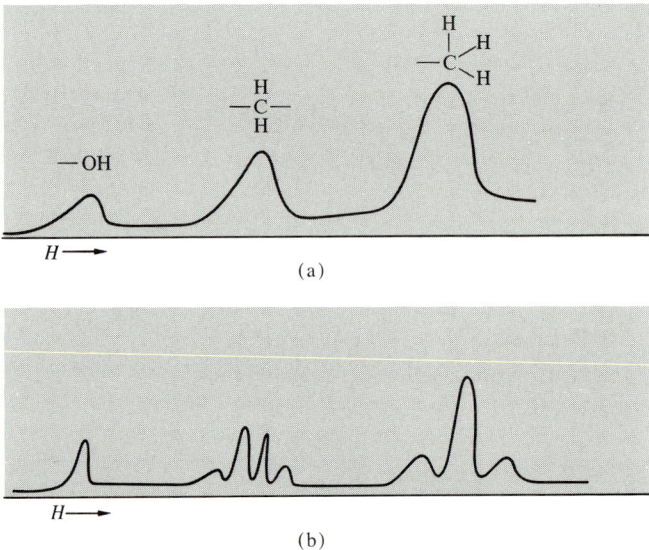

Figure 8–28. (a) Representation of NMR spectrum of ethanol under low resolution. (b) Spectrum of the same compound under high resolution.

absorption lines whose areas in the order of increasing strength of the applied magnetic field have the ratios 1:2:3. The area under an absorption line is proportional to the number of equivalent nuclei involved in the related resonance. Consequently, the smallest peak is assigned to the proton in the hydroxyl group, the next peak to the two protons in the methylene group, and the largest peak to the three protons of the methyl group.

The fact that the resonance for the hydroxyl proton occurs at the lowest value of the applied magnetic field is consistent with the fact that this proton is bonded to an electronegative atom. Because the hydrogen atom is at the positive end of the OH dipole, we conclude that the electrons in the bonding orbital are on the average closer to the oxygen atom, thus leaving the proton relatively bare and unshielded from the external magnetic field.

The electrons in the bonding C–H orbitals of the methyl group are distributed nearly equally between the carbon atom and the three protons, as evidenced by the small polarity of this bond. Consequently these protons are well shielded and require a larger applied field to bring them into resonance than does the hydroxyl proton. The protons in the methylene group have an intermediate degree of shielding because the electron withdrawal by the oxygen atom has some effect on the electron distribution around the methylene carbon atom.

Chemical shifts are very small compared to the strength of the applied magnetic field; they have a magnitude of a few milligauss in a field of 10,000 gauss. Because they vary with field strength, they are usually expressed on a relative basis in parts per million from some resonance peak used as a reference. The single sharp peak for tetramethylsilane (TMS), $(CH_3)_4Si$, which occurs at an exceptionally high field strength, is widely adopted as a reference peak for chemical shifts. When TMS is used as a reference the chemical shift, represented by δ, for a proton in a particular chemical environment is given by the following expression:

$$\delta = \frac{H_{TMS} - H_S}{H_{TMS}} \times 10^6$$

in which H_S is the strength of the applied field at which this proton comes into resonance. A set of typical chemical shifts is given in Table 8-6.

It is clear that in the chemical shift NMR spectroscopy provides us with a remarkably sensitive probe for investigating electron distributions in molecules. The chemical shift is of great value in studying the bonding in even large and complex molecules.

In addition to the chemical shift in molecules containing nonequivalent nuclei, NMR spectra have another valuable feature which can be observed only with spectrometers of high resolution. (This implies that the magnetic field is uniform over the entire sample and constant during a measurement to about one part in 10^7.) In this case many of the lines in the spectrum are split into two or more components, as illustrated by the spectrum of ethanol shown in Figure 8-28b. This phenomenon is known as spin-spin splitting. In contrast to the chemical shift, the separation of the component lines in spin-spin splitting is observed to be independent of the strength of the applied magnetic field.

The spin-spin splitting of a resonance line is attributed to the effect on the nuclei responsible for this line of other, nonequivalent magnetic nuclei in the same molecule. For example, in Figure 8-28b, it can be seen that the resonance line assigned to the methyl group protons is split into a triplet with the areas of

TABLE 8-6   PROTON CHEMICAL SHIFTS

(in parts per million relative to tetramethyl silane)

Compound	Shift
Methyl Protons	
$(CH_3)_4C$	.92
CH_3CN	1.97
CH_3Cl	3.00
CH_3NO_2	4.28
Olefinic Protons	
$(CH_3)_2C{=}CH_2$	4.6
$Cl_2C{=}CClH$	6.45
trans $\begin{smallmatrix}H\quad H\\ C_6H_5C{=}CC_6H_5\end{smallmatrix}$	6.99
Aromatic Protons	
C_7H_8	7.09
C_6H_6	7.27
C_6H_5CN	7.54

the three components in the ratios 1:2:1. Adjacent to the methyl group in the ethanol molecule there is a methylene group with its two equivalent protons. There are three possible ways in which the spins of two protons can be oriented which we can represent as follows:

$$+\tfrac{1}{2}+\tfrac{1}{2} \qquad \begin{matrix}+\tfrac{1}{2}-\tfrac{1}{2}\\ -\tfrac{1}{2}+\tfrac{1}{2}\end{matrix} \qquad -\tfrac{1}{2}-\tfrac{1}{2}$$

Total spin $+1$ 0 -1

In the last row the possible total spins of the two protons are listed. Of course, the designations $+$ and $-$ are purely arbitrary; there is no way of knowing the orientation of the spin of individual protons. The important point is the existence of three different values of the total spin of the methylene protons and the fact that one of these values can be obtained in two different ways. It follows that on the average the magnetic field at the protons in the methyl group may have three different values, one of which will occur twice as often as either of the others. In this way we account for the fact that the methyl proton resonance is split into the observed pattern of three lines.

In an entirely analogous manner we account for the fact that the methylene proton resonance is split into a quadruplet whose components have the relative areas of 1:3:3:1. We see that the three equivalent protons of the methyl group may have their spins oriented in the following ways:

	$+\frac{1}{2}+\frac{1}{2}-\frac{1}{2}$	$+\frac{1}{2}-\frac{1}{2}-\frac{1}{2}$	
$+\frac{1}{2}+\frac{1}{2}+\frac{1}{2}$	$+\frac{1}{2}-\frac{1}{2}+\frac{1}{2}$	$-\frac{1}{2}+\frac{1}{2}-\frac{1}{2}$	$-\frac{1}{2}-\frac{1}{2}-\frac{1}{2}$
	$-\frac{1}{2}+\frac{1}{2}+\frac{1}{2}$	$-\frac{1}{2}-\frac{1}{2}+\frac{1}{2}$	
Total spin $+\frac{3}{2}$	$+\frac{1}{2}$	$-\frac{1}{2}$	$-\frac{3}{2}$

Spin-spin multiplets and their interpretation are as simple as this only if chemical shifts are large compared with the separation of the component lines in a multiplet. If this is not the case, the patterns become quite complex and they must be analyzed on the basis of a detailed theory which is outside the scope of this chapter.

Supplementary Material

Group Theory

A mathematical group consists of a set of elements that are related to each other by some type of combination that is usually called multiplication, although this is often not multiplication in the usual sense of the word. This set must satisfy the following requirements:

   1. The combination, or multiplication, of any two elements must lead to another element in the group. Thus if A and B are any two elements, and AB = C, then C is a member of the group.
   2. The associative law must be obeyed. This means that (AB)C = A(BC). In general the commutative law need not be obeyed. Thus AB may be different from BA.
   3. There must be in the group an identity element I such that IA = A, where A is any member of the group.
   4. Every element A must have an inverse A^{-1}, defined by $A^{-1}A = I$, that is a member of the group.

The set of positive and negative integers including zero forms a group for which group "multiplication" is actually addition. The identity element is zero since A + 0 = A, and the inverse of any integer is its negative since A − A = 0. Notice that these integers do not form a group if group multiplication is algebraic multiplication.

The relation of group theory to the subject of molecular structure is based on the fact that the complete set of symmetry operations for a molecule forms a group, with the symmetry operations as group elements. In this group the multiplication of two symmetry operations is defined as the application of the two operations in succession. The resulting configuration is always one which can be reached by a single symmetry operation that is a member of the group.

The identity operation is an operation that leaves the molecule in its initial configuration; an example is a rotation of 360° about an axis. (The identity operation is not trivial, and it must be included as a group element.) The inverse of a symmetry operation is the operation that returns the molecule to its initial configuration.

The detailed structure of a group can be represented by a group multiplication table which shows the result of multiplying each element of the group by every other element. In such a table, each element of the group appears once and only once in each row and column. As an illustration of a multiplication table we select the table (Table 8-7) for one of the simplest symmetry groups,

TABLE 8-7 MULTIPLICATION TABLE FOR SYMMETRY GROUP C_{2v}

	I	C_2	σ_v	σ_v'
I	I	C_2	σ_v	σ_v'
C_2	C_2	I	σ_v'	σ_v
σ_v	σ_v	σ_v'	I	C_2
σ_v'	σ_v'	σ_v	C_2	I

In this particular group, each element is its own inverse.

C_{2v}, to which belong all molecules with the general formula AB_2 whose nuclei are located at the vertices of an isosceles triangle. These molecules have as their symmetry operations a 2-fold rotation C_2, a reflection σ_v through a mirror plane containing the axis of rotation and perpendicular to the plane of the triangle, and another reflection σ_v' through a mirror plane lying in the plane of the triangle. In addition, there is the identity element I.

Any further discussion of group theory is beyond the scope of this book. The student who goes on to advanced work in physical chemistry will find that this branch of mathematics has many applications to molecular theory, such as the classification of the normal vibrations of a polyatomic molecule as to activity in the infrared or the Raman spectrum, and the classification of molecular wave functions and energy states. These applications are among the most beautiful examples of the relation between mathematics and physical phenomena.

Electron Diffraction by Gases

An important method for the determination of molecular structure makes use of the diffraction of electrons by gas molecules. This method depends on

Determination of Molecular Structure

the wavelike behavior of electrons. It was pointed out on page 177 that experimental evidence for the diffraction of electrons by crystals not only confirmed de Broglie's suggestion that matter has wavelike properties, but also agreed quantitatively with his equation for the associated wavelength, $\lambda = h/mv$.

If a beam of electrons is accelerated through a potential difference of 10,000 volts for example, the associated wavelength, calculated from the de Broglie equation, is 0.12 Å. If such a beam is sent through a very thin foil of metal, the electrons are scattered by the atoms of the metal. Because the wavelength of the electrons is of the same order of magnitude as the interatomic distances in the metal, this scattering results in diffraction. This means that the

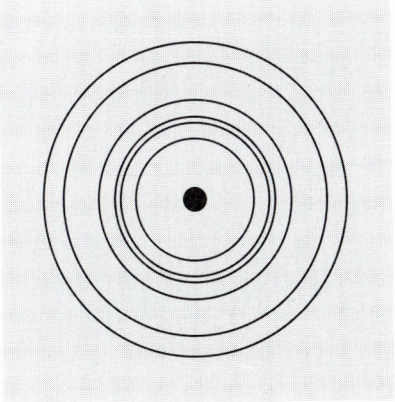

Figure 8–29. Diagrammatic representation of electron diffraction pattern of a metallic foil.

electrons are scattered preferentially at certain angles with respect to the beam because of constructive interference of the electron waves. At other angles the electron waves are canceled by destructive interference. When the electrons transmitted by the foil are allowed to strike a photographic plate, the resulting image consists of a set of concentric rings as shown in Figure 8–29. Each ring corresponds to an angle of diffraction which depends on both the interatomic spacings in the metal and the wavelength of the electrons.

Diffraction also occurs when an electron beam is sent through a diatomic or a polyatomic gas at a low pressure, even though the gas molecules have random orientations with respect to the electron beam. In this case, a photograph of the diffraction pattern shows a set of concentric rings which are much more diffuse than the rings in the diffraction pattern of a metal foil. The theory for the diffraction of electrons by gas molecules was derived by Wierl, who calculated the scattering of electrons by a molecule in a fixed position and then averaged this result over all possible orientations of the molecule.[7] A simplified form of the Wierl equation is the following equation:

$$\frac{I(\theta)}{I_o} = K\left(\sum_{j>i}^{N} \sum_{i=1}^{N} Z_i Z_j \frac{\sin sr_{ij}}{sr_{ij}}\right)$$

In this equation the ratio of intensities $\frac{I(\theta)}{I_o}$ is equal to the fraction of incident electrons scattered at an angle θ, Z_i is the atomic number of the ith atom, s is equal to $\frac{4\pi}{\lambda} \sin \frac{\theta}{2}$, r_{ij} is the distance between any two atoms i and j, and N is the number of atoms in the molecules. The sums are taken over all pairs of atoms.

Any substance that has a vapor pressure of at least a few torr at a moderate temperature can be studied by the electron diffraction method. The apparatus

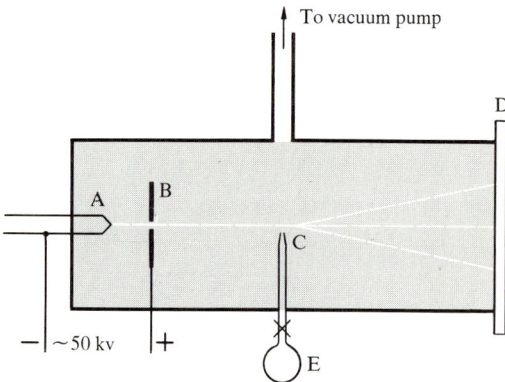

Figure 8–30. Schematic diagram of electron-diffraction apparatus. A, hot cathode source of electrons; B, anode at a positive potential of about 50 kv; C, nozzle for forming a jet of gas; D, photographic plate; E, gas sample.

used is shown schematically in Figure 8–30. The scattering chamber is highly evacuated to remove foreign gas molecules which would produce undesired scattering. The stopcock is opened for a fraction of a second to release a pulse of sample molecules near the inlet tube while a beam of electrons accelerated by a known potential difference passes through the molecules. The electrons scattered by the molecules strike a photographic film or plate. In the image formed after development, the blackness of any area is related to the number of electrons that struck that area. By an appropriate method, the blackness of the image along a line from the center of the pattern to its periphery can be measured and converted to relative intensity of the scattered electrons. In this way the

[7]The basic theory was derived by Debye in 1915 for the scattering of X-rays by a crystalline powder consisting of randomly oriented crystallites. This type of scattering is discussed in Chapter 12.

diffraction pattern can be represented by a graph of relative intensity of scattered electrons versus the parameter s (Fig. 8-31). (Since the de Broglie wavelength

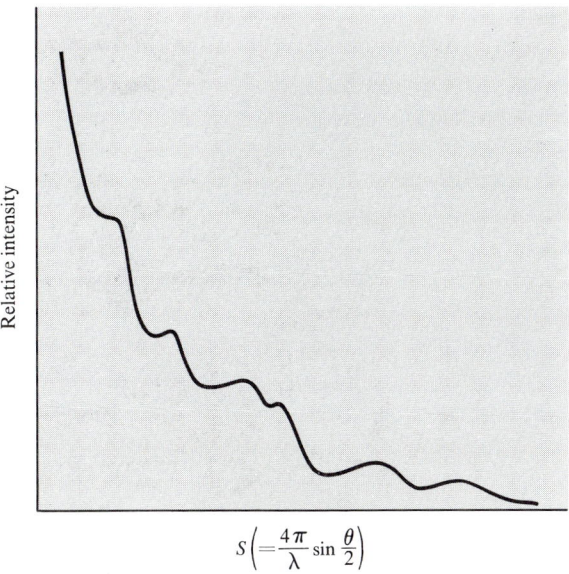

Figure 8-31. Graph of electron diffraction pattern of a polyatomic molecule.

is a constant for a fixed value of the accelerating potential difference, s is a constant times $\sin \theta/2$.)

Unfortunately there is no way of obtaining the desired structural information for a molecule directly from its scattering curve, and thus indirect methods must be used. In the simplest of these, the starting point is a model of the molecule for which approximate values are assigned to the various distances between all of the atoms. Using these values a theoretical scattering curve can be calculated from the Wierl equation, and this curve can be compared with the experimental curve. Usually the two curves are rather different at first, but by systematically varying the distance values a theoretical curve can be obtained that matches the experimental curve. Then the set of interatomic distances used is considered to be the correct one.

If more than a few independent distances must be determined, this trial and error method is time-consuming and subject to error. When there is reason to believe that the molecule under consideration possesses some symmetry elements the problem is greatly simplified because then certain distances are necessarily equal to each other. For example, the CCl_4 molecule has four carbon-to-chlorine distances and six chlorine-to-chlorine distances. From a variety of evidence, however, it is known that this molecule belongs to the point symmetry group T_d, the group of the regular tetrahedron. It follows from the geometry of the tetrahedron that all of the C–Cl distances are the same, and that all of the Cl–Cl distances are equal to $\sqrt{\frac{8}{3}}$ times the C–Cl distance. Thus there is only one independent parameter to be determined, which we can choose to be $r_{\text{C–Cl}}$. In this case the Wierl equation for CCl$_4$ takes the following form:

$$\frac{I(\theta)}{I_o} \propto 4 \times 6 \times 17 \frac{\sin sr_{\text{C–Cl}}}{sr_{\text{C–Cl}}} + 6 \times 17 \times 17 \frac{\sin s\sqrt{\frac{8}{3}} r_{\text{C–Cl}}}{\sqrt{\frac{8}{3}} sr_{\text{C–Cl}}}$$

If a series of values of $sr_{\text{C–Cl}}$ is substituted into the equation above and the resulting values of $I(\theta)/I_o$ are plotted against $sr_{\text{C–Cl}}$, a curve will be obtained that has a number of maxima and minima. When this curve is compared with the experimental scattering curve for CCl$_4$, plotted as a function of s, corresponding maxima and minima in the two curves can be identified. Then for each maximum and minimum, a value of $r_{\text{C–Cl}}$ can be found from the values of $sr_{\text{C–Cl}}$ and s. If the assumption of the tetrahedral model is valid, the set of $r_{\text{C–Cl}}$ distances obtained in this way will cluster around a mean value that is in good agreement with values determined by other methods.

A few additional points are worthy of mention:

1. Because of its low atomic number, hydrogen has very low scattering power and distances to hydrogen atoms are difficult or impossible to determine by electron diffraction.

2. From the experimental diffraction data for a molecule it is possible to calculate values for a function of r known as the radial distribution function, which gives the probability of finding two atoms separated by a distance r. Maxima in this function are observed at values of r corresponding to interatomic distances in the molecule. In this way fairly accurate values for these interatomic distances can be found directly without use of the trial and error method mentioned above.

3. In the early work on electron diffraction by gases the desired diffraction pattern was partially obscured by a background blackening due to electrons that were either not scattered or scattered by a mechanism ("incoherent scattering") that does not involve interference and diffraction. One of the most important advances in recent years is the use of a rotating sector in front of the photographic plate used to record the diffraction pattern. This sector is carefully shaped to reduce the effect of the unwanted electrons, particularly near the center of the pattern, and thus permit the desired diffraction pattern to be obtained in a more quantitative manner than was previously possible.

Supplementary References

G. M. Barrow, *Molecular Spectroscopy*, McGraw-Hill, New York, 1960.

J. C. D. Brand and J. C. Speakman, *Molecular Structure*, Edward Arnold, London, 1960.

P. Debye, *Polar Molecules*, Chemical Catalog Co., New York, 1929, reprinted by Dover Publications.
H. B. Dunford, *Elements of Diatomic Molecular Spectra*, Addison-Wesley, Reading, 1968.
C. P. Smythe, *Dielectric Behavior and Structure*, McGraw-Hill, New York, 1955.
P. J. Wheatley, *Molecular Structure*, 2nd edition, Oxford University Press, London, 1968.

Problems

1. List the symmetry elements and give the point groups for (a) naphthalene; (b) 1,2-dichlorobenzene; (c) 1,3,5-trichlorobenzene; (d) phosphine; (e) chloroform; (f) carbon tetrachloride; (g) ethylene; (h) trans-dichloroethylene; (i) dichloromethane.

2. The molecular polarizability of CCl_4 is 1.01×10^{-23} cm^3/molecule and its vapor pressure at 30°C is 143 torr. Calculate the dielectric constant of CCl_4 vapor in equilibrium with the liquid at 30°C. **Ans.** 1.000576

3. Calculate the dipole moment and the distortion polarizability of chlorobenzene from the following data for the molar polarization of the vapor:

$T(°K)$	373.6	403.7	429.8	476.1	518.4
$P_m(cm^3)$	79.6	75.7	73.1	69.0	66.2

Ans. $\mu = 1.71 D$, $\alpha = 1.25 \times 10^{-23}$ cm^3

4. At 25°C the limiting value at infinite dilution of the molar polarization of *n*-butylamine in benzene is 64.0 cc. The refractive index and density of pure *n*-butylamine are 1.399 and 0.736 g/cc, respectively. Calculate the dipole moment of *n*-butylamine. **Ans.** 1.40*D*

5. Calculate the dipole moment of *n*-amyl fluoride from the following data for its solution in benzene at 25°C:

X_{solute}	0.00723	0.01552	0.2123	0.0000
D	2.302	2.337	2.360	2.273
ρ	0.8726	0.8716	0.8709	0.8735

The refractive index and density of pure *n*-amyl fluoride are 1.3573 and 0.7849 g/cm^3, respectively.

6. The dipole moment of nitrobenzene is 4.22 *D*. Calculate the dipole moment of (a) 1,3-chloronitrobenzene; (b) 1,4-chloronitrobenzene; (c) 1-nitro-3,5-dichlorobenzene. **Ans.** (a) 3.68 *D*; (b) 2.52 *D*; (c) 2.52 *D*

7. The wavenumbers of the pure rotation lines of HBr are represented by the equation $\omega_J = 16.90 J$ cm^{-1}. Calculate (a) the moment of inertia and (b) the internuclear distance for HBr. **Ans.** (a) 3.32×10^{-40} g cm^2; (b) 1.42 Å

8. An absorption line for I^{129}Cl^{35}, due to the transition $J = 0 \rightarrow J = 1$, is observed in the microwave region of the spectrum at a frequency of 6.844×10^9 cycles per second. Calculate (a) the moment of inertia and (b) the internuclear separation for this molecule.

9. The first three lines in the pure rotation spectrum of HF have the following wavenumbers in cm^{-1}: 41.1, 82.2, and 123.2. Calculate the internuclear distance in HF. **Ans.** 0.93 Å

10. HCN has a series of absorption lines in the microwave region of the spectrum, with the frequencies of two adjacent lines having the values 17.73×10^{10} and 26.59×10^{10} cycles per second. Give the quantum numbers of the two states involved in these transitions and calculate the moment of inertia of HCN.

11. Calculate the wavenumbers of the first four rotational lines of DCl^{35}, using a value of 1.28 Å for the internuclear distance.

12. HI has an infrared absorption band with its center at 4.34 microns. (a) Calculate the force constant for this molecule. (b) Calculate the wavelength of the center of the absorption band for DI. **Ans.** (a) 3.12×10^5 dynes/cm; (b) 6.10 μ

13. From the data in Table 8–5, calculate the force constants of (a) Li$_2^7$, and (b) Cl$_2^{35}$.

14. If HBr behaved as a rigid rotor, that is, one with a constant average internuclear separation, some of the lines in its fundamental vibrational absorption band would have the following wavenumbers: 2616.2, 2633.1, 2666.9, 2683.8, and 2700.7 cm^{-1}. (a) For each line assign quantum numbers for the initial and final states of the corresponding transition. (b) Calculate the internuclear separation and the force constant.

15. For acetone a normal mode of vibration that involves the stretching of the carbonyl bond is active in both Raman and infrared spectra. The Raman line associated with this normal mode is found at 5470 Å when excited by incident light whose wavelength is 5000 Å. Calculate the wavelength of the center of the corresponding infrared absorption band. **Ans.** 5.88 μ

16. When the Raman spectrum of CCl$_4$ is obtained, using the mercury arc line at 4358 Å for excitation, Raman lines are observed at the following wavelengths: 4400.0, 4418.8, and 4447.0 Å. Calculate the Raman shift for each of these lines.

17. The absorption spectrum of I_2 vapor consists of a series of bands which terminate in continuous absorption at 4995 Å. I_2 molecules in the upper state dissociate into an atom in its ground state and one with an excitation energy of 21.70 kcal/mole. Calculate the dissociation energy of I_2 in its ground electronic state. **Ans. 35.6 kcal/mole**

18. A particular isotope absorbs energy at the radio frequency of 17.3×10^6 cycles per second when it is in a magnetic field of 10,000 gauss. Calculate the nuclear g factor for this isotope.

19. Calculate the strength of the magnetic field in which protons have a Larmor (precessional) frequency of 6×10^7 cycles per second.
Ans. 14,100 gauss

20. The NMR spectrum of a compound whose empirical formula is C_2H_4O consists of a doublet and a quartet containing 4 lines with relative intensities of 1,3,3,1. The area under the doublet is three times the area under the quartet. Identify the compound and give reasons for your answer.

9

Statistical Mechanics

Introduction

For several chapters we have been concerned with the properties of individual molecules. We have discussed in an introductory way the distribution of electrons in molecules and the various types of energy levels molecules can have. Now we move toward the interpretation of the behavior of matter on a macroscopic scale in terms of molecular properties.

To a certain extent we started on this program in Chapter 5. In that chapter we saw that we could account for some of the properties of gases, notably the pressure, by adopting a very simple molecular model. We also saw, however, that the kinetic theory, as developed on the basis of classical mechanics, leads to incorrect results for the heat capacity of even a diatomic gas (p. 153).

We return to this program with a much better theoretical background than we used in Chapter 5, and we adopt a quite different point of view. As before, we are interested in the behavior of collections of very large numbers of molecules — of the order of Avogadro's number. Because any observed property of matter must be some kind of average of the properties of large numbers of molecules, it is clear that we must depend on statistical methods to obtain the desired averages. For this purpose we need some of the results of probability theory. Actually, some of these results were used implicitly in Chapter 5, and the concept of probability was used in Chapter 6 for the interpretation of quantum-mechanical eigenfunctions. At this point we start with an introduction to the important (and sometimes entertaining!) subject of probability theory.

Probability

The development of the mathematical aspects of probability started with the effort of gamblers to raise their level of success in games of chance, such as cards or dice, in which the ability to calculate odds has a financial value. Any mention of probability always implies the existence of a certain amount of ignorance, for when we can predict exactly the outcome of a situation or the result of a process there is no need to introduce the concept of probability.

If a perfect die is thrown at random, we expect that when it lands any one of its faces will be seen with the same likelihood as any other face. This expectation is expressed by the statement that the probability of the appearance of a particular face is $\frac{1}{6}$. From this point of view we can give the following definition of probability: If a certain event can occur in n equally likely and mutually exclusive ways, n_K of which lead to the result K, then the probability of the occurrence of K is n_K/n. For example, what is the probability of drawing an ace from a well-shuffled deck of cards? A card can be drawn in 52 equally likely ways, of which four result in the drawing of an ace. Consequently, the probability of drawing an ace is $\frac{4}{52}$.

This definition is useful in many circumstances but it has a serious deficiency. How, in general, can we tell that different ways of realizing a certain event are equally likely? Suppose the die that we threw had been loaded without our knowledge; what then is the probability of throwing a five? One way to find out is to throw our die many times and to observe what number appears at each throw. If a five turns up in $\frac{1}{6}$ of the throws, we could say that the probability of throwing a five is $\frac{1}{6}$. Even with a perfect die, however, we might get a run of fives in a series of throws. To reach a reliable conclusion we must throw our die a very large number of times to see what the values of the relative frequencies of the various faces are.

These considerations lead to a second definition of probability: If the relative frequency with which a certain event K occurs in n repeated trials is n_K/n, then $\lim_{n \to \infty} n_K/n$ is the probability of the occurrence of K, provided the limit exists. As an example of the usefulness of this definition, the subject of insurance statistics may be mentioned. No one can predict whether a particular man between the ages of 29 and 30 will live for another year, but the insurance companies can state quite accurately what the probability of his survival is on the basis of a large number of "repeated trials."

The proof that the two probabilities defined above are actually the same is not simple. Here we assume that they are the same, and we will use whichever point of view seems to be appropriate.

Let us take a closer look at the problem of repeated trials. Suppose that we have a bag containing 10 red balls and 5 black balls, all identical except for color. If a ball is taken at random from the bag, what is the probability that it is red? From the first definition above we can say that the answer is $\frac{2}{3}$. Now suppose that this ball is returned to the bag which is then shaken and a ball is drawn again. What is the probability that both of the 2 balls which have been drawn are red? Because the 2 drawings are independent of each other, the probability that the second ball is red is also $\frac{2}{3}$ and the probability that both balls are red is the product of the probabilities of the independent events, namely $\frac{4}{9}$.

What is the probability that on two independent drawings 1 ball is red and 1 ball is black? There are two ways of realizing this event, because the first ball drawn may be either red or black. The probability that the first ball drawn is black and the second one is red is $5/15 \times 10/15 = 2/9$. Likewise the probability that the first ball is red and the second one is black is $10/15 \times 5/15 = 2/9$. Since there are two alternative ways of drawing 1 black ball and 1 red ball the probability of this event is the sum of the probabilities of the two ways that the event can be realized, namely, $2/9 + 2/9 = 4/9$.

The only other possibility in the independent drawings of 2 balls is that both of them should be black, for which the probability is $1/9$. Then the total probability of the three events is $4/9 + 4/9 + 1/9 = 1$, a value which shows that all of the possibilities have been considered.

Instead of continuing to draw balls out of a bag, let us now consider a different physical situation. Imagine that we have a box that is divided by a thin partition into two compartments, A and B, whose open areas expressed as fractions of the total area are a and b. Into this box are tossed at random balls which, though physically identical, are distinguishable. (We will find the question of distinguishability to be of basic significance and we return to it at a later point.)

The probability that a ball tossed into the box will go into compartment A is a, and by hypothesis $a + b = 1$. If 2 balls are tossed into the box, we can express all of the possibilities in the following way:

$$(a + b)^2 = a^2 + 2ab + b^2 = 1$$

We give the successive terms in this equation the following interpretation: a^2 represents the probability that both balls will go into compartment A, and b^2 represents the probability that both balls will go into compartment B. The coefficient 2 before ab indicates that there are two different ways in which one ball goes into A and the other goes into B.

For 3 balls we have

$$(a + b)^3 = a^3 + 3a^2b + 3ab^2 + b^3 = 1$$

This result is consistent with the fact that there are three different ways in which 2 balls can go into either A or B and the other one into the other compartment.

If n balls are tossed into the box, the coefficients in the successive terms of the general equation are given by the coefficients in the binomial expansion which can be written in the following way:

$$(a + b)^n = a^n + \frac{n!}{(n-1)!} a^{n-1}b + \frac{n!}{2!(n-2)!} a^{n-2}b^2 + \cdots$$
$$+ \frac{n!}{r!(n-r)!} a^{n-r}b^r + \cdots + b^n$$
$$= \sum_{r=0}^{r=n} \frac{n!}{r!(n-r)!} a^{n-r}b^r \qquad (9\text{-}1)$$

In this equation $n!$ (n factorial) $= n(n-1)(n-2)\cdots 1$. For consistency it is necessary that $0! = 1! = 1$.

The general term in this equation has the following significance: If a is the probability that a ball goes into compartment A and b is the probability that a ball goes into compartment B, then the probability that the first $n - r$ balls go into A and the remainder go into B is $a^{n-r}b^r$. If, however, we want the probability that *any* $n - r$ balls go into A, then we must calculate the number of different ways that $n - r$ balls can go into A while the remaining r balls go into B.

In order to do this we note that the first ball tossed can be any one of n different choices. Once the first ball is chosen there are $n - 1$ choices for the second, and so on until all of the $n - r$ balls which go into compartment A are chosen. The total number of choices is

$$n(n-1)(n-2) \cdots (n - [n - r - 1])$$

In this calculation, however, we have implied that two ways of choosing the $n - r$ balls are different if they merely involve an interchange of two specific balls. But since by hypothesis we do not care about the order in which the chosen $n - r$ balls go into A, we must divide the total number of choices by the number of ways that the chosen $n - r$ balls can be permuted, or rearranged among themselves, which is $(n - r)!$

Thus the number of different ways that any $n - r$ balls can go into A is given by

$$\frac{n(n-1)(n-2) \cdots (n - [n - r - 1])}{(n - r)!}$$

When we multiply both numerator and denominator of this expression by $r!$ we obtain

$$\frac{n(n-1)(n-2) \cdots (r+1)}{(n-r)!} \frac{r!}{r!} = \frac{n!}{(n-r)!r!}$$

This is exactly the coefficient of the general term in the binomial expansion (Eq. 9–1). We conclude, therefore, that this term represents the probability

Statistical Mechanics

that a particular set of $n - r$ balls will go into A and the remainder will go into B, multiplied by the number of different ways that a set of $n - r$ balls can be obtained from the total number of n balls.

These results can be generalized to include any number of compartments in the box. If the box has K compartments numbered from 1 to K and the probability that a single ball is in compartment K is represented by p_K, the probability that n balls are distributed in such a way that n_1 are in compartment 1 and n_K are in compartment K is given by the following general expression:

$$P_{n_1 \cdots n_K} = \frac{n!}{n_1! n_2! \cdots n_K!} p_1^{n_1} p_2^{n_2} \cdots p_K^{n_K} \tag{9-2}$$

Suppose that the box contains 5 compartments, all of equal size, and that 20 identical balls are tossed into it. What is the probability that 5 balls go into compartment 1, 3 into compartment 2, none in compartment 3, 8 in compartment 4, and 4 in compartment 5? The probability of this distribution is

$$\frac{20!}{5! 3! 0! 8! 4!} \left(\frac{1}{5}\right)^{(5+3+8+4)}$$

Now let us consider two limiting cases for this system which have some bearing on later developments.

First we ask for the distribution of minimum probability. This clearly is the distribution with all of the balls in one compartment, because there is only one way in which this distribution can be realized. Next, we ask for the distribution of maximum probability. The second factor in Equation 9–2, representing the probability of one particular distribution, is not changed by a change in the number of balls in each compartment. Consequently we wish to maximize the first factor. Intuitively, one might expect that the most probable distribution is one with 4 balls in each compartment, and it is easy to prove that this is indeed the case.

For the simplest case we can assume that we have two equal compartments and $2a$ balls (a is any integer). For uniform occupancy the number of ways of realizing this distribution is $\frac{(2a)!}{a! a!}$. Now assume that 1 ball is transferred from one compartment to the other. The number of ways of realizing this distribution is $\frac{(2a)!}{(a+1)!(a-1)!}$. When we divide the first of these expressions by the second, we obtain

$$\frac{(a+1)!(a-1)!}{a!a!} = \frac{(a+1)(a)(a-1)\cdots(a-1)(a-2)\cdots}{a(a-1)(a-2)\cdots a(a-1)(a-2)\cdots}$$
$$= \frac{a+1}{a} > 1$$

If 2 or more balls are transferred, this ratio is even larger. Thus we conclude that the uniform distribution is the most probable.

Boltzmann Distribution Law

After this mathematical digression we come back to the subject of molecules and to the derivation of a fundamental equation, the Boltzmann distribution law. (One special form of this law is expressed by Equation 5–25.) Suppose that we have a system of n noninteracting, identical molecules which contains a certain amount of energy, E. When the system is at equilibrium, how is the energy distributed among the molecules?

The pioneering work on this problem was done in the last century by Boltzmann and by Gibbs, each of whom developed his own approach to a solution. Because their work was done before Planck introduced the quantum theory their efforts were only partially successful. Some of the troublesome points in their theoretical treatments were not clarified until quantum mechanics was developed.

We proceed with the solution to this problem from a quantum point of view. In previous chapters we have learned that a molecule can exist in any one of a set of states characterized by certain eigenfunctions and associated energy levels which we label $\epsilon_0, \epsilon_1, \ldots$. Although we have seen that there are various types of energy levels, such as translational, rotational, vibrational, and electronic levels, for the present we need not specify the nature of the energy levels under consideration.

We now express our problem in the following way: At equilibrium, how are the molecules distributed among the energy levels? In other words, what is the set of occupancy numbers $n_0, n_1, \ldots n_K$ (n_K is the number of molecules in energy level ϵ_K) for the system?

To begin with, we must consider what the word equilibrium means when applied to a system of molecules. The only characteristic of the equilibrium distribution of energy is its independence of time. In other words, the total amount of energy and the number of molecules occupying each energy level remain constant. In the light of the last section we are able to find the most probable distribution, which is also independent of time. This leads us to the first of several basic assumptions which we have to make and which are justified

Statistical Mechanics

by the success of the results obtained. We assume that the equilibrium distribution is the same as the most probable distribution.

Just as in the case of the distribution of balls among compartments (Eq. 9-2), the probability of a distribution of molecules among energy levels is given by the product of (1) the probability of a distribution of particular molecules among the energy levels and (2) the number of different ways this distribution can be realized. (Notice that this implies that the molecules can be distinguished from each other.) In this case, however, we have no way of telling what the probability of a particular distribution is. We therefore make the simplest assumption, which is that all distributions of specific molecules have the same probability. Then the probabilities of all distributions include the same unknown factor.

Because we are interested only in relative values of probability in order to find the distribution of maximum probability, we can ignore the common unknown factor. We need to consider only the number of ways each distribution, specified by a certain set of occupancy numbers, can be realized. This number is called the thermodynamic probability (W) of the distribution. (Although W is called the thermodynamic probability, it is actually not a probability. In general, it is a very large number.)

By analogy with Equation 9-2 we can write the expression for W in the following way:

$$W = \frac{n!}{n_1! n_2! \ldots n_K!} \qquad (9\text{-}3)$$

Our immediate problem is to find the most probable distribution; that is, the occupancy numbers, n_K, of the distribution for which W is a maximum. A small change (δn_K) in the occupancy of any of the energy levels results in a variation (δW) of W. The process of finding the values of the n_K that make W a maximum must be carried out, however, taking into consideration that the total number of molecules and the total energy of the system are constant and that the only change is a shift in the molecules among the energy levels. These requirements are expressed in the following equations:

$$\delta W = 0 \quad \text{(condition for maximum)} \qquad (9\text{-}4)$$
$$\Sigma \delta n_K = 0 \quad \text{(constant } n\text{)} \qquad (9\text{-}5)$$
$$\Sigma \epsilon_K \delta n_K = 0 \quad \text{(constant total energy)} \qquad (9\text{-}6)$$

To carry out the necessary variation we need a more useful way of expressing the factorial of a large number. This way is provided by *Stirling's approximation*, which can be written in several different forms. One of these is the following:

$$n! = n^n e^{-n} (2\pi n)^{\frac{1}{2}}$$

which in logarithmic form becomes

$$\ln n! = n \ln n - n + \tfrac{1}{2} \ln 2\pi n$$

For our purposes the value of n is so large that the first two terms are an adequate approximation.

In order to find the distribution of maximum probability with the use of Stirling's approximation, we find the maximum of the logarithm of W. (This step does not introduce any difficulty because W is a maximum when $\ln W$ is a maximum.) When we take the logarithm of W, the product of factorials in the denominator of Equation 9–3 becomes a sum of logarithms of factorials and we can express $\ln W$ as follows:

$$\ln W = \ln n! - \Sigma \ln (n_K!)$$

With the application of Stirling's approximation this equation becomes

$$\ln W = n \ln n - n - \Sigma(n_K \ln n_K - n_K) \tag{9-7}$$

On taking the variation of both sides of this equation, the result is

$$\delta \ln W = \delta(n \ln n - n) - \delta\Sigma(n_K \ln n_K - n_K)$$

The first term is zero because n is a constant. Because the order of the operators δ and Σ can be interchanged, we can write this equation in the following form:

$$\delta \ln W = -\Sigma \delta n_K \ln n_K + \Sigma \delta n_K$$

When we make use of Eqs. 9–4 and 9–5, this equation becomes

$$\Sigma \delta n_K \ln n_K = 0 \tag{9-8}$$

Although the occupancy numbers are integers they can be treated as continuous variables because we are dealing with very large numbers of molecules, and a few molecules shifted from one energy level to another form only an infinitesimal fraction of the total number. (We have met a similar situation before, p. 128.) In this case the mathematical operation of variation follows the same rules as the operation of differentiation. Applying the rule for differentiating a product to Equation 9–8 we obtain

$$\Sigma(n_K \delta \ln n_K + \ln n_K \delta n_K) = 0$$

Because $\delta \ln n_K = \dfrac{\delta n_K}{n_K}$, this equation finally reduces to

$$\Sigma \ln n_K \delta n_K = 0 \tag{9-9}$$

If the set of variations in the occupancy numbers represented by δn_K could be assigned arbitrary values the only way this equation could be satisfied is for each $\ln n_K$ to be equal to zero individually. But the δn_K cannot be assigned arbitrary values because Equations 9–5 and 9–6 provide two relationships among the δn_K which must be satisfied. A mathematical technique for finding the maxima (or minima) of functions of many variables, subject to given relations among the variables was developed by Lagrange and is known as Lagrange's

Statistical Mechanics

method of undetermined multipliers. To apply this method, each of the Equations 9–5 and 9–6 is multiplied by a constant (independent of K) whose values must be found later. With the constants represented by α and β, these two equations become

$$\alpha \Sigma \delta n_K = 0$$

and

$$\beta \Sigma \epsilon_K \delta n_K = 0$$

When these equations are added to Equation 9–9 the result is

$$\Sigma \ln n_K \delta n_K + \alpha \Sigma \delta n_K + \beta \Sigma \epsilon_K \delta n_K = 0$$

When the terms belonging to the same value of K are grouped together, this equation can be written in the following form:

$$(\ln n_1 + \alpha + \beta \epsilon_1) \delta n_1 + (\ln n_2 + \alpha + \beta \epsilon_2) \delta n_2 + \cdots = 0$$

Because of the inclusion of the undetermined multipliers, α and β, the values of the δn_K can now be considered as independent of each other. It follows that each of the factors in parentheses individually must be equal to zero:

$$\ln n_1 + \alpha + \beta \epsilon_1 = 0$$
$$\ln n_2 + \alpha + \beta \epsilon_2 = 0$$
$$\cdots \cdots \cdots \cdots \cdots \cdots$$
$$\ln n_K + \alpha + \beta \epsilon_K = 0$$

Thus, in general, we can write

$$\ln n_K = -\alpha - \beta \epsilon_K \qquad (9\text{--}10)$$

and

$$n_K = e^{-\alpha} e^{-\beta \epsilon_K} \qquad (9\text{--}11)$$

This equation, which is one form of the Boltzmann distribution law, gives the occupancy numbers of the molecular energy levels for the most probable distribution in terms of the energy magnitudes ϵ_K and the constants α and β. Clearly the next step is the evaluation of these constants. We can eliminate α by summing both sides of Equation 9–11 over K, as follows:

$$\Sigma n_K = n = \Sigma e^{-\alpha} e^{-\beta \epsilon_K}$$
$$= e^{-\alpha} \Sigma e^{-\beta \epsilon_K}$$

Consequently,

$$e^{-\alpha} = \frac{n}{\Sigma e^{-\beta \epsilon_K}} \qquad (9\text{--}12)$$

and the fraction of the total number of molecules in energy level ϵ_K is given by

$$\frac{n_K}{n} = \frac{e^{-\beta \epsilon_K}}{\Sigma e^{-\beta \epsilon_K}} \qquad (9\text{--}13)$$

The quantity represented by $\Sigma e^{-\beta \epsilon_K}$ plays an important role in the further development of our subject. It is so important that it is given a special name, the *molecular partition function*, and a special symbol, q. We will find that when we can evaluate the partition function for a system we can calculate the value of any thermodynamic function for that system.

Before we consider applications of the partition function we need to find the significance of the parameter β. This can be done by expressing some quantity we already know, such as the translational energy of an ideal gas, in terms of the partition function and then noting what β must be for consistency. This is the procedure we used in Chapter 5, page 133. As will be confirmed shortly, $\beta = 1/kT$ where k is Boltzmann's constant.

At this point we generalize our expressions to take into account the fact that in some cases more than one molecular state may be associated with a single energy level. In other words, some eigenfunctions form degenerate sets (see Chapter 6, p. 192) with a degree of degeneracy g_K which is the number of eigenfunctions in the degenerate set associated with the energy level ϵ_K. Introducing the degree of degeneracy into the partition function is equivalent to combining all terms related to the same energy level into a single term. The factor g_K is sometimes called the statistical weight of the related energy level.

When we substitute $1/kT$ for β and include g_K in the molecular partition function it becomes

$$q = \Sigma g_K e^{-\frac{\epsilon_K}{kT}} \tag{9-14}$$

The fraction of the total number of molecules in the energy level ϵ_K when the system is at equilibrium at a temperature T is given by

$$\frac{n_K}{n} = \frac{g_K e^{-\frac{\epsilon_K}{kT}}}{q} \tag{9-15}$$

From this equation it is easy to obtain the ratio of the populations in any two energy levels ϵ_i and ϵ_j:

$$\frac{n_i}{n_j} = \frac{g_i}{g_j} e^{-\frac{(\epsilon_i - \epsilon_j)}{kT}} \tag{9-16}$$

Equations 9-15 and 9-16 are forms of the Boltzmann distribution law.

Thermodynamic Functions in Terms of Partition Functions

The Boltzmann distribution law and the related partition function provide the means for calculating the properties of macroscopic quantities of matter from molecular properties. For this reason the importance of these functions can hardly be overstated.

Statistical Mechanics

The first property we consider is the energy. To obtain the average molecular energy we employ the same definition of the average of a property over a distribution that we have used before (Chapter 5). Applying the Boltzmann distribution function we can write the following expression for the average molecular energy at temperature T:

$$\bar{\epsilon} = \frac{\Sigma \epsilon_K g_K e^{-\frac{\epsilon_K}{kT}}}{\Sigma g_K e^{-\frac{\epsilon_K}{kT}}}$$

The denominator of this fraction is the molecular partition function q and the numerator is equal to $kT^2 \left(\frac{\partial q}{\partial T}\right)_V$. (The use of the partial derivative implies that q is a function of V, the volume of the system.) Consequently we can rewrite the equation for $\bar{\epsilon}$ in the following way:

$$\bar{\epsilon} = \frac{kT^2}{q}\left(\frac{\partial q}{\partial T}\right)_V = kT^2 \left(\frac{\partial \ln q}{\partial T}\right)_V$$

For the molar energy we multiply $\bar{\epsilon}$ by Avogadro's number to obtain:

$$E = \mathfrak{N}kT^2 \left(\frac{\partial \ln q}{\partial T}\right)_V = RT^2 \left(\frac{\partial \ln q}{\partial T}\right)_V \tag{9-17}$$

We cannot evaluate the molar energy from this relation until we have obtained the partition function from the equations for the molecular energy levels. This step is deferred to the next section.

From the equation for the molar energy we can readily find the following general equation for the molar heat capacity at constant volume:

$$C_V = \left(\frac{\partial E}{\partial T}\right)_V = R \frac{\partial}{\partial T}\left[T^2 \left(\frac{\partial \ln q}{\partial T}\right)_V\right] \tag{9-18}$$

The next thermodynamic function to be considered is the entropy. One direct path to this function on a statistical basis is the use of the relation between the entropy of a system and its thermodynamic probability proposed by Boltzmann. In approaching this relation let us recall from Chapter 3 (page 82) that an isolated system is at equilibrium when its entropy is at a maximum. We have identified the equilibrium state with the state of maximum thermodynamic probability. It is reasonable to infer that a relation exists between the entropy of a system at equilibrium and the maximum value of the thermodynamic probability.

To obtain some idea of the nature of this relation we recall that when two identical systems are joined together, the entropy of the combined system is the sum of the entropies of the individual systems; that is, entropy is an extensive property. In contrast, the thermodynamic probability of the combined systems is the product of the thermodynamic probabilities of the individual systems. These statements are consistent with a proportionality between the entropy of a system and the logarithm of its thermodynamic probability, and this is the relation proposed by Boltzmann:

$$S = k \ln W_{max} \tag{9-19}$$

where the use of k implies that we know that the constant of proportionality is Boltzmann's constant. (It turns out that this identity is necessary for consistency.)

We have not derived Equation 9–19 but have merely shown that it is plausible. Assuming that it is valid we can derive an equation for the entropy by expressing $\ln W_{max}$ in terms of the occupancy numbers of the energy levels which are given by the Boltzmann distribution function. From Equation 9–7 we have

$$\ln W_{max} = n \ln n - n - \Sigma(n_K \ln n_K - n_K)_{max}$$
$$= n \ln n - \Sigma(n_K \ln n_K)_{max} \tag{9-20}$$

Now we can use the expression for the occupancy numbers for the most probable distribution we obtained earlier. By writing Equation 9–13 in the following way

$$n_K = \frac{n}{q} e^{-\frac{\epsilon_K}{kT}}$$

and taking the logarithm of both sides, we obtain

$$\ln n_K = \ln \frac{n}{q} - \frac{\epsilon_K}{kT}$$

Substituting this into Equation 9–20 we find

$$\ln W_{max} = n \ln n - \Sigma n_K \left(\ln \frac{n}{q} - \frac{\epsilon_K}{kT} \right)$$
$$= n \ln n - \Sigma n_K \ln n + \Sigma n_K \ln q + \sum \frac{n_K \epsilon_K}{kT}$$

But $\Sigma n_K \epsilon_K$ is the total energy E. Consequently the equation for the entropy can be written in the following way:

$$S = kn \ln q + \frac{E}{T}$$

When E is expressed in terms of the molecular partition function and n is set equal to Avogadro's number, the equation for the molar entropy becomes

$$S = R \ln q + RT \left(\frac{\partial \ln q}{\partial T} \right)_V \tag{9-21}$$

Statistical Mechanics

Although this equation is valid in general, we find in the next section that a correction must be made in calculating the entropy associated with the translational motion of gas molecules. Notice the fact that the entropy depends on the absolute value of the partition function while the energy depends only on the rate of change of the logarithm of the partition function with respect to the temperature.

Now that we have expressed the energy and entropy of a system in terms of its molecular partition function we can express all other thermodynamic functions in an analogous manner. The molar Helmholtz free energy $A = E - TS$ is given by

$$A = RT^2 \left(\frac{\partial \ln q}{\partial T}\right)_V - T\left[R \ln q + RT\left(\frac{\partial \ln q}{\partial T}\right)_V\right]$$
$$= -RT \ln q \tag{9-22}$$

The pressure is given by Equation 3-41

$$P = -\left(\frac{\partial A}{\partial V}\right)_T$$
$$= RT\left(\frac{\partial \ln q}{\partial V}\right)_T$$

From the expressions for A and P we obtain the following form of the Gibbs free energy:

$$G = A + PV = RT\left[-\ln q + V\left(\frac{\partial \ln q}{\partial V}\right)_T\right] \tag{9-23}$$

Evaluation of Partition Functions

In the equation which defines the molecular partition function,

$$q = \sum_K g_K e^{-\frac{\epsilon_K}{kT}} \tag{9-14}$$

ϵ_K stands for all of the energy levels a molecule can have. We can always separate the total molecular energy into translational energy, which is associated with the motion of the molecular center of gravity, and internal energy. In addition, we have seen before that for most molecules it is a very good approximation to express the internal energy of a molecule whose center of gravity is fixed as a sum of rotational, vibrational, and electronic energies, each of which is quantized. On the basis of this approximation we have

$$\epsilon_{tot} = \epsilon_t + \epsilon_r + \epsilon_v + \epsilon_e$$

and the molecular partition function can be expressed in the following forms:

$$q = \Sigma g_t g_r g_v g_e e^{-\frac{\epsilon_t + \epsilon_r + \epsilon_v + \epsilon_e}{kT}}$$

$$= \Sigma g_t e^{-\frac{\epsilon_t}{kT}} \Sigma g_r e^{-\frac{\epsilon_r}{kT}} \Sigma g_v e^{-\frac{\epsilon_v}{kT}} \Sigma g_e e^{-\frac{\epsilon_e}{kT}}$$

$$= q_t q_r q_v q_e$$

We thus reach the conclusion that the molecular partition function can be written as a product of partition functions associated with each of the types of energy levels.

Our immediate objective is the evaluation of each of the factors in the over-all partition function in terms of molecular properties. We start with the translational partition function of an ideal gas, considered as a system of independent particles.

For the evaluation of the translational partition function of an ideal gas we make use of the energy levels of the particle in a box which we derived in Chapter 6 from the Schrödinger equation for this model. For one dimension these energy levels are given by Equation 6–25 which is repeated here:

$$\epsilon_x = \frac{h^2 n_x^2}{8ml^2}$$

where m is the mass of the molecule, l is the length of one edge of the container, and n_x is the translational quantum number for the x coordinate. For one dimension the translational energy levels are nondegenerate, and the factor for the x coordinate in the translational partition function is

$$q_x = \sum_{n_x} e^{-\frac{h^2 n_x^2}{8ml^2 kT}} \qquad (9\text{–}24)$$

To evaluate this sum, let us recall (p. 190) that the order of magnitude of $h^2/8ml^2$ is about $10^{-53}/10^{-21} \sim 10^{-32}$ ergs per molecule. At ordinary temperatures kT is of the order of 10^{-14} ergs per molecule. Consequently for a gas molecule in an ordinary-sized container, $h^2/8ml^2$ is so small compared to kT that a change in n_x by 1 results in only an infinitesimal difference between the corresponding two energy levels. It follows that the translational energy levels can be considered to be distributed continuously and the summation in Equation 9–24 can, without significant error, be replaced by an integration in which n_x is treated as a continuous variable. Then we have

$$q_x = \int_0^\infty e^{-\frac{h^2}{8ml^2 kT} n_x^2} dn_x$$

With the substitution $a = h^2/8ml^2kT$, this integral becomes a standard form:

$$\int_0^\infty e^{-an^2} dn = \frac{1}{2}\sqrt{\frac{\pi}{a}}$$

Statistical Mechanics

Consequently

$$q_x = \frac{1}{2}\left(\frac{8\pi m l^2 kT}{h^2}\right)^{\frac{1}{2}} = \frac{l}{h}(2\pi mkT)^{\frac{1}{2}} \qquad (9\text{--}25)$$

Because $q_t = q_x q_y q_z$ and q_x, q_y, and q_z are identical, for all three dimensions the molecular translational partition function takes the form

$$\boxed{q_t = \frac{V}{h^3}(2\pi mkT)^{\frac{3}{2}}} \qquad (9\text{--}26)$$

in which V, the volume of the system, is written in place of l^3.

Having obtained the translational partition function, we can now calculate the translational contributions to the thermodynamic functions. From Equation 9–17 the molar translational energy is

$$E_t = RT^2 \left(\frac{\partial \ln q_t}{\partial T}\right)_V$$

With

$$\ln q_t = \frac{3}{2}\ln T + \ln \frac{V}{h^3}(2\pi mk)^{\frac{3}{2}}$$

$$E_t = RT^2 \cdot \frac{3}{2}\left(\frac{d \ln T}{dT}\right) = RT^2 \cdot \frac{3}{2}\frac{1}{T}$$

$$= \tfrac{3}{2}RT$$

This, of course, is the equipartition value for the translational energy which is just the result we obtained in Chapter 5 for an ideal gas on a purely classical basis. The identity of the two expressions for the translational energy is not surprising because when we replaced the summation in Equation 9–24 by an integration, and thus assumed that the translational energy is a continuous variable, we left the quantum theory and obtained the classical result.

At this point we can see the significance of the restriction to constant volume in the partial derivative in Equation 9–17. The translational energy levels are functions of the dimensions of the container. Therefore the translational partition function depends on both the volume and the temperature of the system, and in the differentiation of the partition function with respect to the temperature the volume must be held constant.

An alternative way of expressing the molar energy of an ideal gas makes use of the *molar partition function* Q which is the partition function for the whole system of Avogadro's number of molecules. Because by hypothesis the molecules are independent of each other, we might expect that the molar partition function is simply the product of the molecular partition functions of all of the $\mathfrak{N}$ molecules in the system. In symbols,

$$Q = q^{\mathfrak{N}} \quad \text{and} \quad \ln Q = \mathfrak{N} \ln q$$

Then $E = kT^2 \left(\dfrac{\partial \ln Q}{\partial T} \right)_V$. For the translational energy,

$$\ln Q_t = \mathfrak{N} \ln \frac{V}{h^3} (2\pi mkT)^{\frac{3}{2}} \qquad (9\text{–}27)$$

and

$$E_t = \tfrac{3}{2} RT$$

It turns out that we have obtained the correct expression for the molar translational energy from an incorrect expression for the molar translational partition function. Before we consider the entropy which, as we have seen, depends on the absolute value of the partition function, we must make a correction for an assumption we made at the beginning of our derivation of the Boltzmann distribution function. We assumed that the molecules are distinguishable and Equation 9–27 depends on this assumption.

There is, in fact, no way of distinguishing between two gas molecules of the same species. (Isotopic molecules from this point of view belong to different species.) The eigenfunctions for a translating molecule considered as a particle in a box extend throughout the box and only the probability of finding the molecule in a certain region within the container is known. In other words, the eigenfunctions for the translating molecule are delocalized, and interchanging molecules between energy levels (or eigenfunctions) does not lead to a new distribution.

A detailed examination of this problem from the point of view of quantum statistics (either Fermi–Dirac or Einstein–Bose statistics) leads to the conclusion that for all molecular systems under ordinary conditions the correct relation between the molar and molecular translational partition functions is

$$\boxed{Q_t = \frac{q_t^{\mathfrak{N}}}{\mathfrak{N}!}} \qquad (9\text{–}28)$$

The factor $\mathfrak{N}!$ in the denominator is the total number of permutations of $\mathfrak{N}$ molecules, now assumed to be indistinguishable. Then

$$\ln Q_t = \mathfrak{N} \ln q_t - \ln \mathfrak{N}!$$

With the use of Stirling's approximation this becomes

$$\ln Q_t = \mathfrak{N} \ln q_t - (\mathfrak{N} \ln \mathfrak{N} - \mathfrak{N})$$

$$= \mathfrak{N} \ln \frac{q_t}{\mathfrak{N}} + \mathfrak{N}$$

Statistical Mechanics

Writing $\mathfrak{N}$ as $\mathfrak{N} \ln e$, we have

$$\ln Q_t = \mathfrak{N} \ln \frac{q_t e}{\mathfrak{N}}$$

$$= \mathfrak{N} \ln \frac{Ve}{\mathfrak{N} h^3} (2\pi mkT)^{\frac{3}{2}} \quad (9\text{-}29)$$

(Notice that this expression for $\ln Q_t$ leads to the same value of E_t that we obtained before.)

With the corrected expression for $\ln Q_t$ we can now calculate from Equation 9-21 the molar translational entropy of an ideal gas.

$$S_t = k \ln Q_t + kT \left(\frac{\partial \ln Q_t}{\partial T}\right)_V$$

$$= k\mathfrak{N} \ln \frac{Ve}{\mathfrak{N} h^3} (2\pi mkT)^{\frac{3}{2}} + kT \cdot \frac{3}{2} \frac{\mathfrak{N}}{T}$$

$$= R \ln \frac{Ve}{\mathfrak{N} h^3} (2\pi mkT)^{\frac{3}{2}} + \frac{3}{2} R$$

This is one form of the Sackur–Tetrode equation which was obtained semiempirically before the development of quantum mechanics clarified the problem of indistinguishability. We can write this equation in more convenient form by expressing the mass of the molecule in terms of the molecular weight $\left(m = \frac{M}{\mathfrak{N}}\right)$ and combining all of the constants; thus

$$S_t = R \left[\ln V + \frac{3}{2} \ln T + \frac{3}{2} \ln M + \frac{3}{2} \ln \frac{2\pi k}{\mathfrak{N}^{\frac{5}{3}} h^2} + \frac{5}{2}\right]$$

Making use of the ideal gas law we can express the translational entropy as a function of the pressure. With the atmosphere as the unit of pressure, in terms of common logarithms we have

$$S_t = 4.576(-\log P + \tfrac{5}{2} \log T + \tfrac{3}{2} \log M - 0.5053) \text{ cal/mole deg}$$

Because atoms do not have rotational and vibrational energy levels and electronic contributions are negligible except in a few unusual cases, the Sackur–Tetrode equation gives the total entropy of monatomic gases.

From Equation 9-22 we find the translational contribution to the Helmholtz free energy to be

$$A = -kT \ln Q_t$$

$$= -RT \ln \left[\frac{Ve}{\mathfrak{N} h^3} (2\pi mkT)^{\frac{3}{2}}\right]$$

The pressure is given by

$$P = -\left(\frac{\partial A}{\partial V}\right)_T = kT\left(\frac{\partial \ln Q_t}{\partial V}\right)_T$$
$$= RT\frac{d}{dV}\ln V = \frac{RT}{V} \tag{9-30}$$

This result shows that our molar translational partition function is consistent with the ideal gas law.

Next we obtain the Gibbs free energy of translation, as follows:

$$G = kT\left[-\ln Q_t + V\left(\frac{\partial \ln Q_t}{\partial V}\right)_T\right]$$
$$= RT\left\{-\ln\left[\frac{Ve}{\mathfrak{N}h^3}(2\pi mkT)^{\frac{3}{2}}\right] + 1\right\}$$
$$= -RT\ln\left[\frac{V}{\mathfrak{N}h^3}(2\pi mkT)^{\frac{3}{2}}\right]$$
$$= -RT\ln\frac{q_t}{\mathfrak{N}}$$

Before we turn our attention to the partition functions associated with the internal energy of a molecule, it should be emphasized that the correction to the molar partition function for translation expressed in Equation 9–28 has its origin in the fact that the eigenfunctions for the particle in a box are delocalized. This is not true for the eigenfunctions for the internal degrees of freedom. When we made use of the rigid rotor and harmonic oscillator models for a molecule we assumed that the center of gravity of the model was fixed in space. Consequently the rotational and vibrational eigenfunctions are localized, the indistinguishability problem does not arise, and the correction we applied to the translational partition function is not needed for rotational, vibrational, or electronic partition functions.

We can readily dispose of the electronic partition function. For nearly all molecules at ordinary temperatures, ϵ_e for any excited electronic state is so much larger than kT that $e^{-\epsilon_e/kT}$ is essentially zero. Also, for nearly all molecules the ground electronic state ($\epsilon_e = 0$) is nondegenerate and consequently q_e can be taken to have the value 1. There are a few exceptional cases. For O_2, S_2, and other molecules with a triplet ground state, $g_0 = 3$. A small number of molecules, such as NO, have low-lying electronic states for which several terms must be included in the electronic partition function.

We turn next to the rotational partition function. In Chapter 8 we saw that the solution of the Schrödinger equation for the rigid rotor model of a linear molecule leads to the following equation (Eq. 8–12) for the rotational energy levels:

Statistical Mechanics

$$\epsilon_J = \frac{h^2}{8\pi^2 I} J(J+1) \qquad J = 0, 1, 2, \ldots$$

The rigid rotor eigenfunctions are degenerate with a degree of degeneracy equal to $2J+1$. Consequently the molecular partition function for this model is

$$q_r = \sum_{J=0}^{\infty} (2J+1) e^{-\frac{h^2 J(J+1)}{8\pi^2 I k T}}$$

$I = \dfrac{m_1 m_2}{m_1 + m_2} r^2$ = miter nuclear distance reduced mass

For nearly all molecules the moment of inertia is so large that the intervals between rotational energy levels are small relative to kT at ordinary temperatures. In this case it is an adequate approximation to replace the sum by an integral and the approximate rotational partition function can be written as follows:

$$q_r = \int_0^{\infty} (2J+1) e^{-\frac{h^2 J(J+1)}{8\pi^2 I k T}} dJ$$

To carry out this integration we introduce a new variable, X, defined by $X = J^2 + J$; then $dX = (2J+1) dJ$ and the equation for q_r becomes

$$q_r = \int_0^{\infty} e^{-cX} dX = \frac{1}{c}$$

where

$$c = \frac{h^2}{8\pi^2 I k T}$$

Thus

$$q_r = \frac{8\pi^2 I k T}{h^2}$$

This expression for the rotational partition function is applicable to heteronuclear diatomic molecules and to any molecule whose symmetry group is $C_{\infty v}$. For those molecules whose symmetry group is $D_{\infty h}$ it is necessary to divide this expression by two to take into account the fact that these molecules have two indistinguishable positions, one of which can be reached from the other by a rotation of the molecule about an axis perpendicular to the molecular axis and passing through the center of symmetry.

In general, the number of indistinguishable positions of a molecule which can be reached by rigid rotations alone is called the _symmetry number_ and is represented by σ. When this number is introduced into the rotational partition function for linear molecules it becomes

$$q_r = \frac{8\pi^2 I k T}{\sigma h^2} \qquad (9\text{–}31)$$

An alternative way of writing the rotational partition function for linear molecules which makes use of the rotational constant B, defined on page 288 as

$$\frac{h}{8\pi^2 Ic}\ \text{cm}^{-1}$$

is the following:

$$q_r = \frac{kT}{\sigma hcB}$$

For a nonlinear polyatomic molecule with moments of inertia I_A, I_B, I_C, a good approximation for the rotational partition function is found to be

$$q_r = \pi^{\frac{1}{2}}\left(\frac{8kT}{h^2}\right)^{\frac{3}{2}}\frac{(I_A I_B I_C)^{\frac{1}{2}}}{\sigma}$$

The symmetry number for a molecule which belongs to a symmetry group C_{nv} is n; for example, for H_2O and CH_3Cl, $\sigma = 2$ and 3, respectively. Some other values of symmetry numbers are: BCl_3, 6; CH_4, 12; and C_6H_6, 12.

The expression for the rotational partition function given above applies only to molecules which can be treated as rigid rotors. It is not applicable to those molecules, such as ethane, which have internal rotation of one part of the molecule relative to another part. The subject of internal rotation is discussed in more advanced books and is not considered here.

From $\ln q_r = \ln T + \ln \frac{k}{\sigma hcB}$ we can readily obtain the rotational contributions to the thermodynamic functions. Thus the molar rotational energy is

$$E_r = \mathfrak{N}kT^2\frac{d\ln q_r}{dT} = kT^2\frac{d\ln Q_r}{dT} = RT$$

The total derivative is appropriate in this case because Q_r for a particular molecule is a function of T only. Notice that the equipartition value of the rotational energy is obtained, which is consistent with the fact that we carried out an integration to evaluate the partition function.

The molar rotational entropy is given by

$$S_r = k\ln Q_r + kT\frac{d\ln Q_r}{dT}$$

$$= R\ln\frac{kT}{\sigma hcB} + R$$

Statistical Mechanics

The Helmholtz free energy is given by

$$A_r = -RT \ln \frac{kT}{\sigma hcB}$$

Because Q_r is independent of V, the rotational contribution to the Gibbs free energy is the same as A_r.

To evaluate the vibrational partition function for a diatomic molecule, we will employ the energy levels obtained from the solution of the Schrödinger equation for the harmonic oscillator (Eq. 8–16):

$$\epsilon_{v_n} = h\nu(n + \tfrac{1}{2}) \qquad n = 0, 1, 2, \ldots$$

In this case the energy levels are nondegenerate ($g_v = 1$) and the vibrational partition function takes the form:

$$q_v = \sum_{n=0}^{\infty} e^{-\frac{h\nu}{kT}\left(n+\frac{1}{2}\right)} \tag{9-32}$$

For typical values of the vibrational frequency the spacing between the vibrational energy levels is of the order of magnitude of kT at ordinary temperatures. Consequently we cannot replace the summation by an integration as we did in the evaluation of the translational and rotational partition functions. The vibrational partition function, however, takes the following form with $h\nu/kT$ represented by x:

$$q_v = e^{-\frac{x}{2}} \sum_{n=0}^{\infty} e^{-nx}$$

The infinite series in this expression is the geometric series

$$1 + a + a^2 + \cdots \qquad (a = e^{-x})$$

whose sum is known to be $\dfrac{1}{1-a}$. It follows that the equation for q_v is

$$q_v = e^{-x/2}(1 - e^{-x})^{-1}$$
$$= \frac{e^{-h\nu/2kT}}{1 - e^{-h\nu/kT}} \tag{9-33}$$

We can now obtain the molar vibrational energy from

$$E_v = RT^2 \frac{d \ln q_v}{dT}$$

The resulting equation for E_v is

$$E_v = \frac{\mathfrak{N}h\nu}{2} + \frac{\mathfrak{N}h\nu}{e^{h\nu/kT} - 1} \tag{9-34}$$

To derive this equation we can use Equation 9–33 and express E_v as $\Re kT^2 \dfrac{d \ln q_v}{dT} = \dfrac{\Re kT^2}{q_v} \dfrac{dq_v}{dx} \dfrac{dx}{dT}$. On carrying out the differentiations we find

$$\frac{dx}{dT} = -\frac{x}{T}$$

$$\frac{dq_v}{dx} = -\frac{1}{2}e^{-x/2}(1-e^{-x})^{-1} - e^{-x/2}(1-e^{-x})^{-2}e^{-x}$$

$$\frac{d \ln q_v}{dT} = \frac{\dfrac{x}{2T}e^{-x/2}(1-e^{-x})^{-1} + \dfrac{x}{T}e^{-x/2}(1-e^{-x})^{-2}e^{-x}}{e^{-x/2}(1-e^{-x})^{-1}}$$

$$= \frac{x}{2T} + \frac{x}{T}(1-e^{-x})^{-1}e^{-x}$$

$$= \frac{h\nu}{2kT^2} + \frac{\dfrac{h\nu}{kT^2}e^{-h\nu/kT}}{1-e^{-h\nu/kT}}$$

When this is multiplied by $\Re kT^2$ the result is

$$E_v = \frac{\Re h\nu}{2} + \frac{\Re h\nu e^{-h\nu/kT}}{1-e^{-h\nu/kT}}$$

On dividing numerator and denominator of the last term by $e^{-h\nu/kT}$ we obtain Equation 9–34.

The first term in Equation 9–34 is the molar zero-point energy (p. 296). Because the molecules cannot lose this energy, it cannot be transferred in any process in which the molecules participate and it plays no role in any thermodynamic calculation. Consequently it is usually represented by E_0 and subtracted from both sides of Equation 9–34. This equation then becomes

$$\boxed{E_v - E_0 = \frac{\Re h\nu}{e^{+h\nu/kT}-1}} \qquad (9\text{--}35)$$

The use of Equation 9–33 implies that the zero of the energy scale is taken to be the minimum of the harmonic oscillator potential curve (Fig. 8–13). When the zero of the energy scale is the energy of the lowest vibrational level, which is equivalent to omission of the $\tfrac{1}{2}$ in Equation 8–16,

$$\boxed{q_v = (1-e^{-h\nu/kT})^{-1}} \qquad (9\text{--}36)$$

Statistical Mechanics

Let us see what happens to the vibrational energy as the temperature of the system rises. The value of $h\nu/kT$ decreases and when it is sufficiently small $e^{h\nu/kT} \sim 1 + h\nu/kT$ because the higher terms in the series expansion of $e^{h\nu/kT}$ are negligible. The substitution of $h\nu/kT$ for $e^{h\nu/kT} - 1$ in the denominator of Equation 9-35 results in

$$E_v - E_0 = \frac{\mathfrak{N}h\nu}{h\nu/kT} = RT$$

This is the classical equipartition value for the vibrational energy (Chapter 5, p. 153). We see that it is a limiting value that is approached at high temperatures. The same result is obtained by assuming that the vibrational energy levels are distributed continuously and replacing the summation in Equation 9-32 by an integration.

At ordinary temperatures the fact that the vibrational energy levels are quantized cannot be ignored. It follows that the vibrational contribution to the molar energy is significantly less than the equipartition value of RT.

Expressions for the vibrational contributions to the other thermodynamic functions can be readily obtained from the equations given previously. These expressions are:

$$C_{V_v} = \frac{dE_v}{dT} = R \frac{\partial}{\partial T} \left[T^2 \left(\frac{d \ln q_v}{dT} \right) \right]$$

$$= R \left(\frac{h\nu}{kT} \right)^2 \frac{e^{h\nu/kT}}{(e^{h\nu/kT} - 1)^2} \quad (9\text{-}37)$$

$$S_v = R \left[\ln (1 - e^{-h\nu/kT})^{-1} + \frac{h\nu}{kT} (e^{h\nu/kT} - 1)^{-1} \right]$$

$$A_v - E_0 = G_v - E_0 = RT \ln (1 - e^{-h\nu/kT})$$

These equations can be easily generalized to be applicable to polyatomic molecules when the frequencies of the normal modes of vibration are known. Because the normal modes of vibration are independent of each other in the harmonic oscillator approximation, the molecular vibrational energy is obtained as a sum of the energies of the normal modes, each of which is approximated by the harmonic oscillator value (Eq. 8-16). The vibrational partition function is a product of factors, each of the form of Equation 9-36 with one factor for each normal mode. Because it is possible for two or three normal modes to have the same frequency, it is convenient to group the factors with the same frequency. Thus the partition function has the following form:

$$q_v = (1 - e^{-h\nu_1/kT})^{-g_1} (1 - e^{-h\nu_2/kT})^{-g_2} \cdots$$

where g_i is the degeneracy of the energy level associated with the frequency ν_i.

Since each of the thermodynamic functions depends on $\ln q$, it follows that the vibrational contributions to each of them can be expressed as a sum of terms, one for each different vibrational frequency. Before these equations can be evaluated for a particular substance it is necessary that the values of all of the vibrational frequencies and degeneracy factors be known. This implies, in general, that the infrared and Raman spectra of the substance have been measured and analyzed to select the fundamental frequencies (p. 308) in these spectra. It also implies that a correct assignment of frequencies to normal modes of vibration has been made in order to associate each frequency with its degeneracy factor. From the dependence of the vibrational contributions on the frequency it can be seen that the terms with lower frequencies make larger contributions than those of higher frequency and therefore the accuracy of the lower frequency values is particularly important.

Thermodynamic Functions for Diatomic Ideal Gases

In this section we present a summary of the main results of the preceding section, together with some numerical examples. The quantity $h\nu/kT \, (= hc\omega/kT)$ which appears in the vibrational contributions is represented by x. The value of hc/k is 1.4388 cm degree; thus $x = 1.4388 \, \omega/T$. Notice that $\mathfrak{N}h\nu = RTx$.

Molar Partition Function

$$\ln Q = \mathfrak{N}\left[\ln \frac{Ve}{\mathfrak{N}h^3}(2\pi mkT)^{\frac{3}{2}} + \ln \frac{kT}{\sigma hcB} - \ln(1 - e^{-x})\right]$$

$$= \mathfrak{N} \ln \frac{q_t}{\mathfrak{N}} + \mathfrak{N} + \mathfrak{N} \ln q_r + \mathfrak{N} \ln q_v$$

$$= \mathfrak{N}\left(\ln \frac{q_t q_r q_v}{\mathfrak{N}} + 1\right)$$

$$= \mathfrak{N}\left(\ln \frac{q}{\mathfrak{N}} + 1\right)$$

$$\left(\frac{\partial \ln Q}{\partial T}\right)_V = \mathfrak{N}\left(\frac{\partial \ln q}{\partial T}\right)_V; \quad \left(\frac{\partial \ln Q}{\partial V}\right)_T = \mathfrak{N}\left(\frac{\partial \ln q}{\partial V}\right)_T$$

Molar Energy

$$E - E_0 = kT^2 \left(\frac{\partial \ln Q}{\partial T}\right)_V$$

$$= \frac{3}{2} RT + RT + RT \frac{x}{e^x - 1}$$

This equation is valid in the temperature range for which the classical value of the rotational contribution can be used. The lower temperature limit of validity depends on the rotational constant of the molecule and therefore on its moment of inertia. For all diatomic molecules except for H_2, HD, D_2, and HF, this equation can be used at temperatures above about 300°K.

For N_2, $\omega = 2360$ cm^{-1} and the value of x at 298°K is 11.4. The value of e^x is about 10^5 and the vibrational contribution to the molar energy is negligible. This means, of course, that at 298°K essentially all N_2 molecules are in their ground vibrational state. In contrast, for I_2, $\omega = 214.6$ cm^{-1}, x at 298°K is 1.036, and $\dfrac{x}{e^x - 1} = 0.569$. The value of $E - E_0$ for I_2 in the rigid rotor, harmonic oscillator approximation is $3.07 \cdot 1.99 \cdot 298 = 1820$ cal/mole.

Heat Capacity

$$C_V = \left(\frac{\partial E}{\partial T}\right)_V = k\frac{\partial}{\partial T}\left[T^2\left(\frac{\partial \ln Q}{\partial T}\right)_V\right]$$

$$= \frac{3}{2}R + R + R\frac{x^2 e^x}{(e^x - 1)^2}$$

The lower temperature limit for this equation is the same as the lower limit for the energy equation. As the temperature is lowered from a high value, first the vibrational contribution to the heat capacity becomes negligible

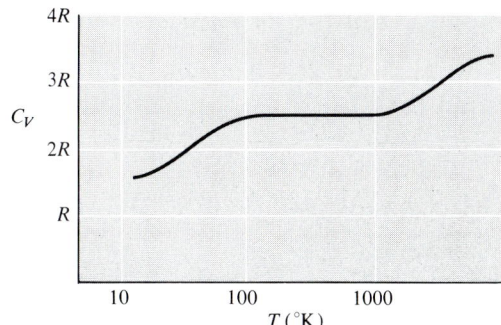

Figure 9-1. Schematic representation of the heat capacity of N_2 as a function of the temperature.

and then the rotational contribution drops below the classical value and approaches zero as a limit for those substances which are still gases. This behavior is illustrated by Figure 9-1 for N_2.

Enthalpy

$$H - E_0 = kT^2\left(\frac{\partial \ln Q}{\partial T}\right)_V + kTV\left(\frac{\partial \ln Q}{\partial V}\right)_T$$

Entropy

$$S = k\left[\ln Q + T\left(\frac{\partial \ln Q}{\partial T}\right)_V\right]$$

$$= R\left[\ln V + \frac{3}{2}\ln T + \frac{3}{2}\ln M + \frac{3}{2}\ln \frac{2\pi k}{\mathfrak{N}^{\frac{5}{3}}h^2} + \frac{5}{2}\right.$$

$$+ \ln T - \ln \sigma - \ln B + \ln \frac{k}{hc} + 1 - \ln(1 - e^{-x}) + \frac{x}{e^x - 1}\bigg]$$

$$= R\bigg[\ln V + \tfrac{5}{2}\ln T + \tfrac{3}{2}\ln M - \ln \sigma - \ln B - \ln(1 - e^{-x})$$

$$+ \frac{x}{e^x - 1} - 4.963\bigg]$$

As in the case of the equations for the energy and the heat capacity, this equation is applicable in the temperature range for which the rotational partition function has its classical value. At temperatures for which x is larger than about 5 the vibrational contributions are negligible. It is assumed that the volume of the system is expressed in cm³.

Helmholtz Free Energy

$$A - E_0 = -kT \ln Q = -RT \ln\left(\frac{q}{\mathfrak{N}} + 1\right)$$

$$= -RT[\ln V + \tfrac{5}{2}\ln T + \tfrac{3}{2}\ln M - \ln \sigma - \ln B - \ln(1 - e^{-x}) - 7.436]$$

Gibbs Free Energy

$$G - E_0 = A - E_0 + PV = A - E_0 + RT$$
$$= -kT \ln Q + \mathfrak{N}kT = -kT(\ln Q - \mathfrak{N})$$
$$= -RT \ln \frac{q}{\mathfrak{N}}$$
$$= -RT[\ln V + \tfrac{5}{2}\ln T + \tfrac{3}{2}\ln M - \ln \sigma - \ln B - \ln(1 - e^{-x}) - 8.436]$$

The Gibbs free energy function $\dfrac{G^\circ - E_0}{T}$ was defined in Chapter 4, p. 117.

Entropy and the Third Law of Thermodynamics

In the last few sections we evaluated the entropy of an ideal gas at equilibrium on the basis of the Boltzmann "principle," $S = k \ln W_{\max}$, and we have identified the statistical entropy defined in this way with the thermodynamic entropy defined in Chapter 3. The connection between these two definitions is so important that we now consider it from some other points of view.

In Chapter 3 the entropy change in a reversible infinitesimal process was

Statistical Mechanics

defined as $dS = dq_{rev}/T$. (Here dq_{rev} represents an infinitesimal amount of heat.) If the volume of the system is constant, $dq_{rev} = C_V \, dT$ and thus $dS = (C_V/T) \, dT$. In terms of the molar partition function we have seen that $C_V = k \frac{\partial}{\partial T}\left[T^2 \left(\frac{\partial \ln Q}{\partial T}\right)_V\right]$ which is equal to $k\left[T^2\left(\frac{\partial^2 \ln Q}{\partial T^2}\right)_V + 2T\left(\frac{\partial \ln Q}{\partial T}\right)_V\right]$.
As we have also seen, the partition function for a particular ideal gas is a function of T and V only. Consequently, we can express the entropy change for a differential temperature change at constant volume in the following way:

$$dS = k\left[T\frac{d^2 \ln Q}{dT^2} + 2\frac{d \ln Q}{dT}\right]dT$$

When this is integrated from $0°K$ to $T°K$ the result is

$$S_T - S_0 = k\left(\ln Q_T - \ln Q_0 + T\frac{d \ln Q_T}{dT}\right) \qquad (9\text{-}38)$$

where Q_T represents the value of the molar partition function at temperature T.

The integration can be carried out in the following way (integration by parts): From the definition of the differential of a product of two functions, $d(uv) = v \, du + u \, dv$, we obtain after integrating $\int v \, du = uv - \int u \, dv$. In our case, set $v = T$,

$$dv = dT, \quad u = \frac{d \ln Q}{dT}$$

and

$$du = \frac{d^2 \ln Q}{dT^2} dT$$

Then

$$\int T\frac{d^2 \ln Q}{dT^2} dT = T\frac{d \ln Q}{dT} - \int d \ln Q$$

and

$$k\int_0^T \left(T\frac{d^2 \ln Q}{dT^2} + 2\frac{d \ln Q}{dT}\right) dT = k\left(T\frac{d \ln Q}{dT} + \ln Q\right)\Big|_0^T$$

$$= k\left(T\frac{d \ln Q_T}{dT} + \ln Q_T - \ln Q_0\right)$$

If Equation 9–38 is to be valid for any value of T, the terms on both sides that depend on T must be equal to each other. Consequently,

$$S_T = k\left(\ln Q_T + T\frac{d \ln Q_T}{dT}\right)$$

This is essentially the same as Equation 9–21. It also follows that $S_0 = k \ln Q_0$. In order to evaluate Q_0 we must recognize that our equations have been derived for a system that consists of independent particles, that is, an ideal gas. Actually, near $0°K$ no substance exists in the gaseous state. Since all substances are solid there is no translational molecular motion. Likewise there is no rotational motion, and only vibrational motion remains. This includes not only the internal vibrations of diatomic or polyatomic molecules but also the vibrational motion of the molecules as units, even if the molecules consist of single atoms as is the case with the crystalline noble gases. (This subject is discussed in Chapter 12.) Then Q_0 is the molar partition function associated with this vibrational motion at $0°K$, and $\ln Q_0 = \mathfrak{N} \ln q_0$.

From the definition of the molecular partition function (setting $\epsilon_0 = 0$),

$$q = g_0 + g_1 e^{-\frac{\epsilon_1}{kT}} + \cdots$$

we see that as $T \to 0$ all terms after the first in this series vanish. This means that all of the molecules are then in the same energy level. Since the ground vibrational energy level is nondegenerate, $g_0 = 1$, $q_0 = 1$, and $S_0 = 0$. Consequently, when the vibrational partition function of an ideal crystal of a substance is evaluated an absolute value of the molar entropy, S_T, can be calculated. This is the statistical basis of the third law of thermodynamics.

Another way of expressing the same conclusion is to return to the Boltzmann definition of entropy, $S = k \ln W_{max}$. If, at $0°K$, all of the molecules in a system are in the same energy state, then there is only one way of realizing this state. It follows that $W_{max} = 1$, and $S_0 = 0$.

One of the obstacles in the way of acceptance of the third law when it was first proposed was the fact that there are some substances which appear not to follow this law. One of these substances is CO for which a discrepancy of 1.10 cal/mole deg between the statistical and calorimetric entropies was found. Detailed studies of this substance have shown that the dipole moment of CO is so small and the two ends of the molecule are so much alike that when CO crystallizes a molecule has almost the same probability of entering the crystal lattice oriented in one direction as in the opposite direction. If we assume that half of the molecules are oriented in one direction and half in the other, the number of ways in which this distribution can be realized is given by

$$W = \frac{\mathfrak{N}!}{\left(\frac{\mathfrak{N}}{2}\right)!\left(\frac{\mathfrak{N}}{2}\right)!}$$

Statistical Mechanics

On application of the Stirling approximation, the result is

$$\ln W = \mathfrak{N} \ln \mathfrak{N} - \mathfrak{N} - 2\left(\frac{\mathfrak{N}}{2} \ln \frac{\mathfrak{N}}{2} - \frac{\mathfrak{N}}{2}\right)$$

$$= \mathfrak{N} \ln \mathfrak{N} - \mathfrak{N} - \mathfrak{N} \ln \frac{\mathfrak{N}}{2} + \mathfrak{N}$$

$$= \mathfrak{N} \ln 2$$

The corresponding entropy is $S = k\mathfrak{N} \ln 2 = 1.38$ cal/deg, a value which is in reasonable agreement with the calorimetric value.

Any other type of randomness in a solid, such as the randomness of structure in a glass, also leads to a nonzero entropy as $T \to 0°K$. Only for the perfectly ordered, ideal crystal can all molecules be in the same energy state and only for such a system can S_0 be set equal to zero.

It may be helpful at this point to return to the analogy of the balls and box compartments discussed early in this chapter. We saw that the distribution for which the number of ways of realization is a minimum, namely 1, is the distribution which is highly ordered because all of the balls are in the same compartment. As balls are shifted to other compartments the number of ways of realizing the distribution increases until it is at a maximum when the distribution of balls throughout the box is uniform.

In the molecular case, as heat is added to a system more and more molecules are shifted to higher energy levels, increasing the thermodynamic probability and thus the entropy. If there were no restriction to a certain amount of energy in the system, W_{max} would correspond to a uniform distribution of molecules among the energy states. This distribution, however, would require an infinite amount of energy and an infinite temperature. For any finite temperature, the most probable distribution of the molecules among the energy states is that given by the Boltzmann distribution law.

Entropy and the Second Law of Thermodynamics

In Chapter 3 the second law was stated in a number of equivalent ways which said in effect that certain kinds of processes are impossible. One form of the second law is the statement that the entropy of an isolated system can never decrease. Now we can examine this statement in the light of the statistical definition of entropy.

Let us consider a simple isolated system which consists of two vessels, 1 and 2, with volumes V_1 and V_2, respectively, separated by a partition. We imagine that initially we have a certain amount, 1 mole for example, of ideal gas in vessel 1 at temperature T. If a hole is formed in the partition we have no doubt about the resulting spontaneous process. The gas will stream through the hole until the pressures in the two vessels are equalized, and the entropy of the system will increase by the amount $\Delta S = R \ln V/V_1$, with $V = V_1 + V_2$.

Once the pressures have become equalized we do not expect that at some later time we will observe that the pressure in vessel 1 is higher than that in vessel 2. According to the second law, this observation is impossible because a transfer of molecules from one vessel to the other is associated with an entropy decrease.

Let us next look at this system from a statistical point of view. After the hole is formed in the partition, the probability that a certain molecule is in V_1 is V_1/V. The probability that m particular molecules are in V_1 while the remaining $\mathfrak{N} - m$ molecules are in V_2 is $(V_1/V)^m(V_2/V)^{\mathfrak{N}-m}$. The probability that *any* m molecules are in V_1 and the remainder are in V_2 is (compare Eq. 9-2)

$$P = \frac{\mathfrak{N}!}{m!(\mathfrak{N}-m)!} \left(\frac{V_1}{V}\right)^m \left(\frac{V_2}{V}\right)^{\mathfrak{N}-m}$$

It is left as an exercise for the student to show that when the molecules are distributed between the two vessels in such a way that the probability is a maximum, $m/(\mathfrak{N}-m) = V_1/V_2$. Since the molecular concentration in vessel 1 is m/V_1 and that in vessel 2 is $(\mathfrak{N} - m)/V_2$, it follows that the molecular concentration, and therefore the pressure, is the same in both vessels.

Now let us imagine that the system contains a very small number of molecules, 100 for example. It is conceivable that at some instant a significant fraction of the molecules would be near the hole with such velocities that these molecules go through the hole. In that case a concentration gradient would be established spontaneously between the two vessels, and the entropy of the system would decrease, in apparent violation of the second law.

What happens when there are, say, 10^{20} molecules in the system? Fluctuations in concentration caused by the passage of a few molecules through the hole can still occur, but these molecules form such a small fraction of the total number that a fluctuation of this type is completely indetectable.

It is possible to calculate the probability of a fluctuation of given magnitude by several methods. One approach is the calculation of the ratio $(P_{max} + \delta P)/P_{max}$ where δP is the change in the probability of the most probable distribution when a small number of molecules, δn, moves from one of the vessels to the other. For the case in which the two vessels have the same volume the following equation can be derived:

$$\ln \frac{P_{max} + \delta P}{P_{max}} = -\frac{(\delta n)^2}{n}$$

(The minus sign is consistent with the expectation that any change from the most probable distribution corresponds to a decrease in probability.) If a fluctuation from equal concentrations occurs in which 0.001% of the molecules

Statistical Mechanics

move from one vessel to the other, a change which would be difficult to observe experimentally, $\delta n = 10^{15}$ and

$$\ln \frac{P_{\max} + \delta P}{P_{\max}} = -\frac{10^{30}}{10^{20}} = -10^{10}$$

The ratio of the probabilities is $e^{-10^{10}}$, a number which, although inconceivably small, is not zero.

The conclusion we can reach from these considerations is that while the second law states that certain processes are impossible, these processes are merely improbable from a statistical point of view, but they are extremely improbable for systems that contain a large number of molecules. For systems that can be manipulated in the laboratory, the probability of a violation of the second law is so small that it is essentially zero.

◈◈◈◈◈◈◈

Another way of expressing this conclusion is the statement that the probability of the most probable distribution is so much greater than the probability of all other distributions that the other distributions are insignificant. This can be shown for the system discussed above, with $V_1 = V_2 = V/2$, by the following calculation of $P_{\max}$:

$$P_{\max} = \frac{n!}{\frac{n}{2}! \frac{n}{2}!} \left(\frac{1}{2}\right)^{\frac{n}{2}} \left(\frac{1}{2}\right)^{\frac{n}{2}}$$

$$\ln P_{\max} = n \ln n - n - 2\left(\frac{n}{2} \ln \frac{n}{2} - \frac{n}{2}\right) + n \ln \frac{1}{2}$$

$$= n \ln n - n \ln \frac{n}{2} + n \ln \frac{1}{2}$$

$$= 0$$

This result says that $P_{\max} = 1$ in spite of the fact that there are an enormous number of other distributions. The statement that $P_{\max} = 1$, exactly, must, of course, be an approximation. It is a consequence of our use of the Stirling approximation.

◈◈◈◈◈◈◈

We can now calculate statistically the entropy change in the isothermal expansion of a mole of gas from V_1 to V. The probability that all $\mathfrak{N}$ molecules are contained in volume V_1 is $(V_1/V)^{\mathfrak{N}}$. From the statistical definition of entropy we have

$$\Delta S_{V_1 \to V} = k \left[\ln P_{\max} - \ln \left(\frac{V_1}{V} \right)^{\mathfrak{N}} \right]$$

$$= R \ln V/V_1$$

This is identical with the value calculated from Equation 3-13.

Equilibrium Constants from Partition Functions

We are now ready to apply our results to chemical reactions involving gases. In Chapter 4 we derived the important relationship between the equilibrium constant for a reaction and its standard Gibbs free energy change expressed by the following equation:

$$\Delta G° = -RT \ln K_P$$

In the calculation of $\Delta G°$ all reactants and products must be in their standard states at temperature T. We limit our treatment to the case in which all of the substances involved are ideal gases. In this case the standard state is usually defined as the state in which the pressure is 1 atm at temperature T (Chapter 3, p. 94); accordingly in the evaluation of the standard free energy from the molecular partition function for the standard state, $q°$, the molar volume at a pressure of 1 atm is used for V in the equation for q.

When $G° - E_0 \left(= -RT \ln \frac{q°}{\mathfrak{N}} \right)$ has been calculated for each reactant and product, there still is a problem to be met before $\Delta G°$ for the reaction can be calculated because the partition function for each substance is calculated with reference to its own ground state as the zero of energy. Before we can add or subtract free energies based on these partition functions, however, they all must be referred to the same energy zero. Since only energy differences are involved, the choice of this zero is arbitrary.

It is often useful to think of the energy of a completely dissociated molecule as an energy zero. This choice is illustrated by consideration of a simple reaction represented by $A + BC \to AB + C$ which, for definiteness, we assume to be exothermic. Each of the diatomic molecules has its own potential energy curve and its own set of energy levels. To establish a relationship between the two sets we can take as our reference energy the energy of the hypothetical state represented by $A + B + C$ in which all molecules are completely dissociated into atoms with negligible translational energy. The energy of this state is represented by the asymptotes of the potential energy curves of both AB and BC (Fig. 9–2) because both asymptotes are related to the same state of complete dissociation. Consequently one of the potential energy curves is shifted along the energy axis until the asymptotes of both curves lie along the same energy line.

Statistical Mechanics

Having found the desired relationship between the two sets of energy levels, we may now refer both sets to the ground energy level ($n = 0$ and $J = 0$) of the reactant molecule, BC. With $\Delta\epsilon_0$ representing the energy difference

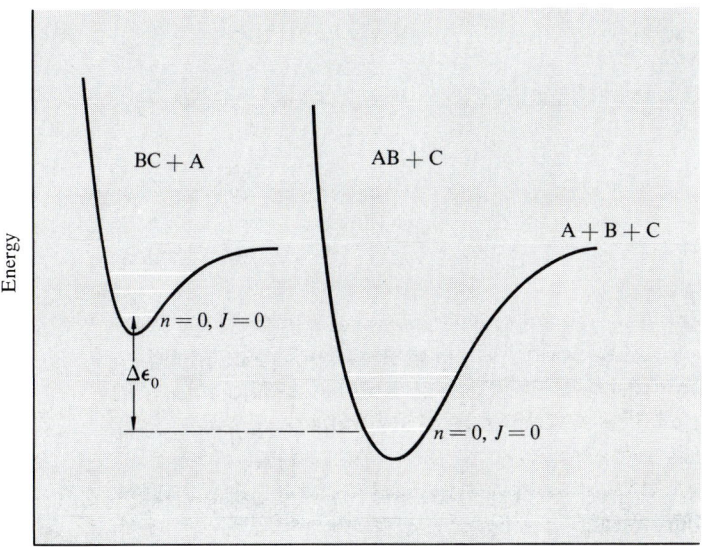

Figure 9–2. Diagram showing relationship of energy levels of BC to the ground state of AB.

between the ground levels of BC and AB, the energy ($\epsilon'_{n,J}$) of any of the levels of AB, referred to the ground level of BC, is given by

$$\epsilon'_{n,J} = \epsilon_{n,J} + \Delta\epsilon_0$$

Then the molecular partition function of AB on the same energy scale, q'_{AB}, can be written in the following forms:

$$q'_{AB} = q_t \sum_n \sum_J e^{-\frac{(\epsilon_{n,J} + \Delta\epsilon_0)}{kT}}$$

$$= q_t e^{-\frac{\Delta\epsilon_0}{kT}} \sum_n \sum_J e^{-\frac{\epsilon_{n,J}}{kT}}$$

Since the double summation over n and J is equal to $q_v q_r$, we have

$$q'_{AB} = (q_t q_v q_r)_{AB}\, e^{-\frac{\Delta\epsilon_0}{kT}}$$

$$= q_{AB}\, e^{-\frac{\Delta\epsilon_0}{kT}}$$

It follows that the difference between the standard free energies of the molecules BC and AB is given by

$$G°_{AB} - G°_{BC} = -RT\left(\ln \frac{q°_{AB}}{\mathfrak{N}} - \ln \frac{q°_{BC}}{\mathfrak{N}} - \frac{\Delta\epsilon_0}{kT}\right)$$

To calculate $\Delta G°$ for the reaction $A + BC \rightarrow AB + C$, we need to include terms for the free atoms whose only contributions come from their translational partition functions. Thus the standard free energy change for the reaction is

$$\Delta G° = G°_C + G°_{AB} - G°_{BC} - G°_A$$
$$= -RT\left(\ln \frac{q°_C}{\mathfrak{N}} + \ln \frac{q°_{AB}}{\mathfrak{N}} - \ln \frac{q°_{BC}}{\mathfrak{N}} - \ln \frac{q°_A}{\mathfrak{N}} - \frac{\Delta\epsilon_0}{kT}\right)$$

When we multiply the numerator and denominator of the last term in this equation by $\mathfrak{N}$ we obtain $\mathfrak{N}\Delta\epsilon_0/RT$. $\mathfrak{N}\Delta\epsilon_0$, which we can represent by ΔE_0, can be interpreted as the heat of reaction when all substances are in their ground energy levels. It is thus the heat of reaction extrapolated to $0°K$, assuming all substances remain ideal gases. It is the same quantity that was introduced in Equation 2–27 as a constant of integration. Since $H = E + RT$ for an ideal gas, $\Delta H_0 = \Delta E_0$.

We can now rewrite the equation above in the following forms:

$$\Delta G° = -RT\left(\ln \frac{q°_C q°_{AB}}{q°_{BC} q°_A} - \frac{\Delta E_0}{RT}\right) = -RT \ln\left(\frac{q°_C q°_{AB}}{q°_{BC} q°_A} e^{-\frac{\Delta E_0}{RT}}\right)$$

From the relation between $\Delta G°$ and K_P we obtain the equilibrium constant in terms of molecular partition functions:

$$K_P = \frac{q°_C q°_{AB}}{q°_{BC} q°_A} e^{-\frac{\Delta E_0}{RT}} \qquad (9\text{–}39)$$

When the expressions for the molecular partition functions are substituted for the q's, we obtain the following equation:

$$K_P = \frac{\frac{V}{h^3}(2\pi m_C kT)^{\frac{3}{2}} \frac{V}{h^3}(2\pi m_{AB} kT)^{\frac{3}{2}} \frac{kT}{\sigma_{AB} hc B_{AB}}(1 - e^{-hc\omega_{AB}/kT})^{-1}}{\frac{V}{h^3}(2\pi m_{BC} kT)^{\frac{3}{2}} \frac{V}{h^3}(2\pi m_A kT)^{\frac{3}{2}} \frac{kT}{\sigma_{BC} hc B_{BC}}(1 - e^{-hc\omega_{BC}/kT})^{-1}} e^{-\frac{\Delta E_0}{RT}}$$

On cancellation of all constant factors in this equation and setting $\sigma_{AB} = \sigma_{BC} = 1$, it reduces to

$$K_P = \left(\frac{m_C m_{AB}}{m_{BC} m_A}\right)^{\frac{3}{2}} \frac{B_{BC}}{B_{AB}} \frac{1 - e^{-hc\omega_{BC}/kT}}{1 - e^{-hc\omega_{AB}/kT}} e^{-\frac{\Delta E_0}{RT}}$$

We see that the equilibrium constant for this reaction depends on the rotational constants, vibrational frequencies, and dissociation energies of the molecules involved. For some reactions all of the data needed to calculate the value of the equilibrium constant at a specified temperature can be obtained from spectroscopic measurements on pure substances.

The reaction chosen as an example above is relatively simple since the total number of molecules does not change during the reaction and the equilibrium constant is dimensionless. Equation 9–39 can be easily generalized to apply to the reaction $aA + bB \rightarrow cC + dD$. In this case the equilibrium constant is given by

$$K_P = \frac{\left(\frac{q_C^\circ}{\mathfrak{N}}\right)^c \left(\frac{q_D^\circ}{\mathfrak{N}}\right)^d}{\left(\frac{q_A^\circ}{\mathfrak{N}}\right)^a \left(\frac{q_B^\circ}{\mathfrak{N}}\right)^b} e^{-\frac{\Delta E_0}{RT}}$$

We can express this equilibrium constant in a form that is more convenient for some purposes by writing

$$\frac{q^\circ}{\mathfrak{N}} = q^* \frac{V}{\mathfrak{N}}$$

in which $q^* = \left(\frac{2\pi mkT}{h^2}\right)^{\frac{3}{2}} q_r q_v$. The value of q^* is independent of pressure or concentration, and only the value of $V/\mathfrak{N}$ depends on the chosen standard state. The expression for the equilibrium constant then becomes

$$K_P = \frac{q_C^{*c} q_D^{*d}}{q_A^{*a} q_B^{*b}} \left(\frac{V}{\mathfrak{N}}\right)^{(c+d)-(a+b)} e^{-\frac{\Delta E_0}{RT}} \qquad (9\text{--}40)$$

For a standard state of 1 atm $\frac{V}{\mathfrak{N}} = \frac{kT}{P}$ where P is the value of the atmosphere expressed in dynes/cm². For a standard state of 1 mole per liter, $V/\mathfrak{N} = 10^3/\mathfrak{N}$, and for a standard state of one molecule per cm³, $V/\mathfrak{N} = 1$.

Supplementary References

N. Davidson, *Statistical Mechanics*, McGraw-Hill, New York, 1962.

M. Dole, *Introduction to Statistical Thermodynamics*, Prentice-Hall, New York, 1954.

T. L. Hill, *Introduction to Statistical Thermodynamics*, Addison-Wesley, Reading, 1960.

W. Kauzmann, *Thermodynamics and Statistics with Applications to Gases*, W. A. Benjamin, New York, 1967.

G. N. Lewis and M. Randall, *Thermodynamics*, 2nd edition revised by K. S. Pitzer and L. Brewer, McGraw-Hill, New York, 1961. Chapter 27.

Problems

1. When a pair of honest dice is thrown, what is the probability of turning up (a) 2 points; (b) 7 points? **Ans.** (a) $\frac{1}{36}$; (b) $\frac{1}{6}$

2. In how many ways can 12 identical but distinguishable balls be distributed into 3 identical boxes with 4 balls in the first box, 5 balls in the second box, and the remainder in the third box? **Ans.** $\frac{12!}{4!5!3!}$

3. 30 identical balls are tossed randomly into a box that is divided into 3 compartments whose available areas have the ratio 1:2:3. (a) What is the probability that 10 balls will go into the smallest compartment, 15 balls will go into the next larger compartment, and 5 balls will go into the largest compartment? (b) What is the distribution of maximum probability? (c) What is the distribution of minimum probability?

4. (a) What is the number of ways of distributing 4 distinguishable balls among 3 identical boxes so that any box contains 2 balls and each of the other boxes contains one? (b) What is the number of ways of distributing 4 distinguishable molecules among 3 energy levels so that 2 molecules are in 1 energy level and 1 molecule is in each of the other 2 energy levels? (c) If the balls (or molecules) are assumed to be indistinguishable, what are the answers to (a) and (b)? **Ans.** (a) 36; (b) 36; (c) 3

5. The center of the fundamental vibration band for HCl is at a wavelength of 3.46μ. What is the ratio of the number of HCl molecules in the first excited vibrational state to the number in the ground vibrational state, if the HCl is at 300°K?

6. The degree of degeneracy of a rotational energy level with quantum number J is $2J+1$. The rotational constant for HF is 20.5 cm^{-1}. Calculate the ratio of the number of HF molecules in the state with $J=3$ to the number in the ground rotational state at a temperature of 300°K. **Ans.** 2.15

7. A certain molecule has an excited electronic energy level 3.50 ev above the ground energy level. What fraction of the molecules are in the excited electron state at equilibrium at a temperature of 1000°K?

8. Calculate the molar entropy of argon at 300°K and 1 atm pressure. **Ans.** 37.0 cal/mole deg

9. Calculate the heat capacity of oxygen at 1000°K given that the fundamental vibrational wavenumber is 1580 cm^{-1} and the moment of inertia is 1.93×10^{-39} g cm^2.

10. Calculate the molar entropy of nitrogen at 298°K and 1 atm, given that the fundamental vibrational wavenumber is 2360 cm^{-1} and the moment of inertia is 1.39×10^{-39} g cm^2. **Ans.** 45.8 cal/mole deg

11. Calculate (a) the free energy function $\dfrac{G° - E_0°}{T}$, and (b) the heat capacity for CO_2 at 300°K and 1 atm pressure, given that the C–O distance is 1.16 Å, and the fundamental vibrational wavenumbers in cm^{-1} are: 1340 (symmetrical stretch), 2350 (asymmetrical stretch), and 667 (bending).
Ans. (a) -43.6 cal/mole deg; (b) 8.90 cal/mole deg

12. Calculate the equilibrium constant at 1000°K for the dissociation $Na_2 = 2Na$, given that the dissociation energy of Na_2 is 0.73 ev, the fundamental vibrational wavenumber is 159 cm^{-1} and the internuclear distance is 3.08 Å.

13. Calculate the equilibrium constant for the reaction $H_2 + D_2 = 2HD$ at 300°K, given that the fundamental vibrational wavenumber for hydrogen is 4400 cm^{-1} and the internuclear distance is 0.74 Å. (The force constant and the internuclear distance are the same for all three isotopic species, and ΔE for this reaction depends on zero-point energies only.) **Ans.** 3.3

14. Given the following data, calculate the value of the equilibrium constant at 500°K for the reaction $H_2 + I_2 = 2HI$.

Molecule	Internuclear distance	Vibrational wavenumber	Dissociation energy
H_2	0.74 Å	4400 cm^{-1}	103.2 kcal/mole
I_2	2.67	215	35.5
HI	1.60	2310	70.5

10

Statistical Mechanics of Real Gases

Introduction

All of the theoretical development presented in the last chapter was based on the rigid rotor, harmonic oscillator molecular model, together with the assumption that the molecules behave as independent particles. If the calculated values are not sufficiently accurate, the error might be the result of either an oversimplified molecular model or the assumption of molecular independence.

It is not a difficult problem to refine the molecular model by applying corrections for centrifugal expansion and anharmonicity, provided the needed spectroscopic data are available. The assumption of the independence of the molecules, however, is a more complicated problem. We have seen that this assumption leads quite directly to the ideal gas equation of state. When it is not adequate to treat a gas as ideal it must be considered as a set of interacting particles. Then the nature and effect of intermolecular forces are of basic importance. The treatment of the problem of real gases requires a completely different point of view than that used in Chapter 9.

Gibbs' Development of Statistical Mechanics

As an introduction to the statistical mechanics of real gases, a brief summary of the theoretical approach developed by Gibbs is presented. This theory is based on classical mechanics and we may anticipate that it is not applicable to the rotational and vibrational degrees of freedom of molecules. Thus our molecular model is a structureless particle of mass m.

The general problem is the one with which Chapter 9 started, namely, the interpretation of the behavior of matter on a macroscopic scale in terms of molecular properties. As before, the system we are interested in contains a number of molecules of the order of Avogadro's number. The properties of this system at equilibrium are some sort of time average over the properties of its molecules. The molecular property of particular significance here is the energy which in Hamiltonian form (Chapter 6, p. 184) is given by the following equation:

$$H = \frac{1}{2m}(p_x^2 + p_y^2 + p_z^2) + U(x,y,z)$$

where p_x, as before, is the x component of the momentum. The kinetic energy is a function of the three momentum components and the potential energy is a function of the three coordinates of the molecular center of gravity.

Because there is no way of knowing the momenta and positions of individual molecules, there is no way of directly obtaining the desired time averages for any system. To overcome this difficulty Gibbs invented the concept of the *ensemble*. An ensemble is an imagined set of an indefinitely large number of replicas of the actual physical system of interest which are assumed to be in thermal equilibrium with each other. If the actual system is too complicated to treat directly, it might seem that doing anything with a large number of such systems is even less of a possibility. Gibbs showed, however, that the average over time for the actual system could be identified with, and replaced by, an instantaneous average over the ensemble, and he developed a way of obtaining this type of average.

In order to indicate how this was done we must introduce the concepts of *phase* and *phase space*. Gibbs defined the phase of a system as a particular assignment of values to all momentum components and positional coordinates of the molecules in the system. (Notice that this use of the word "phase" is different from any previous usage in this book.) If there are n molecules in the system, this implies an assignment of the values of $6n$ variables, the $3n$ momentum components and the $3n$ positional coordinates, which we represent by $p_1, p_2 \ldots p_{3n}$, and $q_1, q_2 \ldots q_{3n}$, respectively.

The simplest example of a phase space is that for a 1-dimensional harmonic oscillator, for which the Hamiltonian function is (Eq. 6–17):

$$H = \frac{p_x^2}{2m} + \frac{kx^2}{2}$$

Because there is only one momentum component and one coordinate for this system, its phase space has only two dimensions, as shown in Figure 10–1. A point in this plane represents a particular value of the momentum and coordinate. During the motion of the harmonic oscillator the momentum and coordinate change continuously, but the total energy remains constant. With the total energy constant, the graph of the Hamiltonian function is an ellipse, with the maximum values of $2p_x$ and $2x$ as the length of its axes, along which the phase point moves.

On the basis of classical mechanics there is no restriction on the value that

Statistical Mechanics of Real Gases

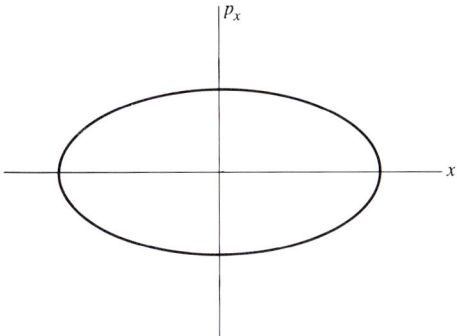

Figure 10-1. Phase space of a harmonic oscillator.

the total energy might have, and therefore the phase point at some instant might be anywhere in the p_x, x plane. From a quantum mechanical point of view, of course, the energy of a harmonic oscillator is limited to the values found from the solution of the Schrödinger equation for this model. This means that the phase point is restricted to move only on the ellipses which correspond to the quantized energy levels. We ignore any such quantum restrictions here, however, because this discussion of real gases is based on classical mechanics.

The phase space of a single molecule is a 6-dimensional space because there are 6 variables. (Although we cannot visualize such a space, the concept is useful as a mathematical device.) This type of phase space is often referred to as μ space (μ for molecule). Correspondingly, the phase space of a system containing n molecules is a $6n$-dimensional space, often called γ space (γ for gas). A single point in this space represents the instantaneous values of all $3n$ momentum components and $3n$ positional coordinates. The momentum components can have any value from $+\infty$ to $-\infty$, while the range of the coordinates is limited by the physical dimensions of the system. As the molecules in the system move about and collide, thus changing their momentum components as well as their positions, the phase point representing this system moves through γ space in accordance with the laws of classical mechanics.

Now we return to Gibbs' ensemble. Each system in the ensemble is imagined to be represented by its own point in γ space. With a large number (N) of systems in the ensemble there are N phase points distributed in some manner throughout phase space, each portion of which is assumed to be accessible to all of the phase points.

A fundamental question then arises. What can be said about the distribution of these N phase points in γ space? More precisely, if we define the element of volume of γ space by analogy with ordinary space as

$$d\phi = dp_1 dp_2 \cdots dp_{3n} dq_1 dq_2 \cdots dq_{3n} \qquad (10\text{--}1)$$

what is the number of phase points in an element of volume of γ space specified by a particular set of values for the $6n$ variables?

If $D(p_1 \ldots p_{3n}, q_1 \ldots q_{3n})$ represents the number of phase points in an element of volume $d\phi$, then a $6n$-fold integration of D over all of γ space must yield the total number of phase points. In symbols,

$$\int \cdots \overset{6n}{\int}_{\text{All } \gamma \text{ space}} D \, d\phi = N$$

If we divide both sides of this equation by N and represent D/N by P, we have

$$\int \cdots \int P \, d\phi = 1 \qquad (10\text{--}2)$$

We can interpret P as a probability; then we can rephrase the question above to read: What is the probability that some system in the ensemble has its phase point in the element of volume $d\phi$?

By a rather abstract mathematical analysis which is not repeated here, Gibbs was led to the conclusions that P is independent of time and that it is an exponential function of the energy of the system. Thus $P = Ae^{-\beta H}$, in which A and β are constants and H, as before, is the energy of the system in Hamiltonian form.

The constant A can be determined by integrating the following equation over all of γ space:

$$P \, d\phi = Ae^{-\beta H} \, d\phi$$

The result is

$$\int \cdots \overset{6n}{\int} P \, d\phi = 1 = A \int \cdots \overset{6n}{\int} e^{-\beta H} \, d\phi$$

and

$$A = \frac{1}{\int \cdots \int e^{-\beta H} \, d\phi}$$

Thus

$$P \, d\phi = \frac{e^{-\beta H} \, d\phi}{\int \cdots \int e^{-\beta H} \, d\phi} \qquad (10\text{--}3)$$

Statistical Mechanics of Real Gases

Gibbs called the denominator of this expression the *phase integral*.

What we have here is the distribution function for the ensemble in thermal equilibrium, which Gibbs called the *canonical ensemble*. By using this distribution function we can obtain averages over the ensemble which, as mentioned above, we identify with the time average for the actual system.

To illustrate this point with the simple system previously employed, let us obtain the average energy of an ideal monatomic gas, represented by $\bar{H}$. Using the same method of averaging we have used before (Chapter 5, p. 132), we have

$$\bar{H} = \frac{\int \overset{6n}{\cdots} \int H e^{-\beta H} \, d\phi}{\int \overset{6n}{\cdots} \int e^{-\beta H} \, d\phi}$$

Before anything further can be done with this equation we must know H as a function of momenta and positional coordinates. In an ideal monatomic gas there are no intermolecular forces and the molecules have kinetic energy of translation only. It follows that H is independent of the positional coordinates of the molecules and it depends on the momenta according to the following equation:

$$H = \sum_{i=1}^{3n} \frac{p_i^2}{2m} \tag{10-4}$$

Then the equation for the average energy becomes

$$\bar{H} = \frac{\int \overset{6n}{\cdots} \int \sum \frac{p_i^2}{2m} e^{-\beta \sum \frac{p_i^2}{2m}} \, dp_1 \ldots dp_{3n} dq_1 \ldots dq_{3n}}{\int \overset{6n}{\cdots} \int e^{-\beta \sum \frac{p_i^2}{2m}} \, dp_1 \ldots dp_{3n} dq_1 \ldots dq_{3n}}$$

Because the integrands of both numerator and denominator are independent of the positional coordinates of the molecules, the $3n$ integrations over these coordinates can be carried out. When the numerator and denominator are divided by the product of these integrals, there remains

$$\bar{H} = \frac{\int \overset{3n}{\cdots} \int \sum \frac{p_i^2}{2m} e^{-\beta \sum \frac{p_i^2}{2m}} \, dp_1 \ldots dp_{3n}}{\int \overset{3n}{\cdots} \int e^{-\beta \sum \frac{p_i^2}{2m}} \, dp_1 \ldots dp_{3n}} \tag{10-5}$$

The right-hand side of this equation can be written as a sum of $3n$ terms, one for each momentum component and all of the same mathematical form. It is sufficient, then, to carry out the integration for any one of the momentum components and multiply the result by $3n$.

To illustrate this procedure, let us obtain the contribution to the average energy from the term involving p_1. In place of $\sum_{i=1}^{3n} p_i^2$ we write $p_1^2 + \sum_{i=2}^{3n} p_i^2$; then Equation 10–5 becomes

$$\bar{H} = \frac{\int \cdots^{3n} \int \left(\dfrac{p_1^2}{2m} + \sum_{i=2}^{3n} \dfrac{p_i^2}{2m} \right) e^{-\dfrac{\beta p_1^2}{2m}} e^{-\beta \sum_{i=2}^{3n} \dfrac{p_i^2}{2m}} dp_1 \ldots dp_{3n}}{\int \cdots^{3n} \int e^{-\dfrac{\beta p_1^2}{2m}} e^{-\beta \sum_{i=2}^{3n} \dfrac{p_i^2}{2m}} dp_1 \ldots dp_{3n}}$$

The average of $\dfrac{p_1^2}{2m}$, which is the average kinetic energy associated with momentum component 1 ($\bar{\epsilon}_1$), is

$$\overline{\dfrac{p_1^2}{2m}} = \frac{\int \dfrac{p_1^2}{2m} e^{-\dfrac{\beta p_1^2}{2m}} dp_1 \int \cdots^{3n-1} \int e^{-\beta \sum_{i=2}^{3n} \dfrac{p_i^2}{2m}} dp_2 \ldots dp_{3n}}{\int e^{-\dfrac{\beta p_1^2}{2m}} dp_1 \int \cdots^{3n-1} \int e^{-\beta \sum_{i=2}^{3n} \dfrac{p_i^2}{2m}} dp_2 \ldots dp_{3n}}$$

When the integrals over $dp_2 \ldots dp_{3n}$ in the numerator and denominator are cancelled, we have

$$\overline{\dfrac{p_1^2}{2m}} = \bar{\epsilon}_1 = \frac{\int \dfrac{p_1^2}{2m} e^{-\dfrac{\beta p_1^2}{2m}} dp_1}{\int e^{-\dfrac{\beta p_1^2}{2m}} dp_1} \tag{10-6}$$

We can convert these integrals into standard forms by the substitution $a = p_1/\sqrt{2m}$. Then $dp_1 = \sqrt{2m}\, da$ and

$$\bar{\epsilon}_1 = \frac{\sqrt{2m} \int a^2 e^{-\beta a^2}\, da}{\sqrt{2m} \int e^{-\beta a^2}\, da}$$

$$= \frac{\dfrac{1}{2\beta}\sqrt{\dfrac{\pi}{\beta}}}{\sqrt{\dfrac{\pi}{\beta}}} = \frac{1}{2\beta}$$

Statistical Mechanics of Real Gases

From our earlier results we found that the average kinetic energy associated with one momentum component is $kT/2$ (Eq. 5-4). It is clear that we have agreement with the present result when we set $\beta = 1/kT$.

◈◈◈◈◈◈◈

From the evaluation of the integrals in Equation 10-5 the average energy of our ideal gas system containing n molecules is given by

$$\bar{H} = \sum_{i=1}^{3n} \bar{\epsilon}_i = \tfrac{3}{2}nkT$$

If the system contains Avogadro's number of molecules then $\bar{H} = E$, the molar translational energy. We thus have obtained the same result we obtained for a monatomic ideal gas from the kinetic theory in Chapter 5 and from the translational partition function (Eq. 9-27).

Equipartition of Energy

We now extend the calculation of the average energy of a large number of molecules to the case of molecules with internal degrees of freedom; that is, molecules with rotational and vibrational energy. In doing this we give a derivation of the generalization known as the law of equipartition of energy, which was mentioned on page 153. We first write the equation for the energy of a diatomic molecule (Eq. 5-28) in Hamiltonian form as follows:

$$H = \frac{p_x^2}{2m} + \frac{p_y^2}{2m} + \frac{p_z^2}{2m} + \frac{p_{\theta_1}^2}{2I} + \frac{p_{\theta_2}^2}{2I} + \frac{kq^2}{2} + \frac{p_q^2}{2\mu}$$

in which p_{θ_1} is the angular momentum associated with rotation about axis 1, $kq^2/2$ is the vibrational potential energy, $q\ (= r - r_e)$ is the displacement of the atoms from their equilibrium positions, p_q is the linear momentum associated with the vibrational motion of the atoms with respect to the center of gravity of the molecule, and μ is the reduced mass.

These seven terms all have the same mathematical form, namely αb^2, in which b is a momentum component or a coordinate and α is a constant. By analogy with Equation 10-6, we can write the average energy associated with a term αb^2 in the following form:[1]

$$\bar{\epsilon}_b = \frac{\int_0^\infty \alpha b^2 e^{-\frac{\alpha b^2}{kT}}\, db}{\int_0^\infty e^{-\frac{\alpha b^2}{kT}}\, db}$$

[1] The limits of these integrals can be taken as either 0 to ∞ or $-\infty$ to $+\infty$, since the value of $\bar{\epsilon}_b$ is the same in both cases.

Evaluation of these integrals leads to the following result:

$$\bar{\epsilon}_b = \frac{kT}{2}$$

We can now state the law of equipartition of energy in its most general form: For each term in the expression of the energy of a molecule which depends on the square of a variable, the contribution to the average molecular energy is $kT/2$, and thus the contribution to the molar energy is $RT/2$. Since there are seven square terms for a diatomic molecule, the molar energy is $7RT/2$ and the corresponding value of the heat capacity at constant volume is $7R/2$. For a nonlinear polyatomic molecule containing N atoms, the number of square terms is $6N - 6$.

We have seen that the law of equipartition of energy has the limited validity which is implied by the fact that its derivation is based directly on classical mechanics.

Relationship of the Phase Integral to the Partition Function

Now let us take another look at the phase integral, the denominator in Equation 10–3. When β is replaced by $1/kT$ and the Hamiltonian function for a monatomic ideal gas (Eq. 10–4) is introduced, the phase integral becomes

$$\int \overset{3n}{\cdots} \int e^{-\frac{1}{2mkT}\Sigma p_i^2}\, dp_1 \ldots dp_{3n} \int \overset{3n}{\cdots} \int dq_1 \ldots dq_{3n}$$

The integrations over the positional coordinates can be carried out quite simply. If a, b, and c are the dimensions of the system, and q_1, q_2, and q_3 are the cartesian coordinates of molecule 1, then

$$\int_0^a \int_0^b \int_0^c dx_1 dy_1 dz_1 = abc = V$$

where V is the volume of the system. The remaining integrals over the coordinates can be treated in the same way, each set of three contributing one factor of V. Consequently

$$\int \overset{3n}{\cdots} \int dq_1 \ldots dq_{3n} = V^n$$

The integrations over the momentum components can be carried out in the same way the denominator of Equation 10–6 was evaluated. The integrals can be written in the form

$$\int e^{-\frac{1}{2mkT}p_1^2}\, dp_1 \int e^{-\frac{1}{2mkT}p_2^2}\, dp_2 \ldots \int e^{-\frac{1}{2mkT}p_{3n}^2}\, dp_{3n}$$

With $\alpha = \dfrac{1}{2mkT}$, the first integral becomes the standard form $\int e^{-\alpha p_1^2} dp_1$, which is equal to $\sqrt{\dfrac{\pi}{\alpha}}$. Consequently

$$\int e^{-\frac{1}{2mkT} p_1^2} dp_1 = (2\pi mkT)^{\frac{1}{2}}$$

and the product of the $3n$ integrals is $(2\pi mkT)^{3n/2}$. Thus the phase integral for one mole of ideal gas has the value

$$[V(2\pi mkT)^{\frac{3}{2}}]^{\mathfrak{N}}$$

The similarity of the phase integral to the molar translational partition function for an ideal gas (Eq. 9–29) should be noticed. The differences are due to the fact that the derivation of the phase integral is based entirely on classical mechanics.

Two corrections, both based on quantum mechanics, must be applied to the phase integral to convert it into a molar partition function. The first of these, which we have discussed before, is the correction for the indistinguishability of gas molecules. As before, this correction consists of a division by $\mathfrak{N}!$.

The second correction is required for consistency with the Heisenberg uncertainty principle (Eq. 6–27). In Equation 10–1 the element of volume of phase space, $d\phi$, was defined and we can write this expression in the following way:

$$d\phi = dp_1 dq_1 dp_2 dq_2 \ldots dp_{3n} dq_{3n}$$

Let us focus on the factors associated with one coordinate, for example $dp_1 dq_1$. According to classical mechanics, these differentials can be allowed to become infinitesimal. In contrast, if we interpret dp_1 and dq_1 as representing the uncertainties, Δp_1 and Δq_1, in p_1 and q_1, then according to the uncertainty principle, $\Delta p_1 \Delta q_1 \sim h$. This requirement leads to the concept that phase space is divided into "cells" and a phase point cannot be located in phase space more precisely than the statement that it is in a particular cell.

Since there are $3n$ factors of the form $dpdq$ included in $d\phi$, the product of all of these factors is h^{3n}. This quantity is the "volume" of each of the cells and thus it is the unit of volume of γ phase space.

Now let us return to Equation 10–2. We interpreted P in this equation as a probability, but this was not strictly correct because $P \, d\phi$ must be dimensionless and $d\phi$ has the dimensions of a volume of phase space. If, however, we define h^{3n} as the unit volume of phase space and express $d\phi$ in terms of this unit, Equation 10–2 becomes

$$\int \overset{6n}{\cdots} \int P \frac{d\phi}{h^{3n}} = 1$$

and P is dimensionless.

When we apply both of the corrections we have described to the general expression of the phase integral for a system containing Avogadro's number of molecules, it becomes the molar translational partition function; thus

$$Q_t = \frac{1}{\mathfrak{N}! h^{3\mathfrak{N}}} \int \cdots^{6\mathfrak{N}} \int e^{-\beta H} d\phi$$

On evaluation of this expression for 1 mole of ideal gas we have

$$Q_t = \frac{1}{\mathfrak{N}! h^{3\mathfrak{N}}} [V(2\pi mkT)^{\frac{3}{2}}]^{\mathfrak{N}}$$

When logarithms are taken, the result is identical with Equation 9–29.

Partition Function for Real Gases

This summary of Gibbs' development of statistical mechanics has been included not from the desire to rederive equations for ideal gases, but rather because this treatment is not limited to ideal gases. Its significance lies in its generality and the fact that it is applicable to real gases. The characteristic feature of a real gas is the existence of intermolecular forces, and these forces are related to an internal potential energy which is represented by $U(q_1 q_2 q_3 \ldots q_{3n})$. This notation implies that the internal potential energy of a system depends on the positions of all of the molecules in the system, and the q's represent any type of positional coordinates.

It follows that the Hamiltonian function can be written in the following form:

$$H = \sum \frac{p_i^2}{2m} + U(q_1 \ldots q_{3n})$$

With $U(q)$ as an abbreviated symbol for the potential energy function, the molar translational partition function for a real gas is

$$Q_t = \frac{1}{\mathfrak{N}! h^{3\mathfrak{N}}} \int \cdots^{6\mathfrak{N}} \int e^{-\frac{1}{kT}\left(\sum \frac{p_i^2}{2m} + U(q)\right)} dp_1 \ldots dq_{3\mathfrak{N}}$$

The integrals over the momentum components are exactly the same as for the ideal gas. Consequently the partition function becomes

$$Q_t = \frac{1}{\mathfrak{N}! h^{3\mathfrak{N}}} (2\pi mkT)^{3\mathfrak{N}/2} \int \cdots^{3\mathfrak{N}} \int e^{-\frac{U(q)}{kT}} dq_1 \ldots dq_{3\mathfrak{N}}$$

The factor involving the integrals over the positional coordinates is usually called the configuration integral, Q_U. If Q_U could be evaluated by carrying out the indicated integrations, all of the thermodynamic properties of a real gas, including the equation of state, could be calculated theoretically in the same

Statistical Mechanics of Real Gases

way as for an ideal gas. The function $U(q)$, however, represents the energy of all of the simultaneous interactions of all of the molecules in the system with each other. Unfortunately, this function is so complicated that no way has been found for expressing it and consequently the configuration integral cannot be evaluated exactly. Since this is the case, the next step is to introduce simplifying assumptions that will permit the calculation of results which are reasonable approximations. Much work has been done on this problem, but the evaluation of the configuration integral is still a subject of research. All that can be done here is to give a descriptive outline of a simplified treatment of the problem.

The first assumption usually made is the assumption of pair-wise interactions, which means that the interaction between 2 molecules is assumed not to be affected by the presence of one or more other molecules in the immediate neighborhood. On the basis of this assumption, the total energy of interaction is the sum of the energies of interaction of all pairs of molecules. It is also assumed that the pair-wise interaction depends only on the distance between the 2 molecules involved. In this case,

$$U(q) = \sum_{\text{all pairs}} u(r_{ij})$$

in which r_{ij} is the distance between molecule i and molecule j and the sum includes each pair just once.

With this approximate expression for the potential energy function, the configuration integral becomes

$$Q_U = \int \cdots \int^{3\mathfrak{N}} e^{-\frac{\Sigma u(r_{ij})}{kT}} dq_1 \ldots dq_{3\mathfrak{N}}$$

$$= \int \cdots \int^{3\mathfrak{N}} \prod_{\text{all pairs } ij} e^{-\frac{u(r_{ij})}{kT}} dq_1 \ldots dq_{3\mathfrak{N}}$$

where the symbol $\prod e^{-u(r_{ij})/kT}$ represents the product of $\dfrac{\mathfrak{N}(\mathfrak{N}-1)}{2}$ exponential factors, one for each pair of molecules.

Next the simplifying assumption is made that the deviation of the gas from ideality is small. This implies that the function $u(r_{ij})$ is small relative to kT over most of the range of r_{ij} and therefore the value of $e^{-u(r_{ij})/kT}$ is close to 1. Because of this fact it is mathematically expedient to define the function $e^{-u(r_{ij})/kT} - 1$, which can be considered as always small, and its integral $\iiint (e^{-u(r_{ij})/kT} - 1) \, dx_j dy_j dz_j$ which will be represented by I. (In this integral molecule i can be assumed to be located at the origin of the coordinate system, and the integration is carried out over the range of the coordinates of molecules j.)

It is possible to express the configuration integral in terms of the integral I, which is the same for all pairs of molecules, according to the following equation:

$$Q_U = V^{\mathfrak{N}}\left(1 + \frac{\mathfrak{N}I}{2V} + \cdots\right)^{\mathfrak{N}} \tag{10-7}$$

The indicated higher terms are negligible when $\mathfrak{N}I/2V$ is small relative to 1, which is true if the deviations from ideality are small. Notice that if $u(r) = 0$, then $I = 0$ and Q_U reduces to its ideal gas value of $V^{\mathfrak{N}}$.

On taking the logarithms of both sides of Equation 10–7 we have

$$\ln Q_U = \mathfrak{N} \ln V + \mathfrak{N} \ln\left(1 + \frac{\mathfrak{N}I}{2V}\right)$$

Because $\mathfrak{N}I/2V$ is small, $\ln(1 + \mathfrak{N}I/2V) \sim \mathfrak{N}I/2V$ and

$$\ln Q_U = \mathfrak{N} \ln V + \frac{\mathfrak{N}^2 I}{2V}$$

Having obtained this expression for Q_U, we can find the theoretical equation of state of our nearly ideal gas by making use of Equation 9–30; thus $P = kT\left(\frac{\partial \ln Q}{\partial V}\right)_T$. Since the only factor in the translational partition function which is a function of V is Q_U, we have

$$P = kT\left(\frac{\partial \ln Q_U}{\partial V}\right)_T = kT \frac{\partial}{\partial V}\left(\mathfrak{N} \ln V + \frac{\mathfrak{N}^2 I}{2V}\right)$$

$$= \frac{\mathfrak{N}kT}{V} - \frac{\mathfrak{N}^2 I kT}{2V^2}$$

$$= \frac{\mathfrak{N}kT}{V}\left(1 - \frac{\mathfrak{N}I}{2V}\right)$$

In a discussion of empirical equations of state for real gases in Chapter 1, it was mentioned that one of the useful and general forms of this equation is the virial equation of state (Eq. 1–7) which is repeated here:

$$P = \frac{\mathfrak{N}kT}{V}\left[1 + \frac{B(T)}{V} + \cdots\right]$$

The quantity $B(T)$ is called the *second virial coefficient*. When we compare the theoretical with the empirical equation we see that the theoretical value of the second virial coefficient is

$$B(T) = -\frac{\mathfrak{N}I}{2}$$

It is significant that we have obtained a theoretical equation of the same mathematical form as the virial equation, but this result is of little use until we can evaluate the integral I. We can simplify the expression of this integral by writing the element of volume in terms of polar coordinates (p. 135). The expression for I then becomes

Statistical Mechanics of Real Gases

$$I = \iiint (e^{-\frac{u(r)}{kT}} - 1) r^2 \, dr \sin\theta \, d\theta \, d\phi$$

When we integrate over the whole range of the angle coordinates, the result is 4π, and the expression for I reduces to

$$I = 4\pi \int_0^\infty (e^{-\frac{u(r)}{kT}} - 1) r^2 \, dr$$

Notice that the upper limit of r is written as infinity. This is a mathematical convenience. Actually the range of r is limited by the size of the container holding the gas, but the value of $u(r)$ decreases so rapidly as r increases that there is no significant error in extending the limit to infinity.

On substitution of the above expression for I, the second virial coefficient takes the form

$$B(T) = -2\pi \mathfrak{N} \int_0^\infty (e^{-\frac{u(r)}{kT}} - 1) r^2 \, dr \tag{10-8}$$

This is as far as we can go without choosing some mathematical form for the potential energy function $u(r)$. The simplest choice for $u(r)$ is that corresponding to the rigid sphere model of a molecule with diameter σ and no attractive forces. In this case the potential energy function is

$$u(r) = \begin{cases} 0 & \text{for } r > \sigma \\ \infty & \text{for } r < \sigma \end{cases}$$

We can separate the integral in Equation 10–8 into two parts, one extending from 0 to σ and the other from σ to ∞. When the appropriate value of $u(r)$ is substituted, the equation for $B(T)$ becomes

$$B(T) = -2\pi\mathfrak{N} \left[\int_0^\sigma (e^{-\infty} - 1) r^2 \, dr + \int_\sigma^\infty (e^0 - 1) r^2 \, dr \right]$$

$$= 2\pi\mathfrak{N} \int_0^\sigma r^2 \, dr$$

$$= \tfrac{2}{3}\pi\mathfrak{N}\sigma^3$$

For the rigid sphere molecular model under consideration we see that the second virial coefficient is positive and independent of the temperature. It is essentially the same as the constant b in the van der Waals equation (Eq. 1–6) if the term a/V^2 is negligible. Since the volume of a sphere of diameter σ is $\pi\sigma^3/6$, the value of $B(T)$ is four times the volume of Avogadro's number of molecules considered as rigid spheres.

To go beyond this oversimplified molecular model requires a consideration of intermolecular potential energy.

Intermolecular Potential Energy Function

The subject of intermolecular potential energy and the related intermolecular forces is one of the most basic parts of the study of the behavior of matter. The existence of liquid and solid phases implies that there are attractive or cohesive forces between molecules, and the small compressibility of these condensed phases shows that strong repulsive forces also act. The deviations from ideality of real gases depend on the same types of forces, which in general are called van der Waals forces.

The van der Waals attractive forces are better known and understood than the repulsive force. Theoretical investigations have resulted in the recognition of the following three types of attractive intermolecular forces:

1. *Dipole-dipole force*. Two electric dipoles, characterized by their permanent electric moment μ (Chapter 8, p. 274), exert a mutual electrostatic force which is a function of their separation (r), their relative orientation, and the magnitude of μ. When this force function is averaged over all orientations to take account of random motion, the result is found to be an attractive force whose corresponding potential energy function is

$$u(r) = -\frac{2\mu^4}{3kTr^6}$$

The negative sign of the potential energy function is consistent with the fact that the dipole-dipole force is an attractive force.

2. *Dipole-induced dipole force*. When two molecules, one with a permanent dipole moment and one without, are adjacent, the electric field of the permanent moment induces an electric moment in the nonpolar molecule. The average electrostatic force between these moments is attractive and it depends on both the magnitude of the permanent moment and the molecular polarizability (Eq. 8–1) of the nonpolar molecule. The corresponding potential energy function is

$$u(r) = -\frac{2\alpha\mu^2}{r^6}$$

3. *London, or dispersion, force*. The two preceding types of attractive force are adequate to account for the behavior of polar gases and mixtures of polar and nonpolar gases. It is clear, however, that they cannot account for deviations from ideality in a pure nonpolar gas such as argon or nitrogen. These deviations were not explained until after the development of quantum mechanics. In 1930, on the basis of an involved calculation using quantum mechanical perturbation theory, London found that when two nonpolar molecules are near each other the instantaneous dipole moment in one of them, resulting from a transient dissymmetry of the electron positions, induces a

moment in the other molecule, and vice versa. His calculations showed that the resultant average force is an attractive force and that the potential energy function can be approximated by

$$u(r) = -\frac{3h\nu_o \alpha^2}{4r^6}$$

In this expression h is Planck's constant and $h\nu_o$ is roughly the ionization potential; that is, the energy required to remove an electron from the molecule. This type of force is usually called the dispersion force because it is related theoretically to the dispersion of the substance, which is the change of refractive index with the frequency of the transmitted radiation.

◈◈◈◈◈◈◈

The equation above implies that the attractive force between two nonpolar molecules increases with the molecular polarizability. This conclusion is approximately confirmed by experimental data. For example, deviations from ideality are much larger for chlorine ($\alpha = 4.50 \times 10^{-24}$ cc/molecule) than for nitrogen ($\alpha = 1.73 \times 10^{-24}$ cc/molecule) when both gases are studied at the same temperature and molecular concentration.

◈◈◈◈◈◈◈

It is remarkable that all three types of attractive force have the same dependence on the intermolecular distance.

In contrast to these equations for the attractive contributions to the intermolecular potential energy, no theoretical equation has been derived so far for the repulsive contribution. Qualitatively, repulsive forces arise during a collision from overlap of the closed electron shells of the molecules, and possibly from nuclear repulsions as well. There is much evidence which indicates that the repulsive potential energy varies with a large inverse power of the internuclear distance but the effect on measurable properties of the value of the exponent is not critical. Experimental data are consistent with a value between 10 and 14, and a value of 12 for this exponent is often used.

The potential energy of a pair of molecules is the sum of terms related to attraction and repulsion. With an attractive term depending on r^{-6} and a repulsive term depending on r^{-12} we can write this sum as follows:

$$u(r) = \frac{A}{r^{12}} - \frac{B}{r^6}$$

where A and B are constants for each type of molecule. An alternative way of writing an expression for the potential energy is

$$u(r) = 4\epsilon \left[\left(\frac{\sigma}{r}\right)^{12} - \left(\frac{\sigma}{r}\right)^{6} \right] \tag{10-9}$$

in which ϵ and σ are characteristic molecular parameters. This potential energy function is usually referred to as the *Lennard–Jones 12-6 potential*. Its graph is given in Figure 10-2.

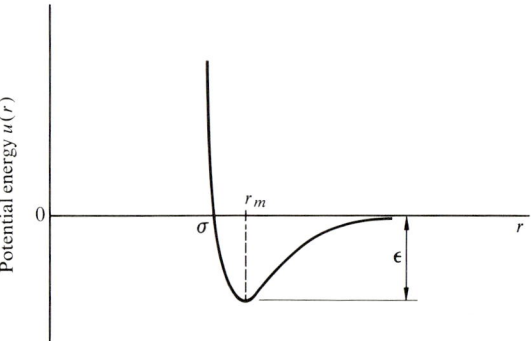

Figure 10-2. Graph of Lennard–Jones 12-6 potential energy function.

We can readily give an interpretation to the parameters σ and ϵ. When $r = \sigma$, $u(r) = 0$; thus σ is the intermolecular distance at which the potential energy of interaction vanishes. It is one type of collision diameter. To find the significance of ϵ let us find the minimum of the Lennard–Jones potential by differentiating it and setting the derivative equal to zero. In this way we obtain

$$\frac{du(r)}{dr} = 4\epsilon(-12\sigma^{12}r^{-13} + 6\sigma^{6}r^{-7}) = 0$$

It follows that $2\sigma^6 = r_m^6$ and $r_m = 2^{\frac{1}{6}}\sigma = 1.122\sigma$, where r_m is the intermolecular distance corresponding to the minimum of the potential energy curve. (This is the distance at which the forces of attraction and repulsion just balance each other.) When we substitute this value of r_m in Equation 10-9 we see that the minimum value of the potential function is

$$u(r_m) = 4\epsilon \left(\frac{\sigma^{12}}{4\sigma^{12}} - \frac{\sigma^6}{2\sigma^6} \right)$$
$$= 4\epsilon \left(\tfrac{1}{4} - \tfrac{1}{2} \right) = -\epsilon$$

This shows that ϵ represents the depth of the potential energy minimum below the value of zero for widely separated molecules.

Now that we have a reasonable expression for the intermolecular potential

Statistical Mechanics of Real Gases

energy, the Lennard–Jones 12–6 function, we can substitute it into the equation for the second virial coefficient (Eq. 10–8) which then becomes

$$B(T) = 2\pi \mathfrak{N} \int_0^\infty \left(1 - e^{-\frac{4\epsilon}{kT}\left[\left(\frac{\sigma}{r}\right)^{12} - \left(\frac{\sigma}{r}\right)^6\right]}\right) r^2 \, dr \qquad (10\text{–}10)$$

In order to obtain the value of the second virial coefficient by carrying out this integration, values of the parameters ϵ and σ must be known. Since there is no way of calculating these values theoretically, it is necessary to give them values which are most nearly consistent with the experimental values of $B(T)$ obtained from P,V,T data. When the best values of ϵ and σ have been calculated in this way, it is found that the theoretical values of $B(T)$ are in agreement with the experimental values over the entire temperature range for which measurements are available.

The most extensive calculations of this type have been carried out by Hirschfelder and his co-workers.[2] They find it convenient to express their results in terms of the dimensionless quantities $T^* = kT/\epsilon$ and $B^*(T) = B(T)/\frac{2}{3}\pi \mathfrak{N}\sigma^3$. When $B^*(T)$ is plotted against T^* for a number of simple gases

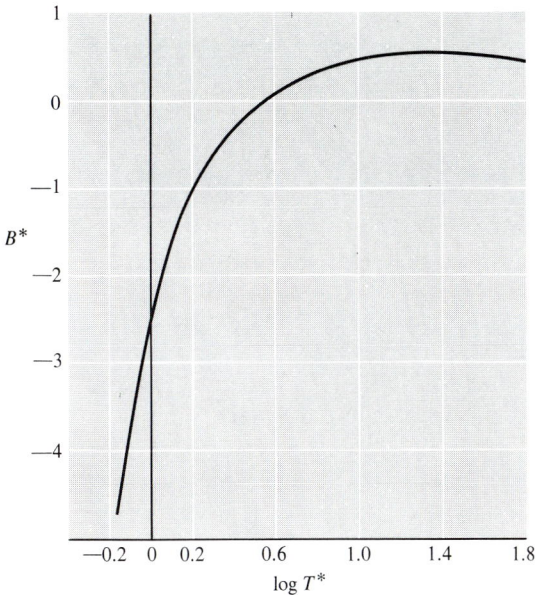

Figure 10–3. "Reduced" second virial coefficient B^* as a function of T^*. (Data from Hirschfelder, Curtiss, and Bird, *loc. cit.*, Table I-B.)

---

[2]Hirschfelder, Curtiss, and Bird, *The Molecular Theory of Gases and Liquids*, John Wiley & Sons, New York, 1954.

it is found that all of the points representing experimental values lie on the same curve (Fig. 10–3).

At temperatures that correspond to a value of T^* less than about 3, the second virial coefficient is negative which indicates that the attractive forces play a dominant role. At higher temperatures $B(T)$ reaches a maximum and then gradually decreases toward a limiting value of 0, the ideal gas value.

The values of ϵ/k and σ for a few gases are listed in Table 10–1. The values of σ listed are in fairly good agreement with collision diameters calculated from

TABLE 10-1 PARAMETERS FOR THE LENNARD-JONES 6-12 POTENTIAL DETERMINED FROM SECOND VIRIAL COEFFICIENTS*

Gas	$\frac{\epsilon}{k}(°K)$	$\sigma(\text{Å})$	B for hard-sphere model $\left(=\frac{2}{3}\pi N\sigma^3\right)$ (cc/mole)
Ne	34.9	2.78	27.10
Ar	119.8	3.41	49.80
N_2	95.1	3.70	63.78
O_2	118	3.46	52.26
CH_4	148.2	3.82	70.16
CO_2	189	4.49	113.9

*Hirschfelder, Curtiss, and Bird, loc. cit., Table 3.6-1.

viscosity data (Table 5–2). Data such as those in Table 10–1 and Figure 10–3 can be used to calculate values of $B(T)$, and thus the compressibility factors of gases at moderate pressures. For example, let us calculate the compressibility factor of methane at 300°K and a pressure of 10 atm. The value of T^* is $300/148 = 2.03$, and from Figure 10–3 the value of $B^*(T)$ is about -0.61. Then $B(T) = -0.61 \times 70.2 \sim -43$ cc/mole. The molar volume at the given conditions is approximately RT/P which is equal to 2460 cc/mole. Then the compressibility factor of methane as given by Equation 1–7 is $Z = 1 - 43/2460 = 0.983$.

Unfortunately, the treatment of real gases in this chapter is applicable only when deviations from ideality are so small that they can be accounted for by the use of the second virial coefficient alone. The treatment of gases at high pressures, for which the higher virial coefficients must be used, is complex and some aspects of the theory are still subjects of research.

Supplementary References

S. M. Blinder, *Advanced Physical Chemistry*, Macmillan, London, 1969; Chapter 22.

N. Davidson, *Statistical Mechanics*, McGraw-Hill, New York, 1962.

J. O. Hirschfelder, C. F. Curtiss, and R. B. Bird, *Molecular Theory of Gases and Liquids*, Wiley, New York, 1954.

Problems

1. Find the second virial coefficient in cm³/mole at 300°K for a gas for which $\epsilon/k = 100°K$ and the hard-sphere molecular diameter is 4.0 Å.

Ans. -9.6 cm²/mole

2. Using Table 10–1 and Figure 10–3, calculate the compressibility factor for CO_2 at a pressure of 5 atm and temperatures of (a) 300°K and (b) 400°K.

3. The value of T^* for which $B^*(T^*) = 0$ is 3.40. Calculate the Boyle point of nitrogen; that is, the temperature at which nitrogen behaves as an ideal gas at moderate pressures. **Ans.** 323°K

11

Reaction Kinetics

Introduction

Except for some aspects of the kinetic theory presented in Chapter 5, this book has not been concerned with time as a variable or with the rates at which processes occur. The major emphasis has been on matter in a state of equilibrium. We have seen how thermodynamics and data on molecular structure can be combined to yield a comprehensive treatment of systems in which chemical equilibrium has been reached. We have also seen how to predict that a certain chemical reaction can or cannot occur in a system in a given state.

Important as thermodynamics and statistical mechanics are for chemistry, they obviously must be supplemented by a treatment of the behavior of matter and the rates of the processes it undergoes when it is not in equilibrium. In this chapter we consider the rates of chemical reactions, a subject which is of major importance not only for its many practical applications, but also for the information it provides on the fundamental problem of how a chemical reaction occurs.

The study of reaction kinetics is in general more difficult than the study of chemical equilibria. The experimental measurements are often vulnerable to subtle errors and the theoretical interpretation of reaction rates is still incomplete. It is not yet possible, except in a very few special cases, to calculate moderately accurately the rate of a reaction from a knowledge of molecular properties, as equilibrium constants are calculated. In fact, the details of the sequence of events resulting in a chemical reaction are still subjects of research. This does not mean that little has been accomplished in the century during which reaction rates have been studied. Much has been done, but much remains for the future.

The dependence of the rate of a chemical reaction on the concentrations of the reacting substances began to be recognized a little over a hundred years ago. The first generalization on this point was made in 1864 by Guldberg and Waage who stated that a reaction rate is proportional to the product of the active masses of the reactants. By the term "active mass" they meant the quantity we call concentration. Subsequent work has confirmed their generalization for some reactions, but it has been found that the dependence of reaction rate on reactant concentration is usually more complicated than this.

Empirical Description of Reaction Rates

The measurement of a reaction rate requires that concentration changes be determined as a function of time. Basically, this determination is a problem in quantitative analysis, to be solved either by conventional analytical methods or by some indirect method. If a conventional method, such as titration, is employed, it is usually necessary to stop or "freeze" the reaction at definite times to make sure that any additional concentration change during completion of the analysis is negligible.

Any property of the reacting system which can be measured without disturbing it and which can be related unambiguously to its composition may be utilized in kinetic studies. Among the wide variety of such properties that have been employed by ingenious investigators are the absorption spectrum in an appropriate spectral range, refractive index, electrolytic conductance, and the rotation of the plane of transmitted plane polarized light. For those gas phase reactions during which the total number of gas molecules changes, a measurement of the total pressure of the reacting system provides sufficient information.

Reacting systems are classified as homogeneous if all of the reactants are in the same phase, or heterogeneous if the reactants are not in the same phase. At first, we limit our attention to homogeneous systems which are held at constant temperatures.

Let us consider a general reaction whose equation is

$$aA + bB \rightarrow cC$$

We can express the rate of this reaction in terms of the rate of change with time of any of the three concentrations, $-\frac{d[A]}{dt}$, $-\frac{d[B]}{dt}$, or $\frac{d[C]}{dt}$. (The concentration of a substance is represented by its chemical symbol in square brackets.) The three rate expressions are related by the stoichiometric coefficients in the following way:

$$-\frac{1}{a}\frac{d[A]}{dt} = -\frac{1}{b}\frac{d[B]}{dt} = \frac{1}{c}\frac{d[C]}{dt}$$

For example, if $a = 2$ and $b = 1$, the rate of disappearance of B is half the rate of disappearance of A.

For many (but not all!) reactions it has been found that the rate as a func-

Reaction Kinetics

tion of the concentrations of the reactants can be represented by an equation of the form

$$-\frac{d[A]}{dt} = k_r[A]^m[B]^n$$

When it is possible to write a rate equation in this form, the *order of the reaction* is defined as the sum of the exponents, $m + n$. For simple reactions the exponents and the order are found to be small positive integers, but for other types of reactions the exponents and the order may be nonintegral. The exponent for the concentration of any one of the reactants is said to be the order of the reaction with respect to that substance.

The factor k_r in the rate equation is called the specific reaction rate, or simply the *rate constant*. Although it is independent of concentrations it is a function of the temperature. Its dimensions are (concentration)$^{1-(m+n)}$ (time)$^{-1}$ and thus its numerical value depends, except for the case of a first-order reaction, on the units used for both concentration and time. The concentration unit used most commonly is the mole/liter.

In general, there is no relationship between the exponents (m and n) in the rate equation and the stoichiometric coefficients (a and b) in the chemical equation. In other words, it is not possible to predict the dependence of the rate of a reaction on the concentrations of the reactants from the chemical equation for the reaction. One of the best illustrations of this point is the following pair of gas-phase reactions:

$$H_2 + I_2 = 2HI$$
$$H_2 + Br_2 = 2HBr$$

These equations appear quite similar, but the kinetics of the two reactions are completely different. The first reaction is a second-order reaction while the second reaction follows a complicated rate law without a definite order. The order of a reaction is an empirical quantity which can be found only from experimental data for concentrations at measured time intervals.

Determination of the Order of a Reaction

When the differential equations representing the reaction rates corresponding to different orders are integrated, the resulting equations give the concentration of a reactant or product as a function of time. If one of these equations for a particular order is found to represent adequately the experimental data for a certain reaction, then the order of this reaction has been determined.

We now carry out the needed integrations of a few of the simpler differential rate equations.[1]

[1] Zero-order reactions with rate independent of reactant concentrations are possible in heterogeneous systems but not in the type of system being considered here.

First-Order Reactions

For a first-order reaction represented by $A \rightarrow B + C$, the rate equation is

$$-\frac{d[A]}{dt} = k_1[A]$$

On separating the variables in this equation, it becomes

$$-\frac{d[A]}{[A]} = -d \ln [A] = k_1 \, dt \tag{11-1}$$

When both sides are integrated indefinitely, the result is

$$\ln [A]_t = -k_1 t + \text{constant}$$

The constant of integration is easily found from the initial condition. If the concentration of A at time zero is $[A]_o$, then the constant is $\ln [A]_o$ and the integrated equation can be written in the following equivalent forms:

$$\ln [A]_o - \ln [A]_t = k_1 t; \quad [A]_t = [A]_o e^{-k_1 t}; \quad \log \frac{[A]_o}{[A]_t} = \frac{k_1 t}{2.3}$$

A graph of $\log [A]_t$ versus t is a straight line with a slope equal to $-k_1/2.3$. The value of k_1 is independent of the concentration units employed.

If Equation 11-1 is integrated between the limits $[A]_1, t_1$ and $[A]_2, t_2$ the result is

$$\log \frac{[A]_1}{[A]_2} = \frac{k_1}{2.3}(t_2 - t_1)$$

This equation implies that if the concentrations are measured at equal time intervals, for a first-order reaction the ratios of the concentrations at the beginning and at the end of each interval are constant.

It is sometimes convenient to express the rate equation in terms of a concentration change such as the change in the concentration of A at time t, which in this case is the same as the concentration of B formed at time t. Representing by a the initial concentration of A and by x the concentration change at time t, we can write the first-order rate equation in the following way:

$$\frac{dx}{dt} = k_1(a - x) \tag{11-2}$$

When the variables are separated and the equation is integrated between the limits $x = 0$, $t = 0$, and x, t, the result is

$$\log \frac{a}{a - x} = \frac{k_1 t}{2.3}$$

The time ($t_{\frac{1}{2}}$) required for the reaction of half of the initial amount of reactant A is of particular significance for first-order reactions. When half of A has reacted, $a - x = a/2$ and

$$t_{\frac{1}{2}} = \frac{2.30}{k_1} \log 2 = \frac{0.693}{k_1}$$

It is evident from this equation that $t_{\frac{1}{2}}$, which is called the half-life of reactant A, is independent of the initial amount (or concentration) of A. It is interesting to note that just this behavior is observed for the disintegration of all radioactive isotopes, for which the half-life is a characteristic property.

TABLE 11-1   DECOMPOSITION OF DI-TERTIARY BUTYL PEROXIDE AT 154.6°C*

Time (min)	Total Pressure (torr)	$a - x$	$k_1 x\ 10^4\ (\text{sec}^{-1})$
0	169.3		
3	189.2	159.3	3.39
6	207.1	150.4	3.19
9	224.4	141.7	3.30
12	240.2	133.8	3.18
15	255.0	126.4	3.16
18	269.7	119.1	3.31
21	282.6	112.6	3.12

*J. H. Raley, F. F. Rust, and W. E. Vaughan, J. Am. Chem. Soc. 70, 88 (1948). The values of the total pressure have been corrected for a small pressure of nitrogen used to propel the reactant into the reaction vessel.

Experimental data for the rate of decomposition of di-tertiary butyl peroxide into acetone and ethane are given in Table 11-1. This gas-phase reaction is represented by the following equation:

$$(CH_3)_3COOC(CH_3)_3 \rightarrow 2\ CH_3COCH_3 + C_2H_6$$

The volume of the reaction vessel was kept constant and consequently the concentration of each substance was proportional to its partial pressure. Because the number of molecules changes as the reaction proceeds, measurements of the total pressure of the system provided all of the data needed for a calculation of the rate constant.

If P, a, and x represent respectively the total pressure, the initial pressure of di-tertiary butyl peroxide, and the change in this pressure at time t, then $P = (a - x) + 3x$ because 3 moles of products are formed for each mole of reactant consumed. It follows that $x = \dfrac{P - a}{2}$, and $a - x = \dfrac{3a - P}{2}$. The values of the first-order rate constant in sec^{-1} were calculated for successive three-minute intervals from the equation $k_1 = \dfrac{2.3}{3 \times 60} \log \dfrac{(a-x)_i}{(a-x)_f}$, in which the subscripts i and f refer to the initial and final value for each interval. It can be seen that the values of k_1 in the last column of Table 11–1 are indeed constant within reasonable limits. The rate data are plotted in two different ways in Figures 11–1 and 11–2.

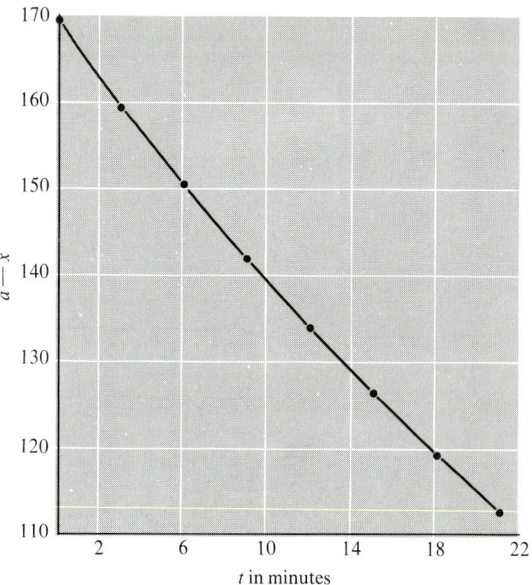

Figure 11–1. Graph of $a - x$, the partial pressure of di-tertiary butyl peroxide, for the decomposition reaction at 154.6°C.

Second-Order Reactions

There are two different types of second-order reactions which must be considered separately. One of these types is a second-order reaction of one compound, A → products, for which the rate equation is

$$-\frac{d[A]}{dt} = k_2[A]^2$$

When the variables are separated and the integrations are carried out the result is

$$-\int \frac{d[A]}{[A]^2} = k_2 \int dt$$

$$\frac{1}{[A]} = k_2 t + \text{constant}$$

Reaction Kinetics

Again evaluating the constant of integration from the initial condition, we have

$$\frac{1}{[A]_t} - \frac{1}{[A]_o} = k_2 t \tag{11-3}$$

When concentrations are expressed in moles/liter and time in seconds, the units of k_2 are liters mole^{-1} sec^{-1}. For a gas phase reaction, when partial pressures in torr are used as a measure of concentration the units of k_2 are torr^{-1} sec^{-1}.

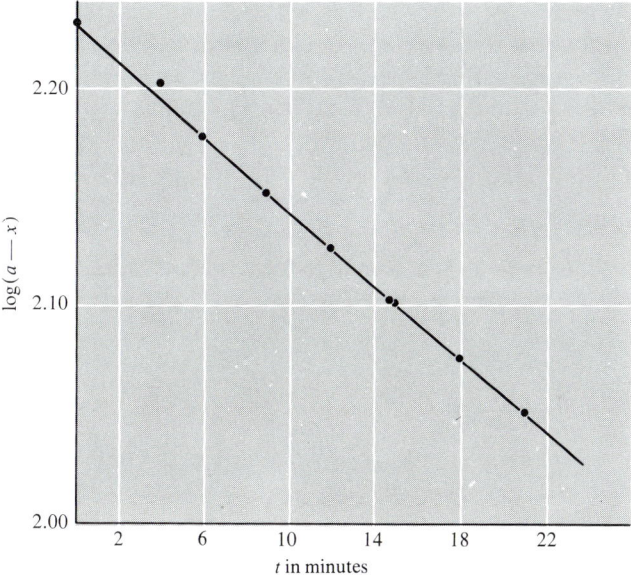

Figure 11–2. Graph of log $(a - x)$ for the decomposition of di-tertiary butyl peroxide at 154.6°C. The linearity of the plotted points is evidence that this reaction follows a first-order rate law. The rate constant is proportional to the negative of the slope of the graph line.

When the concentration of the reactant is reduced to half of its initial value,

$$\frac{1}{[A]_t} - \frac{1}{[A]_o} = \frac{1}{[A]_o}$$

and

$$t_{\frac{1}{2}} = \frac{1}{k_2[A]_o}$$

Thus in this case the half-life of the reactant is inversely proportional to its initial concentration.

For the other type of second-order reaction the general chemical equation is A + B → products and the reaction is of first order with respect to each reactant. The rate equation can be written in either of the two equivalent forms

$$-\frac{d[A]}{dt} = -\frac{d[B]}{dt} = k_2[A][B]$$

or

$$\frac{dx}{dt} = k_2(a - x)(b - x) \tag{11-4}$$

where x is the number of moles/liter of either A or B which has reacted in time t, and a and b are the initial concentrations of A and B, respectively.

On separating the variables in Equation 11-4 and integrating, we have

$$k_2 t = \int \frac{dx}{(a - x)(b - x)} + \text{constant}$$

When $a \neq b$, the integral is a standard form whose value is $\frac{1}{b-a} \ln \frac{b-x}{a-x}$. The constant of integration evaluated from the initial condition ($x = 0$ at $t = 0$) is $\frac{1}{b-a} \ln \frac{b}{a}$. It follows that

$$k_2 t = \frac{1}{b - a}\left(\ln \frac{b - x}{a - x} - \ln \frac{b}{a}\right)$$

$$= \frac{1}{b - a} \ln \frac{a(b - x)}{b(a - x)} \tag{11-5}$$

TABLE 11-2 DIMERIZATION OF BUTADIENE AT 326°C*

Time (min)	Total Pressure (torr)	$a - x$	$k_2 \times 10^5$ (torr^{-1}min^{-1})
0	632.0		
6.12	606.6	581.2	2.29
12.18	584.2	536.4	2.37
17.30	567.3		
29.18	535.4	438.8	2.44
42.50	509.3	386.6	2.32
60.87	482.8		
90.05	453.3	274.6	2.23
119.00	432.8		
176.67	405.3	178.6	2.29

*W. E. Vaughan, J. Am. Chem. Soc. 54, 3863 (1932). The k_2 values are calculated for successive time intervals.

Reaction Kinetics

For a second-order reaction between two reactants, a graph of $\ln \dfrac{b-x}{a-x}$ versus time is a straight line.

If the initial concentrations of the two reactants are the same, Equation 11–5 cannot be used. For this special case, however, $(a-x)(b-x) = (a-x)^2$ and the differential equation is the same as that for the case of a single reactant. The integrated equation is the same as Equation 11–3, which in the present notation becomes

$$\frac{1}{a-x} - \frac{1}{a} = k_2 t$$

In Table 11–2, data are given for the rate of dimerization of butadiene, according to the equation $2C_4H_6 \rightarrow C_8H_{12}$. From the constancy of the second-order rate constant, it is apparent that this reaction follows a second-order rate law. The same conclusion can be reached from Figure 11–3, in which $10^3/a - x$ is plotted against the time.

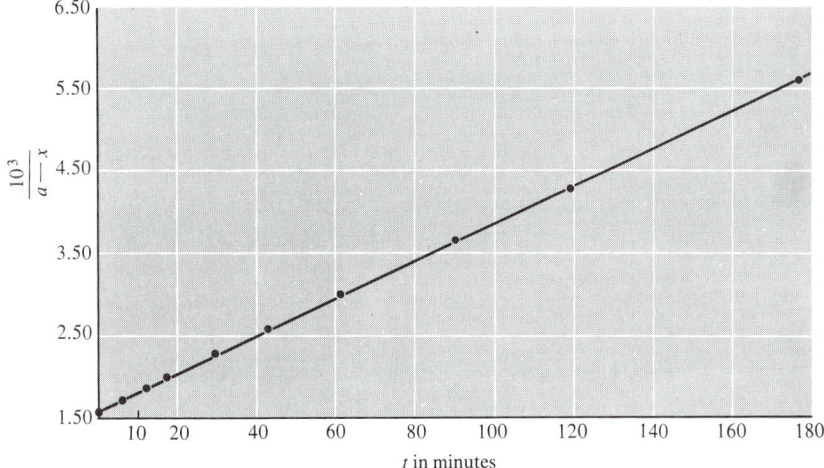

Figure 11–3. Graph of $\dfrac{10^3}{a-x}$ vs time for the dimerization of butadiene at 326°C.

Pseudo First-Order Reactions

When for the second-order reaction, $A + B \rightarrow$ products, discussed above, one of the reactants is present in large excess, the concentration of this reactant will undergo relatively little change during the course of the reaction, and it can be treated as a constant. This constant can be combined with the second-order rate constant to form a pseudo first-order constant. Notice that if $b \gg a$, then $b - x \sim b$ and Equation 11-4 reduces to the first-order equation (Eq. 11-2). This type of pseudo order is found for a reaction carried out in solution when the solvent is also a reactant. A typical example is the hydrolysis of an ester in aqueous solution according to the equation

$$R_1C\begin{matrix}O-R_2\\ \\O\end{matrix} + H_2O \rightarrow R_1C\begin{matrix}O-H\\ \\O\end{matrix} + R_2OH$$

Incidentally, the technique of adding a large excess of one reactant to a reacting system is sometimes used to unravel the kinetics of a complex reaction because it eliminates the concentration of this reactant as a variable.

Third-Order Reactions

There are very few reactions which are definitely known to be of third order and consequently we omit this case, which is treated in specialized books on reaction kinetics.

Having obtained the integrated form for a number of rate equations, we turn to the methods used for the determination of reaction order. Among these methods are the following:

1. *Trial and error.* This is the method used most frequently. A reasonable assumption is made for the order and the corresponding integrated rate equation is checked with the experimental data by calculating the rate constant for a number of time intervals. If the calculated rate "constant" is actually constant, then the assumed order is correct. A variant of this method is the graphing against time of the appropriate functions of the experimental concentration corresponding to various orders. The function which yields a straight-line plot corresponds to the correct order. The graphical technique is useful for following changes in order which sometimes occur as a reaction proceeds.

2. *Half-life method.* If the dependence of the half-life of a reactant on its initial concentration is known the order may often be inferred.

3. *Differential method.* It is sometime advantageous to study the dependence of the approximate rate $\Delta x/\Delta t$ on reactant concentrations by varying the initial concentration of one reactant at a time.

4. *Isolation or "swamping" method.* This refers to the technique, mentioned above, of adding a large excess of a reactant to remove its concentration as a variable in the rate equation.

This list by no means exhausts the methods which have been used to find reaction orders, but it does include the methods used most frequently.

Parallel and Opposed Reactions

The reacting systems discussed so far have been assumed to be simple in the sense that only one reaction is occurring. Unfortunately, for most reacting systems a variety of complications can exist. One type of complication in gas-phase reactions is the effect of the walls of the container on the reaction rate. Since, however, we are limiting this discussion to homogeneous reactions, we ignore possible wall effects.

Another type of complication results from the fact that a given set of reactants can often react in more than one way at the same time. When this occurs, the rate equation may appear to be quite complex, particularly if the parallel reactions are of different orders.

If the parallel reactions are $A \rightarrow B + C$ and $A \rightarrow D + E$ and both are of first order with rate constants k' and k'', respectively, then the rate of disappearance of A is given by the equation

$$-\frac{d[A]}{dt} = k'[A] + k''[A] = k[A]$$

where $k = k' + k''$. In terms of the concentration of A the rate equation appears to be one corresponding to a simple first-order reaction. The integration of this rate equation yields $[A]_t = [A]_o e^{-kt}$. The rate equation for the formation of B, however, is

$$\frac{d[B]}{dt} = k'[A]$$

After substitution of the expression given above for [A], this equation becomes

$$d[B] = k'[A]_o e^{-kt}\, dt$$

and its integrated form is

$$[B]_t = \frac{k'[A]_o}{k}(1 - e^{-kt})$$

Similarly,

$$[D]_t = \frac{k''[A]_o}{k}(1 - e^{-kt})$$

Thus,

$$\frac{[B]_t}{[D]_t} = \frac{k'}{k''}$$

We see that the concentration ratio of the products of these parallel reactions is equal to the ratio of the individual rate constants.

If a reaction carried out in a closed system is chemically reversible, as soon as significant amounts of products have been formed they react to form the reactants. (For this reason, emphasis in kinetic studies is often placed on the initial rates of reactions.) The observed reaction rate is then the difference between the rates of the forward and the reverse reactions. For the reversible reaction $A + B \underset{k_r}{\overset{k_f}{\rightleftarrows}} C + D$, which is assumed to be of second order in both forward and reverse directions, the rate equation is

$$-\frac{d[A]}{dt} = k_f[A][B] - k_r[C][D]$$

When the reacting system initially contains only A and B, the net reaction rate decreases with time from its initial maximum value to zero; at that time,

$$k_f[A][B] = k_r[C][D]$$

and

$$\frac{k_f}{k_r} = \frac{[C][D]}{[A][B]} = K_{eq}$$

Because k_f and k_r are independent of concentrations, this is also true of K_{eq}, the equilibrium constant expressed here in terms of concentrations. (Compare Chapter 4, p. 102.) The treatment of equilibrium from the reaction rate point of view was first carried out by Guldberg and Waage in connection with their kinetic studies. Their result, expressed in the equation above, is sometimes called the law of mass action. As mentioned in Chapter 4, this derivation of the expression for K_{eq} clearly depends on the assumption of simple rate laws for both the forward and reverse reactions, in contrast to the thermodynamic approach, which is completely general.

When the equilibrium constant for a reversible reaction is known, the differential and integrated forms of the rate equation can be expressed in terms of either the equilibrium constant or the equilibrium concentrations of reactants. Equations have been derived for various combinations of orders of the forward and reverse reactions; for example, the forward reaction might be of second order while the reverse reaction is of first order.

It will be sufficient here to treat the simplest case, in which both the forward and reverse reactions are of first order. Consider the reaction

$$A \underset{k_r}{\overset{k_f}{\rightleftarrows}} B$$

with the equilibrium constant, $K = k_f/k_r$, assumed to be known. The rate equation is

$$-\frac{d[A]}{dt} = k_f[A] - k_r[B]$$

If no B is present initially, $[B] = [A]_o - [A]$ after the reaction starts. When $[B]$ and k_r in the rate equation are replaced by their equals, we have

$$-\frac{d[A]}{dt} = k_f[A] - \frac{k_f}{K}([A]_o - [A])$$

$$= k_f\left([A] + \frac{[A]}{K} - \frac{[A]_o}{K}\right)$$

$$= k_f\left\{\left(\frac{K+1}{K}\right)[A] - \frac{[A]_o}{K}\right\}$$

After separation of the variables, this equation becomes

$$k_f\,dt = -\frac{d[A]}{\left(\frac{K+1}{K}\right)[A] - \frac{[A]_o}{K}}$$

$$= -\frac{K}{K+1}\,d\ln\left\{\left(\frac{K+1}{K}\right)[A] - \frac{[A]_o}{K}\right\}$$

When this equation is integrated between the limits $[A]_o$ at $t = 0$ and $[A]_t$ at t, the result after simplification is

$$k_f t = \frac{K}{K+1}\ln\left\{\frac{[A]_o K}{[A]_t(K+1) - [A]_o}\right\} \tag{11-6}$$

We can express the integrated equation in terms of the equilibrium concentration of A by noting that $K = [B]_e/[A]_e$ and $[B]_e = [A]_o - [A]_e$. Consequently $K = ([A]_o - [A]_e)/[A]_e$ and $K + 1 = [A]_o/[A]_e$. On substitution of these expressions the integrated rate equation takes the following form:

$$k_f t = \frac{[A]_o - [A]_e}{[A]_o}\ln\frac{[A]_o - [A]_e}{[A]_t - [A]_e}$$

Still another form for this equation is obtained when the relation $(K+1)/K = (k_f + k_r)/k_f$ (derived from $K = k_f/k_r$) is used to eliminate K from Equation 11-6. The result is

$$(k_f + k_r)t = \ln\frac{[A]_o - [A]_e}{[A]_t - [A]_e}$$

Thus the first-order constants k_f and k_r can be evaluated individually from measurements of the concentration of A as a function of time when the initial and the equilibrium concentrations of A are known.

Consecutive Reactions

It has already been pointed out that relatively few reactions are kinetically simple. The rates of many complex reactions have been interpreted on the basis that the complex reaction is actually a series of consecutive, simple reactions in which the product of one reaction is the reactant for another. In some cases the intermediate substances are formed in so high a concentration that they can be identified and their concentrations can be determined quantitatively. In most cases, however, the intermediates react so rapidly that the series of steps by which the over-all reaction takes place can only be inferred. A series of steps by which a reaction is believed to occur is usually called the mechanism of the reaction.

The differential equation representing the over-all rate of a series of consecutive reactions, some of which may be reversible, is in general quite complicated, and special computational techniques are needed for its solution. Here we consider only the example of a two-step reaction with each step of first order and irreversible, represented by the following equation:

$$A \xrightarrow{k_1} B \xrightarrow{k_2} C$$

We assume that initially only A is present.

The rate equations for this reaction form a coupled set of differential equations which we can write in the following form:

$$\frac{d[A]}{dt} = -k_1[A]$$

$$\frac{d[B]}{dt} = k_1[A] - k_2[B]$$

$$\frac{d[C]}{dt} = k_2[B] \tag{11-7}$$

Integration of the first equation yields

$$[A]_t = [A]_0 e^{-k_1 t}$$

When this expression for [A] is introduced into the second equation, we obtain

$$\frac{d[B]}{dt} = k_1[A]_0 e^{-k_1 t} - k_2[B] \tag{11-8}$$

The integration[2] of this equation between the limits $[B]_0 = 0$ at $t = 0$ and $[B]_t$ at t results in

$$[B]_t = \frac{[A]_0 k_1}{k_2 - k_1}(e^{-k_1 t} - e^{-k_2 t}) \tag{11-9}$$

To find the concentration of C as a function of time, we note that according to the assumed chemical equation a molecule of B or a molecule of C is formed for every molecule of A which reacts. Consequently

[2] This integration can be carried out by using $e^{-k_2 t}$ as an integrating factor.

Reaction Kinetics

$$[A]_t + [B]_t + [C]_t = [A]_o$$

and
$$[C]_t = [A]_o - [A]_t - [B]_t$$

Substituting for $[A]_t$ and $[B]_t$ the expressions found above, we have

$$[C]_t = [A]_o - [A]_o e^{-k_1 t} - \frac{[A]_o k_1}{k_2 - k_1}(e^{-k_1 t} - e^{-k_2 t})$$

$$= [A]_o \left[1 - \frac{(k_2 - k_1)e^{-k_1 t} + k_1(e^{-k_1 t} - e^{-k_2 t})}{k_2 - k_1} \right]$$

$$= [A]_o \left(1 - \frac{k_2 e^{-k_1 t} - k_1 e^{-k_2 t}}{k_2 - k_1} \right)$$

A graph of the concentrations of A, B, and C as a function of time is given in Figure 11–4. The concentration of A decreases exponentially, the concentra-

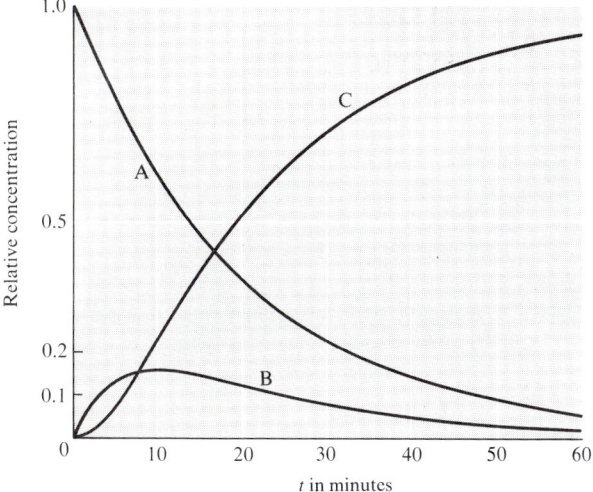

Figure 11–4. Dependence on time of relative concentrations for consecutive first-order reaction A → B → C. $k_1 = 0.05$ min^{-1}, $k_2 = 0.20$ min^{-1}.

tion of B rises to a maximum and then decreases, and the concentration of C rises slowly at first, then more rapidly, and approaches the initial concentration of A as a limit. The somewhat indefinite time interval during which the product of a complex reaction, such as C, is experimentally indetectible is often called the induction period of the reaction.

If in a series of consecutive reactions one of the rate constants is very much smaller than any of the others (as is often the case), then the over-all rate of the reaction is limited by the rate of this slowest reaction, the "bottleneck," which is called the rate-determining step. In the example above, if $k_2 \gg k_1$ B reacts about as fast as it is formed, and its formation is the rate-determining step. Then Equation 11–9 for the concentration of B as a function of time simplifies to the following:

$$[B]_t = \frac{[A]_o k_1}{k_2}(e^{-k_1 t} - e^{-k_2 t})$$

After the reaction has proceeded for a time t such that $t \gg 1/k_2$ then $e^{-k_1 t} \gg e^{-k_2 t}$, and this equation becomes

$$[B]_t = \frac{[A]_o k_1}{k_2} e^{-k_1 t} \tag{11–10}$$

The corresponding equation for [C] is

$$[C]_t = [A]_o(1 - e^{-k_1 t}) \tag{11–11}$$

This is the equation for the first-order formation of C directly from A with k_1 as the rate constant.

Equation 11–11 can be obtained from Equations 11–7 and 11–8 by setting $d[B]/dt = 0$ in Equation 11–8, which implies that the concentration of B is constant. Because we assumed that k_1/k_2 in Equation 11–10 is very small, in this case the concentration of B is never large, and the assumption that it reaches a constant value soon after the start of the reaction is a good approximation. This *steady-state* (or stationary-state) *approximation* is useful in the study of reactions which are so complex that the use of exact equations is impractical. An important example of such a complex reaction is the chain reaction.

Chain Reactions

It was pointed out in an earlier section that the reaction $H_2 + I_2 \rightarrow 2HI$ has been shown to be a second-order reaction in both directions. When, however, the rate of the apparently closely related reaction $H_2 + Br_2 \rightarrow 2HBr$ was studied by Bodenstein and Lind, they found the surprising result that the rate of formation of HBr is expressed by the following empirical equation:

$$\frac{d[HBr]}{dt} = \frac{k_r[H_2][Br_2]^{\frac{1}{2}}}{1 + k'_r \frac{[HBr]}{[Br_2]}} \tag{11–12}$$

This equation implies that in the initial stages of the reaction, before significant amounts of HBr have been formed, the reaction order is 3/2. The formation of HBr decreases the reaction rate. This rate decrease by a reaction product or other substance is known as inhibition.

More than a decade after the work of Bodenstein and Lind, several

Reaction Kinetics

groups of investigators simultaneously proposed a theoretical interpretation of this complicated rate equation in terms of a mechanism which consisted of a series of steps called a *reaction chain*. Since that time many other complex reactions have been shown to be chain reactions.

The rate of formation of HBr is accounted for as the over-all rate of the following steps, each with its own rate constant:

$$Br_2 \xrightarrow{k_1} 2Br \qquad \text{chain initiation}$$
$$Br + H_2 \xrightarrow{k_2} HBr + H \qquad \text{chain propagation}$$
$$H + Br_2 \xrightarrow{k_3} HBr + Br \qquad \text{chain propagation}$$
$$H + HBr \xrightarrow{k_4} H_2 + Br \qquad \text{chain inhibition}$$
$$2Br \xrightarrow{k_5} Br_2 \qquad \text{chain breaking}$$

The reaction starts with the thermal dissociation of a few Br_2 molecules into atoms. These atoms react with H_2 molecules to form HBr and reactive H atoms which propagate the chain. When one of these H atoms reacts with a Br_2 molecule, another molecule of HBr is formed as well as another Br atom which can react with H_2 to form more HBr. If, however, a H atom reacts with a HBr molecule it lowers the net rate of formation of HBr. (Notice that reaction 4 is the reverse of reaction 2.) The chain is terminated when two Br atoms combine to form Br_2, which is the reverse of reaction 1.

We now proceed to derive a theoretical equation for the rate of formation of HBr which has the same mathematical form as Equation 11-12. On the basis of the chain mechanism we can write the following rate equation:

$$\frac{d[HBr]}{dt} = k_2[Br][H_2] + k_3[H][Br_2] - k_4[H][HBr] \tag{11-13}$$

The difficulty with this equation is the fact that it contains the concentrations of the H and Br atoms, which are too low to be detectable. Although it is conceivable that these concentrations might be calculated from a solution of the set of five simultaneous differential equations, it is more realistic to make use of the steady-state approximation discussed in the preceding section. Consequently we set $d[H]/dt = d[Br]/dt = 0$. This leads to the following pair of simultaneous equations:

$$\frac{d[H]}{dt} = 0 = k_2[Br][H_2] - k_3[H][Br_2] - k_4[H][HBr] \tag{11-14}$$

and

$$\frac{d[Br]}{dt} = 0 = 2k_1[Br_2] - k_2[Br][H_2] + k_3[H][Br_2] + k_4[H][HBr] - 2k_5[Br]^2 \tag{11-15}$$

The factors of 2 in the first and last terms on the right-hand side of the equation above are due to the fact that the rate constants of reactions 1 and 5 are conventionally written in terms of Br_2 molecules. Thus

$$-\frac{d[Br_2]}{dt} = k_1[Br_2] = \frac{1}{2}\frac{d[Br]}{dt}$$

and

$$\frac{d[Br_2]}{dt} = k_5[Br]^2 = -\frac{1}{2}\frac{d[Br]}{dt}$$

When we subtract Equation 11–14 from Equation 11–13 we obtain

$$\frac{d[HBr]}{dt} = 2k_3[H][Br_2] \qquad (11\text{–}16)$$

On solving Equation 11–14 for [H] we obtain

$$[H] = \frac{k_2[H_2][Br]}{k_3[Br_2] + k_4[HBr]} \qquad (11\text{–}17)$$

To find an expression for [Br] we add Equations 11–14 and 11–15. The result of this step is

$$2k_1[Br_2] = 2k_5[Br]^2$$

Consequently,

$$[Br] = \left(\frac{k_1}{k_5}[Br_2]\right)^{\frac{1}{2}} \qquad (11\text{–}18)$$

(Notice that $k_1/k_5 = K_{eq}$ for the dissociation of Br_2.) Now we can substitute Equation 11–18 in Equation 11–17 to eliminate [Br]. Then

$$[H] = \frac{k_2[H_2]}{k_3[Br_2] + k_4[HBr]}\left(\frac{k_1}{k_5}[Br_2]\right)^{\frac{1}{2}}$$

Substitution of this equation in Equation 11–16 yields

$$\frac{d[HBr]}{dt} = \frac{2k_3[Br_2]k_2[H_2]\left(\frac{k_1}{k_5}\right)^{\frac{1}{2}}[Br_2]^{\frac{1}{2}}}{k_3[Br_2] + k_4[HBr]}$$

Finally we can simplify this equation by dividing numerator and denominator by $k_3[\text{Br}_2]$. The resulting equation is

$$\frac{d[\text{HBr}]}{dt} = \frac{2k_2 \left(\dfrac{k_1}{k_5}\right)^{\frac{1}{2}} [\text{H}_2][\text{Br}_2]^{\frac{1}{2}}}{1 + \dfrac{k_4[\text{HBr}]}{k_3[\text{Br}_2]}}$$

When this theoretical equation is compared with the empirical equation of Bodenstein and Lind (Eq. 11–12), the similarity is evident. In fact, the two equations are identical if the constants in the empirical equation are interpreted as combinations of the rate constants for the steps in the reaction chain:

$$k_r = 2k_2 \left(\frac{k_1}{k_5}\right)^{\frac{1}{2}} \quad \text{and} \quad k'_r = \frac{k_4}{k_3}$$

The reactions used above as steps in the HBr chain mechanism are not the only reactions that could take place between Br_2, H_2, HBr, Br, and H. For example, reactions such as $\text{H}_2 \rightarrow 2\text{H}$ or $\text{Br} + \text{HBr} \rightarrow \text{Br}_2 + \text{H}$ might be thought to be possible steps in the chain mechanism. Careful consideration has shown, however, that although these reactions do occur their rates are much too low to affect the rate of formation of HBr.

One of the characteristic features of chain reactions in general is the fact that from a relatively small number of reactive intermediates (chain-carriers) produced in the initiation reaction, many product molecules are formed. The length of the chain, defined by the number of product molecules formed per chain-carrier produced in the initiation reaction, clearly depends on the relative rates of the chain-propagating and the chain-breaking steps in the mechanism. In many reacting systems the important chain-breaking steps are heterogeneous reactions of the chain-carriers at the wall of the container or reactions of the chain-carriers with impurities in the reactants. In these cases the chain length and the reaction rate are markedly dependent on the relative area and the condition of the wall of the container and the purity of the reactants.

The mechanism for the formation of HBr, which involves atoms as the chain-carriers, is not the only kind of chain mechanism which has been employed for the interpretation of complex reactions. Another type of chain reaction is known in which the chain-carriers are free radicals, and many reactions of organic molecules have been shown to be of this type. Still another type of chain reaction is one which has a branching-chain mechanism. This means that in at least one of the steps in the mechanism, more than one chain-carrier is formed for every chain-carrier removed by the reaction.

An important reaction which is interpreted on the basis of a branching-chain mechanism is the reaction $2\text{H}_2 + \text{O}_2 \rightarrow 2\text{H}_2\text{O}$. This reaction is extremely complex and certain of its aspects are not entirely clear yet, even though a large number of investigators have worked on this problem for many years. A mechanism that is widely accepted is the following:

1. $H_2 + O_2 \rightarrow HO_2 + H$ initiation
2. $H_2 + HO_2 \rightarrow H_2O + OH$ propagation
3. $OH + H_2 \rightarrow H_2O + H$ propagation
4. $H + O_2 \rightarrow OH + O$ branching
5. $O + H_2 \rightarrow OH + H$ branching
6. $\left. \begin{array}{l} HO_2 \\ OH \\ H \end{array} \right\} + \text{wall} \rightarrow \text{products}$ termination

In the last step chain-carriers are removed by reactions at the wall of the reaction vessel. In the fourth and fifth steps, the reaction of one chain-carrier leads to the formation of two more. As a result of this multiplication, the number of chain-carriers and therefore the rate of the reaction can rise extremely rapidly and thus lead to an explosion. Whether an explosion actually occurs under given conditions depends on the rate of formation of chain-carriers relative to the rate of their removal by the termination steps.

It is interesting to note that a similar type of branching-chain mechanism is responsible for nuclear explosions. When a U^{235} nucleus absorbs a slow neutron it can undergo fission (that is, it splits into two nuclei of roughly equal mass). In this process it emits, on the average, between two and three neutrons with high kinetic energy which comes from the conversion of mass into energy. When these neutrons, which are the chain-carriers in this process, are slowed down (moderated) by collisions with nuclei, they can be absorbed by U^{235} to propagate the chain. If the number of neutrons increases at a rate higher than that at which they escape from the mass of uranium or are consumed by other processes, an explosion results.

Dependence of Reaction Rate on Temperature

It has been known for a long time that a temperature rise greatly increases the rate of most chemical reactions. A generalization often quoted states that the rates of many reactions are approximately doubled for each 10° rise in temperature above room temperature. The dependence of reaction rates on temperature is of great practical value for the control of reaction rates by adjustment of the temperature of the reacting system. In addition, this temperature dependence sometimes provides important clues to the mechanism of a reaction. In discussing the effect of temperature we need to consider only simple reactions, for we have seen that complex reactions can be interpreted as combinations of simple reactions.

The quantity in the rate equation that is temperature dependent is the rate constant. Arrhenius suggested in 1889, largely on the basis of analogy with the van't Hoff equation for the temperature coefficient of the equilibrium constant in terms of concentration ($d \ln K_c/dT = \Delta E/RT^2$), p. 115, that the temperature coefficient of the rate constant may be represented by

$$\frac{d \ln k_r}{dT} = \frac{E_a}{RT^2} \tag{11-19}$$

in which the energy E_a is called the *activation energy*.

If we write the corresponding equation for the reverse reaction as $d \ln k_r'/dT = E_a'/RT^2$ and subtract this equation from Equation 11–19 we have

$$\frac{d \ln k_r}{dT} - \frac{d \ln k_r'}{dT} = \frac{E_a - E_a'}{RT^2}$$

The left-hand side of this equation can be written as

$$\frac{d(\ln k_r - \ln k_r')}{dT} = \frac{d \ln (k_r/k_r')}{dT} = \frac{d \ln K_c}{dT}$$

because, as we have seen, $K_c = k_r/k_r'$. It follows that the difference between the activation energies for the forward and reverse reactions equals ΔE, the heat of reaction at constant volume.

When Equation 11–19 is integrated, assuming that E_a is independent of the temperature, the result is

$$\ln k_r = -\frac{E_a}{RT} + \ln A$$

or

$$k_r = Ae^{-\frac{E_a}{RT}} \tag{11-20}$$

The constant of integration, A, is usually called the pre-exponential factor.

Many reactions have been found to follow the Arrhenius law over moderate temperature ranges. For a typical reaction, $\ln k_r$ is plotted as a function of $1/T$ in Figure 11–5. For some reactions which have been studied over a wide temperature range it is found that the "constant" A varies slightly with the temperature. In a temperature range in which E_a is known and constant, the value of the rate constant, k_1, at temperature T_1 may be calculated from its known value, k_2, at temperature T_2 from the integrated form of Equation 11–19,

$$\ln k_1 = \ln k_2 - \frac{E_a}{R}\left(\frac{1}{T_1} - \frac{1}{T_2}\right)$$

Conversely, it is evident that E_a can be calculated from values of the rate constant at two temperatures.

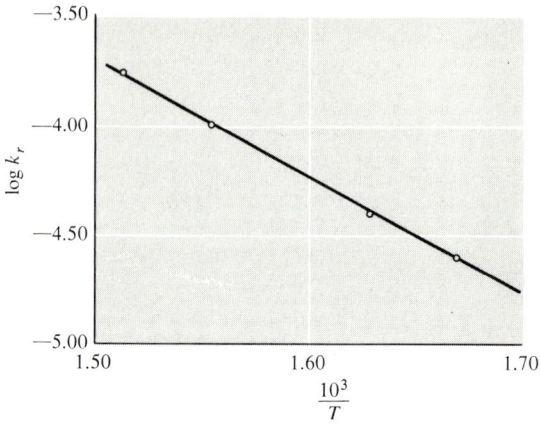

Figure 11–5. Graph of log k_r vs $1/T$ for the dimerization of butadiene. The slope of the graph line is proportional to the Arrhenius activation energy which in this case is equal to 24.7 kcal/mole.

Collision Theory of Bimolecular Gas Reactions

Up to this point in this chapter we have been discussing the empirical description of reaction rates, and we have seen how the rates of complex reactions can be interpreted in terms of combinations of simpler reactions. Before proceeding further we introduce a useful term, the *molecularity* of a simple (or elementary) reaction, such as one inferred to be part of the mechanism of a complex reaction. The molecularity is the number of molecules, atoms, or other reacting species, thought to participate in a simple reaction. Its value can be only 1, 2, or possibly 3. It may or may not be the same as the order of the reaction. In contrast to the order of a reaction, the molecularity is a theoretical quantity.

The reactions that are believed to be unimolecular are mostly decompositions and isomerizations of organic compounds; for example, the isomerization of cyclopropane to propylene. Many of the bimolecular reactions are dimerizations and free radical recombinations. For a termolecular reaction to occur it is necessary that three reactive molecules participate in a single step. This is a relatively improbable process, and few gas reactions are believed to occur by a termolecular mechanism.

Now we turn to one of the basic problems in chemistry. Precisely what happens when a chemical reaction occurs? We know that certain kinds of molecules disappear and other kinds take their place. How can we account for this transformation and predict its rate? As was indicated in the introduction to this chapter, we do not yet have completely satisfactory answers to these profound questions but we will consider some of the answers that have been obtained.

Reaction Kinetics

The earliest theory of chemical reactions is the *collision theory*. It is evident that two substances can react with each other only when they are in contact. Since the substances are composed of molecules, it is reasonable to expect that the molecules of these substances must come in "contact" or, in other words, the molecules must collide. Because our knowledge of molecular collisions is more complete for the gas phase than for the liquid phase, for the present we restrict our attention to gas phase reactions.

If a reaction takes place whenever two molecules collide, we should be able to calculate the rate of the reaction by calculating the rate at which bimolecular collisions occur. In Chapter 5, as part of the development of the kinetic theory, we obtained the following expression for the number of bimolecular collisions per second per cm³ among the molecules of one species:

$$Z = 2N^2\sigma^2 \left(\frac{\pi kT}{m}\right)^{\frac{1}{2}} \quad (5\text{-}20)$$

If the reaction under consideration involves two different gases, A and B, we need to calculate the rate of bimolecular collisions between unlike molecules. This rate is given by the following expression:

$$Z_{AB} = N_A N_B \sigma_{av}^2 \left(\frac{8\pi kT}{\mu}\right)^{\frac{1}{2}} \quad (11\text{-}21)$$

in which σ_{av} is defined as $\frac{\sigma_A + \sigma_B}{2}$ and μ is the reduced mass $\left(= \frac{m_A + m_B}{m_A m_B}\right)$. In terms of the molecular weights (M_A and M_B) of A and B, the collision number is given by

$$Z_{AB} = N_A N_B \sigma_{av}^2 \left[\frac{(M_A + M_B) 8\pi RT}{M_A M_B}\right]^{\frac{1}{2}}$$

In the special case in which A and B are the same species, $\sigma_{av} = \sigma$, $\mu = m/2$, and Equation 11-21 reduces to Equation 5-20 when account is taken of the indistinguishability of the collision partners by a division by 2.

When appropriate values of molecular properties under laboratory conditions are introduced into the expression for Z_{AB}, this number turns out to be extremely large. Let us take as an example the reaction of H_2 with I_2 at 700°K with each gas at atmospheric pressure. In this case $N_{H_2} = N_{I_2} \sim 10^{19}$ molecules/cc, $\sigma_{H_2} \sim 2.2$ Å, $\sigma_{I_2} \sim 4.6$ Å and $\sigma_{av} \sim 3.4$ Å. Then

$$Z = 10^{38}(3.4 \times 10^{-8})^2 \left[\frac{(254 + 2) \times 8 \times 3.14 \times 8.31 \times 10^7 \times 700}{2 \times 254}\right]^{\frac{1}{2}}$$

$$= 10^{38} \times 1.16 \times 10^{-15} \times 8.58 \times 10^5 \sim 10^{29} \text{ collisions/cc sec}$$

Since there are $\sim 10^{19}$ collisions/sec for $\sim 10^{19}$ molecules of each kind, each molecule makes $\sim 10^{10}$ collisions in 1 second with molecules of the other kind.

If a reaction occurred at every collision between unlike molecules, the whole reaction would be over in $\sim 10^{-10}$ second. (A similar conclusion would be reached for any other bimolecular gas reaction because only the values of σ and μ would change.)

This calculated rate is in obvious disagreement with observed reaction rates. In addition, the temperature dependence of the collision number is in marked disagreement with the observed sensitivity of the reaction rate, represented by the Arrhenius equation, to temperature change. Consequently we must conclude that not all collisions result in reaction.

Effective collisions must be "violent" enough to produce the breaking and forming of chemical bonds that is the essential feature of a reaction. Thus, it is reasonable to expect that only those molecules which have sufficient energy undergo reactive collisions. In fact, the name "activation energy" given to the quantity E_a in the Arrhenius equation suggests that only activated, energy-rich molecules can react.

The form of the Arrhenius equation would appear to be related to the exponential dependence on temperature and molecular kinetic energy in the Maxwell distribution law for kinetic energy (Eq. 5-16). Let us examine this idea a little more closely. The distribution law just referred to gives the fraction of the molecules with kinetic energy between ϵ and $\epsilon + d\epsilon$. From this equation we can calculate the fraction of molecules with kinetic energy greater than some specified value. Even a brief consideration of a bimolecular collision, however, leads to the conclusion that it is not necessary for *one* of the collision partners to have *all* of the activation energy itself. From the point of view of an observer at the center of mass of the colliding molecules it makes no difference how the kinetic energy is divided between the partners. It is only the relative kinetic energy along the line of molecular centers which is significant in a reactive collision.

A detailed analysis of the dynamics of bimolecular collisions leads to the result that the number of collisions per cc per second between unlike molecules whose relative kinetic energy along the line of centers is greater than ϵ_a, represented by Z'_{AB}, is given by

$$Z'_{AB} = Z_{AB} e^{-\frac{\epsilon_a}{kT}}$$

We assume that Z'_{AB} also gives the rate of reactive collisions between A and B, and so we can equate it to the rate at which the molecular concentration of either A or B changes. Thus

$$-\frac{dN_A}{dt} = Z'_{AB}$$
$$= N_A N_B \sigma_{av}^2 \left[\frac{(M_A + M_B) 8 \pi RT}{M_A M_B} \right]^{\frac{1}{2}} e^{-\frac{\epsilon_a}{kT}} \frac{\text{molecules}}{\text{cc sec}} \quad (11\text{-}22)$$

Reaction Kinetics

Now we can obtain a theoretical expression for the rate constant. With [A] expressed in moles/liter, $[A] = 10^3 N_A / \mathfrak{N}$. In place of the rate equation

$$-\frac{d[A]}{dt} = k_2 [A][B]$$

we can write

$$-\frac{10^3}{\mathfrak{N}} \frac{dN_A}{dt} = k_2 \frac{10^6}{\mathfrak{N}^2} N_A N_B$$

Solving for k_2, we have

$$k_2 = -\frac{\mathfrak{N}}{10^3 N_A N_B} \frac{dN_A}{dt}$$

When $\dfrac{dN_A}{dt}$ is replaced by its equal in Equation 11–22, the resulting equation is

$$k_2 = \frac{\mathfrak{N} \sigma_{av}^2}{10^3} \left[\frac{(M_A + M_B) 8\pi RT}{M_A M_B} \right]^{\frac{1}{2}} e^{-\frac{E_a}{RT}} \frac{\text{liters}}{\text{mole sec}} \qquad (11\text{–}23)$$

From a comparison of Equation 11–23 with Equation 11–20 we can write the following theoretical equation for the pre-exponential factor A:

$$A = \frac{\mathfrak{N} \sigma_{av}^2}{10^3} \left[\frac{(M_A + M_B) 8\pi RT}{M_A M_B} \right]^{\frac{1}{2}}$$

The concentration unit implied in this equation is the mole per liter. We can identify the Arrhenius activation energy with the relative kinetic energy along the line of centers of two colliding molecules that is required for a reaction between these molecules to occur. (Although the theoretical pre-exponential factor depends on $T^{\frac{1}{2}}$, this quantity varies so slightly for a moderate temperature change that this temperature dependence is usually not important.)

Our theoretical equation can be tested with experimental rate data when values are found for σ_{av} and E_a. For this purpose let us again use the H_2–I_2 reaction for which the value of E_a has been found to be about 40,000 cal/mole. Substituting the appropriate values, for $T = 700°K$ we find

$$A = 6.95 \times 10^5 \times 8.58 \times 10^5$$
$$= 6.0 \times 10^{11} \text{ liter/mole sec}$$

It follows that the value of the rate constant is given by

$$k_2 = 6.0 \times 10^{11} e^{-\frac{40{,}000}{2 \times 700}}$$
$$= 0.22 \text{ liter/mole sec}$$

At this temperature the experimental value found by Bodenstein is 0.064 liter/mole sec. Considering the uncertainty in the value of the effective collision diameter and, particularly, the activation energy, this agreement of theory with experimental results is quite satisfactory.

Unfortunately, there are very few reactions for which the same type of agreement with experimental results is found. In many cases the observed reaction rate is much smaller than the calculated one, sometimes by a factor as high as 10^5 for the reactions of fairly complicated molecules. It is not difficult to see the source of at least some of the discrepancy. Our model for a reacting molecule is a rigid sphere with no internal energy. By using a spherical model we have ignored the dependence of the effectiveness of a collision on the relative orientation of the collision partners, although we know that only one part of a molecule, such as the halogen in an alkyl halide, may be involved in the reaction.

In addition, by treating the activation energy as though it were entirely related to translation we have ignored any effect of rotational and vibrational energy on the reaction rate. Consequently we must expect the collision theory in its simplest form to be applicable only to the reactions of very simple molecules.

Attempts have been made to generalize the collision theory by introducing another factor, the steric factor p, into the equation for the bimolecular rate constant to take account of the orientation requirement for reaction. Then this equation becomes

$$k_2 = pAe^{-\frac{E_a}{RT}} \qquad (11\text{--}24)$$

This step has little significance, however, for it has not been found possible to calculate the value of such a steric factor theoretically.

Unimolecular Reactions

When we attempt to apply the collision theory to simple reactions which follow a first-order rate equation and are presumably unimolecular, another problem arises. By definition, in a unimolecular reaction only one molecule is believed to participate in the rate-determining step. The half-life of the reactant is independent of its pressure, and thus its collision rate. On the other hand, unimolecular reactions have the Arrhenius dependence on temperature. Although a certain fraction of the molecules have a relatively high speed and kinetic energy, there is no reason to believe that this translational energy is available for the breaking of chemical bonds that is necessary for a dissociation or isomerization reaction to occur without collisions. What, then, is the source of the activation energy?

In contrast to chemical reactions, the spontaneous disintegration of radioactive nuclei, which also proceeds by a "unimolecular" mechanism, has a rate which is completely independent of the temperature. The disintegrating nuclei

Reaction Kinetics

have sufficient internal energy for the "reaction" to occur. No external source of energy is needed.

At one time the problem of the activation energy in unimolecular reactions was thought to be a serious objection to the collision theory. Then simultaneously and independently Lindemann and Christiansen in 1921 proposed a mechanism which accounted for the experimental observations. The key point in this mechanism is the interpretation of a unimolecular reaction as two consecutive steps: a bimolecular activation step followed by a unimolecular reaction.

Consider the reaction $A \rightarrow B + C$. Let us represent by A^* an activated molecule, one which has sufficient energy to react. Applying the collision theory, we must assume that activated molecules are formed in bimolecular collisions of A molecules and we also assume that this process is reversible; that is, an activated molecule can become deactivated by colliding with an ordinary molecule and losing its activation energy. The rate constants of the forward (activation) and reverse (deactivation) processes are represented by k_{2f} and k_{2r}, respectively. Now we assume that there is a certain probability that an activated molecule can react to form B and C and that this step is unimolecular with a rate constant k_1. We can represent these processes in the following way:

$$A + A \underset{k_{2r}}{\overset{k_{2f}}{\rightleftarrows}} A + A^* \\ \downarrow k_1 \\ B + C$$

The concentration of A^* at any time clearly depends on the competing processes of deactivation and reaction, and we can write the following equation for the net rate of formation of A^*:

$$\frac{d[A^*]}{dt} = k_{2f}[A]^2 - k_{2r}[A^*][A] - k_1[A^*]$$

The concentration of activated molecules can be expected to be always small and therefore we can make use of the steady-state approximation discussed earlier. In this case, because $d[A^*]/dt = 0$ it follows that

$$k_{2f}[A]^2 = k_{2r}[A^*][A] + k_1[A^*]$$

Solving for $[A^*]$, we obtain

$$[A^*] = \frac{k_{2f}[A]^2}{k_{2r}[A] + k_1}$$

The rate of formation of B is given by $\frac{d[B]}{dt} = k_1[A^*]$ and consequently

$$\frac{d[B]}{dt} = \frac{k_1 k_{2f}[A]^2}{k_{2r}[A] + k_1}$$

This rate law clearly has no well-defined order. There are, however, two special cases of particular interest. If the pressure of A is so high that $k_{2r}[A] \gg k_1$, then the rate law becomes

$$\frac{d[B]}{dt} = \frac{k_1 k_{2f}}{k_{2r}}[A]$$

and the formation of B is a first-order reaction. In this case, the rate of collisional deactivation of A is relatively high and the unimolecular reaction is the rate-determining step.

If the pressure of A is so low that $k_{2r}[A] \ll k_1$, the formation of B has the following second-order rate law:

$$\frac{d[B]}{dt} = k_{2f}[A]^2$$

Collisional deactivation is so slow under these conditions that essentially every activated molecule reacts to form products and the bimolecular activation process is the rate-determining step, although the reaction is still unimolecular.

Thus we see that the Lindemann–Christiansen mechanism accounts for the first-order and unimolecular reaction at high pressures of A in a way that is consistent with the collision theory. In addition, it predicts that at sufficiently low pressures of reactant the reaction should change from first order to second order. This mechanism was verified by the observation that for several carefully studied reactions this predicted change in order with reactant pressure actually occurs.

The concept of activation by collision as a process distinct from reaction can be checked in another way. For simplicity we have assumed that the reacting system initially contained only pure reactant, and that activation and deactivation processes involved collisions of reactant molecules only. This is an unnecessary restriction. The reacting system might contain an inert gas as well as the reactant, and then the activation and deactivation processes could occur by collisions of reactant with inert gas molecules. The Lindemann–Christiansen mechanism is still applicable to this system and the effect of inert gases is found to be in reasonable agreement with the theory.

In spite of the success of the Lindemann–Christiansen mechanism in explaining some aspects of unimolecular reactions, many basic questions remain. For example, in what form is the energy of activation of the activated molecules? As mentioned previously, it is not in the form of translational energy, and the effect of rotational energy is probably small. Vibrational energy is the remaining possibility. The idea that the activation energy in a unimolecular reaction is in the form of vibrational energy is quite plausible for we would expect that, by analogy to the dissociation of a diatomic molecule, if enough energy could be put into one of the normal modes of vibration of the activated molecule, at least one bond would break.

Reaction Kinetics

The problem then involves both the transfer of kinetic energy into vibrational energy during an activating collision and the distribution of the vibrational energy among the $3N - 6$ normal modes of vibration. Since certain dissociation products are formed in the reaction, there must be a significant probability that enough energy will concentrate in the appropriate mode of vibration to break a particular bond and thus lead to the formation of products. The calculation of this probability, which is the key to the calculation of the unimolecular rate constant k_1, is still a matter for research, and this is where we must leave the subject of unimolecular reactions.

Transition State Theory

We have seen that the collision theory of bimolecular gas reactions makes possible a useful but crude picture of the reaction process, and it permits rather good calculations of the rate constant for the reactions of some simple molecules when the activation energies are known. This theory, however, leaves much to be desired. It does not provide a way for calculating activation energies theoretically. It does not account for the effect of the orientation of collision partners or the role that internal energy might play in a reaction. It does not provide any information on the details of a reactive collision.

The development of quantum mechanics has made possible another theoretical approach to chemical reactions which in principle permits the understanding and quantitative treatment of the reaction process. Let us illustrate this point of view with the bimolecular reaction of A_2 and B_2 to form AB as an example. We may assume that we have a considerable amount of information about A_2, B_2, and AB molecules as stable individual systems.

Now let us picture the sequence of events as a molecule of A_2 and a molecule of B_2 approach each other from a large distance. The first interaction is a weak van der Waals attraction as a result of the London dispersion force mentioned on p. 386. As the molecules approach each other still more closely, van der Waals repulsion becomes dominant and the molecules may separate after an elastic "collision."

If, however, the molecules have enough relative kinetic energy, they may approach each other so closely that another type of interaction occurs. In this interaction the mutual perturbation is so great that the four atoms can be considered to be bonded together in a single system which is called an *activated complex*. The system is said to be in a *transition state* because the energy-rich activated complex is unstable. It may dissociate in such a way that A_2 and B_2 molecules are reformed and then go their separate ways. But, more importantly, the complex may dissociate into two AB molecules; in this event, a chemical reaction has occurred. We can illustrate these steps in the following way:

$$
\begin{array}{c} \text{A—A} \\ + \\ \text{B——B} \end{array}
\quad \rightleftarrows \quad
\begin{array}{c} \text{A---A} \\ \diagup \quad \diagdown \\ \text{B-------B} \\ \text{Activated} \\ \text{complex} \end{array}
\quad \rightarrow \quad
\begin{array}{c} \text{A} \quad \text{A} \\ | \; + \; | \\ \text{B} \quad \text{B} \end{array}
$$

The dashed lines are used to represent interactions between atoms in the activated complex.

In Figure 11–6 we have a schematic graph of the energy of the 2-molecule system plotted against an abscissa, called the *reaction coordinate*, which is a qualitative measure of the progress of the reaction. In this graph the transition state lies at the top of an energy barrier separating products from reactants.

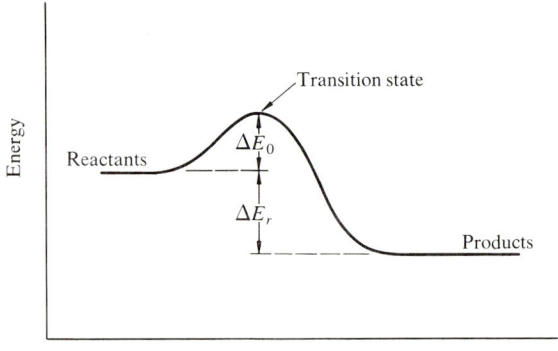

Figure 11–6. Energy of a reacting system as a function of the progress of the reaction. ΔE_0 is the activation energy of the forward reaction and ΔE_r is the difference in energy between products and reactants. The reaction is depicted as exothermic.

The difference between the energy of the activated complex and the energy of the reactants is interpreted as the activation energy. If the energy values along this curve could be calculated it would mean that a giant step would have been taken toward a complete theory of chemical reactions. To calculate the energy of the reacting system we can turn to the method we used in Chapter 7 for the calculation of the energy of stable molecules, namely, the solution of the Schrödinger equation for the system.

It is not difficult to write the Schrödinger equation for a system of four nuclei and the associated electrons. If this equation could be solved we could calculate the energy for any configuration of the nuclei. We have seen, however, that the solution of the Schrödinger equation for even a simple molecule is a formidable task and the ground state energies obtained are more or less rough approximations. The problem of the transition state is much more complicated than this, and it is not surprising that up to now even semiquantitative calculations have been carried out for only a few of the simplest reactions.

Reaction Kinetics

One of the reactions which has been studied quantitatively is the following reaction: $D + H_2 \rightarrow DH + H$. Here only three nuclei and three electrons are involved. Since each nucleus requires three coordinates to fix its position, the system of three nuclei has nine coordinates, or degrees of freedom, which can be divided in the following way (compare Chapter 8, p. 302): three coordinates are needed to locate the center of mass of the activated complex with respect to a coordinate system in space and three coordinates specify its orientation with respect to the coordinate axes. This leaves three internal coordinates which may be interpreted as the three internuclear distances.

Calculations of the energy have shown that the configuration of this system with the lowest energy is the linear configuration. In this case we can represent the course of the reaction in the following way:

$$
\begin{array}{cccc}
\overset{r_1}{\overbrace{\quad\quad}}\,\overset{r_2}{\overbrace{\quad\quad}} & \overset{r_1\;\;r_2}{\overbrace{\quad\quad}} & \overset{r_1}{\overbrace{\quad\quad}}\,\overset{r_2}{\overbrace{\quad\quad}} \\
\underset{D}{\bullet\!\rightarrow} \quad \underset{H_2}{\bullet\;\bullet} \;\rightarrow\; \underset{\text{Activated complex}}{\bullet\text{--}\bullet\text{--}\bullet} \;\rightarrow\; \underset{DH}{\bullet\;\bullet} \quad \underset{H}{\bullet\!\rightarrow}
\end{array}
$$

By the use of suitable approximation methods the energy of this system has been calculated as a function of the two distances r_1 and r_2. The calculated values can be represented as a surface in a 3-dimensional graph by plotting the energy as a function of the two distances. They can also be represented by a contour diagram in two dimensions as shown in Figure 11–7. The plateau at large values of both r_1 and r_2 represents the energy of the three separated atoms. When r_1 is large and r_2 is small, the system consists of an H_2 molecule with the D atom far away. This region of the contour diagram represents a valley whose cross-section is the potential energy curve for the H_2 molecule.

As r_1 decreases while r_2 is constant, the energy of the system at first is constant and then begins to rise toward a maximum at nearly equal values of r_1 and r_2. At large values of r_2 and small values of r_1 the system consists of a DH molecule with an H atom at a distance. This is represented by another valley whose cross-section is the potential energy curve for DH. As r_2 decreases, this valley rises toward the maximum mentioned above. This maximum is now seen to be a saddle-shaped pass through the energy barrier which separates the two valleys. The energy value at the center of the pass relative to the energy of the reactant molecules is interpreted as the activation energy.

The reacting system can be roughly represented by a ball with large initial value of r_1 which starts to roll from right to left. If the ball does not have enough energy to climb to the pass, it rises only part way and then rolls back down into the valley; no reaction has taken place. If the ball has just enough energy to climb to the pass, which corresponds to the formation of the activated complex, it still can roll back to the initial valley, but it also can continue on, enter the next valley, and roll downhill to the floor of this valley. In this case a reaction has taken place.

The dashed line in the diagram represents the reaction path, and the distance along this line is the reaction coordinate. (In Figure 11–6 the reaction coordinate has an analogous significance.)

We have described the potential energy diagram for the simplest reaction system, and for this system we have shown only the simplest reaction path in which the reactant and product molecules are in their lowest energy states.

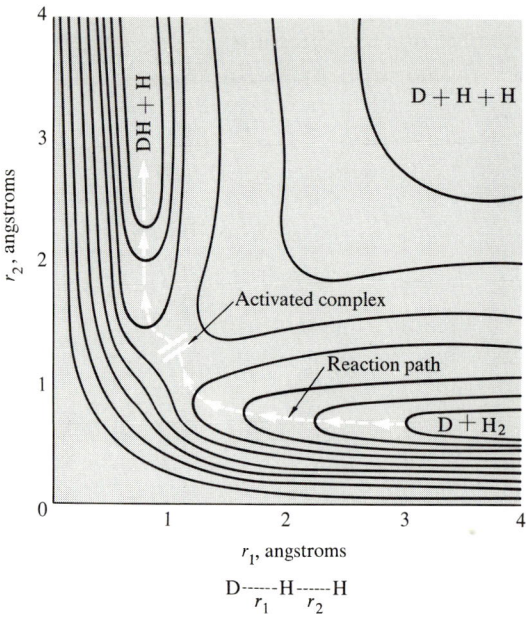

Figure 11–7. Contour diagram of the energy of the 3-particle system D- - -H- - -H as a function of the distances r_1 and r_2.

(The ball mentioned above need not roll along the floors of the two valleys.) Nevertheless the study of this case is of great value because it provides a framework for the more qualitative consideration of complicated reactions for which adequate calculations cannot yet be carried out.

We will see how this framework is used in the transition state theory[3] of reaction rates which has as its objective the calculation of rate constants without the use of empirical parameters. This theory had its origin in the work of Wigner and Pelzer in 1932. It was extended by Eyring and Polanyi in 1935, and it has been applied extensively by Eyring and others to the study of the rates of many physical processes as well as chemical reactions.

[3] The general theory in which the potential energy is calculated at least semiquantitatively is often called the theory of absolute reaction rates.

Let us consider a bimolecular reaction represented by

$$A + B \rightarrow C^{\ddagger} \rightarrow \text{products}$$

$C^{\ddagger}$ represents the activated complex; the superscript on this symbol is used to designate all quantities related to activated complexes. We now proceed to a derivation of a theoretical expression for the rate constant, k_2, of this reaction at a temperature T. As a start, we assume that the reaction rate is the rate of passage of the reacting systems over the pass in the energy barrier which separates the products from the reactants. This rate, in turn, depends on the product of the concentration of activated complexes, $[C^{\ddagger}]$, and the average rate at which the complexes travel along the reaction path over the pass and become reaction products.

The average rate of crossing of the pass can be expressed by $\bar{v}_x^{\ddagger}/\delta$, where $\bar{v}_x^{\ddagger}$ is the average velocity of the activated complexes in the positive direction along the reaction path and δ is a length which can be interpreted as the length of the pass. (We will soon see that a definition more precise than this is not needed.) We can now write the theoretical rate law for our reaction in the following form:

$$-\frac{d[A]}{dt} = [C^{\ddagger}]\frac{\bar{v}_x^{\ddagger}}{\delta}$$

In a more general formulation of this theory another factor, called the transmission coefficient and usually represented by κ, is included in the theoretical rate law. The transmission coefficient is defined as the fraction of the activated complexes at the top of the pass which cross it and then go on to form products. Its value cannot be calculated in general but it is usually so close to 1 that we can ignore it.

We can also express the rate law in terms of the second-order rate constant as $-d[A]/dt = k_2[A][B]$. Consequently we can equate these two forms of the rate law, as follows:

$$k_2[A][B] = [C^{\ddagger}]\frac{\bar{v}_x^{\ddagger}}{\delta}$$

Solving for k_2, we obtain the following theoretical equation for the second-order rate constant:

$$k_2 = \frac{[C^{\ddagger}]}{[A][B]}\frac{\bar{v}_x^{\ddagger}}{\delta} \qquad (11\text{–}25)$$

In this equation we have three unknown quantities, $[C^‡]$, $\bar{v}_x^‡$, and δ, that cannot be evaluated directly. An indirect procedure must be used and we now undertake its development.

First we focus on the average velocity $\bar{v}_x^‡$. In Chapter 5 we discussed the method for obtaining averages of quantities by use of a distribution function. We now assume that the distribution function for the velocities of activated complexes along the reaction path is the Maxwell distribution law for velocities in one dimension (Eq. 5–12). According to this equation the fraction of activated complexes with velocity $v_x^‡$ along the reaction path is given by

$$\frac{dn_x}{n} = \left(\frac{m^‡}{2\pi kT}\right)^{\frac{1}{2}} e^{-\frac{(mv_x^2)^‡}{2kT}} dv_x^‡$$

in which $m^‡ (= m_A + m_B)$ and $\frac{1}{2}m^‡ v_x^{‡2}$ are the mass and the kinetic energy of the activated complex in its motion along the reaction path. Then the average velocity in the forward direction is obtained by multiplying this distribution function by $v_x^‡$ and integrating over the positive values of $v_x^‡$ as follows:

$$\bar{v}_x^‡ = \left(\frac{m^‡}{2\pi kT}\right)^{\frac{1}{2}} \int_0^\infty v_x^‡ e^{-\frac{(mv_x^2)^‡}{2kT}} dv_x^‡$$

$$= \left(\frac{kT}{2\pi m^‡}\right)^{\frac{1}{2}}$$

When we substitute this value for $\bar{v}_x^‡$ in Equation 11–25 we obtain

$$k_2 = \frac{[C^‡]}{[A][B]} \frac{1}{\delta} \left(\frac{kT}{2\pi m^‡}\right)^{\frac{1}{2}}$$

Next we turn our attention to the unknown concentration $[C^‡]$. We assume that the activated complexes are in equilibrium with reactants A and B. (It has been shown in a more detailed and complicated derivation of transition state theory that this assumption is not essential, but rather is a matter of mathematical convenience.) The first factor in the equation above is just the expression for the equilibrium constant, $K_{eq}^{‡\prime}$, for the formation of activated complexes, expressed in terms of concentrations. Thus we can write this equation in the following form:

$$k_2 = K_{eq}^{‡\prime} \frac{1}{\delta} \left(\frac{kT}{2\pi m^‡}\right)^{\frac{1}{2}} \tag{11–26}$$

In Chapter 9 we saw how equilibrium constants can be expressed in terms of partition functions for standard states (Eq. 9–40) and we express $K_{eq}^{‡\prime}$ in this way. If concentrations are given in number of molecules per cm³, the appropriate standard state is a concentration of 1 molecule/cm³ and the quantity $V/\mathfrak{N}$ in Equation 9–40 is set equal to 1. Representing the partition function

Reaction Kinetics

for the activated complex in this standard state by $q_‡^{o\prime}$, we can express the equilibrium constant in the following way:

$$K_{eq}^{‡\prime} = \frac{q_‡^{o\prime}}{q_A^o q_B^o} e^{-\frac{\Delta E_0^‡}{RT}} \qquad (11\text{–}27)$$

where $\Delta E_0^‡$ is the difference in energy per mole between the activated complex and the reactant molecules, all in their lowest energy states.

Before we can go any further we must find the value of $q_‡^{o\prime}$. To arrive at this value we assume that the activated complex is essentially like a molecule except for one characteristic feature. This feature is a degree of freedom which in an ordinary molecule would be a normal mode of vibration, but which for an activated complex is a translation along the reaction path.

This distinction can be clarified by consideration of the linear triatomic activated complex discussed earlier for the $D + H_2$ reaction. If this triatomic system were a stable molecule, as we saw in Chapter 8, one of its normal modes of vibration would be the asymmetric stretching vibration (Fig. 8–17) and this mode of vibration would continue indefinitely. Since this system is an activated complex, however, as the D atom approaches the closer H atom the other H atom moves away and continues to move in the same direction to complete the reaction. The following diagram illustrates this difference:

Molecule   Activated complex

Thus we see that a vibrational degree of freedom in a stable molecule is replaced by a translational degree of freedom in an activated complex.

From another point of view we can say that the distance δ at the top of the energy barrier is so small that the change in potential energy over this distance is essentially zero. A linear motion with constant potential energy is a translation.

Taking this unusual degree of freedom of the activated complex into consideration, we can separate the partition function into a product of two factors, one for the 1-dimensional translation along the reaction path and the other for the remaining factors in the partition function. In symbols,

$$q_‡^{o\prime} = q_{‡\,\text{trans}}^o q_‡^o \qquad (11\text{–}28)$$

For $q^\circ_{\ddagger\,\text{trans}}$, by analogy with the partition function for a 1-dimensional translation (Eq. 9–25) we can write the following expression:

$$q^\circ_{\ddagger\,\text{trans}} = \frac{(2\pi m^\ddagger kT)^{\frac{1}{2}}\delta}{h}$$

in which δ, as before, represents the length of the pass and therefore the length within which the activated complex is contained.

When we substitute this expression for $q^\circ_{\ddagger\,\text{trans}}$ in Equation 11–28 and then introduce the resulting expression for $q^{\circ\prime}_\ddagger$ into Equation 11–27, this equation becomes

$$K^{\ddagger\prime}_{eq} = \frac{q^\circ_\ddagger}{q^\circ_A q^\circ_B} \frac{(2\pi m^\ddagger kT)^{\frac{1}{2}}\delta}{h} e^{-\frac{\Delta E^\ddagger_0}{RT}}$$

Substitution of this equation in Equation 11–26 leads to the following equation for the bimolecular rate constant:

$$k_2 = \frac{q^\circ_\ddagger}{q^\circ_A q^\circ_B} \frac{(2\pi m^\ddagger kT)^{\frac{1}{2}}\delta}{h} \cdot \frac{1}{\delta}\left(\frac{kT}{2\pi m^\ddagger}\right)^{\frac{1}{2}} e^{-\frac{\Delta E^\ddagger_0}{RT}}$$

This equation reduces to

$$k_2 = \frac{kT}{h}\left(\frac{q^\circ_\ddagger}{q^\circ_A q^\circ_B} e^{-\frac{\Delta E^\ddagger_0}{RT}}\right) \qquad (11\text{–}29)$$

We have thus reached our objective, the theoretical equation for the rate constant k_2. The factor kT/h, which includes physical constants only, has the dimensions of frequency. At ordinary temperatures its value is about 10^{13} sec^{-1}. All of the variables that are characteristic for a particular reaction are included in the factor in parentheses. Because this factor has the form of an equilibrium constant, Equation 11–29 is often written in the very simple form

$$k_2 = \frac{kT}{h} K^\ddagger \qquad (11\text{–}30)$$

Before we consider the evaluation of $K^\ddagger$ in general let us apply Equation 11–30 to the simplest bimolecular reaction in which 2 atoms, A and B, combine to form the diatomic molecule AB.

Although the calculation of the rate of this reaction is a useful test of the theoretical equation, actually a bimolecular association of two atoms does not occur. (The transmission coefficient in this special case is zero.) This follows from the fact that if a molecule AB were formed from atoms it would have so much vibrational energy that it would dissociate into atoms again. In contrast, for a 3-particle collision, A + B + M, where M could be A, B, or a foreign molecule, the excess energy can be carried off by the third particle. Bimolecular association reactions of molecules or free radicals can occur because in this case the excess energy of the newly formed molecule can be distributed among a number of its normal modes of vibration.

In order to evaluate $K^{\ddagger}$ for the reaction $A + B \to AB$ we need to obtain the partition function for the activated complex $(AB)^{\ddagger}$. We can consider this complex as similar to a molecule except for the absence of the vibrational degree of freedom. Thus the activated complex has 3 degrees of translational freedom and 2 degrees of rotational freedom. For the corresponding partition functions we use Equation 9–26 and Equation 9–31, respectively. Then the partition function for the complex is

$$q_{\ddagger}^{0} = V \left(\frac{2\pi m^{\ddagger} kT}{h^2} \right)^{\frac{3}{2}} \frac{8\pi^2 I^{\ddagger} kT}{h^2}$$

In this equation, $m^{\ddagger} = m_A + m_B$, $I^{\ddagger}$ is the moment of inertia of the activated complex, and $V = 1$ cm³ because of our choice of standard state.

The atoms A and B have only translational degrees of freedom and their partition functions are identical except for the mass factor. Consequently,

$$K^{\ddagger} = \frac{q_{\ddagger}^{0}}{q_A^0 q_B^0} e^{-\frac{\Delta E_0^{\ddagger}}{RT}}$$

$$= \frac{\left[\frac{2\pi(m_A + m_B)kT}{h^2} \right]^{3/2} \frac{8\pi^2 I^{\ddagger} kT}{h^2}}{\left(\frac{2\pi kT}{h^2} \right)^3 m_A^{3/2} m_B^{3/2}} e^{-\frac{\Delta E_0^{\ddagger}}{RT}}$$

We can simplify this equation by noting that $I^{\ddagger} = \mu^{\ddagger} r_{AB}^2$, where r_{AB} is the distance between the atoms in the complex and $\mu^{\ddagger} \left(= \frac{m_A m_B}{m_A + m_B} \right)$ is the reduced mass of the complex. When these substitutions are made and common factors are cancelled, this equation becomes

$$K^{\ddagger} = h \left[\frac{8\pi(m_A + m_B)}{m_A m_B kT} \right]^{\frac{1}{2}} r_{AB}^2 \, e^{-\frac{\Delta E_0^{\ddagger}}{RT}} \text{ cm}^3/\text{molecule}$$

Now we can readily obtain the theoretical equation for the rate constant of the reaction under consideration:

$$k_2 = \frac{kT}{h} K^{\ddagger}$$

$$= \left[\frac{8\pi(m_A + m_B)kT}{m_A m_B} \right]^{\frac{1}{2}} r_{AB}^2 \, e^{-\frac{\Delta E_0^{\ddagger}}{RT}} \text{ cm}^3/\text{molecule sec}$$

On comparing this equation for k_2 with the equation derived on the basis of the collision theory (Eq. 11-23), we find a remarkable similarity in form, although some of the quantities have different significances in the two theories. Instead of σ_{av}, the average collision diameter, we have r_{AB}, the internuclear distance in the activated complex; and instead of E_a, the Arrhenius activation energy, we have $\Delta E_0^\ddagger$, the difference between the energy of the activated complex and the ground state energies of the reactants. Even when these differences are taken into account, the agreement between the two theories, based on entirely different concepts, is satisfying.

When, however, an attempt is made to apply Equation 11-30 to the calculation of the rates of realistic reactions, formidable difficulties arise. In general, the Schrödinger equation for the reacting system cannot be solved to a sufficient degree of approximation, and consequently potential energy surfaces comparable to the one represented in Figure 11-7 cannot be obtained. It follows that $\Delta E_0^\ddagger$ for the activated complex cannot at present be calculated from theory.

On the other hand, it is possible to calculate fairly reliable values of the pre-exponential factor of Equation 11-29 for some reactions. This factor involves the partition functions of the reactants and the activated complex. The partition functions of many molecules of even moderate complexity are adequately known, but the calculation of the partition function of the activated complex, $q_\ddagger^0$, requires values for its vibrational frequencies and interatomic distances which are not measurable.

TABLE 11-3 COMPARISON OF THEORETICAL AND EXPERIMENTAL PREEXPONENTIAL FACTORS FOR SOME BIMOLECULAR REACTIONS*

Reaction	Preexponential factor × 10^{-12} (cc/mole sec)	
	Calculated	*Experimental*
$NO + O_3 \rightarrow NO_2 + O_2$	0.44	0.80
$NO_2 + O_3 \rightarrow NO_3 + O_2$	0.14	5.9
$NO_2 + F_2 \rightarrow NO_2F + F$	0.12	1.6
$NO_2 + CO \rightarrow NO + CO_2$	6.0	12
$2NO_2 \rightarrow 2NO + O_2$	4.5	1.8
$NO + NO_2Cl \rightarrow NOCl + NO_2$	0.84	0.83

*D. R. Herschbach, H. S. Johnston, K. S. Pitzer, and R. E. Powell, J. Chem. Phys. 25, 736 (1956).

Reaction Kinetics

Under these circumstances it is necessary to assume that the activated complex can be treated as a molecule with a missing degree of vibrational freedom (the one involving the reaction coordinate). Then, with reasonable estimates of the structural parameters of the activated complex which are obtained by analogy with actual molecules, $q_{\ddagger}^o$ can be evaluated. In this way values of the pre-exponential factor have been calculated for a number of reactions of simple molecules. The resulting values (Table 11–3) agree in general with those obtained from measured reaction rates within a factor of 10, and many are considerably better than this. Considering the fact that there is appreciable uncertainty in some of the experimental data, this agreement represents a real achievement.

Closely related to the transition state theory is what might be called a pseudo-thermodynamic approach to reaction rates. It was mentioned earlier that the factor $K_{\ddagger}$ in Equation 11–30 has the form of an equilibrium constant. By analogy with an actual equilibrium constant (Eq. 4–5), we can relate $K_{\ddagger}$ to a "free energy of activation," $\Delta G_{\ddagger}^{\circ}$. Thus we can write[4]

$$\Delta G_{\ddagger}^{\circ} = -RT \ln K_{\ddagger}$$

and

$$K_{\ddagger} = e^{-\Delta G_{\ddagger}^{\circ}/RT}$$

Again by analogy with thermodynamics we can make use of the relationship $\Delta G_{\ddagger}^{\circ} = \Delta H_{\ddagger}^{\circ} - T\Delta S_{\ddagger}^{\circ}$, in which we have introduced the enthalpy (or "heat") of activation and the entropy of activation. Then we can express the rate constant k_r in the following ways:

$$k_r = \frac{kT}{h} K_{\ddagger} = \frac{kT}{h} e^{-\frac{\Delta G_{\ddagger}^{\circ}}{RT}}$$

$$= \frac{kT}{h} e^{-\frac{\Delta H_{\ddagger}^{\circ}}{RT}} e^{\frac{\Delta S_{\ddagger}^{\circ}}{R}} \tag{11–31}$$

On taking the logarithm of both sides of this equation and differentiating with respect to T, we obtain

$$\frac{d \ln k_r}{dT} = \frac{\Delta H_{\ddagger}^{\circ}}{RT^2} + \frac{1}{T} = \frac{\Delta H_{\ddagger}^{\circ} + RT}{RT^2}$$

From Equation 11–19

$$\frac{d \ln k_r}{dT} = \frac{E_a}{RT^2}$$

[4]Strictly speaking, the equilibrium constant related by thermodynamics to ΔG° is K_P, the equilibrium constant in terms of partial pressures. Since the use of thermodynamic equations here is not rigorous, we ignore the difference between K_P and K_C, the equilibrium constant in terms of concentrations.

Consequently, $\Delta H_\ddagger^\circ = E_a - RT$. Neither $\Delta H_\ddagger^\circ$ nor E_a is the same as the energy difference $\Delta E_0^\ddagger$ introduced in Equation 11–27, but the differences among these three quantities are not large compared to the uncertainty in their values.

We now equate the expression for the rate constant in Equation 11–31 with the one in Equation 11–24, based on the collision theory:

$$\left(\frac{kT}{h} e^{\frac{\Delta S_\ddagger}{R}}\right) e^{-\frac{\Delta H_\ddagger}{RT}} = (pA) e^{-\frac{E_a}{RT}}$$

Since $\Delta H_\ddagger \sim E_a$, the 2 pre-exponential factors are roughly equal; thus

$$\frac{kT}{h} e^{\frac{\Delta S_\ddagger}{R}} \sim pA$$

This approximate equality leads to some understanding of the magnitude of the steric factor, p, which, as mentioned earlier, cannot be calculated on the basis of the collision theory. We conclude that the steric factor is related to the entropy of activation.

The numerical value of $\Delta S_\ddagger$ depends on the chosen standard state. This, in turn, depends on the unit of concentration used in the rate constant which is usually the mole per liter. For a standard state concentration of one mole per liter, the pre-exponential factor of a bimolecular rate constant is roughly $10^{10} - 10^{11}$ liter/mole sec (p. 417). If the value of p for a reaction is about 1, since kT/h is about 10^{13} sec^{-1},

$$e^{\frac{\Delta S_\ddagger}{R}} \sim 10^{10} \times 10^{-13} \sim 10^{-3}$$

Thus $\Delta S_\ddagger \sim 2 \ln (10^{-3}) \sim -13.8$ cal/mole deg. If $p < 1$, $\Delta S_\ddagger$ is lower than this value (larger in absolute value).

A negative entropy of activation corresponds qualitatively to an increase in order and a loss of excited degrees of freedom in the activated complex relative to the reactant molecules. A large negative entropy of activation, corresponding to a relatively highly ordered activated complex, is correlated with a small value of the steric factor. Even rough estimates of the entropy of activation are sometimes useful in comparing the reaction rates of similar compounds and they can throw some light on the reaction mechanism.

It was mentioned at the beginning of this section that the transition state theory in its most complete form provides the possibility of a detailed treatment of chemical reactions and a quantitative calculation of absolute reaction rates. This possibility has not been realized so far because of computational difficulties. Activation energies have been calculated from theory for only a few simple gas reactions and even then with simplifying assumptions. As mentioned above, the pre-exponential factor for the rate constant has been calculated for a number of reactions by evaluation of partition functions with reasonable agreement with experimental data. It can be said, however, that up to now the main usefulness of the transition state theory to reaction kinetics has not been the quantitative results obtained, but rather the background of concepts in terms of which reaction rates can be correlated and reaction mechanisms can be elucidated.

Reaction Rates in Solutions

Although the theories of reaction rates discussed in the last few sections were developed for gas reactions, it turns out that they are also applicable to many reactions involving dissolved substances. The collision theory is based on a calculation of the rate of bimolecular collisions of independent molecules in the gas phase, and there is no reason to believe that this calculation has meaning for the liquid phase. Nevertheless, there are reactions for which the rate constant and the activation energy are almost the same when the reactants are dissolved in a number of solvents as when they are in the gas phase. This fact implies that for these cases the pre-exponential factors are also nearly the same.

The observed rates can be interpreted by the hypothesis that when two reactant molecules encounter each other in a solution they are hindered from separating after an unreactive collision because they are surrounded by a "cage" of solvent molecules. Consequently they have the opportunity to make many collisions before separating. These multiple collisions compensate for the relatively slow diffusion of the reactant molecules towards each other in the liquid phase.

In general, reactions carried out in polar solvents do not occur in the gas phase. For these reactions the solvent usually has a marked effect on the reaction rate. To a considerable extent these solvent effects can be reasonably accounted for in terms of the nature of the activated complex and its interaction with the solvent. In the case of reactions which involve ions as reactants or products, electrostatic interactions are particularly important.

Catalysis

There are many reactions whose rates can be accelerated by the presence of small amounts of a substance which does not appear to be a reactant because it can be recovered quantitatively when the reaction is completed. In some cases the reaction rate can be increased by a very large factor. A dramatic example is the reaction of H_2 and O_2. At room temperature a mixture of these gases does not undergo a detectable reaction, but if a small piece of spongy platinum is introduced into the system the reaction occurs with explosive violence.

The acceleration of a reaction rate by a substance which does not appear to enter into the reaction was called *catalysis*[5] by Berzelius in 1835. It has been the subject of a large amount of research, both because of interest in the nature of catalytic activity and because the acceleration of the rates of many reactions has great economic value. In some catalyzed reactions the role of the catalyst has been well established. In many others, in spite of great efforts the mechanism of catalytic activity has not yet been determined.

Careful investigation has shown that the equilibrium constant of a reversible reaction is not changed by the presence of a catalyst.[6] This conclusion is consistent with the fact that the catalyst can be recovered without chemical change. Because the free energy of the catalyst is not changed by the reaction, the free energy change of the reaction is not changed by the catalyst. From the relationship of the forward and reverse reaction rates to the equilibrium constant it follows that a catalyst which increases a reaction rate also increases the rate of the reverse reaction by the same factor.

The general explanation of catalytic activity is based on the concept that the catalyst in some way provides a reaction path for which the activation energy is considerably lower than that of the uncatalyzed reaction. In many cases of homogeneous catalysis, in which the catalyst and the reacting system are in the same phase, the catalyst can be shown to participate in the formation of a reactive intermediate which may be a molecule, a free radical, or an ion. This reactive intermediate then reacts to form products and regenerate the catalyst. In many reactions the steady-state concentration of the reactive intermediate is so low that its presence can only be inferred.

In heterogeneous catalysis, sometimes called contact catalysis, the catalyst is a solid. Although the detailed interpretation of this type of catalysis varies with the kind of reaction and the nature of the catalyst, it is believed that the first step is the formation of a layer of reactant molecules on the surface of the catalyst, a process known as *adsorption*. The adsorbed molecules are held at the surface of the solid by the unbalanced intermolecular forces present there. At least some of the adsorbed molecules are modified by reaction with some component of the surface or by distortion due to surface forces, in such a way that reactive intermediates are formed which then react to form product molecules. These product molecules are then desorbed and diffuse away, to be replaced by fresh reactant molecules.

The effectiveness of a solid catalyst depends on both the nature and the extent of its surface. For this reason the solids used as catalysts are often those which have large numbers of pores and fissures to increase the useful area. For example, a typical catalyst may have a surface area of the order of 10^6 cm^2

---

[5]It is possible for a small amount of a substance which apparently does not enter into a reaction to decrease the reaction rate significantly. This phenomenon is known as inhibition or negative catalysis.

[6]The few experimental results which appear to contradict this statement involved such large amounts of catalyst that they are accounted for in terms of an interaction that is similar to a solvent effect.

per gram of catalyst. Small amounts of a foreign substance may act as catalyst "poisons" because they are strongly adsorbed and thus reduce the surface area available for reaction. In fact, if product molecules are adsorbed too strongly they can act as poisons for the reaction which produces them.

The identification of the reactive intermediates on a catalyst surface is very difficult. Although many ingenious investigations have been carried out in this field, few firm conclusions have been reached. The development of solid catalysts is still largely empirical and is as much an art as a science.

Photochemical Reactions

All of the reactions considered so far in this chapter are included in the class of thermal reactions. The word "thermal" refers to the fact that the source of the activation energy for these reactions is basically the kinetic energy of the molecules in motion which is distributed among the molecules at the temperature of the reacting system according to the Maxwell distribution law.

There are other classes of reactions in which the activation energy is supplied from a source external to the reacting system in the form of electromagnetic radiation or a beam of particles such as electrons, protons, or α particles. The study of reactions of this general type is often called radiation chemistry. Part of this broad subject is *photochemistry*, the study of reactions resulting from the absorption of visible or ultraviolet radiation. This, itself, is an extensive subject which is expanding rapidly. The discussion here is limited to some of its basic concepts, with emphasis on the rates of photochemical reactions.

The fact that light can have a marked effect on some chemical reactions has been known for a long time. As far back as 1818, Grotthus and, somewhat later, Draper concluded that only the light which is absorbed by a reactant is effective in producing a photochemical reaction. Although this statement, which is often called the first law of photochemistry, seems obvious today it was far from obvious in 1818. The photochemical consequences of the absorption of light were not understood at all until the development of the quantum theory and its application to molecular spectroscopy. A modern version of the first law of photochemistry is the statement that a photochemical reaction always involves the production of electronically excited molecules by the absorption of visible or ultraviolet radiation. (At this point it would be helpful to review the section on electronic spectroscopy in Chapter 8.)

A central problem in photochemistry is the nature and fate of the electronically excited molecules. Processes directly involving these excited molecules are called primary processes in contrast to secondary processes which include all subsequent transitions or reactions. A number of primary processes have been identified. Among them are the following:

1. The excited molecule may radiate its excitation energy and make a transition to its ground electronic state.

2. The excited molecule may collide with an unexcited molecule and transfer to it a small part of its excitation energy, while making a radiationless transition to the lowest vibrational level of the excited electronic state. It will then make a transition to its ground electronic state while emitting a photon whose frequency is lower than that of the photon it originally absorbed. This secondary process is called fluorescence.

3. The excited molecule may transfer essentially all of its excitation energy to a molecule of another species which then reacts. This process is called a sensitized photochemical reaction.

4. The excited molecule may undergo a chemical reaction. This reaction may be a unimolecular dissociation or isomerization, or it may involve unexcited molecules of the same or a different species.

We are concerned here with only the last process in which the source of the activation energy is the absorbed photon. This idea was first stated by Einstein in 1912 as an application of his concept of the photon. His statement, which is sometimes called the second law of photochemistry, is the following: Every absorbed photon activates one molecule in the primary process of a photochemical reaction. Thus the activation of 1 mole of reactant requires Avogadro's number of photons. The radiant energy associated with this number of photons is called the einstein. Its value in ergs is given by $\mathfrak{N}h\nu$ (or $\mathfrak{N}hc/\lambda$) where ν and λ represent the frequency and the wavelength of the radiation. Thus for radiation of wavelength λ in angstrom units, this energy is given by

$$\frac{6.02 \times 10^{23} \times 6.62 \times 10^{-27} \times 3.00 \times 10^{10}}{10^{-8}\lambda} = \frac{1.20 \times 10^{16}}{\lambda} \text{ ergs}$$

$$= \frac{2.86 \times 10^{5}}{\lambda} \text{ kcal}$$

For light of wavelength 5000 Å, 1 einstein represents an energy of 57.2 kcal.

Einstein's photochemical law does *not* imply that one molecule reacts for each photon absorbed. On the one hand, in the first two primary processes mentioned above absorption does not lead to reaction; on the other hand, the secondary processes may include a chain reaction in which many thousands of molecules react for each photon absorbed. The ratio of the number of molecules reacting to the number of photons absorbed is called the *quantum yield* (ϕ) of the reaction.[7] The magnitude of the quantum yield of a photochemical reaction, which can range from near zero to 10^6 or more, provides useful information about its mechanism.

[7] When several competing processes occur simultaneously, a quantum yield for each process can be defined. The quantum yield defined above is the gross, or over-all, quantum yield.

In order to investigate the kinetics and determine the quantum yield of a photochemical reaction it is necessary to have a quantitative measure of the number of photons absorbed by the reacting system during a measured time interval. This type of measurement is in general neither easy nor direct. We describe it for a simplified case.

Let us imagine that our reacting system is contained in a vessel or cell of length b which is equipped with flat transparent windows (Fig. 11-8). A collimated monochromatic beam of radiation of an appropriate wavelength λ is incident on the entrance window. Part of this radiation is absorbed by the reacting system in the cell and the remainder (neglecting reflection losses at

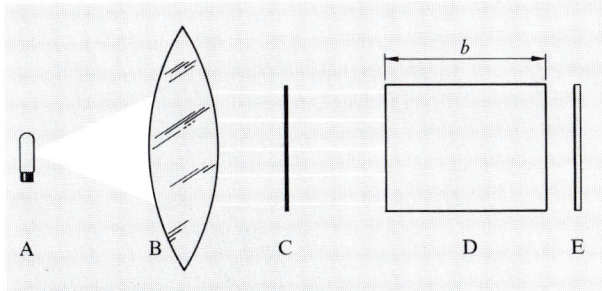

Figure 11-8. Schematic diagram of an apparatus for photochemical studies. A is a source of light, B is a collimating lens, C is a device for selecting a narrow range of wavelengths (either a monochromator or an optical filter), D is the reaction vessel usually immersed in a thermostat, and E is a detector which may be a photoelectric cell or a thermopile.

the windows) is transmitted through the cell and reaches a detector whose output signal s is proportional to the rate at which photons are incident on it.

The fraction of the radiation incident on the cell that is transmitted and reaches the detector is given by *Beer's law*[8] which can be expressed in several equivalent ways, among which are the following:

$$\frac{s}{s_0} = e^{-\alpha_\lambda bc} \quad \text{and} \quad \ln \frac{s_0}{s} = \alpha_\lambda bc \qquad (11\text{-}32)$$

in which s_0 is the detector signal when there is no absorbing substance in the cell, c is the concentration of the absorbing reactant, and α_λ is a property of the absorbing substance called the absorption coefficient. The subscript λ indicates that α_λ is a function of the wavelength. With b expressed in cm and c expressed in moles/cm³, the units of α_λ are cm²/mole.

[8] This law is also associated with the names of Bouguer and Lambert who did the pioneering work in the measurement of the absorption of light by matter.

The following oversimplified derivation of Beer's law serves to point out the significance of the absorption coefficient. Let us say that the cross-section of the cell is S cm² and that the incident beam has a uniform intensity I, defined as the number of photons crossing a unit area perpendicular to the beam in 1 second. Then the total number of photons incident on the entrance window per second is SI. Imagine that the beam traverses the cell for a distance x and then continues for an infinitesimal distance dx. The volume of this portion of the cell is $S\,dx$ and it contains $\mathfrak{N}cS\,dx$ molecules. If the cross-section area of a molecule is a, the total target area presented to the photons by this number of molecules is $a\mathfrak{N}cS\,dx$.

Now assume that the fraction of the photons absorbed in traversing the distance dx, represented by $-d(SI)/SI_x\,(=-dI/I_x)$, is equal to the ratio of the molecular target area to the cross-section of the cell. This equality is expressed by

$$-\frac{dI}{I_x} = \frac{a\mathfrak{N}cS\,dx}{S} = a\mathfrak{N}c\,dx$$

When this equation is integrated between the limits $x = 0$ and $x = b$, the result is

$$\ln \frac{I_o}{I_b} = a\mathfrak{N}cb \qquad (11\text{--}33)$$

in which I_o and I_b are the intensities of the incident beam and the transmitted beam, respectively. Since a is a molecular cross-section for absorption, the quantity $\mathfrak{N}a$, which is equal to the absorption coefficient α_λ, can be interpreted as a molar cross-section in consistency with its units of cm²/mole.

That Equation 11–33 is another form of Beer's law is evident from a comparison with Equation 11–32. Since the fraction of the incident radiation transmitted through the cell is a dimensionless quantity, it can be expressed in terms of a number of equivalent ratios. For example, the detector signal in Equation 11–32 by hypothesis is proportional to SI and therefore to I, since S is a constant for a particular cell. Consequently, $s/s_o = I/I_o$. In addition, if I is multiplied by $h\nu$, it becomes the energy transported per second per unit cross-section area of the beam in ergs/sec cm². Energy transport or conversion per unit time is power; thus the fraction of the incident radiation transmitted can be expressed as the ratio of the radiant power transmitted by the cell (P) to the radiant power incident on the cell (P_o). Still another form of Beer's law which is commonly used in spectrophotometry is expressed in terms of common logarithms. The quantity $\log P_o/P$ is called the absorbance and is represented by A. Thus Beer's law is

$$A = \epsilon_\lambda bc$$

in which c is expressed in moles per liter and ϵ_λ ($= \alpha_\lambda/2.303$) is called the molar absorptivity.

For photochemical investigations it is not sufficient to know the fraction of the incident radiation that is absorbed by the reacting system. The important quantity is the number of photons or number of einsteins absorbed per second. To obtain this number the detector must be carefully calibrated to obtain the relation between the detector signal and the number of einsteins per second of radiation at a definite wavelength reaching the detector. The details of this calibration are beyond the scope of our discussion here.

Assuming that the detector is properly calibrated and that the extent of photochemical reaction is determined by an appropriate method of quantitative analysis, we can express the quantum yield of a typical decomposition, or photolysis, reaction represented by $R \rightarrow$ products in the following way:

$$\phi = \frac{\text{number of moles of } R \text{ decomposed per second}}{\text{number of einsteins absorbed per second}}$$

The number of einsteins absorbed per second is equal to $S(I_o - I)$ in which the intensity of the radiation is expressed in einsteins/cm² sec.

From Beer's law

$$I = I_o e^{-\alpha b[R]}$$

and it follows that

$$S(I_o - I) = SI_o(1 - e^{-\alpha b[R]}) \tag{11-34}$$

The quantity $\alpha b[R]$ is often so small that $1 - e^{-\alpha b[R]}$ can be approximated by $\alpha b[R]$. In this case

$$\phi = \frac{\text{number of moles of } R \text{ decomposed per second}}{SI_o \alpha b[R]}$$

Since Sb is the volume of the reaction cell, and the number of moles of R divided by the cell volume is the concentration of R in moles/cm³, we have

$$\phi = \frac{\text{change in } [R] \text{ per second}}{I_o \alpha [R]}$$

When the rate of change of $[R]$ is expressed by $-\dfrac{d[R]}{dt}$, the photochemical rate law is

$$-\frac{d[R]}{dt} = \phi I_o \alpha [R]$$

❖❖❖❖❖❖❖

The other limiting case of Equation 11–34 occurs when $ab[R]$ is so large that all of the incident radiation is absorbed in the cell. Then $-\dfrac{d[R]}{dt} = \phi I_o$, a rate law that is of zero order in R. If $ab[R]$ has an intermediate value, a complicated situation results because the reaction rate varies with distance from the entrance window and so does $[R]$.

❖❖❖❖❖❖❖

We now apply some of these relationships to a relatively simple photochemical reaction that has been thoroughly investigated, the photolysis of HI by ultraviolet radiation at room temperature. The principal result is that two molecules of HI are decomposed for every photon absorbed; that is, 2 moles/einstein. The interpretation of this result is based on the following mechanism:

$$\begin{array}{ll} HI + h\nu = H + I & (1) \\ H + HI = H_2 + I & (2) \\ \underline{I + I(+M) = I_2(+M)} & (3) \\ 2HI + h\nu = H_2 + I_2 & \end{array}$$

Reaction 1 is the primary process which in this case is a dissociation into atoms. The conclusion that every photon absorbed leads to dissociation is consistent with the fact that HI has a continuous absorption spectrum because this implies that the excited electronic state of HI is unstable (see Fig. 8–23c). Reaction 3 in which I atoms are removed must involve a 3-body collision, as mentioned earlier (p. 428). The third body can be a gas molecule or the wall of the reaction cell. Other conceivable reactions, such as $I + HI \rightarrow I_2 + H$, have been shown to be so slow that they are insignificant.

Another photochemical reaction that has been carefully studied is the reaction of H_2 and Br_2 to form HBr. We considered the analogous thermal reaction on page 409 and saw that it has a chain mechanism. It turns out that the photochemical reaction has the same mechanism as the thermal reaction except for the chain initiation step which is $Br_2 + h\nu \rightarrow 2Br$. Although the quantum yield of this primary process is probably close to 1, the overall quantum yield at room temperature is quite small, in spite of the chain mechanism. This observation is explained by the fact that one of the chain steps, $Br + H_2 \rightarrow HBr + H$, has a rather large activation energy and a small rate at room temperature. Since this reaction competes for the Br atoms with the chain terminating step, the recombination of Br atoms, the overall quantum yield is small.

Reaction Kinetics

In contrast, the photochemical reaction $H_2 + Cl_2 \rightarrow 2HCl$ has a quantum yield near 10^6. The mechanism of this reaction is analogous to the reaction of H_2 and Br_2. The difference between the quantum yields is ascribed to the fact that the chain step $Cl + H_2 \rightarrow HCl + H$ has a much smaller activation energy and a much higher rate than the corresponding bromine reaction and thus utilizes the Cl atoms very effectively in forming HCl.

Much of the extensive work in photochemistry being carried out at present is concerned with the photochemical reactions of organic compounds. In some cases these reactions lead to the synthesis of compounds which cannot be obtained, or obtained as easily, by other synthetic methods. The most important photochemical reaction is undoubtedly the reaction in which CO_2 and H_2O are converted into carbohydrates in the green leaves of plants by absorption of sunlight.

Supplementary References

S. W. Benson, *Foundations of Chemical Kinetics*, McGraw-Hill, New York, 1960.

J. G. Calvert and J. N. Pitts, *Photochemistry*, Wiley, New York, 1966.

A. A. Frost and R. G. Pearson, *Kinetics and Mechanisms* Wiley, New York, 1961.

W. C. Gardiner, Jr., *Rates and Mechanisms of Chemical Reactions*, W. A. Benjamin, New York, 1969.

K. J. Laidler, *Chemical Kinetics*, McGraw-Hill, New York, 1965.

Problems

1. The vapor phase decomposition of ethylene oxide according to the equation $C_2H_4O \rightarrow CH_4 + CO$ was studied at 414°C with the following results:

t(min)	0	7.0	12.0	18.0
P_{tot}(torr)	119.0	130.7	138.2	146.4

(a) Show that this is a first-order reaction. (b) Calculate the rate constant at 414°C. (c) Calculate the half-life of the ethylene oxide. (d) Calculate the time required to reach a total pressure of 200 torr.

Ans. (b) 0.0145 min^{-1}; (c) 48 min; (d) 79 min

2. A gaseous compound decomposes completely into gaseous products according to the equation $A \rightarrow B + C$. When the reaction is carried out at 200°C the following data are obtained:

t(sec)	0	180	360
P_{tot}(torr)	35.2	46.3	53.9

(a) Show that this is a first-order reaction. (b) How long a time is required to reach a total pressure of 61.6 torr? (c) What fraction of A is decomposed after 7 min?

3. At 600°C, phosphine decomposes according to the equation $4PH_3(g) \rightarrow P_4(g) + 6H_2(g)$. When phosphine was introduced into a flask containing inert gas at 600°C the following total pressures were observed at the indicated time intervals:

t(sec)	0	60	120	∞
P_{tot}(torr)	129.30	150.11	155.65	157.30

(a) Show that this decomposition is a first-order reaction. (b) Calculate the rate constant of the reaction at 600°C and the half-life of phosphine.

Ans. 0.023 sec^{-1}, 30 sec

4. Complete Table 11-2 by calculating the missing values of the rate constant.

5. At 500°C dimethyl ether decomposes according to the equation $(CH_3)_2O \rightarrow CH_4 + H_2 + CO$. At this temperature the following data were obtained:

t(sec)	0	390	777	1195	3155
P_{tot}(torr)	312	408	488	562	779

Determine the order of the reaction and calculate the rate constant at 500°C.

Ans. First order, 4.3×10^{-4} sec^{-1}

6. From the following data for the decomposition of acetaldehyde into methane and carbon monoxide:

t(sec)	0	42	190	480
P_{tot}(torr)	360	394	474	554

(a) Find the order of the reaction. (b) Calculate the rate constant at the temperature at which the data were taken.

7. Show that the half-life of a reactant in a reaction that is of order n with respect to this reactant ($n > 1$) is given by

$$t_{\frac{1}{2}} = \frac{1}{k(n-1)} \frac{2^{n-1} - 1}{a^{n-1}}$$

in which a is the initial concentration of the reactant.

8. A compound X undergoes an isomerization reaction which is reversible and first order in both directions. At a certain temperature the rate constant in the forward direction is found to be 9.2×10^{-4} sec^{-1}. After a long time the

composition of the reacting system is constant at 20 mole % X. Calculate the rate constant for the reverse reaction.

9. For the consecutive first-order irreversible reactions $X \xrightarrow{k_1} Y \xrightarrow{k_2} Z$, the values of k_1 and k_2 are 0.75 min^{-1} and 0.25 min^{-1}, respectively, at a certain temperature. If this reaction is started with pure X at a concentration of 1.0 mole/liter, (a) how long a time will be required for the concentration of Y to reach a maximum? (b) What will be the maximum concentration of Y? (c) What will be the composition of the reacting system after a time interval of 10 min? **Ans.** (a) 2.2 min; (b) 0.58 mole/liter; (c) [X] = 0, [Y] = 0.12, [Z] = 0.88 mole/liter

10. A first-order decomposition reaction is observed to have the following rate constants at the indicated temperatures:

k(min^{-1}) × 10^5	2.46	45.1	576
temp (°C)	0	20	40

(a) Calculate the activation energy. (b) Calculate the value of the rate constant at 30°C. **Ans.** (a) 23.3 kcal; (b) 1.68 × 10^{-3} min^{-1}

11. The value of the rate constant in sec^{-1} for the decomposition of N_2O_5 is 3.46 × 10^{-5} at 25°C, 4.98 × 10^{-4} at 45°C, and 4.87 × 10^{-3} at 65°C. Calculate (a) the activation energy for this reaction; (b) the pre-exponential factor.

12. The half-time for a first-order decomposition reaction at 377°C is 363 min and its energy of activation is 52.0 kcal/mole. (a) What fraction of the molecules at 377°C have sufficient energy to react? (b) Calculate the time required for the compound to be 75% decomposed at 450°C.
Ans. (a) 4.0 × 10^{-18}; (b) 12.5 min

13. Calculate the activation energy for a reaction whose rate doubles for a 10° rise above 20°C.

14. The activation energy for a reaction is 28.0 kcal/mole. At what temperature will the rate constant have twice the value it has at 600°C?

15. 2 gases, with molecular weights of 30 and 60 respectively, undergo a bimolecular reaction with an activation energy of 27.5 kcal/mole. Calculate the rate constant at 100°C for this reaction, using a value of 4.0 Å for the average collision diameter. **Ans.** 1.9 × 10^{-5} liter/mole sec

16. The rate constant of a first-order reaction is represented by the equation $k = 4.3 \times 10^{13} e^{-\frac{25,000}{RT}}$ sec^{-1}. Calculate (a) the value of the rate constant at 100°C; (b) the entropy of activation.

17. The reaction considered in the preceding problem can be carried out in the presence of a catalyst that lowers the entropy of activation by 2 cal/mole degree and lowers the activation energy by 5.0 kcal/mole. Calculate the ratio of the rate constant of the catalyzed reaction to that of the uncatalyzed reaction at 100°C. **Ans. 330**

18. When monochromatic radiation of wavelength λ was sent through an absorption cell 5.0 cm long, containing a 0.10 molar solution of an absorbing substance in a nonabsorbing solvent, 23.0% of the incident radiation was absorbed. Calculate the molar absorptivity, ϵ_λ. **Ans. 0.228 liter/mole cm**

19. Absorption of ultraviolet radiation decomposes acetone according to the equation $(CH_3)_2CO + h\nu \rightarrow C_2H_6 + CO$, and the quantum yield of this reaction at 2800 Å is 0.20. A sample of acetone absorbs monochromatic radiation at 2800 Å at the rate of 7.50×10^4 ergs/sec. Calculate the rate of formation of CO. **Ans. 3.5×10^{-9} mole/sec**

20. A gaseous compound X is photochemically decomposed by ultraviolet radiation. This compound was introduced into an absorption cell 10.0 cm long and 25.0 cm² in cross-section, at a pressure of 0.50 atm and a temperature of 31°C. When a uniform beam of monochromatic radiation with wavelength 3000 Å which filled the cell was sent through the compound, 5.1% of the incident radiation was absorbed. After irradiation for 10.0 min, analysis showed that 1.6% of the X initially present had decomposed. A calibrating measurement showed that the intensity of the beam transmitted by the empty cell was 0.020 watt/cm². Calculate (a) the absorption coefficient, α_λ, of X at 3000 Å; (b) the number of einsteins absorbed by the compound; (c) the quantum yield of this reaction.

12

Solids

Introduction

In this chapter and Chapter 13 we focus our attention on systems which consist of a single pure substance in a condensed phase. Condensed phases are characterized by their high density and low compressibility compared to the values of these properties for the gas phase. The values of these properties for a condensed phase indicate that in such a phase the molecules (or ions) are relatively close together, and we may anticipate the conclusion that the properties of a condensed phase are largely determined by intermolecular forces.

Solids are distinguished from liquids by their rigidity; that is, their resistance to change of shape under the influence of external forces. Solids can be divided into two broad categories: crystals and amorphous substances. (There are some borderline cases which we do not discuss here.) The outstanding characteristics of a crystal are its sharp melting point and its flat faces and sharp edges which, in a well-developed crystal, are usually arranged symmetrically. These properties are the result of a very high degree of internal order which extends throughout the crystal. Amorphous substances, such as glasses, do not have this long-range order. In many ways they are more closely related to liquids than to crystalline solids and we defer their consideration to Chapter 13.

Because the properties of crystals are closely related to their structure, we first discuss the subject of crystalline structure and its determination, and then turn to a consideration of some of the properties of crystals.

Geometry and Symmetry of Crystals: Crystallography

The symmetrical arrangements of flat faces and sharp edges of naturally occurring crystals have been a source of pleasure and wonder since very early times.

Apparently the first quantitative measurements on crystals were made in the seventeenth century when N. Stensen (or Steno, which is the Latin form of his name) measured the angles between the faces of quartz crystals having different shapes. He found that in spite of the differences in shape, the angles between corresponding faces were always the same. Later when crystals of many other substances were observed, each was found to have characteristic angles between faces. The statement that the interfacial angles are characteristic for each kind of crystal is sometimes called the *first law of crystallography*.

A particular kind of crystal often exists in a variety of shapes which are superficially different. For example, ice crystals in the form of snow flakes can be observed in a wide variety of shapes. Today we know that differences in the shape of a crystallized substance depend on the conditions under which the crystals were grown.

At about the same time as Stensen's work, Robert Hooke speculated that all observed forms of crystals could be pictured as consisting of regular arrangements of small spherical particles. More than a century later Hooke's idea was extended by Hauy, who became interested in the regular shape of the fragments formed when a crystal of calcite ($CaCO_3$) was accidentally shattered. From his observations on this and other crystals he concluded that their plane faces and straight edges could be accounted for on the basis of a stacking of "unit" particles whose shape was related to the external shape of the crystal. He found that the orientation of any face of a crystal could be expressed in a simple manner by using three crystal edges that meet in a point as a set of reference axes.

Hauy's observations led to the concept of a crystal as an ordered 3-dimensional arrangement of atoms, molecules, or ions, which is called a *crystal lattice*. From a geometric point of view a lattice is a set of points for which the arrangement about any point is the same as it is about any other point. (We can avoid consideration of the points near the boundaries of the lattice by assuming that the lattice extends indefinitely in all directions.) If we start at any particular point in such a lattice and move in a certain direction, parallel to one of the reference axes, for example, at a certain distance we will find an identical lattice point. On proceeding an equal distance further, another lattice point will be found and so on throughout the lattice.

The simplest example of a lattice is the 1-dimensional lattice formed by a set of points arranged at equal distances along a straight line. The only pa-

Solids

rameter in this case is the minimum repeat distance. In a 2-dimensional planar lattice, the parameters are two basis vectors which give the repeat distances along two axes and the angle between these axes. From the two basis vectors a unit pattern, or unit "cell," can be constructed with which a planar lattice can be formed by repeated translations of the unit cell, in much the same way that a piece of wallpaper can be produced by repeated translations of a unit design.

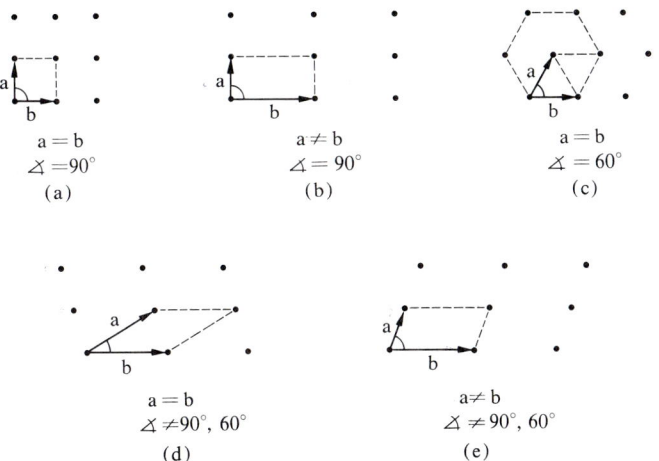

Figure 12–1. Five types of planar lattice.

The two repeat distances in a planar lattice may be the same or different, and the angle may be 90°, 60°, or any value other than these two. These possibilities lead to the existence of just five possible types of planar lattices, each characterized by its unit cell. These five types of planar lattices and their unit cells are shown in Figure 12–1.

It should be noticed that for some lattices the unit cell can be chosen in more than one way. For example, in Figure 12–1c the unit cell may be a parallelogram, an equilateral triangle, or a regular hexagon with a lattice point in its center. The first two of these are called *primitive unit cells*, which means that all lattice points are at the corners of unit cells.

In a 3-dimensional space lattice there are three basis vectors, and the parameters are three repeat distances and three angles. Depending on the value of these parameters, there are seven possible lattice types which can be constructed from primitive unit cells. Accordingly, crystals can be classified into seven *crystal systems*, whose names and definitions are given in Table 12–1. In 1850 Bravais showed that when nonprimitive unit cells, in which there are lattice points at the centers of the cells or at the centers of some of the faces, are

TABLE 12-1   THE SEVEN CRYSTAL SYSTEMS*

System	Unit Cell	Examples
Cubic	$a = b = c$ $\alpha = \beta = \gamma = 90°$	NaCl, diamond
Tetragonal	$a = b \neq c$ $\alpha = \beta = \gamma = 90°$	TiO_2, Sn (white)
Orthorhombic	$a \neq b \neq c$ $\alpha = \beta = \gamma = 90°$	S (rhombic), $BaSO_4$
Monoclinic	$a \neq b \neq c$ $\alpha = \gamma = 90°, \beta \neq 90°$	S (monoclinic), $CaSO_4 \cdot 2H_2O$
Rhombohedral (or trigonal)	$a = b = c$ $\alpha = \beta = \gamma \neq 90°$	$CaCO_3$ (calcite)
Hexagonal	$a = b \neq c$ $\alpha = \beta = 90°, \gamma = 120°$	SiO_2, graphite
Triclinic	$a \neq b \neq c$ $\alpha \neq \beta \neq \gamma \neq 90°$	$K_2Cr_2O_7$

*In this table the symbol $\neq$ should be interpreted to mean "not necessarily equal because of symmetry." There are some border-line crystals which appear to be more symmetrical than they actually are because certain parameters have nearly the same value. K. J. Mysels, *J. Chem. Ed.* **34**, 40 (1957).

included, there are 14 possible space lattice types. These lattice types, now called *Bravais lattices*, are illustrated in Figure 12-2. One of these Bravais lattices can be assigned to every crystal.

Another way of classifying crystals is based on the symmetry elements which can be observed for crystals with well-developed faces. These symmetry elements, which are listed in Table 8-1, include rotation axes, reflection planes, rotation-reflection axes, and the center of symmetry. It was pointed out in Chapter 8 that certain combinations of these symmetry elements, or the associated symmetry operations, form point symmetry groups.

An important difference between the symmetry of a molecule and the symmetry of a crystal is the result of the relationship between the symmetry of the crystal as a whole and the symmetry of the unit cell from which the crystal lattice can be constructed. Because the crystal lattice must occupy all of the space in the crystal, the only rotational axes that can be present in crystals are those of order 2, 3, 4, or 6; no such limitation exists for molecules. It can be shown that axes of other orders are inconsistent with the space-filling requirement.

The point symmetry elements for crystals can be combined into point

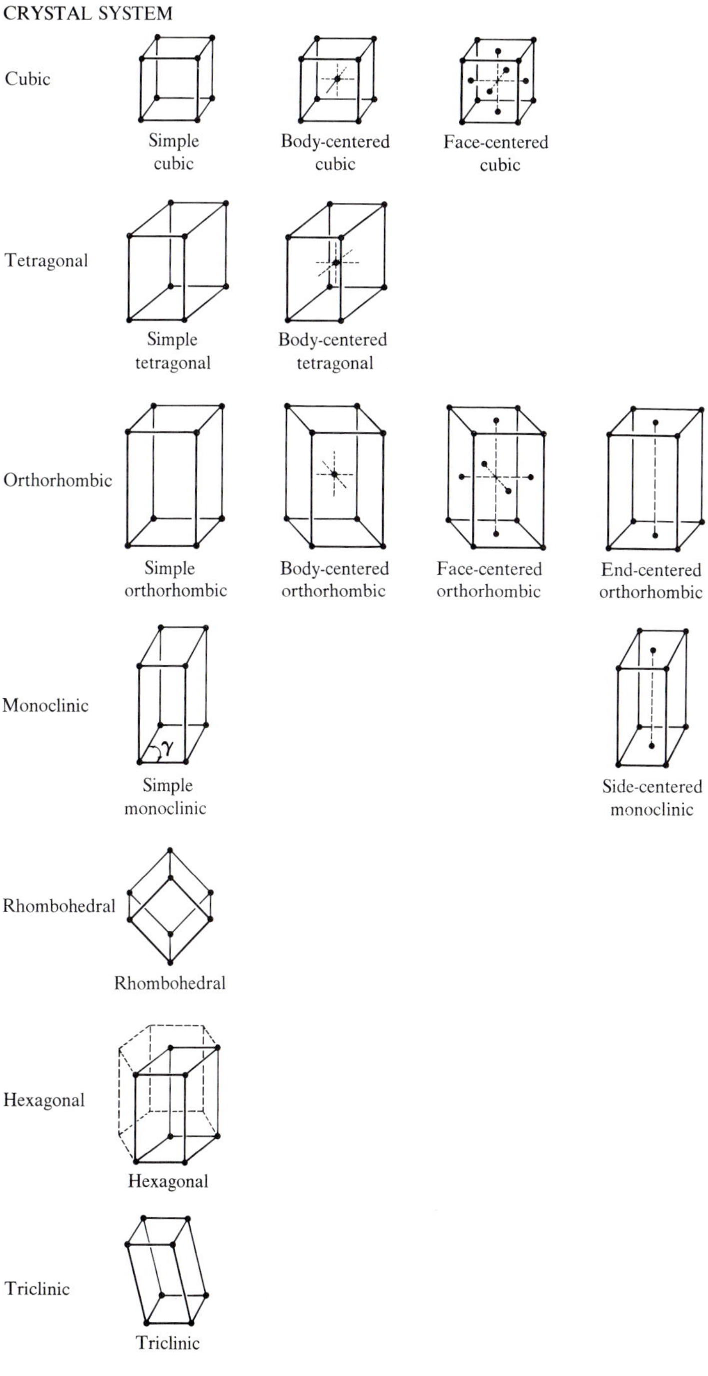

CRYSTAL SYSTEM

Cubic
- Simple cubic
- Body-centered cubic
- Face-centered cubic

Tetragonal
- Simple tetragonal
- Body-centered tetragonal

Orthorhombic
- Simple orthorhombic
- Body-centered orthorhombic
- Face-centered orthorhombic
- End-centered orthorhombic

Monoclinic
- Simple monoclinic
- Side-centered monoclinic

Rhombohedral
- Rhombohedral

Hexagonal
- Hexagonal

Triclinic
- Triclinic

Figure 12–2. The fourteen Bravais lattices.

symmetry groups in just 32 ways. On the basis of its symmetry group, a crystal can be classified into one of 32 *crystal classes*.

A crystal possesses another type of symmetry which a molecule does not have, namely the translational symmetry of the space lattice. If we imagine that a unit cell is moved in the direction of one of the basis vectors by a distance equal to the repeat distance in that direction, the cell then occupies a position that is indistinguishable from its initial position. Consequently for a space lattice, or a crystal, the unit translation also is a symmetry element.

For a complete description of the symmetry of a crystal the translational symmetry elements must be combined with the point symmetry elements into combinations which are called *space groups*. On a purely mathematical basis it can be shown that there are just 230 different space groups. Every crystal belongs to one of these space groups which completely specifies its symmetry. One of the major steps in the determination of the structure of a crystal is its assignment to a space group.

The combination of the symmetry elements of point groups with translations leads to the existence of two types of symmetry elements which are characteristic of crystals. One of these is the *glide plane*, for which the symmetry operation is a reflection followed by a translation parallel to the mirror plane. The other symmetry element is the *screw axis* of order 2, 3, or 4. The associated symmetry operation is a rotation by 180°, 120°, or 90°, respectively, followed by a translation along the rotational axis.

In order to discuss the structure of a crystal we need a way to describe the orientation of planes passing through the lattice points of the crystal. Every face of a crystal is such a plane, which passes through a large number of lattice points. The orientation of a lattice plane is described by considering the intercepts of the plane on the three basis vectors of the lattice. It follows from the geometry of the lattice that the ratio of each of the intercepts, h′, k′, and l′, to the unit length, a, b, and c, of the corresponding basis vector must be an integer or a ratio of integers. This statement is called the *law of rational indices*.

In a crystal the orientation of a lattice plane can be specified by giving the three intercept ratios h′/a, k′/b, and l′/c along the three basis vectors. It is customary in crystallography, however, to use for this purpose the reciprocals of these three ratios, multiplied by the smallest number that makes them all integers. The three integers, represented by (hkl), are called the *Miller indices* of the plane. Thus if the intercept ratios are h′/a = 2, k′/b = $\frac{1}{3}$, l′/c = $\frac{1}{2}$,

Solids   451

the Miller indices of this plane are (164). If a plane is parallel to one of the basis vectors, its intercept on this vector is at infinity and the corresponding Miller index is 0.

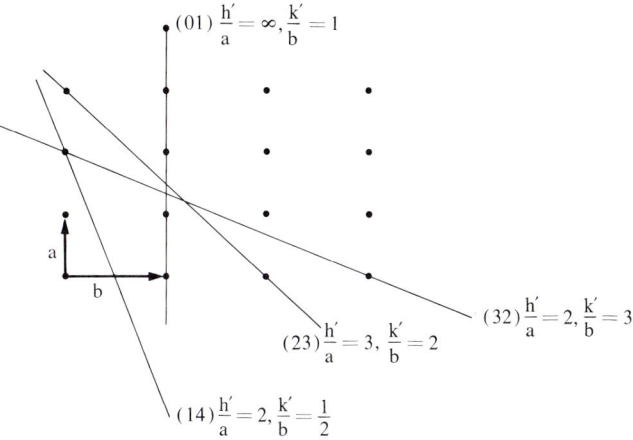

Figure 12–3. Miller indices for lines in a 2-dimensional lattice.

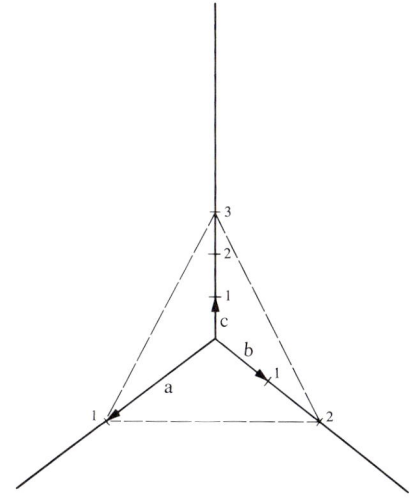

Figure 12–4. A plane whose Miller indices are (632) in an orthorhombic lattice.

The use of Miller indices to specify the orientation of lines in a 2-dimensional lattice is illustrated in Figure 12–3. The orientation of a plane in an orthorhombic lattice is shown in Figure 12–4, and some planes in cubic lattices are shown in Figure 12–5.

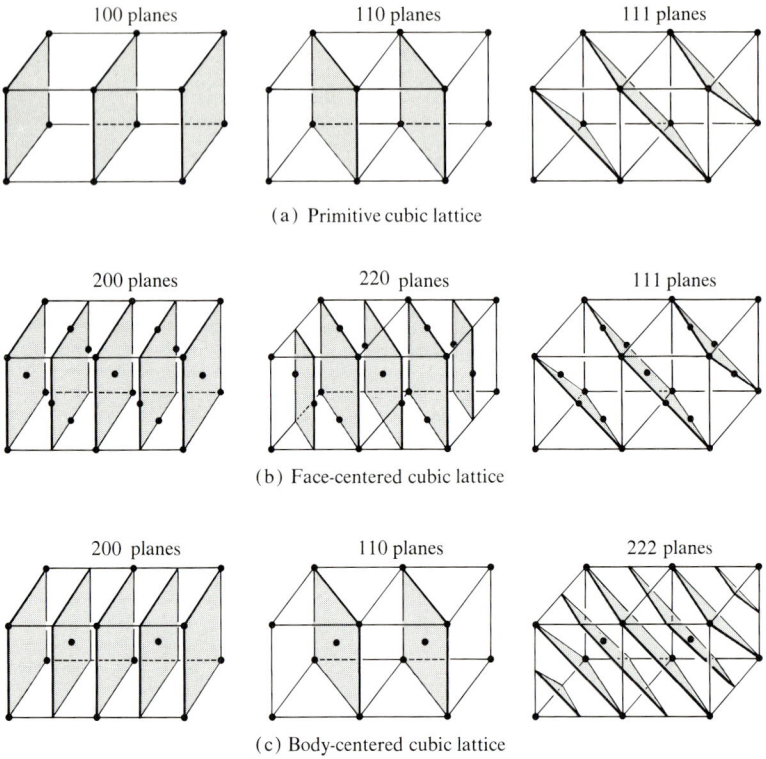

Figure 12-5. Some planes in cubic lattices.

Diffraction of X-Rays by Crystals

By the end of the nineteenth century a considerable amount of information had been obtained about a wide variety of crystals. The experimental data, however, were limited to the results of a few kinds of physical measurements, mainly the measurement of interfacial angles. There was no way of reaching definite conclusions about the internal arrangement of the particles (atoms, molecules, or ions) of which a crystal is composed.

This situation was changed radically in 1912 when von Laue realized that the distances between particles in a crystal are of the same order of magnitude as the wavelengths of X-rays. He predicted that an array of these particles arranged in a lattice should act like a 3-dimensional diffraction grating, and that when a beam of X-rays is passed through a crystal, a diffraction pattern would be observed. The experiment was carried out by Friedrich and Knipping, who obtained a diffraction pattern which completely verified von Laue's pre-

Solids

diction. This result opened up the possibility of determining the details of the internal arrangement in a crystal from the interpretation of its diffraction pattern. The subject of X-ray crystallography has been growing in scope and refinement ever since.[1]

The formation of a diffraction pattern of the type predicted by von Laue depends on the passage through a crystal of a beam of X-rays with a continuous distribution of wavelengths. Soon after the announcement of the observation of X-ray diffraction, W. H. Bragg and his son W. L. Bragg developed another experimental procedure in which they measured the "back-scattering" of monochromatic X-rays by a crystal. Because the diffraction patterns obtained by the Bragg method are easier to interpret than those obtained by the Laue method, we limit our discussion to the Bragg method.

Let us imagine that a monochromatic beam of X-rays is directed at a face of a crystal with an atom at each lattice point of a simple lattice (Fig. 12-6). The incident X-rays excite the atoms in the crystal which emit secondary X-rays in all directions. The secondary waves emitted by the atoms in a single

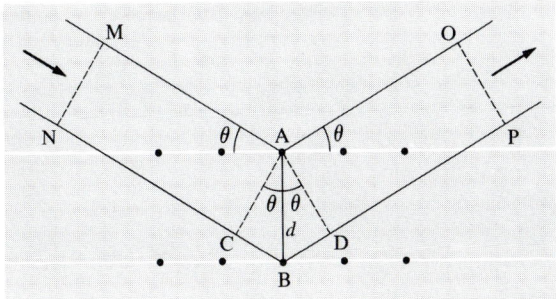

Figure 12-6. "Reflection" of an X-ray beam from lattice planes.

lattice plane combine to form a scattered beam which is at the same angle with respect to the scattering plane as the incident beam. Because this behavior is analogous to the reflection of light by a plane mirror (angle of incidence equal to angle of reflection), this scattering process is often referred to as a reflection of the X-ray beam, although this usage is not strictly correct.

As the incident X-ray beam penetrates the crystal, it produces scattered beams from many parallel lattice planes. These scattered beams interfere with each other constructively, and thus reinforce each other, whenever the difference in path length between beams originating in different lattice planes is an integral multiple of the X-ray wavelength. This condition imposes a

[1] In 1912 the nature of X-rays was not completely certain. The observation of a diffraction pattern by Friedrich and Knipping not only provided information about crystal structure but it also established beyond question that X-rays have wave properties.

restriction on the angle θ at which X-rays of a given wavelength, λ, can be "reflected" by a set of lattice planes with a particular lattice spacing, d.

In order to calculate the magnitude of crystal lattice spacings and thus find the size of the unit cell, we need a relationship between θ, λ, and d. The equation which relates these quantities was first derived by W. L. Bragg. We derive it with the use of Figure 12–6, which represents the scattering from two atoms, A and B, in adjacent lattice planes. The lines AC and AD are perpendicular to NB and BP, respectively. The difference in the path lengths NBP and MAO is CB + BD, which is equal to 2 CB. This path difference equals $n\lambda$ in which, as mentioned above, n must be an integer. Because the angles CAB and BAD are both equal to θ, CB = BD = $d \sin \theta$. Therefore,

$$n\lambda = 2d \sin \theta$$

This is the *Bragg equation* for X-ray diffraction.[2]

In the Bragg equation, n represents the order of diffraction. From this equation it can be seen that the angle θ for a certain value of n and d is the same as the first-order diffracted beam for a set of planes with spacing d/n. It is conventional in the use of the Bragg equation to take n as equal to 1 and to consider the beams of higher orders of diffraction as though they come from planes with correspondingly reduced spacing, whether or not such planes actually exist. For example, the second order of diffraction from the (111) planes of a lattice with spacing d occurs at the same angle as the first-order diffraction from a parallel set of planes with spacing $d/2$, whose Miller indices are (222). Similarly, the third-order diffraction line is ascribed to a plane whose indices are (333).

To apply the Bragg technique, an apparatus (spectrometer or diffractometer) is used (Fig. 12–7) in which a well-defined beam of monochromatic X-rays is produced. This is accomplished by passing the radiation emitted by an X-ray tube through a set of slits and a filter which transmits only the radiation corresponding to the most intense emission line (the $K\alpha$ line) of the target of the X-ray tube. To change the wavelength of the X-rays, an X-ray tube with a different target metal must be used. The most commonly used target metals and their characteristic wavelengths are copper, 1.541 Å, molybdenum, 0.709 Å, and chromium, 2.290 Å.

The crystal to be studied is mounted in the X-ray beam in such a way that the angle θ between a crystal face and the beam can be varied while the angle between the crystal face and the direction to an X-ray detector is kept always equal to θ. The detector, which may be a Geiger counter, a scintillation counter, or an ionization chamber, provides an electrical signal which is proportional

---

[2] The Bragg equation is similar to, but different from, the optical equation for the diffraction of light by a plane ruled diffraction grating, $n\lambda = d \sin \theta$. In the optical equation θ is the angle between the diffracted beam and the perpendicular to the plane of the grating, and d is the distance between the lines in the grating.

Solids

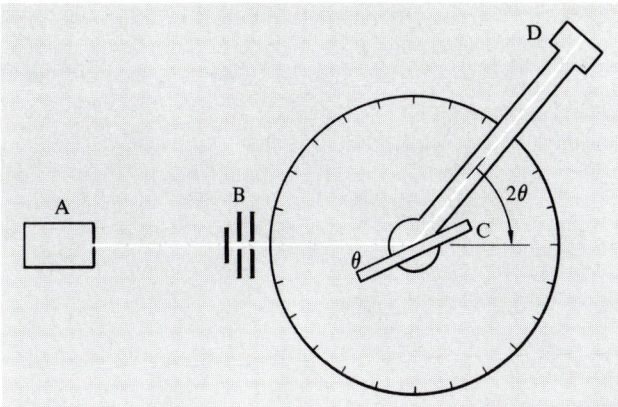

Figure 12-7. Schematic diagram of an X-ray spectrometer. A is the X-ray source, B is the filter and slit system for defining the X-ray beam, C is the crystal, and D is the X-ray detector.

to the intensity of the incident X-rays. As θ is varied, the detector signal goes through a maximum whenever this angle reaches a value which is given by the Bragg equation. In this way the value of θ which corresponds to the maximum signal can be measured.

From the known value of λ and the observed value of θ the spacing of the lattice planes parallel to the crystal face can be calculated. To find the d values for other sets of planes, the diffraction patterns from faces parallel to these planes must be obtained.

Powder Method

The Bragg technique requires the use of fairly large single crystals with flat faces carefully oriented in a number of specific directions. Not all crystalline substances can be easily obtained in the needed form and, in addition, the Bragg technique is very time-consuming. A simpler approach, known as the *powder method*, was developed independently by Debye and Scherrer (1916) and Hull (1917). In this method the sample under investigation consists of a finely divided crystalline powder which is contained in a small thin-walled tube of glass or plastic, or is attached by some adhesive to a fiber. The X-ray beam is sent through the powdered crystal and the diffracted beams are either photographed or detected as a function of the angle from the incident beam by one of the detectors mentioned above.

The particles in the crystalline powder are very small crystals, called crystallites, which are oriented at random directions. Because of this randomness, some of the crystallites are oriented in such a way that a particular lattice plane, the (100) plane, for example, will have the Bragg angle with respect to the primary beam. Each such crystallite will emit a diffracted beam at an angle of 2θ with respect to the primary beam (Fig. 12-8). Since these crystallites are oriented at random except for the common angle θ with respect to the primary beam, the beams diffracted by the (100) planes form the surface of a cone.

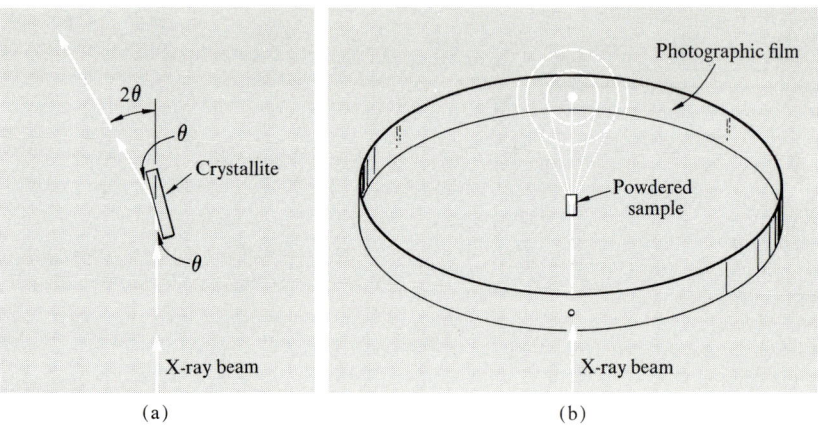

Figure 12–8. (a) Reflection from a crystallite oriented by chance at the appropriate angle with respect to the incident X-ray beam; (b) diagram showing X-ray diffraction by a powdered sample.

If this cone of X-rays is intercepted by a photographic plate whose plane is perpendicular to the primary beam, the resulting image on the photographic plate is a circle.

In a similar way, there are crystallites oriented at the Bragg angles for the other lattice planes, each of which has its own value of the interplanar spacing. Consequently the diffraction pattern recorded on the photographic plate consists of a set of concentric circles which is different for each kind of crystal. From the radii of these circles and the distance from the photographic plate to the sample, the angle of diffraction can be calculated.

❖❖❖❖❖❖

In an actual diffraction camera (Fig. 12-9) a strip of photographic film bent into a cylindrical surface is used. Holes are usually punched in the film at those positions where the incident X-ray beam enters and the undeviated portion of this beam leaves the camera. The sample, which is mounted on the axis of the cylinder, is often rotated about this axis to insure randomness in the orientation of the crystallites. When the exposed and developed film is

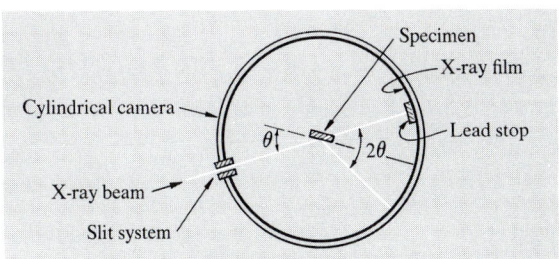

Figure 12–9. An X-ray diffraction camera.

placed on a flat surface, the image of the diffraction pattern (Fig. 12–10) consists of a set of pairs of circular arcs arranged symmetrically around a point which is the intercept of the undeviated incident beam on the film. From the distance y between a pair of related lines on the film and the radius of the cylindrical camera r the angle of diffraction can be readily calculated from the relationship $\theta = 360y/4\pi r$.

In order to obtain sharp diffraction lines of uniform thickness the crystallites in the powdered crystal must have an average dimension of a few microns. If the crystallites are too large, relatively few of them will contribute to a

Figure 12–10. X-ray powder diffraction pattern for NiO. (Courtesy of Professor L. L. Pytlewski.)

diffraction line which then is a discontinuous set of spots. If the crystallites are too small the diffraction lines become broadened. This broadening is a consequence of the fact that as the size of a crystallite decreases, the number of its lattice planes and thus the extent of the orderly arrangement also decreases. It is possible to estimate the average size of very small crystallites from measurements of the broadening of their diffraction lines.

The powder method is useful for the crystal systems that have only one or two lattice parameters to be determined. This includes the cubic, tetragonal, hexagonal, and rhombohedral systems. We restrict our attention to the simplest case, the cubic system with three possible Bravais lattices which are based on primitive (P), face-centered (F), or body-centered (B) unit cells.

From the observed diffraction pattern of a cubic crystal we can find its lattice type and the edge length of the unit cube a. For this purpose we need the following geometrical relationship[3] between a and d_{hkl}, interplanar spacing for planes with Miller indices h, k, and l:

$$d_{hkl} = \frac{a}{\sqrt{h^2 + k^2 + l^2}} \tag{12-1}$$

When this expression for d_{hkl} is substituted into the Bragg equation, the resulting equation for $\sin \theta$ is

$$\sin \theta_{hkl} = \frac{\lambda}{2a} \sqrt{h^2 + k^2 + l^2} \tag{12-2}$$

The value of $\sin \theta_{hkl}$ for each observed diffraction line must satisfy this equation. We must, however, assign values of the Miller indices to the diffraction lines before we can use this equation to calculate a. In order to index the lines, that is, find the correct values of h, k, and l, we will predict the diffraction pattern for each type of cubic lattice. From a comparison between the predicted and observed diffraction patterns we will be able to index the lines unambiguously and thus make a correct choice of lattice type.

We first predict the diffraction pattern for a primitive cubic lattice. Since a is a constant for a particular kind of crystal, we can rewrite Equation 12-2 in the form

$$\sin^2 \theta_{hkl} = K(h^2 + k^2 + l^2)$$

By assigning consecutive integral values (0, 1, 2, . . .) to h, k, and l we can calculate a series of values of $\sin^2 \theta$ and a series of values of d_{hkl} from Equation 12-1. Some of these values are listed in Table 12-2. Notice that $\sin^2 \theta$ cannot have the value $7K$ because there is no way in which the integer 7 can be written in the form $h^2 + k^2 + l^2$. (This is also true of the integers 15, 23, 28, and so on.)

The predicted diffraction pattern thus consists of a set of six lines which are equally spaced when plotted against $\sin^2 \theta$, followed by a gap and then another series of lines. The observation of such a set of diffraction lines shows directly that the crystal under study has a primitive cubic lattice. It then is

---

[3] This is a special case of the more general relationship

$$\frac{1}{d_{hkl}^2} = \frac{h^2}{a^2} + \frac{k^2}{b^2} + \frac{l^2}{c^2}$$

which is applicable to a lattice with orthogonal basis vectors a, b, and c. It can be derived by calculating the distance from some lattice point to the nearest plane whose Miller indices are h, k, and l, using the 3-dimensional form of the theorem of Pythagoras.

easy to assign to each line the correct values of h, k, and l. From the measurement of θ for any one of these lines, the cube edge length a can be calculated from the equation

$$a = \frac{\lambda}{2 \sin \theta_{hkl}} \sqrt{(h^2 + k^2 + l^2)} \qquad (12\text{--}3)$$

If the indexing of the lines has been done correctly, the same value of a must be found from all of the values of $\sin \theta_{hkl}$.

TABLE 12-2   INTERPLANAR DISTANCES FOR A PRIMITIVE CUBIC LATTICE

hkl	100	110	111	200	210	211	220	221 300	310
$\sin^2 \theta_{hkl}$	K	2K	3K	4K	5K	6K	8K	9K	10K
d_{hkl}	a	$\dfrac{a}{\sqrt{2}}$	$\dfrac{a}{\sqrt{3}}$	$\dfrac{a}{2}$	$\dfrac{a}{\sqrt{5}}$	$\dfrac{a}{\sqrt{6}}$	$\dfrac{a}{2\sqrt{2}}$	$\dfrac{a}{3}$	$\dfrac{a}{\sqrt{10}}$

The patterns of lines expected from face-centered and body-centered lattices are different from the pattern just described, and they are different from each other. In the case of a body-centered lattice, Figure 12-5 shows that all of the atoms lie on (110) planes and the corresponding diffraction line will be relatively intense. Only half of the atoms, however, lie in the (100) planes and the remainder lie in the (200) planes which are located halfway between adjacent (100) planes. As a consequence of this fact, at the Bragg angle for "reflection" from the (100) planes the X-rays scattered by these planes are out of phase with those scattered by the (200) planes, destructive interference occurs, and the diffraction line corresponding to the (100) planes is absent. Conversely, at the Bragg angle for reflection from the (200) planes all scattered X-rays are in phase and a strong diffraction line results.

A situation similar to that for the (100) planes exists with respect to the (111) planes. It can be shown in general that for a body-centered cubic lattice all diffraction lines for which h + k + l is an odd integer must be absent.

For the face-centered cubic lattice, only half of the atoms lie in the (100) and the (110) planes and the corresponding diffraction lines are absent. All atoms lie in (111), (200), and (220) planes and the corresponding diffraction lines are strong. In general for this type of lattice the values of h, k, and l must all be even (considering 0 as even) or all odd for strong diffraction lines to be produced.

The predicted diffraction patterns for the three types of cubic lattice are shown schematically in Figure 12–11. The differences between them clearly indicate the usefulness of missing reflections in distinguishing between different lattice types. In general the search for missing reflections is an important step in the determination of crystal structures. In the simplest cases described above, it is sufficient for a complete structure determination.

As an example of this procedure let us consider the powder diffraction pattern of crystalline silver, which is known from optical examination to

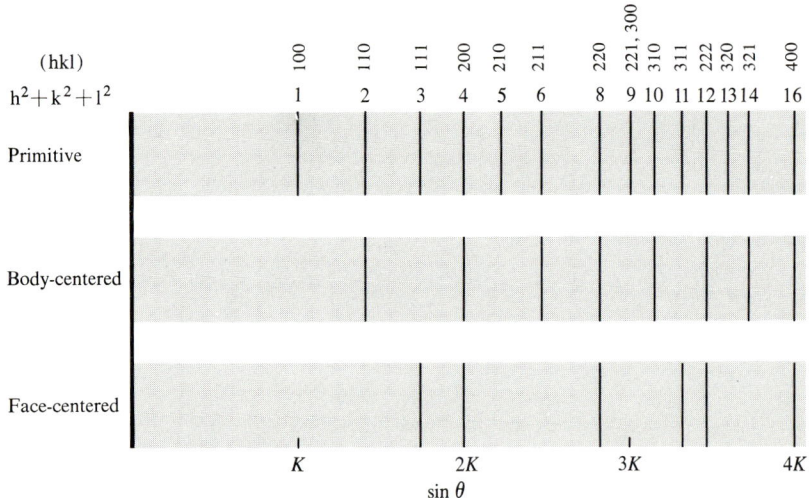

Figure 12–11. Predicted positions of diffraction lines for the three cubic lattices as a function of sin θ. The same value of the edge length of the unit cube is used for all three lattices.

belong to the cubic system. The Bragg angles, using copper $K\alpha$ X-rays with $\lambda = 1.541$ Å, for the first six diffraction lines, and the corresponding sin θ values are as follows:

θ	19.08°	22.17°	32.26°	38.74°	40.82°	49.00°
sin θ	0.3268	0.3773	0.5338	0.6257	0.6536	0.7547

The problem now is to index these lines and to calculate the edge length of the unit cube. One way to accomplish this is to compare the ratios of the sin θ values given above with those predicted for the structures in Figure 12–11. From this figure we see that the values of sin θ for the first three diffraction lines of a primitive or a body-centered lattice are in the ratio $1:\sqrt{2}:\sqrt{3}$, while the ratio for a face-centered lattice is $\sqrt{3}:2:2\sqrt{2}$. A comparison with the data for silver shows unambiguously that it has a face-centered cubic lattice, and the indexing of the diffraction lines follows from this conclusion.

Solids

The edge length, a, of the unit cell can now be calculated from each of the values of $\sin \theta$, and they all, of course, yield the same value for a. When a value of 0.3268 for $\sin \theta$ is substituted into Equation 12–3, we obtain

$$a = \frac{\sqrt{3} \times 1.541}{2 \times 0.3268} = 4.086 \text{ Å}$$

The distance between (111) planes, d_{111}, is then

$$\frac{4.086}{\sqrt{3}} = 2.359 \text{ Å}$$

A useful check on the correctness of a proposed structure is a comparison of the crystal density calculated for this structure with the measured density value. The density can be calculated from the mass contained in a unit cell and its volume. For a cubic lattice, if M is the molecular weight (or formula weight) of the substance in question and n is the number of molecules per unit cell, the mass per unit cell is $nM/\mathfrak{N}$, and the value of the density is given by

$$\rho = \frac{nM}{a^3 \mathfrak{N}} \tag{12–4}$$

To use this expression we must know the value of n. In a face-centered cubic lattice with atoms at the lattice points, each atom in the center of a face is shared between two unit cells. Since there are six such atoms their contribution to the unit cell in question is $6 \times \frac{1}{2} = 3$ atoms. Each corner atom is shared among eight unit cells. There are eight corner atoms and their contribution to the unit cell is $8 \times \frac{1}{8} = 1$ atom. Consequently $n = 3 + 1 = 4$. In contrast, for a primitive cubic lattice, $n = 1$ atom/unit cell, and for a body-centered cubic lattice $n = 2$ atoms/unit cell.

With the value of n for a face-centered lattice and the value of a found above for the silver lattice, the calculated density of silver is

$$\rho = \frac{4 \times 107.9}{(4.086)^3 \times 10^{-24} \times 6.023 \times 10^{23}}$$
$$= 10.50 \text{ g/cc}$$

This value is very close to the measured density of silver, a result which confirms the choice of a face-centered lattice for silver.

◊◊◊◊◊◊◊

If any four of the quantities in Equation 12–4 are known the fifth can be readily calculated. This equation was valuable in the early days of X-ray crystallography because at that time there was no method for the measurement of X-ray wavelengths. By finding the structure of NaCl from the observed pattern of its diffraction lines, W. L. Bragg was able to use the corresponding value of n and the measured density to calculate an approximate value of a for this crystal. With this value of a and the Bragg equation he was able to calculate fairly accurate wavelengths for the X-rays he used. Later, better wavelength values were obtained by the use of ruled diffraction gratings at grazing incidence.

When accurate values of λ became known, it was possible to obtain accurate values of a for a number of crystals. With these accurate values of a, Equation 12–4 was applied to the evaluation of Avogadro's number. For example, the diamond lattice is known to be cubic with 8 carbon atoms in the unit cell. From precise measurements the edge length of the unit cube is calculated to be 3.5668 Å. The density of diamond is 3.5184 g/cc and the atomic weight of carbon is 12.011. From these data[4] the calculated value of Avogadro's number is

$$\mathfrak{N} = \frac{8 \times 12.011}{(3.5668)^3 \times 10^{-24} \times 3.5184}$$
$$= 6.0236 \times 10^{23}$$

In this way some of the most accurate values of Avogadro's number have been obtained.

◈◈◈◈◈◈

In this section we have considered the application of the powder method to the simplest types of crystals, namely, those with cubic lattices having identical atoms at the lattice points. Such crystals actually form a very small fraction of all crystals since most crystals do not belong to the cubic system. In addition, for most crystals the crystal lattice does not consist of an array of atoms but rather an array of molecules or compound ions. The powder method is not adequate for the determination of the structure of these crystals and other, more complicated, methods must be used.

Determination of Crystal Structure by X-Ray Diffraction

The goal of the complete determination of the structure of a crystal is a statement of the location of each atom in the unit cell. These atomic locations are conveniently described by using the basis vectors of the space lattice as a coordinate system for the unit cell. As simple examples of this method for specify-

[4] T. Batuecas, *Nature*, **173**, 345 (1954).

ing the positions of atoms, we can use the cubic lattices discussed in the last section.

In the primitive cubic lattice we have seen that there is only one atom per unit cell, and we can take the location of this atom as the origin of the coordinate system; thus the coordinates of this atom are 0,0,0. In the body-centered unit cell there are two atoms, whose coordinates are 0,0,0 and $\frac{1}{2},\frac{1}{2},\frac{1}{2}$. Finally, in the face-centered unit cell there are four atoms, whose coordinates are 0,0,0, $\frac{1}{2},\frac{1}{2},0$, $\frac{1}{2},0,\frac{1}{2}$, and $0,\frac{1}{2},\frac{1}{2}$.

For these three cases the atomic coordinates are fixed by the lattice type. This is not true for any other lattice type, however, and the general problem of the determination of atomic coordinates is far from simple. A detailed discussion of this problem is beyond the scope of this book and our treatment here is limited to a brief description of the methods used for structure determination.

Because of the limitations of the powder method, mentioned above, single crystal techniques, in general, must be used to obtain diffraction patterns. These techniques usually involve rotation or oscillation of the crystal about selected axes. In this way the various lattice planes in the crystal are brought to their Bragg angles while the diffracted beams are photographed or measured with an X-ray detector. The experimental data consist of the angular positions and relative intensities of a large number, sometimes several thousand, of diffracted beams.

The structure of the crystal is then obtained by what is essentially a trial and error method. A reasonable structure is proposed on the basis of whatever information is available, and the theoretical diffraction pattern for this structure is calculated and compared with the experimental data. Although the agreement may be poor at first, by systematically varying the parameters of the structure the calculated diffraction pattern will eventually agree with the observed pattern in both the positions and the intensities of all of the diffracted beams.

When this result has been achieved, the locations of all atoms in the unit cell are known with a high degree of reliability. Then all the distances between atoms, whether the atoms are in the same molecule or not, are known and bond lengths and angles can be calculated. More data on bond lengths and angles have been obtained in this way than by all other techniques combined, and these data have been of great importance in the development of structural chemistry.

The scattering of X-rays by atoms, which is the prerequisite for X-ray diffraction, is basically the result of the interaction of the X-rays with the electrons of the atoms. The scattering power of atoms for X-rays depends on the number of electrons in the atom, and is roughly proportional to the atomic number. Consequently, in the diffraction pattern of a crystalline compound, the scattering from planes of atoms of relatively high atomic number predominates and the coordinates of heavy atoms can usually be determined with greater precision than those of light atoms. Conversely, the scattering power of hydrogen atoms is so low that their positions cannot be determined accurately and they often cannot be determined at all but must be inferred from the positions of the heavier atoms to which they are bonded.

The development of neutron diffraction has overcome this difficulty. Because neutrons are neutral they do not interact with electrons. They do, however, interact with atomic nuclei, and this interaction is about the same for all nuclei. As a result of this interaction, neutrons are scattered like waves whose length is the deBroglie wavelength (h/mv) and diffraction from a crystal lattice can occur. If a crystal contains hydrogen atoms (or protons) their contribution to the neutron diffraction pattern is comparable to that of other atomic nuclei.

For example, crystalline UH_3 contains H atoms combined with atoms of a very high atomic number. The contribution of the H atoms to the X-ray diffraction pattern of this crystal is completely masked by the scattering from the uranium atoms. In contrast, from the neutron diffraction pattern of UH_3 the positions of the H atoms also can be found and a complete structure determination can be carried out.

Types of Crystal Structures

In this section a few structures are discussed as representative of the many thousands that have been studied.

1. *Packed spheres.* Since Hooke proposed that all crystals are composed of identical spherical particles packed in a regular array, many people have considered the possible ways that spheres can be packed together. The various structures were well known before the development of X-ray crystallography made it possible to relate them to known crystal structures in which the lattice points are occupied by atoms.

One way to pack spheres is to start with a layer in the form of a square array (Fig. 12-12) and then add identical layers with each sphere directly above its nearest neighbor in the layer below. The unit cell of this structure is a primitive cube and each sphere is in contact with six nearest neighbors. This

Figure 12-12. Packed spheres in a simple cubic lattice.

structure is relatively open since only 52% ($\pi/6$) of the total volume is occupied by the spheres. No crystalline element has been found to have this structure which is the least favorable of all regular arrays of spheres for bonding interactions between the atoms.

Another possible method of packing spheres is obtained by starting with the same first layer as above and then placing a sphere in each of the hollows formed by sets of spheres in the bottom layers (Fig. 12-13). By adding other

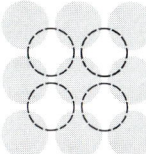

Figure 12-13. Two layers of a body-centered cubic lattice.

layers in the same way, a body-centered cubic lattice is formed in which each sphere has eight nearest neighbors and 68% of the total volume is occupied. About 20 of the metallic elements have this type of structure.

A body-centered cubic lattice of spheres does not fill space to the greatest extent possible. To obtain a structure that is packed as closely as possible, the first layer must be modified (Fig. 12-14) by surrounding each sphere with

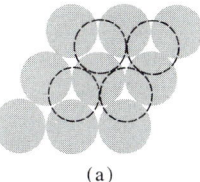 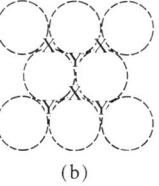

(a)   (b)

Figure 12-14. (a) First and second layers of packed spheres in cubic close-packing. (b) Two sets of trigonal depressions in second layer, one set marked by X, the other by Y. Occupation of one set by spheres of the third layer leads to hexagonal close-packing and occupation of the other set leads to a face-centered cubic lattice.

six nearest neighbors in the same layer. Then the second layer is formed by placing spheres in the trigonal depressions formed by three first-layer spheres in contact.

There are now two nonequivalent sets of trigonal depressions in the second layer, and the third layer must occupy one set or the other. In one case, the spheres of the third layer lie directly over the spheres in the first layer. Because this structure can be viewed as a set of hexagonal prisms in contact, it is called the hexagonal close-packed structure. When, on the other hand, the spheres of the third layer are placed in the other set of depressions, there are no spheres directly below the third-layer spheres. If the array is extended by placing the fourth-layer spheres directly over the first-layer spheres and continuing other layers analogously, the result is cubic close-packing for which the unit cell is a face-centered cube.

In each type of close-packed structure, hexagonal or cubic, each sphere has 12 nearest neighbors and 74% of the total volume is occupied by the spheres. Most metals crystallize in one of the close-packed structures as do the noble gases, which form face-centered cubic lattices.

2. *Simple ionic structures: a. Sodium chloride structure.* A crystal of NaCl was used by the Braggs in their first work on X-ray diffraction and the structure of this crystal was the first one determined by X-ray crystallography. A schematic drawing of the powder pattern of NaCl is shown in Figure 12–15.

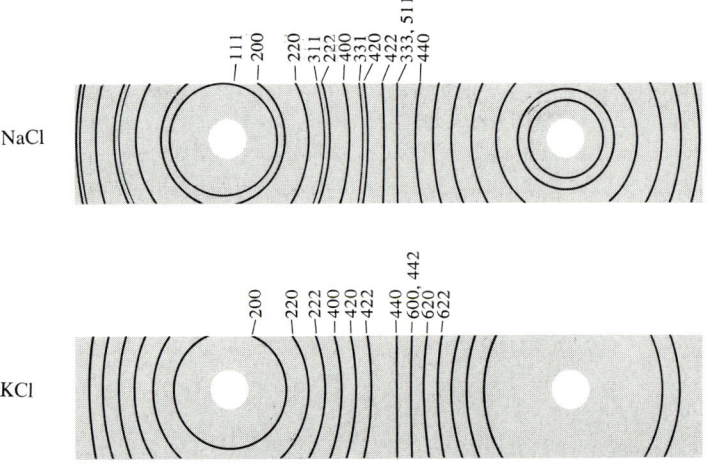

Figure 12–15. X–ray powder patterns of NaCl and KCl.

The indicated indexing of the diffraction lines, which leads to a single value for the edge length of the unit cube, is consistent with a face-centered cubic lattice. From the relative intensities of the lines it can be concluded that the NaCl structure consists of two interpenetrating face-centered lattices with each ion of a given kind surrounded by 6 ions of the other kind as nearest neighbors, as shown in Figure 12–16.

In the NaCl crystal lattice the (111) planes contain only Na^+ ions or Cl^- ions. At the Bragg angle for which the diffracted beams from the Na^+ ion planes

Solids

alone would reinforce each other, they are exactly out of phase with the beams from the Cl⁻ ion planes and destructive interference occurs. If the two kinds of ions had the same scattering power for X-rays the interference would be complete and there would be no (111) diffraction line. Because the Cl⁻ ion has a larger number of electrons than the Na⁺ ion, the interference is not complete and a weak (111) line results. At the angle corresponding to the (222) diffraction line, the beams from the Na⁺ planes and the Cl⁻ planes are in phase and a strong (222) line is observed. The weakness of the (333) line and the relative strength of the (444) line are accounted for on the same basis.

It is interesting to compare the powder pattern of NaCl with that of KCl, which is also shown in Figure 12-15. At first glance the pattern for KCl looks like that of a primitive cubic lattice since there are six lines followed by a gap. Actually, however, KCl has a face-centered cubic lattice like that of NaCl. The resemblance to a primitive lattice is a consequence of the fact that K⁺ ions and Cl⁻ ions have the same number of electrons and thus have the

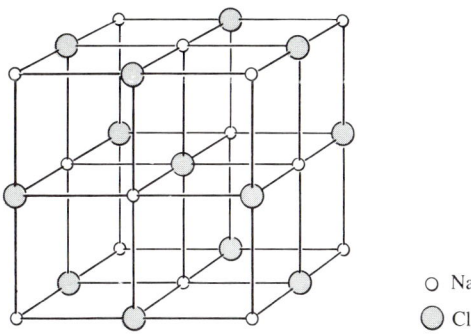

Figure 12-16. Structure of NaCl.

same scattering power for X-rays. The unit cell of the apparently primitive lattice formed by considering K⁺ and Cl⁻ ions as identical has half the edge length of the actual face-centered unit cell, and the indexing of the KCl lines in Figure 12-15 is consistent with this structure. Incidentally, the resemblance of the KCl diffraction pattern to that for a primitive cubic lattice is direct evidence that the scattering centers are ions and not atoms.

Many ionic compounds with the general formula AB form crystals with the face-centered NaCl structure.

b. Cesium chloride structure. In the series K⁺, Rb⁺, Cs⁺ the number of filled electron shells and the radius of the ion increases. Although RbCl has the NaCl structure, CsCl has a structure that can be described as two interpenetrating primitive lattices with each ion of one kind surrounded by eight nearest neighbors of the other kind at the corners of a cube. The change in lattice type between RbCl and CsCl is attributed to the fact that the ratio of the radius of the positive ion to that of the negative ion is larger in CsCl than in RbCl.

3. *Diamond structure.* The diffraction pattern of diamond indicates, as mentioned above, a face-centered cubic lattice with 8 atoms per unit cell. This structure can be pictured as consisting of two interpenetrating face-centered lattices (Fig. 12-17) in which the unit cell has 8 atoms at the corners of the cube,

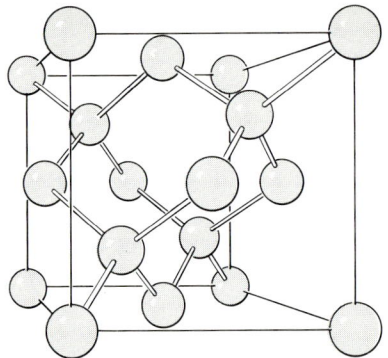

Figure 12-17. Face-centered cubic unit cell of the diamond lattice.

6 atoms in the face centers, and 4 atoms within the cube. Each C atom is surrounded by 4 C atoms at tetrahedral angles. The C–C distance is 1.54 Å, the same as that found in ethane and all other saturated hydrocarbons.

All of the other Group IV elements, except Pb, crystallize with the diamond structure, which is also found in SiC, one form of ZnS, and AlP and other compounds of Group III and Group V elements. In each of these compounds, each atom of one kind is surrounded by four of the other kind at tetrahedral angles.

4. *Layered structures.* Many crystals are known whose structures consist of planar networks, or sheets, of atoms with relatively large distances between the sheets. For example, graphite has this type of structure. The graphite lattice consists of sheets of fused hexagons of C atoms (Fig. 12-18), with each atom bonded to three others with an interatomic distance of 1.42 Å. This distance is close to the C–C bond length in aromatic hydrocarbons. The distance between the sheets of C atoms is found to be 3.35 Å, which is considerably larger than the distance between chemically bonded C atoms. A number of the silicate minerals related to mica have a similar type of layered structure. Crystals with this structure can be cleaved relatively easily along planes parallel to the layers.

Solids

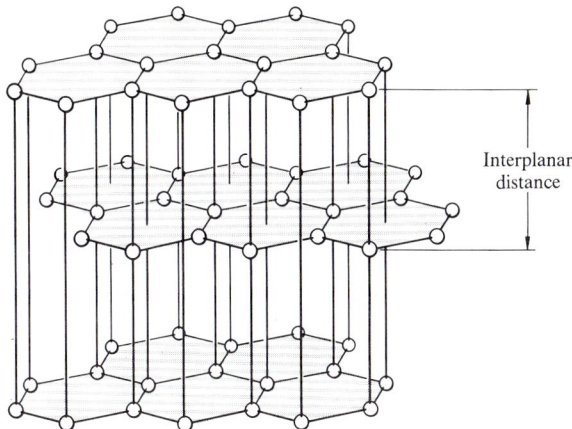

Figure 12-18. Graphite lattice with layered structure.

Physical Properties of Crystalline Substances

Before we turn to the interpretation of some of the properties of crystals, we first ask a basic question. Why does a substance at equilibrium at a certain temperature and external pressure exist in the form of a crystal with a particular type of structure? Thermodynamics supplies a general answer to this question: the Gibbs free energy of this structure must be lower than that for any other type of structure or state under the given conditions. This answer, however, is so general that it is not very informative.

From a statistical point of view the central problem in accounting for the stability of a crystal is the evaluation of its partition function, for we have seen that all equilibrium properties can be calculated from the partition function. The molecules in a crystal, however, interact with each other so strongly that the partition function depends on the quantum states of the crystal as a whole. To obtain information about these quantum states requires in principle the solution of the Schrödinger equation for this crystal. Even the writing of this equation in complete form cannot be done at present. Although some progress has been made in the development of partition functions for simplified models of crystal lattices, it is not yet possible to predict on a theoretical basis the type of lattice in which a particular substance will crystallize.

In view of the complexity of the general approach to the stability of crystals, we turn to the more modest task of accounting for some of the observed properties of crystals, assuming that we know their structure. Historically, the first property to be reasonably well understood is the heat capacity, whose study played an important role in the development of our knowledge of the solid state even before the discovery of X-ray diffraction.

Heat Capacity of Crystals

Early in the nineteenth century, the limited experimental data available at the time for the heat capacity of solid elements, mostly metals, were summarized by Dulong and Petit in the following generalization: the heat capacity of a solid element is 6 cal/deg per gram atomic weight. This law was helpful in the determination of the atomic weights of some of the elements. It was soon found, however, that some of the solid elements, notably carbon in the form of diamond, have atomic heat capacities considerably less than 6 cal/deg at room temperature. Later it was found that the atomic heat capacities of all elements when measured at sufficiently low temperatures are below the Dulong and Petit value and, in fact, approach a value of 0 as $T \to 0°K$.

Boltzmann was able to account for the value of 6 cal/deg on the basis of a simple model for a crystalline element and the law of equipartition of energy (Chapter 5, p. 153). Consider a system which consists of Avogadro's number of atoms arranged in a 3-dimensional lattice. Because of the rigidity of the crystal the possibility of translational motion of the atoms can be excluded. Since there is no evidence for rotational motion, the only remaining type of motion is vibration. The crystal can then be considered as a set of $\mathfrak{N}$ identical 3-dimensional harmonic oscillators, each vibrating about a fixed equilibrium position. The restoring force for a particular atom results from its interaction with its neighbors.

For the motion of one atom in one direction, the Hamiltonian function is the harmonic oscillator function which we have seen before (Eq. 6–17). According to the law of equipartition of energy the average kinetic energy associated with this motion is the same as the average potential energy, and each is equal to $kT/2$. Consequently for the $3\mathfrak{N}$ vibrational degrees of freedom the total energy is $3\mathfrak{N}kT = 3RT$, and the atomic heat capacity is $3R$ which is approximately 6 cal/deg.

Although this classical theory yields the law of Dulong and Petit, it can account neither for any heat capacity value less than $3R$ nor for the fact that the heat capacity of a crystal is a function of the temperature. The difficulty here is closely related to the difficulty with the classical theory for the heat capacity of a diatomic gas (Chapter 5, p. 153). It was resolved in an analogous manner when Einstein in 1907 recognized that the classical theory was not applicable to this oscillator model for a crystal. He made use of the quantum theory to derive a vibrational partition function which is essentially the same as Equation 9–36. With the assumption that all of the harmonic oscillators have the same frequency, ν, he obtained the following molar vibrational partition function for this model of a crystal:

$$Q = (1 - e^{-\frac{h\nu}{kT}})^{-3\mathfrak{N}} \qquad (12-5)$$

Solids

From this partition function the vibrational contributions to all of the thermodynamic properties of the model can be calculated, as we have seen before. In particular, the energy is (Eq. 9–35)

$$E - E_o = \frac{3h\mathfrak{N}\nu}{e^{\frac{h\nu}{kT}} - 1}$$

and the heat capacity at constant volume is (Eq. 9–37)

$$C_V = 3R \left(\frac{h\nu}{kT}\right)^2 \frac{e^{\frac{h\nu}{kT}}}{(e^{\frac{h\nu}{kT}} - 1)^2} \qquad (12\text{--}6)$$

This expression reduces to $C_V = 3R$, the classical value, for temperatures so high that $h\nu \ll kT$. At sufficiently low temperatures, $h\nu \gg kT$ and $C_V \to 0$ as $T \to 0°K$.

◊◊◊◊◊◊◊

To obtain these limiting values let us represent $h\nu/kT$ by x; then

$$C_V = 3Rx^2 \frac{e^x}{(e^x - 1)^2}$$

$$= \frac{3Rx^2 e^x}{e^{2x} - 2e^x + 1}$$

When the numerator and denominator are divided by e^x we have

$$C_V = \frac{3Rx^2}{e^x + e^{-x} - 2}$$

If x is small relative to 1, $e^{\pm x} = 1 \pm x + \frac{x^2}{2} + \cdots$. When this expression is substituted for $e^{\pm x}$ in the denominator, its value becomes x^2 and it follows that $C_V = 3R$. When x is large relative to 1, e^x is so large that $e^{-x} - 2$ in the denominator is negligible; also, as x increases, e^x increases more rapidly than x^2 and C_V approaches 0 as a limit.

◊◊◊◊◊◊◊

In order to compare the theoretical values of the heat capacity calculated from Equation 12–6 with experimental data for a particular crystalline substance,[5] it is necessary to select an appropriate value for the frequency ν, which must be treated as an adjustable parameter. A value of ν is chosen for which the theoretical curve of heat capacity versus temperature has the closest fit to the experimental points. As shown schematically in Figure 12–19, the portion of

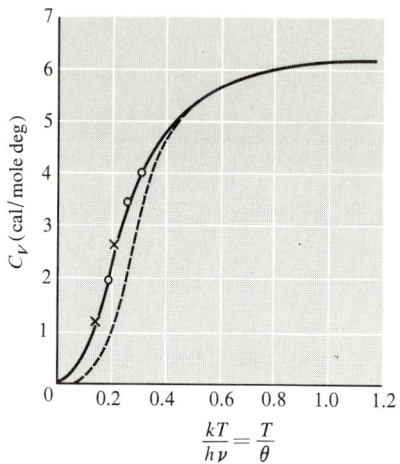

Figure 12–19. Heat capacity of crystalline solids as a function of the temperature. The solid line is obtained from the Debye equation (Eq. 12–8) and the dotted line is the graph of the Einstein equation (Eq. 12–6).

the curve corresponding to higher temperatures can be fitted quite well but the deviations in the lower-temperature portion far exceed experimental errors. This discrepancy is not surprising because there is no reason to believe that the oscillators all have the same frequency, and thus the Einstein model is oversimplified.

A few years after the publication of Einstein's work on crystals, Debye in 1912 proposed a more complicated model of a crystal which led to improved results. In this model there is a continuous distribution of vibrational frequencies from zero up to a maximum value that is characteristic of a particular crystalline substance. The equation derived by Debye for the heat capacity of a crystal reduces for very low temperatures to the following simple form:

$$C_V = \alpha T^3 \qquad (12\text{–}7)$$

[5] A careful comparison must take into account the fact that the experimental heat capacities are measured at constant pressure, while the theory gives an expression for the heat capacities at constant volume. The difference between these heat capacities is given by the thermodynamic equation (Eq. 3–28)

$$C_P - C_V = T \left(\frac{\partial P}{\partial T}\right)_V \left(\frac{\partial V}{\partial T}\right)_P$$

This difference is usually small for crystals and it is negligible at low temperatures.

Solids

in which α for each substance is a characteristic constant which must be calculated from experimental data. The Debye T^3 law has proved to be of great importance because, as we have previously seen (Chapter 4, p. 112), it is used to extrapolate C_V values to 0°K in the calculation of molar entropies by use of the third law of thermodynamics.

◈◈◈◈◈◈◈

We do not present here a complete derivation of the Debye equation for the heat capacity of crystals but only outline the ideas on which it is based. To introduce these ideas let us consider a 1-dimensional (linear lattice) model because it shows some of the main features of a 3-dimensional model, while permitting easy visualization. Our model (Fig. 12–20) consists of a large

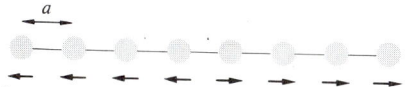

Figure 12–20. Linear lattice with equilibrium interatomic distance a. The normal mode of vibration with the lowest frequency is indicated schematically.

number, n, of atoms arranged with uniform interatomic distance, a, along a line which we can take as the x axis. Between each adjacent pair of atoms there is a massless spring with force constant k, and the center of gravity of the model is fixed in space.

From a physical point of view this 1-dimensional model for a crystal is equivalent to a giant linear molecule that contains n atoms. As we have seen in Chapter 8, the vibrational motion of this model can be described in terms of a set of n normal modes of vibration. (Strictly speaking, there are $n-2$ normal modes because we exclude translation and rotation of the model, but the error is insignificant when n is very large.)

Each of the normal modes of vibration has a different frequency. In the normal mode with the lowest frequency we can imagine that half of the atoms vibrate as a unit (all atoms in phase) against the other half. This type of vibration is equivalent to a longitudinal acoustic vibration with wavelength na. The remaining normal modes have successively higher frequencies, corresponding to shorter wavelengths, up to a maximum frequency, ν_m, for a normal mode in which each atom vibrates against its nearest neighbors.

We can obtain the energy of this model by expressing the average energy corresponding to each frequency and then summing over the whole set of frequencies for the normal modes. Because there are a large number of frequencies in the range from 0 to ν_m, the differences between successive frequencies are very small and the summation can be replaced by an integration over the frequency range. In order to do this, however, it is necessary to find the distribution function $g(\nu)$ for the frequencies of the normal modes; that is, the number of normal modes with frequencies between ν and $\nu + d\nu$. Then the integration can be carried out and the resulting energy expression can be differentiated with respect to the temperature to obtain C_V.

For the actual, 3-dimensional crystal lattice the problem of calculating the frequency distribution of the normal vibrations is very complicated. Debye assumed that the distribution function is the same as the one which had been derived for the elastic vibrations of a continuous solid medium. The number of such vibrations with frequency ν is given by the following expression:

$$g(\nu)\, d\nu = \frac{12\pi V}{c^3} \nu^2\, d\nu$$

in which V is the volume of the solid and c is an average of the velocities of transverse and longitudinal waves in the solid. Debye obtained an expression for the maximum frequency ν_m by setting the integral over the distribution function from 0 to ν_m equal to the total number of normal modes of vibration, $3\mathfrak{N}$; thus

$$\int_0^{\nu_m} \frac{12\pi V}{c^3} \nu^2\, d\nu = 3\mathfrak{N}$$

It follows that

$$\frac{12\pi V}{c^3} \frac{\nu_m^3}{3} = 3\mathfrak{N}$$

and

$$\frac{12\pi V}{c^3} = \frac{9\mathfrak{N}}{\nu_m^3}$$

With the use of this equality the distribution function can be expressed in terms of ν_m as follows:

$$g(\nu)\, d\nu = \frac{9\mathfrak{N}}{\nu_m^3} \nu^2\, d\nu$$

We can now calculate the total vibrational energy of the crystal by multiplying the average energy corresponding to a particular frequency (Eq. 9–35) by the number of normal modes that have this frequency and integrating this product over the range of frequencies from 0 to ν_m. Thus we obtain the following equation:

$$E - E_o = \frac{9\mathfrak{N}}{\nu_m^3} \int_0^{\nu_m} \frac{h\nu}{e^{h\nu/kT} - 1} \nu^2\, d\nu$$

In a similar way by use of the expression for the contribution to the heat capacity corresponding to a single frequency (Eq. 9–37) we can obtain the following equation for the total heat capacity:

$$C_V = \frac{9\mathfrak{N}k}{\nu_m^3} \int_0^{\nu_m} \left(\frac{h\nu}{kT}\right)^2 \frac{e^{h\nu/kT}}{(e^{h\nu/kT} - 1)^2} \nu^2 \, d\nu \tag{12-8}$$

This equation can be written in a slightly simpler form by using x as a symbol for $h\nu/kT$ and introducing a parameter which is known as the Debye characteristic temperature, defined by the relation $\theta_D = h\nu_m/k$. The upper limit of x is $h\nu_m/kT$ which is equal to θ_D/T. When these substitutions are made, Equation 12–8 becomes

$$C_V = 9R \left(\frac{T}{\theta_D}\right)^3 \int_0^{\theta_D/T} \frac{x^4 e^x}{(e^x - 1)^2} \, dx$$

The integral in the above equation cannot be evaluated exactly but it can be accurately approximated by numerical or graphical methods. When this is done, the values of θ_D for a number of simple crystals are found by fitting the theoretical heat capacities to experimental data. Then, as predicted by the above equation, the measured heat capacities plotted as a function of T/θ_D all fall on the same graph line. The success of this treatment can be seen from Figure 12–19 which includes data for elements as different as carbon (diamond) with $\theta_D = 1890°K$ and lead with $\theta_D = 90°K$.

As is evident from the graph the limiting value of C_V for $T \gg \theta_D$ is $3R$. The limiting form for the equation for C_V at temperatures much lower than θ_D can be shown to be

$$C_V = \frac{12}{5} \pi^4 R \left(\frac{T}{\theta_D}\right)^3$$

Because all quantities on the right-hand side except the temperature are constants for a particular crystal, we can write this equation in the form of Equation 12–7, $C_V = \alpha T^3$, with $\alpha = 12\pi^4 R/5\theta_D^3$. This equation is in excellent agreement with experimental data at temperatures lower than θ_D, even for crystals which are more complex than those of the elements, but the agreement is significantly less good at higher temperatures. The deficiencies of Debye's theory of the heat capacity of crystals result from the fact that it replaces the actual distribution of vibrational frequencies by an oversimplified distribution function which is obtained by ignoring the structure of crystals.

Other Properties of Crystals

Properties of crystals other than the heat capacity are discussed most conveniently in terms of types of crystals, with the basis of classification being the kind of bonding that holds the "particles" together in the crystal lattice. We will distinguish the following five types of crystals: (1) metallic, (2) ionic, (3) covalent, (4) molecular, and (5) hydrogen-bonded. (This classification is not exact and many crystals belong to more than one type.) At this point we discuss some of the characteristic properties of each of these types.

1. *Metallic crystals.* The outstanding properties of metallic crystals are their high electrical and thermal conductivities and their reflectance of light. We have seen that metals, with few exceptions, crystallize in a close-packed or body-centered structure with 12 or 8 nearest neighbors, respectively. The existence of a structure that can be represented as an array of packed spherical particles implies that the forces between the particles are independent of angles and therefore depend only on the distance between the particles. Also basic for the understanding the properties of metals is the fact that the atoms of typical metallic elements all have a relatively small number (usually 1 or 2) electrons in their valence shell orbitals.

These facts and many others have led to a model of a metallic crystal as an array of spherical positive ions immersed in a "sea" of electrons whose wave functions extend throughout the crystal. In a real sense we can say that the entire metallic crystal is a giant molecule, and it is useful to think of the electrons as occupying delocalized molecular orbitals.

To make this point of view a little more definite, let us use the simplified model of a linear lattice and consider how such a lattice might be formed from atoms of a typical metal such as sodium. Two atoms of Na which approach each other to a sufficiently small distance begin to interact when their $3s$ atomic orbitals overlap significantly (Chapter 7, p. 232). As a result of this interaction molecular orbitals are formed which can be expressed approximately as linear combinations of the two $3s$ atomic orbitals. From the two atomic orbitals two molecular orbitals can be constructed, one bonding, the other antibonding, each of which can be occupied by two electrons of opposite spin. In this way, as we have seen before, we can account for the existence of stable Na_2 molecules in the gas phase.

Let us next imagine that three Na atoms are placed on a line with their separations about the same as that between the Na nuclei in a crystal of Na. (That these three atoms do not form a stable structure need not be of concern.) In this case three molecular orbitals can be formed from the combination of the three $3s$ atomic orbitals (Fig. 12–21). Now let us imagine that n Na atoms, with n of the order of Avogadro's number, are arranged on a line with uniform spacing. We can say that n decentralized molecular orbitals are formed, of which half are bonding and the other half are antibonding.

Solids

Each of the n molecular orbitals is associated with a different energy level. Because of the very large number of energy levels for our linear lattice, the difference between any two successive levels is infinitesimal and the entire set of n energy levels is said to form an *energy band*. (There is a rough analogy with the translational energy levels of a mole of ideal gas. In that case also we saw

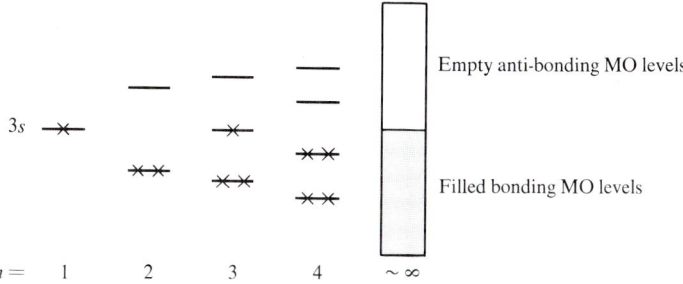

Figure 12–21. Schematic representation of energy levels of a linear lattice of n Na atoms.

(Chapter 9, p. 348) that the energy levels lie so close together that they form a practically continuous energy band.)

When we turn from the linear lattice model to the stable 3-dimensional structure of an actual crystal, the concept of energy bands retains its validity. The band model for the energy levels of a crystal underlies much of the theoretical background of solid state physics. Although we arrived at this model by way of a molecular orbital approach, it can be developed from other points of view which we do not discuss here.

We have mentioned the overlap of the Na 3s orbitals. There are also, however, electrons in 1s, 2s, and 2p atomic orbitals. Because the 1s orbitals are concentrated near the Na nucleus and the electrons in these orbitals are partially shielded from external influences by the electrons in the 2s shell, the 1s orbitals and the corresponding energy levels are affected only slightly by the formation of the crystal. The 2s and 2p orbitals are affected to a greater extent, and the sharp energy levels of the isolated Na atoms are broadened in the crystal into narrow energy bands which are separated by gaps in which there are no energy levels.

Let us now picture what happens if we add n electrons to an array of Na$^+$ ions arranged as in the Na crystal. The electrons will begin to occupy the molecular orbitals of lowest energy, corresponding to the bottom of the band of energy levels. According to the Pauli exclusion principle, two electrons will go into each molecular orbital. This process will continue until the n electrons occupy completely the lowest $n/2$ orbitals, leaving the upper half of the energy band unfilled. This is shown schematically in Figure 12–22a.

We can now account for the high electrical conductivity of Na in the following way: The flow of current in an electrical conductor under the influence of a small potential difference between the ends of the conductor implies that

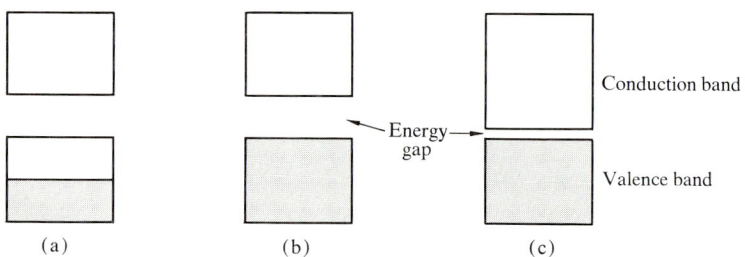

Figure 12–22. Schematic representation of energy bands of (a) a typical metal with the lowest band half-filled, (b) a typical insulator with the filled lower band separated from the conduction band by a relatively large energy gap, (c) an intrinsic semiconductor with an energy gap so small that a few electrons can make a transition from the filled band to the conduction band.

the electrons which carry the current have gained both some momentum and kinetic energy from the electric field. In order for an electron to increase its energy it must make a transition from one energy level to another. In a crystal of Na the electrons whose energy levels lie significantly below the top of the filled portion of the energy band cannot absorb energy from the field because they cannot make transitions to slightly higher energy levels since these higher levels are already occupied by other electrons.

The electrons initially in levels at the top of the filled part of the energy band, however, can make transitions to the unfilled levels which lie directly above the filled levels. As a result of their absorption of energy, these electrons are accelerated by the electric field and move through the crystal, thus forming an electric current.

In contrast, the band structure for a crystal which is nonmetallic and an electrical insulator is shown in Figure 12–22b. In this case there are enough electrons to fill all the available levels of the lowest band, often called the valence band, and there is a considerable difference in energy between the top of this band and the bottom of the lowest unfilled band, the conduction band.

Consequently, no low-energy transition at all can occur, and there is no electric current when a small potential difference is applied to the crystal.

Let us next consider a Group II metal, such as Mg, which has two $3s$ electrons per atom. In this case, for a crystal containing n atoms there are n molecular orbitals formed from $3s$ atomic orbitals. Since these molecular orbitals are completely filled by the $2n$ valence electrons, it appears that Mg should be an insulator. Detailed studies for this case, however, have shown that the band of energy levels formed from combinations of the $3p$ atomic orbitals partially overlaps the band formed from the $3s$ orbitals. Consequently, electrons at the top of the $3s$ band can make transitions into the overlapping $3p$ band and so contribute to the electric current. Thus Mg is an electrical conductor and it is a metal, although its properties and those of the other Group II elements are not as typically metallic as those of the Group I elements or those of Cu, Ag, or Au whose atoms have a single s electron outside filled electron shells.

We do not discuss the other properties of metals beyond pointing out that (1) the high thermal conductivity and reflectance can be accounted for on the basis of the band model sketched above; and (2) the mechanical properties of metals (tensile strength, malleability, and ductility) can be interpreted as a consequence of the close-packed structures of metallic lattices.

A few supplementary remarks may be of interest. For all metals the electrical resistance decreases (and the conductance increases) with falling temperature. For a perfect single crystal, as the temperature approaches 0°K the theoretical value of the resistance approaches zero. Any change that disturbs the perfect periodicity of the lattice raises the resistance. As the temperature is raised, the increasing vibrational amplitude of the positive ions tends to lower the lattice periodicity. Introduction of foreign atoms, as in the formation of an alloy, has the same effect. Likewise, any defect in the structure of the crystal increases the resistance.

Another approach to the theory of metals ignores the periodicity of the lattice and treats the electrons in a metal as a gas with wave functions obtained by analogy with an ideal gas from the solutions of the Schrödinger equation for the particle in a box. The electron gas, however, does not follow Maxwell-Boltzmann statistics but Fermi-Dirac quantum statistics. Although the electron gas model is crude, it has been of use in understanding some of the properties of metals.

2. *Covalent crystals.* Covalent crystals are usually brittle and hard, and they have high melting points. Some of them, diamond and quartz in particular, are very good electrical insulators.

Let us use diamond as an example of a covalent crystal. In the description of the diamond structure (p. 468) it was pointed out that each C atom has four nearest neighbors at tetrahedral angles and that the C–C distance is the same as in ethane. It can be concluded that the bonding of the C atoms in diamond is quite similar to the bonding in a saturated chain of C atoms. We can describe this bonding by saying that a diamond crystal is a giant molecule in which the C atoms have sp^3 hybridization, and the bonds are formed by overlap of these tetrahedral hybrids to form localized bonding molecular orbitals, each occupied by two electrons. (In other language, we can say that covalent bonds are formed between C atoms by shared pairs of electrons.)

The description above implies the major differences between metallic and covalent crystals. In covalent crystals there are chemical bonds between atoms and there are definite angles between these bonds. All valence electrons are bound tightly to relatively small regions between the bonded atoms. In terms of the energy band model, for a covalent crystal there is a gap between the valence band and the conduction band. For diamond this energy gap is quite large (5.33 ev) which is consistent with the fact that diamond is one of the best electrical insulators known. It is also consistent with the fact that a diamond without impurities is transparent in the visible and near ultraviolet regions of the spectrum.

For the other Group IV elements, which have the diamond structure, the energy gap decreases with increasing number of filled electron shells as follows: Si, 1.14 ev; Ge, 0.67 ev; and Sn, 0 ev. The case of Ge is particularly interesting because it is a semiconductor. This means that at ordinary temperatures it has a measurable conductance which, however, is far less than that of any of the typical metals. As is characteristic of semiconductors, the conductance of Ge increases with rising temperature, in contrast to metallic conductance. This behavior can be attributed to the fact that the energy gap is so small (Fig. 12–22c) that at moderate temperatures some electrons can acquire enough thermal energy to make the transition from the valence band to the conduction band, and thus contribute to an electric current. The fraction of the electrons with the required energy increases with rising temperature and consequently the conductance increases.

Pure Ge is an example of an intrinsic semiconductor; that is, a pure substance which is a semiconductor. The addition of an impurity has a marked effect on the conductance of a semiconductor, especially if the impurity atoms have a number of valence electrons that is different from that of the major component. In effect, the presence of the impurity introduces additional energy levels into the gap between the valence and the conductions bands and thus raises the conductance. Crystals of Ge and Si which have been "doped" with traces of Group III or Group V elements are widely used in transistors and other electronic devices.

3. *Ionic crystals.* Ionic crystals are good insulators with fairly high melting points. They are easily fractured or cleaved along lattice planes. We have seen that simple ionic crystals, such as those of the alkali halides, belong to the cubic system. An ionic crystal can be thought of as a giant molecule in which

there are no specific bonds between ions and thus no definite bond angles. The ions behave much like charged spheres which are packed in a way that maximizes the electrostatic interaction while maintaining over-all electrical neutrality.

The electrostatic interaction between ions obeys Coulomb's law. Because of the simplicity of this law, ionic crystals were the first for which the bonding was understood. Even before the development of quantum mechanics, the *cohesive energy* of ionic crystals could be calculated. The cohesive energy of a crystal is defined as the energy required to separate the crystal into a set of particles (atoms, molecules, or ions, depending on the type of crystal) which are so far apart that they exert negligible forces on each other. This energy is related to the sublimation energy of the crystal. Its value depends on the intermolecular forces holding the particles in the crystal lattice.

The ideas underlying the calculation of the cohesive energy of ionic crystals are now outlined. (For details, see page 485.) For an ionic lattice to be in stable equilibrium, the net attractive force must be balanced by a repulsive force between neighboring ions due to the interaction of filled shells of electrons. The small compressibility of a crystal shows that this repulsive force, and the potential energy associated with it, must increase very rapidly as the separation of the ions decreases below its equilibrium value.

The expression for the potential energy of two singly-charged ions of opposite sign at an interionic distance r, considering their energy at infinite separation as zero, consists of two terms. One of these is the coulombic term, $-\epsilon^2/r$, with ϵ representing the charge. The negative sign indicates that this term corresponds to an attractive force between the ions. The second term, which represents the repulsion energy mentioned above, may be written as a large inverse power of r of the form br^{-n}. (Compare the Lennard–Jones potential energy function, Eq. 10–9.) The larger the value of n, the more closely does the ion approach a rigid sphere model.

Having an expression for the potential energy of a pair of ions, we can calculate the potential energy for an actual crystal by summing our expression over all pairs of ions for which the interaction is significant. Because the repulsion term decreases very rapidly with increasing values of r, this term is summed only for those pairs of ions that are nearest neighbors.

To illustrate the procedure just described, we use a linear lattice consisting of ions of alternating sign separated by a constant distance a (Fig. 12–23).

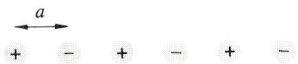

Figure 12–23. Linear ionic lattice.

The potential energy of any particular ion in the lattice is given by a sum of terms which includes the attraction and repulsion contributed by its nearest neighbors, $2(-\epsilon^2/a + ba^{-n})$, plus the coulombic repulsion contribution $(2\epsilon^2/2a)$ of its two next nearest neighbors with the same sign, plus the attraction contribution $(-2\epsilon^2/3a)$ of its next, next nearest neighbors, and so on. Thus the total potential energy, U, of a single ion in a linear lattice is given by

$$U = 2[ba^{-n} - \frac{\epsilon^2}{a}(1 - \tfrac{1}{2} + \tfrac{1}{3} - \tfrac{1}{4} + \cdots)]$$

The total potential energy of the entire lattice is obtained by multiplying this expression by the number of ions in the lattice, neglecting the fact that the contributions of the few ions near the ends of the lattice are slightly modified.

In an actual crystal whose structure has been determined, the relative position of each of the ions is known. For example, in a face-centered cubic lattice each ion has six nearest neighbors of opposite sign at a distance of $a/2$ and 12 next nearest neighbors of the same sign at a distance of $a/\sqrt{2}$, and so on, where a is the edge length of the unit cell (Fig. 12–16). Thus a potential energy expression for each type of lattice can be written. In order to evaluate this expression it is necessary to find the values of the constants b and n.

The value of n can be found, or at least approximated, by deriving a relationship between n and the coefficient of compressibility, which has been measured for many crystals. The values of n calculated from this relationship for the alkali halides are in the range 8–13 and tend to increase with the size of the alkali ion. These values are in the same range as those calculated for the repulsive forces between gas molecules (Chapter 10, p. 387).

To obtain the value of b we make use of the requirement that for the lattice at equilibrium the potential energy must be a minimum with respect to change in lattice dimensions. This means that $dU/da = 0$. Accordingly the potential energy function for the lattice is differentiated with respect to a, the resulting function is equated to zero, and this equation is solved for b.

It is interesting to note that the value of b calculated in this way for KCl is fairly close to the value it has for the interaction of two argon atoms. This is to be expected because the K$^+$ and Cl$^-$ ions each have the same number of electrons as the argon atom, and the repulsive forces between nearest neighbors in the KCl crystal and the atoms in argon gas are due to interactions between the same types of filled electron shells.

With reasonable values of n and b known, the potential energy change can be calculated for the formation of a mole of compound in the form of an ionic crystal, from $\mathfrak{N}$ positive and $\mathfrak{N}$ negative ions, all of which are initially far apart. This energy change is taken to be numerically equal to the molar cohesive energy, E_c, of the ionic crystal.

A comparison of the calculated value of the cohesive energy for a crystal with an experimentally determined value of this energy would be highly desirable, but this would require a measurement of the energy required for the process, represented by MX(c) → M⁺(g) + X⁻(g), in which a mole of a crystalline compound of the metal M and the halogen X is converted into a gas consisting of M⁺ ions and X⁻ ions at such large interionic distances that the electrostatic forces between the ions are negligible. No method is known for carrying out such a measurement directly.

Born and Haber, however, devised an independent though indirect way of calculating the cohesive energy of a crystal by considering a series of processes whose net effect is the desired process mentioned above. The processes used by Born and Haber form a thermodynamic cycle which can be represented in the following way:

		Energy change
1.	M⁺(g) + X⁻(g) → MX(c)	$E_c \; (= -\Delta U)$
2.	MX(c) → M(c) + ½X₂(g)	$-\Delta H_{form}$
3.	M(c) → M(g)	ΔH_{sub}
4.	½X₂(g) → X(g)	ΔH_{dis}
5.	M(g) → M⁺(g) + $\mathfrak{n}\epsilon$	I_M
6.	X(g) + $\mathfrak{n}\epsilon$ → X⁻(g)	A_X

The first step is the reverse of the process mentioned above, whose energy we want to find. Step 2 is the reverse of the formation of a mole of crystalline MX from the elements in their stable states. Step 3 is the sublimation of a mole of crystalline M and step 4 is the dissociation of half a mole of gaseous X_2 into atoms. The energy of ionization of M atoms, I_M, in step 5 is called the ionization potential, and the energy of attachment of electrons to X atoms in step 6 is called the electron affinity, A_X.

Because these processes form a complete cycle, it follows from the first law of thermodynamics that the sum of all of the energy changes must equal zero. Consequently we can calculate the energy change for step 1 if we can evaluate the energy changes for the other five steps. The energy changes for steps 2, 3, and 4 can be found from calorimetric measurements or by less direct methods. The ionization potential and the electron affinity can be obtained from spectroscopic measurements. Thus it is possible to compare theoretical values of the cohesive energy with values calculated from the Born-Haber cycle.

Table 12–3 presents a comparison for a few of the crystals whose cohesive energies have been evaluated by both methods. It can be seen that the values obtained by these completely different methods are in good agreement for crystals that are completely ionic, such as the alkali halides. In those cases, such as the silver halides, for which there is other evidence for the existence of a significant covalent contribution to the cohesive energy, the theoretical values are lower than those calculated from the Born-Haber cycle.

4. *Molecular Crystals.* As a general rule, molecular crystals have low

melting points and relatively high vapor pressures. They are often soft and easily deformable. Nearly all crystalline organic compounds are of this type. All of the evidence available is consistent with a structure in which molecules containing normal chemical bonds are held in a lattice by relatively weak van der Waals forces. X-ray crystallography and absorption spectroscopy show that the interatomic distances and molecular vibration frequencies are modified only slightly by the formation of a molecular crystal.

TABLE 12-3   LATTICE ENERGIES OF IONIC CRYSTALS (KCAL/MOLE)*

Crystal	Born-Haber Cycle	Theoretical Calculation
LiF	241.2	242.0
LiCl	198.2	200.2
NaCl	183.8	184.4
KCl	166.8	167.5
CaF_2	624	622
AgCl	216	199
Ag_2O	714	585

*Data quoted by T. C. Waddington, *Advances in Inorganic Chemistry and Radiochemistry*, page 157, Volume I, Academic Press, New York, 1959.

The simplest molecular crystals are those of the noble gases. Except for helium, which differs from the others in that it cannot be solidified at atmospheric pressure, they all have the cubic close-packed structure. With the use of the Lennard–Jones potential function, whose parameters for each gas are obtained from the second virial coefficient (Chapter 10, p. 389), values can be calculated for the cohesive energy of these crystals which are in good agreement with the experimental heats of sublimation.

5. *Hydrogen-bonded crystals.* Crystals of this type are rather similar to molecular crystals except for the fact that the molecules are held together by hydrogen bonds. In such a bond a proton can be considered bound to two electronegative atoms (Chapter 7, p. 259). The hydrogen bonds have an important effect on the way the molecules are arranged in the crystal. They are found in salt hydrates, organic compounds, such as carboxylic acids, containing OH or NH groups, and in the crystals of a large number of compounds of biochemical interest.

The outstanding example of a hydrogen-bonded crystal is ice, in which each oxygen atom is surrounded by four others at tetrahedral angles with hydrogen bonds between them. Such a structure is relatively open, a fact which accounts for the unusually low density of ice.

Crystal Defects

We have been discussing the structures of crystals as though these structures consisted of perfectly regular arrays of molecules or ions. Actually, a perfect crystal is a rarity, if indeed, it exists at all. Almost all actual crystals are polycrystalline mosaics and contain a variety of defects. Some crystalline properties are greatly affected by the presence of imperfections or impurities. For example, it was pointed out that the electrical conductivity of metals at low temperatures (p. 479) and of semiconductors (p. 480) varies greatly with the number and type of impurity atoms.

With many solids, particularly metals, the mechanical properties such as tensile strength and malleability are extremely dependent on crystal defects. This is also true of the coefficient of diffusion. When a crystal grows from a melt or a saturated solution, the rate of growth of a face is determined to a large extent by the defects in the face.

Because of its importance both theoretically and practically, the study of crystal defects is a major topic in solid state physics and chemistry. We cannot pursue this topic any further, but it is discussed in several of the references listed on page 488.

Supplementary Material

Born Theory of Cohesive Energy of Ionic Crystals

The potential energy of two ions whose charges have the same magnitude, $\pm z$, is assumed to be represented by

$$U(r) = \pm \frac{z^2}{r} + \frac{b}{r^n}$$

where r is the interionic distance and b and n are constants to be determined.[6] The sign of the first term is $+$ if the two ions are identical and $-$ if they have charges of opposite sign.

To obtain the total potential energy of a mole of ionic compound in the form of a crystal the first term in the equation above must be summed for all pairs of ions (nearest neighbors, next-nearest neighbors, and so on) while the second term is summed over nearest neighbors only, because the repulsive

[6]Quantum mechanical treatment of the overlap between filled electron shells leads to an exponential function for the potential energy of repulsion. Because the form of the function has little effect on the calculated results, we use the inverse power function which is somewhat simpler to manipulate.

force is significant only for ions that are "in contact." We can express the molar potential energy in the following way:

$$U = \frac{1}{2}\left[\underbrace{\sum \pm \frac{z^2}{r_{ij}}}_{\text{all pairs}} + \underbrace{\sum \frac{b}{r_{ij}^n}}_{\text{nearest neighbors}}\right]$$

(The factor $\frac{1}{2}$ is included to count each pair only once.)

The evaluation of the first term is simplified by the fact that the number of neighbors of each type that an ion has is determined by the lattice type. For a cubic lattice all interionic distances can be expressed in terms of a single parameter that might be, for example, the edge length of the unit cube, a. The sum can then be computed, once and for all, for each lattice type. The resulting number, represented by A, is called the Madelung number after the man who first evaluated these sums. The value of A is 3.495 for the face-centered NaCl lattice and 2.035 for the body-centered CsCl lattice. The molar potential energy is then given by

$$U = -\mathfrak{N}A\frac{z^2}{a} + \frac{B}{a^n} \qquad (12\text{-}9)$$

This equation contains two constants, B and n, whose value must be found by using additional information. The constant B can be evaluated from the requirement that the crystal lattice is at equilibrium when a has the particular value a_o. Consequently, the potential energy must be at a minimum with respect to change in lattice dimensions and thus $(dU/da)_{a=a_o} = 0$. On differentiating U with respect to a, we obtain

$$\frac{dU}{da} = \frac{\mathfrak{N}Az^2}{a^2} - \frac{nB}{a^{n+1}}$$

When the right-hand side is equated to zero and a is set equal to a_o, solution of this equation yields

$$B = \frac{\mathfrak{N}Az^2}{n}a_o^{n-1}$$

When this expression is substituted for B in Equation 12–9 it becomes

$$U = -\mathfrak{N}Az^2\left(\frac{1}{a} - \frac{a_o^{n-1}}{na^n}\right) \qquad (12\text{-}10)$$

With $a = a_o$, the equilibrium value of the molar potential energy then is given by

$$U_e = -\frac{\mathfrak{N}Az^2}{a_o}\left(1 - \frac{1}{n}\right) \qquad (12\text{-}11)$$

Solids

To evaluate this expression it is necessary to find the value of the parameter n. Because n is a repulsion constant and because the compressibility of an ionic crystal depends on the repulsive forces between neighboring ions, it is reasonable to expect that the value of n can be calculated from measured compressibilities. The coefficient of compressibility, κ, is defined by

$$\kappa = -\frac{1}{V}\left(\frac{\partial V}{\partial P}\right)_T$$

To relate κ to the potential energy equation we can start from the thermodynamic equation of state, Equation 3–27, which is completely general.

$$\left(\frac{\partial E}{\partial V}\right)_T = T\left(\frac{\partial P}{\partial T}\right)_V - P$$

If T is set equal to $0°K$ and then the resulting equation is differentiated with respect to V, we obtain

$$\left(\frac{\partial^2 E}{\partial V^2}\right)_{T=0} = -\left(\frac{\partial P}{\partial V}\right)_{T=0}$$

Now two approximations are made. We assume that (1) the thermodynamic energy E is all in the form of potential energy at $T = 0°K$ and (2) this equation can be used with negligible error at temperatures above $0°K$. With these approximations, it follows that

$$\frac{1}{\kappa} = V\left(\frac{d^2 U}{dV^2}\right)_{T=0} \tag{12-12}$$

In Equation 12–9 U is given as a function of a, the edge length of the unit cell. To find d^2U/dV^2 we make use of the mathematical identity

$$\frac{d^2 U}{dV^2} = \frac{d^2 U}{da^2}\left(\frac{da}{dV}\right)^2 + \frac{d^2 a}{dV^2}\left(\frac{dU}{da}\right) \tag{12-13}$$

The second term in this equation equals zero because of the equilibrium condition. To evaluate the first term, we need the relationship between the molar volume, V, and a. If there are j molecules (or ion pairs) per unit cell, the molar volume is given by $V = \mathfrak{N}a^3/j$ and consequently $da/dV = j/3\mathfrak{N}a^2$.

By differentiating Equation 12–10 twice with respect to a and then setting $a = a_o$, we obtain

$$\left(\frac{d^2 U}{da^2}\right)_{a=a_o} = -\mathfrak{N}Az^2\left(\frac{2}{a_o^3} - \frac{n+1}{a_o^3}\right)$$

$$= \frac{\mathfrak{N}Az^2}{a_o^3}(n-1)$$

When this expression is substituted into Equation 12–13, the result is

$$-\frac{d^2U}{dV^2} = \frac{j^2}{9\mathfrak{N}^2 a_0^4} \frac{\mathfrak{N} A z^2}{a_0^3}(n-1)$$

On substituting this result into Equation 12–12, we obtain

$$\frac{1}{\kappa} = \frac{\mathfrak{N} a_0^3}{j} \frac{j^2}{9\mathfrak{N}^2 a_0^4} \frac{\mathfrak{N} A z^2}{a_0^3}(n-1)$$

and

$$\frac{1}{\kappa} = \frac{jAz^2}{9a_0^4}(n-1)$$

It follows that

$$n = \frac{9a_0^4}{jAz^2\kappa} + 1$$

For many ionic crystals the value of n is approximately 9.

With n, A, and a_0 known, the value of U_e can be calculated from Equation 12–11. The values listed in the last column of Table 12–3 were obtained by a refined version of this method.

Supplementary References

N. B. Hannay, *Solid-state Chemistry*, Prentice-Hall, Englewood Cliffs, New Jersey, 1967.
C. Kittel, *Introduction to Solid State Physics*, 3rd edition, Wiley, New York, 1966.
D. E. Sands, *Introduction to Crystallography*, W. A. Benjamin, New York, 1969.
P. J. Wheatley, *Molecular Structure*, 2nd edition, Oxford University Press, London, 1968.

W. G. Gehman, "Standard Ionic Crystal Structures." *J. Chem. Ed.* **40,** 54 (1963).
R. J. Sime, "Some Models of Close Packing." *J. Chem. Ed.* **40,** 61 (1963).

Problems

1. A certain face of a crystal intercepts the a, b, and c axes at relative distances of $\frac{1}{2}$, 2, and ∞. What are the Miller indices of this face?   **Ans. (410)**

2. When a certain crystal was studied by the Bragg technique, using X-rays of wavelength 2.29 Å, an X-ray reflection was observed at an angle of 23°20′.

(a) What is the corresponding interplanar spacing? (b) When another X-ray source was used, a reflection was observed at 15°26'. What was the wavelength of these X-rays? **Ans.** (a) 2.89 Å; (b) 1.54 Å

3. Gold has a cubic face-centered lattice with an edge length of the unit cube of 4.07 Å. Calculate the diffraction angles of the first three lines in the diffraction pattern when copper X-rays of wavelength 1.54 Å are used.
Ans. 19°5', 22°13', 32°21'

4. Index all of the diffraction lines of silver given on page 480. What is the distance between the (222) planes?

5. The unit cell of aluminum is a cube with edge length 4.05 Å. The density of aluminum is 2.70 g/cm³. (a) What is the structure of aluminum crystals? (b) What is the smallest interatomic distance in aluminum?
Ans. (a) fcc; (b) 2.86 Å

6. Chromium forms body-centered cubic crystals with the edge length of the unit cube equal to 2.875 Å. Calculate its density.

7. Copper crystallizes in the cubic system. X-rays of wavelength 0.588 Å are reflected from a copper crystal at angles of 8°8', 9°24', and 13°18'. Find (a) the structure of the copper crystals; (b) the edge length of the unit cube; (c) the density of copper.

8. Molybdenum forms body-centered cubic crystals whose density is 10.3 g/cm³. Calculate (a) the edge length of the unit cube; (b) the distances between the (110) and between the (111) planes.
Ans. (a) 3.14 Å; (b) 2.22 Å, 1.81 Å

9. Use the data given below to find the type of cubic lattice for each of the following elements:

1. Fe $a = 2.86$ Å, $\rho = 7.86$ g/cm³
2. V $a = 3.01$ Å, $\rho = 5.96$ g/cm³
3. Pd $a = 3.88$ Å, $\rho = 12.16$ g/cm³

10. Calculate the fraction of the total volume occupied by spheres in contact in (a) a primitive cubic lattice; (b) a face-centered cubic lattice.

11. The edge length of the unit cube of NaCl is 5.63 Å. If the first diffraction line of NaCl is observed at an angle of 20°38', what is the wavelength of the X-rays used? **Ans.** 2.29 Å

12. When X-rays of wavelength 0.709 Å are reflected from a KCl crystal, the first reflection peak is observed at $\theta = 6°30'$. Calculate the edge length of the unit cube and the density of KCl.

13. The edge length of the unit cube of diamond is 3.567 Å and this cube contains 8 carbon atoms. (a) Calculate the closest distance between carbon atoms, assuming them to be spheres in contact. (b) Calculate the fraction of the total volume that is occupied by carbon atoms.   **Ans.** (a) 1.54 Å; (b) 0.34

14. The molar heat capacity of silver at 20.0°K is 0.390 cal/deg. Calculate (a) the Debye characteristic temperature; (b) the heat capacity of silver at 10.0°K.

15. Calculate the electron affinity of the Cl atom from the following data, given in kcal/mole except where noted: lattice energy of NaCl, 184; ΔH_f of NaCl(c), -98.2; ΔH of sublimation of metallic Na, 25.9; ΔH_f of Cl atoms, 28.9; ionization potential of Na atoms, 5.12 ev.   **Ans.** 3.78 ev

16. Calculate the lattice energy of KCl from the following data, given in kcal/mole except where noted: ΔH_f of KCl(c), -104.2; ΔH of sublimation of metallic K, 21.5; ΔH_f of Cl atoms, 28.9; electron affinity of Cl atoms, 3.78 ev; ionization potential of K atoms, 4.32 ev.

17. Calculate the lattice energy of KCl in kcal/mole from the following data: $a_o = 6.28$ Å; n (exponent of repulsion term) = 9.   **Ans.** 164 kcal/mole

13

Liquids

Introduction

In earlier chapters the development of a detailed and comprehensive theory of gases was based to a large extent on the concept of the random chaotic motion of independent molecules. In Chapter 12 we saw that the theoretical treatment of crystalline solids depends greatly on the simplification resulting from their high degree of order and the absence of translational motion in crystals.

When we come to the study of liquids a serious problem arises because liquids have neither the randomness of molecular motion which facilitates the treatment of gases, nor the completely ordered and symmetrical arrangement which characterizes crystalline solids. Since the enthalpy change when a crystal melts is always positive, the entropy change on melting is also a positive quantity. This implies that there is a loss of order when a crystal melts. The liquid state is thus intermediate between the complete order of the crystalline state and the complete disorder of the gaseous state. Because of this fact, the development of a molecular theory of liquids has encountered formidable difficulties. There is at present no completely satisfactory theory of liquids in spite of strenuous efforts in this direction. Consequently, any treatment of liquid properties must be more descriptive than the treatment of gases and crystals, and greater reliance must be placed on the use of thermodynamic relationships involving measured quantities than on molecular theory.

We first consider what is known about the structure of liquids, then turn to the current status of theories of the liquid state, and finally discuss some of the characteristic properties of liquids.

Structure of Liquids

When we compare the properties of a substance at temperatures just below and just above its melting point, we find that most of these properties change very little. For example, the molar volume of nearly all substances increases only about 10% on melting. Also the entropy of melting is usually much smaller than the entropy of vaporization. These facts suggest that the difference in structure between crystals and liquids may be one of degree and that liquids should have at least some vestige of the orderly arrangement of crystals.

This point of view is supported by observations of the diffraction of X-rays by liquids. Let us use the diffraction pattern of liquid argon, Figure 13–1, as an example. We choose a monatomic liquid because the diffraction pattern for diatomic or polyatomic liquids has complicating features resulting from the presence of nearly constant interatomic distances in the molecules.

In contrast to a typical crystalline powder pattern with its many sharp

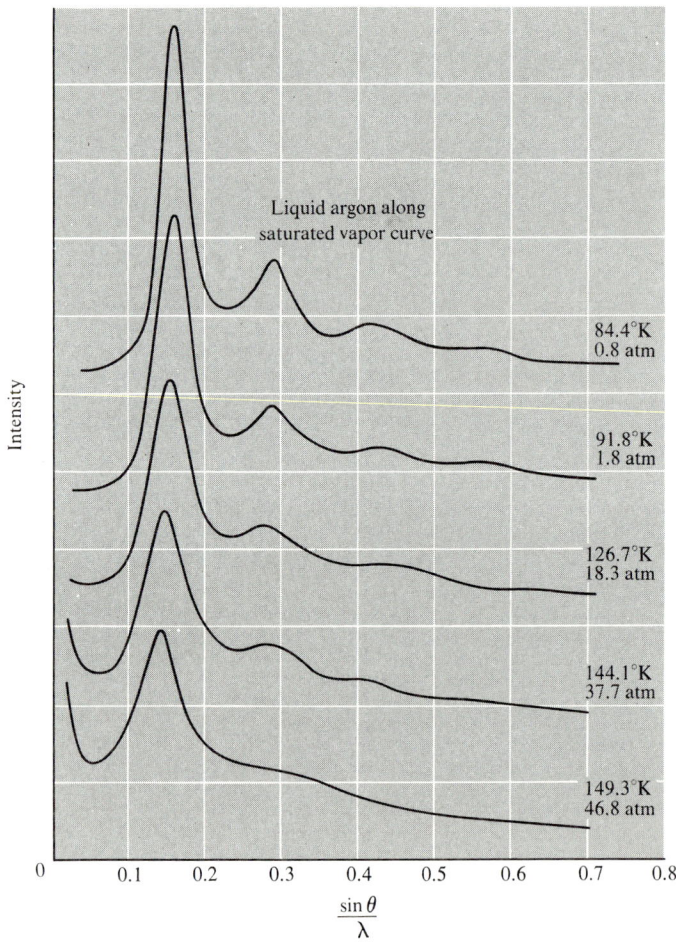

Figure 13–1. Graph of X-ray diffraction pattern of liquid argon at a series of temperatures. The ordinate of each curve is shifted relative to the preceding curve.

diffraction lines, the pattern of liquid argon just above its melting point shows only a very few broad maxima and minima which become broader and more diffuse as the temperature is raised. The presence of these maxima and minima is highly significant, however, for they would not be present if the molecular arrangement in the liquid were completely random.

In order to interpret the smeared-out appearance of the liquid diffraction pattern let us recall that in the discussion of the powder method (p. 457) it was pointed out that as the crystallite size and thus the extent of the orderly arrangement decreases, the diffraction lines become broadened. Consequently the appearance of the liquid pattern is attributed to a partially ordered structure for liquids. The presence of only a few maxima and minima is consistent with the presence about a typical molecule of order which extends only a few multiples of the mean intermolecular distance in the liquid. The absence of long-range order is a basic characteristic of liquids, and in this respect liquids resemble gases.

The X-ray diffraction pattern of a liquid is not a consequence of any definite arrangement of the atoms (or molecules); instead, it represents a time average over all of the positions of the atoms resulting from their translational motion. It is possible, however, to use this pattern for the calculation of the *radial distribution function*,[1] $4\pi r^2 N(r)$, which gives the number of atoms whose centers are at a distance r from the center of any atom in the liquid. For a crystal at 0°K the radial distribution function would consist of a series of lines at values of r equal to the distances between nearest neighbors in the lattice, next nearest neighbors, and so on. As the temperature of the crystal is raised, the lines broaden into narrow bands because of the vibrational motion of the atoms, with the area under each band proportional to the number of atoms at that distance. Thus for crystalline argon, which has a face-centered structure, the first band, whose r value is 3.82 Å, has an area that is consistent with the presence of 12 nearest neighbors (Fig. 13–2e). The second band, with an r value of 5.43 Å, results from the presence of six next nearest neighbors.

It can be seen in Figure 13–2a that the radial distribution function for liquid argon just above its melting point has a maximum at very nearly the same value of r as the distance between nearest neighbors in crystalline argon. From the area under this maximum for liquid argon just above its melting point, the number of nearest neighbors is calculated to be about 10.5. The difference between 12 and 10.5 has important implications for the theory of liquids, and we will return to this point later.

As the temperature of a liquid is raised its radial distribution function, as

---

[1] The radial distribution function is related to a function $g(r)$, known as a correlation function, which gives the probability of finding the center of an atom in a volume element, $\sin \theta \, d\theta \, d\phi \, r^2 \, dr$, at the terminus of a vector r whose origin is at the center of any selected atom. When this function is integrated over the angles θ and ϕ (p. 135) and then is multiplied by the average molecular concentration $\bar{N}$, the result is $4\pi r^2 N(r)dr$. If we imagine a spherical shell of radius r and thickness dr with the selected origin at its center, $4\pi r^2 N(r)dr$ represents the number of atoms whose centers lie in this spherical shell.

calculated from its X-ray diffraction pattern, undergoes several characteristic changes (Fig. 13–2b and c). The maxima shift slightly toward larger r values and become even broader and shallower. Evidently the degree of order decreases with rising temperature.

In discussing the structure of liquids we have emphasized thus far the relationship between the liquid state and the crystalline state. We should not, how-

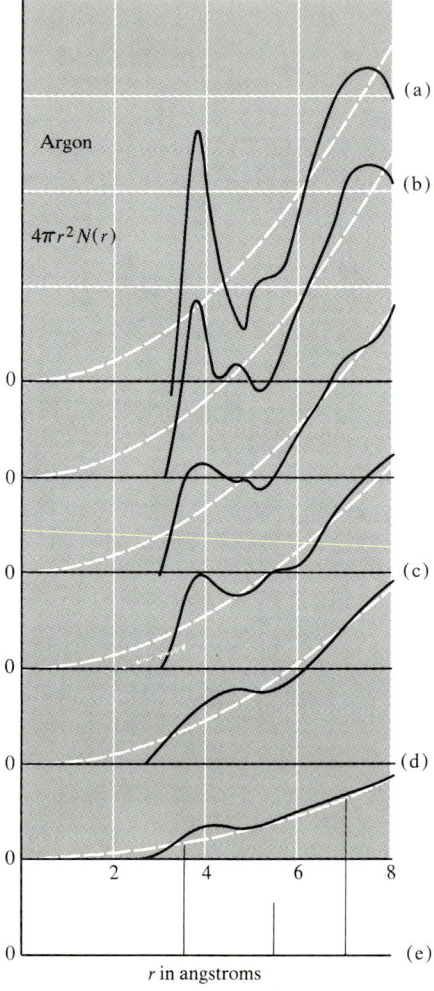

Figure 13–2. Graphs of the radial distribution function of argon at a series of temperatures. For each curve the rising dotted line is the graph of $4\pi r^2$. (a), (b), and (c) refer respectively to liquid argon at 84°K and 0.8 atm, 92°K and 1.8 atm, 149°K and 46.8 atm; (d) is a graph for gaseous argon at 149°K and 43.8 atm; and (e) shows the position of the diffraction "bands" in crystalline argon.

Liquids

ever, overlook the relationship between liquids and the gaseous state. The fact that both liquids and gases are fluids is one obvious point of similarity, but their relationship is much deeper than this, particularly near the critical state. In Chapter 1 it was pointed out that if the temperature of a gas having a pressure higher than the critical pressure is dropped from just above the critical temperature to a temperature just below, a transition from the gaseous phase to the liquid phase occurs without any discontinuous change in properties. This fact strongly implies a similarity in structure between a liquid and a dense gas, at least at temperatures not far from the critical temperature. This conclusion is supported by observations of the X-ray diffraction pattern of argon at temperatures near its critical temperature. The diffraction patterns and the corresponding radial distribution functions for the liquid and vapor phases show little difference (Fig. 13–2c and d).

Theories of the Liquid State

In order to develop a theory on the basis of which the properties of liquids can be interpreted and calculated in terms of molecular properties, it seems necessary to make the maximum use of what is known about the gaseous and crystalline states. The first even partially successful approach to such a theory was made by van der Waals in connection with the development of his equation of state for gases (p. 20). By introducing two molecular constants, one related to intermolecular attractive forces and the other to repulsion forces, he obtained an equation of state that is fairly successful in accounting for the deviations from ideality of real gases at moderate pressures. In addition, his equation accounts for the existence of a critical state and roughly predicts the condition under which a gas would condense into a liquid. The van der Waals equation, however, has not led to further understanding of the liquid state.

From a statistical point of view, the theoretical goal is a partition function for the liquid state, considered as a system of interacting molecules that have both kinetic and potential energies. In Chapter 10 we saw that the total molar partition function, Q_{tot}, for such a system can be written as a product of two factors, $Q_{tot} = Q_t Q_u$. The first factor (p. 382),

$$Q_t = \frac{1}{\mathfrak{N}!} \left[\frac{(2\pi mkT)^{\frac{1}{2}}}{h} \right]^{3\mathfrak{N}}$$

is related to the translational kinetic energy, expressed as a function of the momentum, for the system of $\mathfrak{N}$ molecules. This factor can be evaluated without difficulty.

The configuration integral Q_u is defined by

$$Q_u = \int \cdots \int^{3\mathfrak{N}} e^{-U(q)/kT} \, dq_1 \ldots dq_{3\mathfrak{N}}$$

in which $U(q)$ represents the potential energy of interaction of all of the molecules among themselves, expressed as a function of the molecular coordinates $q_1 \ldots q_{3\mathfrak{N}}$. Q_u has the dimensions of (volume)$^{\mathfrak{N}}$. (For an ideal gas, $U(q) = 0$, and $Q_u = V^{\mathfrak{N}}$, where V is the volume of the gas.)

In Chapter 10 we needed to use approximations in evaluating Q_u for a real gas in order to obtain the second virial coefficient (Eq. 10–10) in terms of molecular parameters. The method of approximation used there depends on the assumption that the deviations from ideality are small. In other words, the effect of intermolecular forces is assumed to be small except when two molecules come close together. As the pressure and density of a gas are raised to relatively large values, however, this assumption becomes decreasingly appropriate. It is not valid at all for a liquid in which the molecules are always interacting with each other and the effect of intermolecular forces is predominant.

It is clear that another theoretical approach to the evaluation of Q_u for a liquid is needed. In one type of theory attention is focused on the structure of a liquid as inferred from the X-ray diffraction results mentioned above. In such a "lattice" or "cell" theory the assumption is made that much of the orderly arrangement of the molecules in the crystalline state is retained on melting. The liquid is assumed to have a partially disordered lattice structure in which most of the molecules are confined to a cell or cage formed by their nearest neighbors.

In a version of the theory called the "hole" theory, the increment of volume gained on melting (the free volume) is assumed to be distributed in the form of holes or cavities of roughly molecular size. Thus an average molecule will have a smaller number of nearest neighbors in the liquid than it has in the crystalline state. This assumption is consistent with the average number of nearest neighbors calculated from the radial distribution function. The hole theory is helpful because it provides a model of liquids that is easily visualized and that at least qualitatively accounts for their properties.

Unfortunately, none of these approximate methods for evaluating the partition function has proved to be successful from a quantitative point of view, although much effort has been expended.

Eyring and his co-workers have developed an interesting theoretical approach to the liquid state, which he calls the *significant structure theory*, in which some of the features of a compressed gas theory and a disordered lattice theory have been combined. In this theory the molecular arrangement and the number of nearest neighbors for a certain fraction of the molecules are assumed to be essentially the same as in the corresponding crystal. The molecules in this fraction can vibrate with the same frequencies as those in the crystal and are said to be solid-like.

For each of the other molecules there is on the average at least one hole in its immediate neighborhood, corresponding to a vacant lattice site. A molecule which is located next to a hole may acquire enough thermal energy to move into the hole and thus carry out a translational motion. Such a molecule with a translational degree of freedom can be considered as gas-like. The holes are pictured as moving in the opposite direction from the translating molecules. Eyring uses the term "fluidized vacancies" to refer to the holes moving about randomly in the liquid.

In order to divide the molecular degrees of freedom into vibrational and translational degrees of freedom, the assumption is made that the fractions of solid-like and gas-like molecules are given by V_s/V and $(V - V_s)/V$, respectively, with V_s and V representing the molar volumes of the crystalline solid and the corresponding liquid.

The next step in the theoretical development is the writing of an expression for the liquid partition function on the basis of the significant structure theory, Q_{ss}. Without going into the details we can say that Q_{ss} is formulated as a product of two factors, one corresponding to the solid-like and the other to the gas-like molecules. The factor for the solid-like molecules depends mainly on the Einstein partition function for the vibrational degrees of freedom of crystals (Eq. 12–5), and the factor for the gas-like molecules depends on the ideal gas partition function for the translational degrees of freedom. When the volume of the system equals V_s, Q_{ss} reduces to the Einstein partition function for the crystalline form of the substance in question. As the volume of the system becomes much larger than V_s, Q_{ss} approaches the ideal gas partition function.

To evaluate Q_{ss} for a monatomic element, such as one of the noble gases, three properties of the elements in the crystalline state are needed: the heat of sublimation, the Einstein characteristic temperature, and the molar volume. In addition, the molar volume of the liquid state at the melting point must be known. From this type of partition function a variety of properties have been calculated with rather good agreement with experimental data. For other classes of liquids (diatomic and polyatomic molecules, metals, and fused salts) additional assumptions must be made, but again good agreement is obtained with measured properties.

In another type of theoretical treatment of the liquid state emphasis is placed on the radial distribution function mentioned previously, rather than on the partition function. It is possible to express the equation of state and the thermodynamic properties of a liquid in terms of the radial distribution function. Unfortunately, experimentally determined radial distribution functions (Fig. 13–2) are not accurate enough to be used for this purpose. Exact equations have been derived for this function but no general solutions of these equations have been obtained so far. Recent work, however, on approximation methods applied to a model of a liquid composed of rigid spherical molecules has led to equations from which a number of thermodynamic properties of liquefied noble gases have been calculated in good agreement with experimental data.

Properties of Liquids

Viscosity

One of the important properties that liquids share with gases is the ability to flow under the influence of unbalanced forces. It was pointed out in Chapter 5, page 142, that a force must be applied to a fluid to cause it to flow and that the resistance to flow is called the viscosity. It was also mentioned that Newton derived the following equation for the force, the viscous drag, acting on a surface in a fluid undergoing laminar flow:

$$f = \eta A \frac{dv}{dz} \qquad (13\text{--}1)$$

A is the area of the surface, dv/dz is the rate of change of flow velocity in a direction perpendicular to the surface, and η, the viscosity coefficient, is a property of the fluid.

The rate at which a liquid flows through a tube under a constant driving force depends on the value of the viscosity coefficient. This relationship provides one of the methods for the measurement of η. The basic equation, known as Poiseuille's equation, is

$$\eta = \frac{\pi R^4 t \, \Delta P}{8VL}$$

in which R is the radius of the tube and L is its length, V is the volume of liquid which flows through the tube in time t, and ΔP is the difference in the pressures on the ends of the tube.

Poiseuille's equation is a direct consequence of Eq. 13–1. When a liquid, assumed to be incompressible, undergoes steady laminar flow in a tube, the linear velocity of flow, v, decreases from a maximum value at the axis of the tube to a value of 0 in a layer adjacent to the wall of the tube. The rate of change of velocity with radial distance from the axis, r, is given by $-dv/dr$ (Fig. 13–3).

Let us picture a cylinder of liquid with radius r and length L, concentric

Liquids

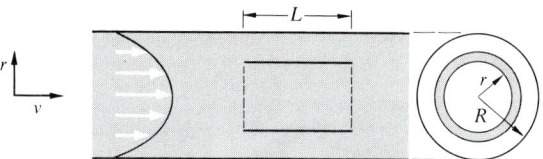

Figure 13-3. Laminar flow in a tube.

with the tube. The area of the cylindrical surface is $2\pi rL$, and according to Newton's law the viscous drag on this surface is

$$f = -\eta 2\pi rL \frac{dv}{dr}$$

Because the liquid is flowing steadily without acceleration, this drag force must be numerically equal to the driving force which is given by the product of the pressure difference and the area of the cross-section of the cylinder. Consequently,

$$\pi r^2 \Delta P = -\eta 2\pi rL \frac{dv}{dr}$$

and thus

$$dv = -\frac{\Delta P}{2L\eta} r \, dr$$

To obtain the flow velocity as a function of r we integrate this equation and obtain

$$v = -\frac{\Delta P}{4L\eta} r^2 + C$$

The constant of integration is evaluated from the condition that $v = 0$ when $r = R$. Then $C = \Delta P R^2 / 4L\eta$, and we have

$$v = \frac{\Delta P}{4L\eta}(R^2 - r^2)$$

(This equation implies that the graph of v as a function of r, the flow profile, is a parabola as shown in Figure 13-3.)

If we focus on the area between two concentric circles of radius r and $r + dr$, given by $2\pi r \, dr$, the volume of liquid flowing through this area in time t is given by

$$dV = vt \cdot 2\pi r \, dr$$

When we integrate this equation from $r = 0$ to $r = R$ we obtain the total volume of liquid which flows through a cross-section of the tube in time t; thus,

$$V = \int_0^R 2\pi vt r \, dr$$

On introduction of the expression for v above, this becomes

$$V = 2\pi t \frac{\Delta P}{4L\eta} \int_0^R (R^2 - r^2) r\, dr$$

$$= 2\pi t \frac{\Delta P}{4L\eta} \left(\frac{R^4}{2} - \frac{R^4}{4}\right)$$

$$= \frac{\pi t R^4 \Delta P}{8L\eta}$$

When this equation is solved for η, Poiseuille's equation is obtained.

It should be noted that Poiseuille's equation applies only to the laminar flow (parallel stream-lines) of fluids which obey Newton's law of viscosity. With increasing flow rate or increasing tube radius, at some point there is an onset of turbulence which produces a drastic effect on flow conditions.

Many different types of apparatus have been devised for the measurement of viscosity. One of the most commonly used types is some form of the Ostwald viscometer (Fig. 13-4). With this apparatus the time is measured for a fixed

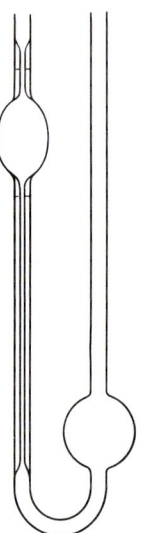

Figure 13-4. Ostwald viscometer.

volume of liquid to flow through a vertical capillary tube as the liquid level falls between two reference marks. The driving force is the net force of gravity resulting from the difference in the levels, Δl, of the liquid in the two reservoirs.

In this case $\Delta P = \rho g \Delta l$, where g is the acceleration of gravity. Because Δl decreases as the liquid flows, the Ostwald viscometer is not convenient for absolute measurements of the viscosity. It is almost always used for relative measurements with some liquid of known viscosity, often water, as a reference. Then an unknown viscosity coefficient is obtained from the following equation:

$$\eta = \eta_r \frac{\rho t}{\rho_r t_r}$$

where ρ_r and t_r are the density and efflux time of the reference liquid.

Another type of apparatus, known as the falling ball viscometer, is an application of Stokes' law of viscous flow. Stokes showed that if a sphere of radius r moves with constant velocity v through a liquid whose viscosity coefficient is η, the drag force on the sphere is given by

$$f = 6\pi\eta r v \qquad (13\text{-}2)$$

If a sphere falls vertically through a liquid, the net force on it in the downward direction is $(m - m_l)g$, where m and m_l are respectively the masses of the sphere and the liquid it displaces, and g is the acceleration of gravity.

When the sphere falls at a constant speed the forces acting upward and downward are numerically equal. Since m and m_l are equal to the respective densities times the volume of the sphere, we have the following equations:

$$\tfrac{4}{3}\pi r^3 (\rho - \rho_l) g = 6\pi\eta r v$$

and

$$\eta = \frac{2r^2(\rho - \rho_l)g}{9v}$$

By measuring the time required for the sphere to fall between two reference marks and thus obtaining v, the viscosity coefficient can be calculated when the values of r and the densities are known. This method is used mostly for liquids with relatively high viscosities.

The viscosity coefficient has been determined as a function of the temperature for many liquids. Without exception the viscosity falls with rising temperature and the temperature dependence for "normal" liquids without hydrogen bonding is well represented by the following empirical equation:

$$\eta = A e^{\frac{B}{RT}} \qquad \text{or} \qquad \ln \eta = A' + \frac{B}{RT}$$

in which B has the dimensions of energy per mole. The same relationship is sometimes expressed in terms of the fluidity, ϕ, which is defined as the reciprocal of the viscosity coefficient; thus,

$$\phi = \frac{1}{\eta} = \frac{1}{A} e^{-\frac{B}{RT}} \qquad (13\text{-}3)$$

This temperature dependence has some interesting implications for the theory of viscous flow in liquids. In the first place one can conclude that the mechanism of flow must be different for gases than for liquids, since the viscosity of a gas increases with rising temperature while that of a liquid decreases. The concept of momentum transfer by bimolecular collisions, which was used successfully with gases, is not applicable here.

In addition, Equation 13–3 has the same mathematical form as the Arrhenius equation for reaction velocity (Eq. 11–20). This suggests that it might be possible to interpret the quantity B as some kind of energy of activation for viscous flow. This interpretation is quite consistent with the concept of fluidized vacancies, which provides a simple mechanism for the flow of a liquid in which the molecules are packed nearly as closely as they are in a crystal.

Let us assume that a molecule which is near a hole must "push" its neighbors out of the way to enter the hole. This assumption is equivalent to saying that the molecule must pass over an energy barrier in order to occupy the hole. In the absence of external forces there is no preferred direction in the liquid and the molecules can move into holes in any direction with equal probability. When a driving force is applied to the liquid, the potential energy on the "upstream" side of the energy barrier is raised and that on the "downstream" side is lowered. A molecule can lower its potential energy by moving in the direction of the applied force, and motion into a hole in this direction is more probable than in the opposite direction. The rate of flow of the liquid depends on the net rate at which the molecules pass over the energy barrier in the downstream direction.

The height of the energy barrier is thus identified with the molecular energy of activation, ϵ_a, for viscous flow. Assuming thermal equilibrium in the liquid, we can say that the fraction of the molecules which have at least this much energy is given by the Boltzmann factor, $e^{-\epsilon_a/kt}$ ($= e^{-B/RT}$). In this way the exponential dependence of the viscosity (or fluidity) on the temperature is accounted for.

◈◈◈◈◈◈

If we had a complete theory of liquid viscosity, we could calculate the preexponential factor and the activation energy for viscous flow from molecular properties. With the help of simplifying assumptions such calculations have been carried out, mainly by Eyring and his co-workers, with moderate success. The energy of activation appears to be related to the energy required to form a hole in the liquid and this, in turn, is related to the heat of vaporization.

The mechanism of flow described above is undoubtedly oversimplified. That the basic concept has some validity is indicated, however, by an empirical equation obtained by Batschinski in 1913, which can be written in the following form:

Liquids

$$\frac{1}{\eta} = \phi = \frac{V_l - V}{C}$$

in which C is a constant characteristic of the liquid, V_l is its molar volume under the conditions used for the viscosity measurement, and V is a volume which is very close to the molar volume of the crystalline form of the same substance. This equation implies that the fluidity is proportional to the free volume of the liquid and thus to the concentration of the holes it contains.

Surface Tension

It is a familiar observation that small droplets of mercury on glass and small droplets of water on a piece of paraffin wax assume a shape that is very nearly spherical. This shape is the result of a characteristic property of liquids known as the *surface tension* whose effects can be observed whenever a liquid is in contact with its vapor or some other gas.

The surface tension, γ, is defined as the force in a liquid surface perpendicular to any line of unit length in the surface. The common unit of surface tension is the dyne/cm. Another property closely related to surface tension is the *surface energy* which is the energy required to form a unit area of a surface from molecules initially in the interior of the liquid. The surface energy in ergs/cm² is numerically equal to the surface tension.

This equality can be demonstrated by imagining that a film of liquid is held in a wire frame with one movable side (Fig. 13–5). The force due to surface

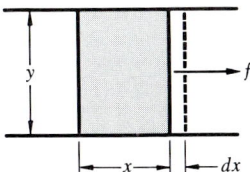

Figure 13–5. Liquid film held in wire frame with one movable side.

tension, which tends to reduce the value of the length x, is $2y\gamma$ because the film has two surfaces. When an equal and opposite force, f, is applied to the movable wire the surface is in equilibrium. If the magnitude of f is increased slightly and the length x is increased by Δx, the work done is $f \Delta x$ which is equal to $2y\gamma \Delta x$. This amount of work leads to the formation of new surfaces of total area $2y \Delta x$. The work done per unit area, which is equal to the surface energy, is then equal to γ ergs/cm².

The proportionality of the energy of a surface to its area provides an explanation of the spherical shape of the droplets mentioned above. The droplets must assume a shape that minimizes the energy of their surfaces, and of all geometrical figures the sphere has the smallest surface-to-volume ratio. Thus we see that the surface of a liquid behaves like a stretched membrane. A liquid film in the form of a hollow sphere is similar to a spherical rubber balloon, and the surface, like the rubber, exerts a pressure directed toward the center of the sphere. It follows that at equilibrium the pressure inside the sphere must be greater than the outside pressure.

The excess pressure ($\Delta P = P_{\text{inside}} - P_{\text{outside}}$) due to the spherical surface depends on the surface tension and the radius, r, of the sphere. The energy of the surface is $2\gamma(4\pi r^2)$; the factor 2 takes into account the fact that the film has two surfaces. If the radius of the sphere increases by dr, the change in the total energy of its surface is given by $16\gamma\pi r\, dr$. This increase in the energy of the surface comes from the work $\Delta P\, dV$ done against surface tension in enlarging the sphere. Since V is the volume of the sphere, $dV = 4\pi r^2\, dr$, and we have

$$16\gamma\pi r\, dr = \Delta P \cdot 4\pi r^2\, dr$$

It follows that $\Delta P = 4\gamma/r$. For a single surface, such as that of a gas bubble in a liquid, $\Delta P = 2\gamma/r$ and

$$\gamma = \frac{r\,\Delta P}{2} \tag{13-4}$$

One of the methods used for the determination of the surface tension of a liquid depends on a measurement of the maximum pressure that can exist in a bubble in the liquid. When a bubble of inert gas is formed at the tip of a tube which is immersed in a liquid to a depth h (Fig. 13–6), the pressure of the gas is balanced by the sum of the hydrostatic pressure at the tip and the pressure due to surface tension.[2] This equality can be represented by

$$P_g = \rho g h + \frac{2\gamma}{r}$$

in which r is the radius of curvature of the bubble and ρ is the density of the liquid.

As the gas pressure is increased the bubble grows and the radius of curvature decreases until it equals the radius of the tip. The bubble then has a hemispherical shape. With further increase of the gas pressure, the bubble separates from the tip and the gas pressure drops. From a measurement of the maximum gas pressure and the radius of the tip, the surface tension of the liquid can be calculated using the equation above.

Another method for determining surface tensions depends on the difference

[2]Thus $\Delta P = P_g - \rho g h$; this implies that the liquid is under atmospheric pressure and that the gas pressure is measured with an open-end manometer.

Liquids

between the level of the curved surface (called the *meniscus*) formed inside a vertical tube of small diameter dipping into a liquid and the level of the surface outside the tube. Whether the meniscus rises or falls is determined by the angle

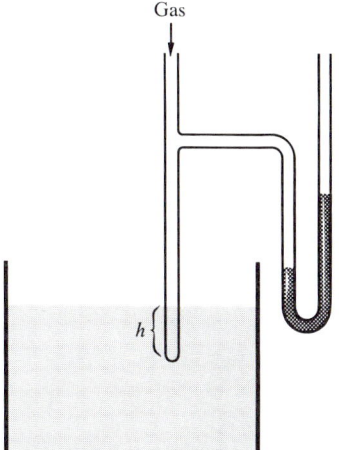

Figure 13-6. Determination of surface tension by the maximum bubble pressure method.

in the liquid between the liquid surface and the wall of the tube. This angle, called the *contact angle* (θ), depends on the relative values of the cohesive intermolecular forces in the liquid and the adhesive forces between the molecules of the liquid and those of the solid with which it is in contact.

If the adhesive forces are larger than the cohesive forces, the contact angle is less than 90° (Fig. 13-7). In this case a drop of liquid placed on a horizontal

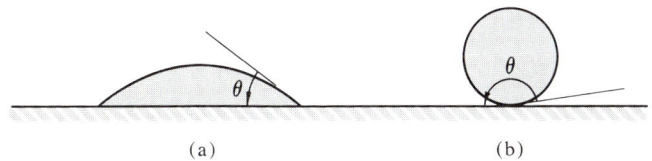

Figure 13-7. (a) Contact angle $\theta < 90°$, (b) $\theta > 90°$.

surface of the solid will spread, and the liquid is said to wet the solid. In the limiting case the value of θ is 0° and a drop cannot be formed because the liquid will spread over the available solid surface until a nearly monomolecular layer is formed. This is the behavior of water on a clean glass surface.

In contrast, if the adhesive forces are much smaller than the cohesive forces, the value of θ approaches 180°, the liquid does not wet the surface, and instead of spreading a drop of liquid will assume a nearly spherical shape. This is the case for mercury on a clean glass surface.

Now let us return to the vertical tube partially immersed in a vessel of liquid and let us assume that the contact angle is less than 90°. In this case the meniscus rises above the flat surface of the liquid outside the tube (Fig. 13-8). The

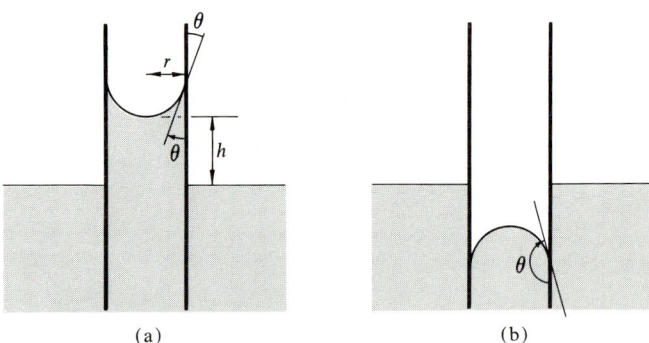

Figure 13-8. (a) Capillary rise ($\theta < 90°$) and (b) depression ($\theta > 90°$) of a liquid in a tube.

upward force f_u that causes the rise is the vertical component of the force due to surface tension. This component acts along the edge of the liquid whose length is $2\pi r$, where r is the radius of the tube. Thus $f_u = 2\pi r \gamma \cos \theta$. At equilibrium this force is balanced by the downward force of gravity on the column of liquid, f_d, which is equal to the mass of this column times the acceleration of gravity. Since the mass of the liquid is given by $\rho \pi r^2 h$, where h is the height of the column, it follows that

$$2\pi r \gamma \cos \theta = \rho \pi r^2 h g$$

and also,

$$\gamma = \frac{\rho r h g}{2 \cos \theta} \tag{13-5}$$

For water in contact with clean glass, the value of $\cos \theta$ is very close to 1.

If the value of θ is greater than 90°, $\cos \theta$ is a negative number and a depression of the meniscus results with the concave side in the downward direction (Fig. 13-8b). This depression is observed with mercury in a glass tube.

Equation 13-5 can also be derived from 13-4. Assume that the meniscus, whose radius of curvature is R, is momentarily at the same level as the flat liquid surface outside the tube (Fig. 13-9). (This is not a condition of equilib-

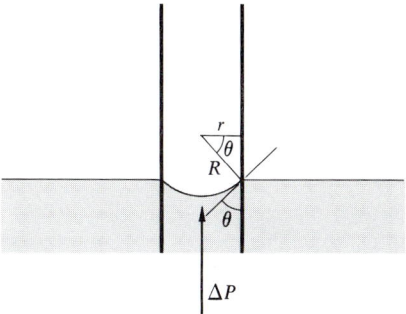

Figure 13-9. Nonequilibrium condition in which the meniscus inside the tube is at the same level as the liquid surface outside the tube.

rium.) Then the pressure on the liquid just under the meniscus is less than the pressure on the outer flat surface by the pressure ΔP exerted upward by the meniscus. The meniscus will then rise until the pressures are equalized. At this time $\Delta P = \rho h g$, and from Equation 13-4 we have

$$\Delta P = \frac{2\gamma}{R} = \rho h g$$

From Figure 13-9 it can be seen that $R = r/\cos \theta$. When this substitution is made in the equation above and the resulting equation is solved for γ, Equation 13-5 is obtained.

Other methods are also used for the measurement of surface tension but the ones mentioned are sufficient for our purpose.

When the surface tension of a liquid is measured over a temperature range it is found that it invariably decreases with rising temperature. One of the empirical equations that have been found to represent the data reasonably well is the following equation proposed by McLeod:

$$\gamma = C(\rho_l - \rho_v)^4$$

in which C is a constant for each substance and ρ_v is the density of the vapor in equilibrium with the liquid, whose density is ρ_l. This equation implies that γ approaches zero as the temperature approaches the critical temperature of the substance because ρ_v then approaches ρ_l. This is to be expected for when the densities of the liquid and vapor phases become equal, the surface disappears.

Now we turn to the interpretation of surface tension from a molecular point of view. It is clear that a molecule in a surface exists in an unusual situation because the cohesive forces exerted by its neighbors in the liquid are not balanced by the attractive force of the vapor molecules. When a molecule moves from the body of the liquid into the surface, its number of nearest neighbors decreases and some "bonds" are broken. Thus the energy of the molecules in the surface is higher than that of the molecules in the body of the liquid and this fact accounts for the surface energy defined earlier.

The magnitude of the surface energy (and surface tension) depends on the strength of the intermolecular forces. The liquids with relatively weak intermolecular forces have low surface tensions, and conversely. For example, the cohesive forces in water are relatively large, as we have seen, because of extensive hydrogen bonding, and the surface tension of water is exceptionally large.

Amorphous Solids and Glasses

We have seen that the rigidity and characteristic shapes of crystals are associated with a high degree of internal order. For liquids, the absence of rigidity and characteristic shape resulting from fluidity is associated with a low degree of internal order. Intermediate between crystals and liquids is a class of solids which have rigidities comparable to that of some crystals, but no characteristic shape. This is the class of amorphous solids and glasses.

When some amorphous solids are heated the only change is chemical decomposition. Many others, however, do not decompose with rising temperature but instead gradually lose their rigidity. They become liquids whose viscosities are very high at first and then drop rapidly as the temperature rises. A solid of this type has a rather indefinite "softening range," rather than a sharp melting point. When the resulting liquid is cooled its viscosity increases and other properties change continuously, with no sign of a discontinuity such as is observed at a freezing point. A solid that shows this type of behavior is called a *glass*.

If an X-ray diffraction pattern of a glass (or of any amorphous solid) is obtained, it looks very much like the diffraction pattern of a liquid. Consequently we can conclude that although the viscosity of a glass-forming liquid increases to an immeasurably high value with falling temperature, the internal structure retains the randomness characteristic of a liquid. This conclusion is consistent with the observation that when the heat capacity of a glass is measured down to very low temperatures, the entropy value calculated from the experimental data is abnormally high. (This is direct evidence that the third law of thermo-

Liquids

dynamics (Chapter 4, p. 112) is not applicable to glassy samples.) Because glasses have the internal disorder associated with liquids, they are usually considered to be supercooled liquids.

Among the types of liquids which tend to form glasses on cooling are mixtures of certain oxides often including silica, hydrogen-bonded liquids such as glycerol, and liquids containing molecules with very little symmetry. In order for crystallization to occur the molecules must move to fit into a growing crystal lattice. If the intermolecular forces, and thus the viscosity, in the liquid are high, or if a particular orientation of the molecules is needed to form a crystal lattice, the disorder in the liquid may be "frozen-in" before crystallization has had enough time to occur.

Another type of substance which yields liquid-like X-ray diffraction patterns includes some polymers of high molecular weight which consist of long chains of atoms intertwined among each other. Some of these polymers, such as polyethylene, contain both crystalline and amorphous regions whose proportions depend on the conditions under which the polymer was formed and on its rate of cooling. One of the more interesting polymers is natural rubber (polyisoprene). When a series of X-ray diffraction patterns is obtained for a rubber sample which is stretched to an increasing extent, the patterns consist of discrete spots which grow in intensity with elongation of the sample. This is evidence for partial crystallization of the sample by the alinement of the long molecular chains resulting from the stretching.

Liquid Crystals

In the preceding section we discussed briefly the behavior of matter in a state in which it has the rigidity of a solid but the internal structure of a liquid. In contrast, many compounds can exist in the form of a liquid which has to some extent the orderly structure of a crystal. Such a liquid is often called a *liquid crystal;* more precisely, the compound is said to be in a *mesomorphic* state or phase.

The compounds that can exist in a mesomorphic state are in general polar organic compounds with "stiff," moderately long chains. A fairly typical example is para-*n*-hexylbenzoic acid whose structural formula is

$$H_{13}C_6-\bigcirc-C\begin{matrix}\nearrow O \\ \searrow OH\end{matrix}$$

When crystals of such a compound are heated, at a reproducible temperature a transition takes place to a turbid, cloudy liquid which, in some cases, has a very high viscosity and in others, a viscosity in the range of typical ordinary liquids. On further heating, at some characteristic temperature this mesomorphic phase changes sharply to a clear liquid.

The interpretation of this behavior is suggested by the observation that a mesomorphic phase is *anisotropic;* that is, many of its properties have different values in different directions. This is not true of ordinary liquids but it is a phenomenon observed with most crystals. Investigations by a variety of methods, including X-ray diffraction and the transmission of polarized light, have led to the conclusion that, although the formation of a mesomorphic phase involves a loss of the 3-dimensional lattice structure of the crystal, a considerable amount of orderly arrangement remains. Because the molecules are polar, the intermolecular attractive forces are relatively strong. As a result of these forces, the molecules tend to line up parallel to each other over regions that are large on a molecular scale. When the mesomorphic phase "melts," this residual order disappears.

Supplementary References

G. W. Gray, *Molecular Structure and the Properties of Liquid Crystals*, Academic Press, New York, 1962.

J. O. Hirschfelder, C. F. Curtiss, and R. B. Bird, *Molecular Theory of Gases and Liquids*, Wiley, New York, 1954.

H. Eyring and R. P. Marchi, "Significant Structure Theory of Liquids." *J. Chem. Ed.*, **40**, 562 (1963).

R. F. Kruh, "Diffraction Studies of the Structure of Liquids." *Chem. Rev.*, **62**, 319 (1962).

Problems

1. 135.0 cc of a liquid flow through a capillary tube 8.56 cm long with an internal diameter of 1.10 mm, under a constant head of 1.00 atm, in 64.2 sec. What is the viscosity of the liquid at the temperature at which the measurements were made? **Ans.** 0.202 poise

2. How long will it take for 100 cc of benzene at 25°C to flow through a tube 15.0 cm long with an internal diameter of 2.0 mm, under a constant head of 2.00×10^4 dynes/cm²? The viscosity of benzene at 25°C is 6.10 millipoise.
 Ans. 11.5 sec

3. When an Ostwald viscometer was calibrated with water at 20°C the flow time was 57.4 sec. For water at 20°C, $\eta = 10.05$ millipoise and $\rho = 0.998$ g/cm³. With *n*-octane at 20°C the flow time in the same viscometer was 43.7 sec. For *n*-octane at 20°C $\rho = 0.704$ g/cm³. Calculate the viscosity of *n*-octane at 20°C.

Liquids

4. A brass sphere (ρ = 8.55 g/cm³) 5.5 mm in diameter falls through a distance of 22.8 cm in a viscous liquid (ρ_l = 1.17 g/cm³) in 12.3 sec. What was the viscosity of the liquid at the temperature of the experiment?   **Ans.** 65.5 poise

5. The viscosity of diethyl ether in millipoise is 2.84 at 0°C, 2.33 at 20°C, and 1.97 at 40°C. (a) Calculate the activation energy of diethyl ether for viscous flow. (b) Calculate its viscosity at 60°C.

6. The maximum bubble pressure observed when a tube of radius 5.0 mm was immersed in a liquid (ρ = 0.918 g/cm³) to a depth of 1.50 cm below the liquid surface was 1.10 torr. Calculate the surface tension of the liquid.

7. Water at 25°C was observed to rise in a clean glass tube until the meniscus was at a height of 27.4 mm above the bulk surface. What was the radius of the glass tube? The surface tension of water at 25°C is 72.0 dynes/cm.

8. When a clean glass tube with a radius of 1.20 mm was immersed in a pool of mercury at 20°C, the meniscus was depressed 6.05 mm below the surface of the pool. Calculate the surface tension of mercury at 20°C (ρ = 13.6 g/cm³).

9. A fritted glass filter can hold mercury without leakage if the holes in the glass are small enough. What is the radius of the largest hole of circular cross-section if the filter just holds a column of mercury 2.0 cm high at 20°C. (Use data from the preceding problem.)   **Ans.** 3.6×10^{-2} cm

14

Phase Transitions and Equilibria for Pure Substances

Introduction

Practically all processes of interest to chemists involve transitions from one phase to another. No matter how much we might know about gaseous, liquid, and solid phases by themselves, our information would be quite incomplete without some understanding of the conditions under which different phases can coexist in equilibrium and the dynamics of phase change. These two topics are the analogs of the topics of chemical equilibrium and the kinetics of chemical change. For both phase change and chemical change, the equilibrium aspects are easier to treat both experimentally and theoretically in a quantitative manner, largely because the powerful tool of thermodynamics is applicable to all types of equilibria.

In this chapter we start by deriving a general relationship first obtained by Gibbs which is known as the *Gibbs phase rule*. Then we consider phase equilibria for pure substances, and finally present briefly some of the results of the study of the mechanism of phase change.

Phase Equilibrium

Our criterion for phase equilibrium is the one developed in Chapter 3 and applied to chemical equilibria in Chapter 4, namely, for a system to be at equilibrium at constant temperature and pressure, the Gibbs free energy must be at a minimum, and therefore $dG_{T,P} = 0$ for any infinitesimal process. Let us apply this criterion to a closed system at constant T and P in which a pure substance is present in more than one phase, and let us take as our infinitesimal process a transfer of dn moles of the substance from a phase in which its chemical potential is μ_a to a phase in which it is μ_b. Then the free energy change for this process is (compare Eq. 3-59):

$$dG_{T,P} = -\mu_a\, dn + \mu_b\, dn$$

For equilibrium to exist between these two phases, $dG_{T,P} = 0$ and it follows that

$$\mu_a\, dn = \mu_b\, dn$$

Therefore,

$$\mu_a = \mu_b$$

This conclusion is completely general. No matter how complex a system might be and no matter how many phases and components it contains, when this system is at equilibrium the chemical potential of any component in any phase must be equal to the chemical potential of this component in any other phase in which it is present. If this were not so, a spontaneous transfer of this substance from one phase to another could occur and consequently the system would not be in equilibrium.

The Phase Rule

Before we proceed to a derivation of the *phase rule* we need to define rather precisely the two terms, "number of components" and "number of degrees of freedom." By the *number of components*, C, in a system we mean the smallest number of substances from which the entire system can be formed. If the substances in the system do not react with each other, the number of components is simply the number of different substances in the system. If, however, a reaction can occur, according to our definition the number of components is less than the number of substances present. For example, a system containing NH_3, HCl, and NH_4Cl in arbitrary proportions at an elevated temperature is considered to be a 2-component system because it can be formed from NH_4Cl and either NH_3 or HCl, depending on which is present in the larger amount. If the numbers of moles of NH_3 and HCl in the system are equal, the system has only one component because it could be formed from NH_4Cl alone.

By the *number of degrees of freedom*, F, of a system, sometimes called the variance of the system, we mean the number of independent intensive variables which must be specified in order to determine the state of the system at equilibrium. As mentioned earlier (p. 3), the state of a system which consists of one component in one phase is determined by two intensive variables; thus $F = 2$, and the system is said to be *bivariant*. We choose these variables to be the temperature and the pressure of the system.

When more than one component is present in a single phase the composition of the system must be specified, as well as the temperature and the pressure. To specify the composition requires $C - 1$ intensive variables which we take to be mole fractions. There are only $C - 1$ independent compositional variables because when $C - 1$ mole fractions are given, the remaining one is fixed by the requirement that the sum of all of the mole fractions is 1.

Phase Transitions and Equilibria for Pure Substances

If there are $\mathcal{P}$ phases in the system, each with $C - 1$ compositional variables, the total number of compositional variables is $\mathcal{P}(C - 1)$. Including T and P, which have the same values for all phases, the total number of variables is then $\mathcal{P}(C - 1) + 2$. In general, the free energy of the system and therefore the chemical potentials of the components in the system are functions of all of these variables. When the system is at equilibrium, however, not all of these variables are independent. This fact is a mathematical consequence of the thermodynamic requirement that for each component the chemical potential must have the same value in all phases. For each component this requirement results in $\mathcal{P} - 1$ equalities between the chemical potentials, and for all C components there are then $C(\mathcal{P} - 1)$ equalities. For example, for two components (1 and 2) in three phases (a, b, and c), we have the following equalities:

$$\mu_{1a} = \mu_{1b} = \mu_{1c}$$
$$\mu_{2a} = \mu_{2b} = \mu_{2c}$$

As shown below, each of these equalities reduces the number of independent variables by 1. It follows that the number of independent variables, which is the number of degrees of freedom, is given by

$$F = \mathcal{P}(C - 1) + 2 - C(\mathcal{P} - 1)$$
$$= C - \mathcal{P} + 2 \tag{14-1}$$

This equation is known as the Gibbs phase rule.

For a simple example of an application of the phase rule let us take a system consisting of one component distributed between a liquid and a vapor phase. In this case our criterion for equilibrium is expressed by the following equality:

$$\mu_l(T,P) = \mu_v(T,P)$$

The number of degrees of freedom is given by $F = 1 - 2 + 2 = 1$, and thus the system is univariant. This implies that the equilibrium pressure (in this case, the vapor pressure) is a function of the temperature only, or vice versa, in agreement with observation.

To show that in general an equality between functions of the same variables reduces the number of independent variables by 1, we may consider two functions of the same two independent variables, $f(x,y)$ and $g(x,y)$, which are equal over a range of the variables; thus

$$f(x,y) = g(x,y)$$

This equation can be solved for one of the variables in terms of the other variable. In other words, because of the equality only one of the variables, either x or y, is independent in the range in which the equation applies.

From a geometrical point of view, the graph of a function of two independent variables, such as $f(x,y)$, is a surface in 3-dimensional space (Fig. 14–1). The graph of $g(x,y)$ is another surface in the same space. Equality of these functions for a range of values of x and y implies that the two surfaces intersect

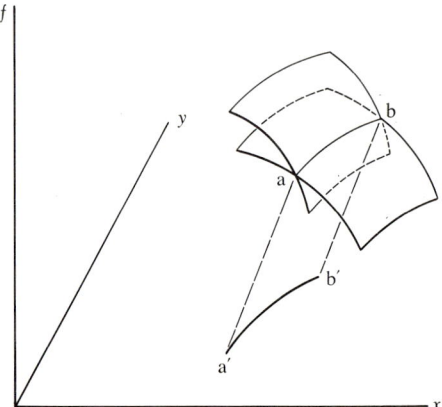

Figure 14–1. Graph showing two surfaces intersecting in a line ab. The line a'b' is the projection of the line ab in the x,y plane.

in this range. Such an intersection is a line whose projection on the x,y plane is the graph of y as a function of x, or conversely.

If a third function is equal to the other two,

$$f(x,y) = g(x,y)$$
$$f(x,y) = h(x,y)$$

these two equations in two variables can be satisfied simultaneously only by a particular value of x and a particular value of y; consequently, *neither* variable is independent. Our results here can be generalized by the statement that each equality between functions of the same set of variables reduces the number of independent variables by 1.

Liquid-Vapor and Solid-Vapor Equilibria

We have just seen that when a pure substance exists in both a liquid and a vapor phase at equilibrium, according to the phase rule $F = 1$ and the system is *univariant*. The vapor pressure is then a characteristic function of the temperature which can be represented by a line in a phase diagram with T and P as coordinates (line CD in Fig. 14–2). This line is a coexistence curve which divides the phase diagram into two regions, each of which represents a 1-phase

Phase Transitions and Equilibria for Pure Substances

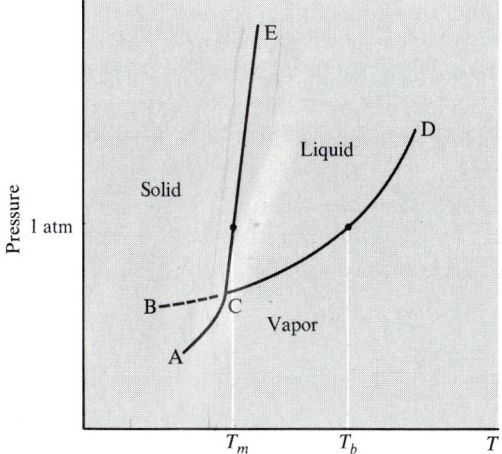

Figure 14–2. Schematic phase diagram for a 1-component system. C is a triple-point and D is the critical point. BC, which is an extension of the vapor pressure curve CD, is the vapor pressure curve of supercooled liquid. AC is the sublimation curve of the solid and CE is the melting curve. T_m is the melting point at 1 atm and T_b is the normal boiling point.

system with $F = 2$ that is all liquid in one case and all vapor in the other. The establishment of equilibrium between vapor and liquid restricts the possible combinations of T and P values to those which lie on the coexistence curve.

We found the form of this coexistence curve when we derived the Clapeyron equation (Eq. 3–52) and applied it to a system consisting of a pure liquid in equilibrium with its vapor. By assuming that the vapor behaves as an ideal gas and that the molar volume of the liquid can be neglected in comparison to the molar volume of the vapor, we obtained the differential form of the Clausius-Clapeyron equation (Eq. 3–53):

$$\frac{d \ln P}{dT} = \frac{\Delta H_v}{RT^2}$$

in which ΔH_v is the molar heat (or enthalpy change) of vaporization.

In Chapter 3 we obtained an expression for the vapor pressure as a function of the temperature (Eq. 3–55) by integrating the above equation under the assumption that ΔH_v is independent of the temperature. In order to carry out this integration on a more general basis we need an expression for ΔH_v as a function of the temperature. This expression is obtained by integration of the Kirchhoff equation (Eq. 2–25), which is

$$d(\Delta H) = \Delta C_P \, dT$$

Since the process under consideration here is vaporization, ΔC_P in this case is the difference between the heat capacities at constant pressure of the liquid and the vapor. Integration of the Kirchhoff equation requires, in turn, an expression for ΔC_P as a function of the temperature.

Because a theoretical value for ΔC_P cannot be calculated at present, we must turn to empirical equations which represent the experimental data for the two heat capacities. It turns out that in most cases it is a good approximation to take ΔC_P as a constant, independent of the temperature. Then the integration of the Kirchhoff equation leads to

$$\int d(\Delta H_v) = \Delta C_P \int dT + \Delta H_o$$

and

$$\Delta H_v = \Delta C_P T + \Delta H_o$$

in which ΔH_o is a constant of integration which must be evaluated from a known value of ΔH_v at some temperature.

Now we can substitute this expression for ΔH_v in Equation 3-53 and carry out the integration as follows:

$$\int d \ln P = \int \frac{\Delta H_o + \Delta C_P T}{RT^2} dT$$

and thus we obtain

$$\ln P = -\frac{\Delta H_o}{RT} + \frac{\Delta C_P}{R} \ln T + \text{constant}$$

This equation is often written in terms of common logarithms in the following form:

$$\log P = -\frac{a}{T} + b \log T + c$$

in which a, b, and c are constants for each substance. It is found that experimental vapor pressure data are well represented by empirical equations of this form. Equation 3-55 is a special case of this equation with ΔC_P (and therefore b) equal to zero.

If we had an adequate molecular theory of the liquid state, this theory would provide an expression for the vapor pressure as a function of the temperature and molecular properties. It then would not be necessary to rely so heavily on experimental measurements. In the absence of such a theory, the thermodynamic relationships between molar properties are indispensable.

From the point of view of the kinetic theory, equilibrium between a liquid and its vapor implies that the rate at which vapor molecules strike the liquid surface and enter the liquid must be equal to the rate at which they evaporate from

Phase Transitions and Equilibria for Pure Substances 519

the surface. In Chapter 5 we derived an equation (Eq. 5–17) from which we can calculate the number of vapor molecules that strike a unit area of surface each second as a function of the vapor pressure. Since condensation is an exothermic process and there is no activation energy, we can expect that each vapor molecule which reaches the surface enters the liquid. Unfortunately, there is no direct way of calculating the rate at which liquid molecules leave the surface and enter the vapor phase, and we cannot use the equality of the rates of condensation and evaporation at equilibrium to calculate the vapor pressure.

Even though our information about the liquid state is incomplete, we can give a molecular interpretation of the form of the relation between P, T, and ΔH_v in Equation 3–53. We saw in Chapter 13 that a molecule in a liquid is in contact on the average with a certain number of nearest neighbors. As a result of these interactions the potential energy of the molecule is less than if it were isolated. When an average molecule moves from the interior of the liquid to the vapor phase work must be done against the attractive intermolecular forces. Consequently the potential energy of the molecule increases by a certain amount, say ϵ_p.

From a statistical point of view the equilibrium distribution of molecules between the liquid and vapor phases is given by the Boltzmann distribution law. Thus if N_l and N_v represent the molecular concentrations in the liquid and vapor phases, respectively, we have the following relationships:

$$\frac{N_v}{N_l} = e^{-\frac{\epsilon_p}{kT}} = e^{-\frac{\Delta E_p}{RT}}$$

and

$$\ln \frac{N_v}{N_l} = -\frac{\Delta E_p}{RT}$$

If the vapor behaves ideally $N_v = P/kT$, and the equation above can be written in the following form:

$$\ln P = \ln kN_l - \frac{\Delta E_p}{RT} + \ln T$$

Because the temperature coefficient of expansion of liquids is small, N_l does not change greatly with a change in the temperature. When the assumption is made that N_l is a constant, differentiation of the equation above with respect to temperature leads to

$$\frac{d \ln P}{dT} = \frac{\Delta E_p}{RT^2} + \frac{1}{T}$$

$$= \frac{\Delta E_p + RT}{RT^2}$$

This equation has the same form as Equation 3–53 with $\Delta H_v \sim \Delta E_p + RT$. We thus see that although it is not yet possible to calculate values of ΔE_p theoretically, the molecular and thermodynamic approaches are mutually consistent.

Dependence of Vapor Pressure on Droplet Size

In Chapter 13 it was pointed out that the molecules in the surface of a liquid exist in a situation which is different from that of the molecules in the interior of the liquid. Up to now we have ignored the effect of surface molecules on the vapor pressure of the liquid. Usually this neglect of the surface molecules is completely justified, for when the liquid is in a container as a single mass the fraction of the molecules in the surface layer is very small. If the liquid, however, is divided into minute droplets the neglect of the surface is not justified because a significant fraction of the molecules is present in surface layers. The pressure of the vapor in equilibrium with the droplets is higher than the equilibrium vapor pressure of the bulk liquid. The dependence of the equilibrium pressure of the droplets on their radius and on the surface energy of the liquid is given by the following equation, usually called the Kelvin equation:

$$\ln P = \ln P_o + \frac{2M\gamma}{RT\rho r} \qquad (14\text{–}2)$$

in which r is the radius of the droplet, P_o is the vapor pressure of the bulk liquid, and M, γ, and ρ are its molecular weight, surface free energy (numerically equal to the surface tension), and density, respectively.

The effect of the surface is not large, as can be seen from the example of water droplets with a radius of 0.1μ (10^{-5} cm) at 25°C. With a γ value of 72 dynes/cm the equation above for this case becomes

$$\ln \frac{P}{P_o} = \frac{2 \times 18 \times 72}{8.32 \times 10^7 \times 298 \times 1.0 \times 10^{-5}} \sim 0.010$$

Consequently $P/P_o \sim 1.01$, and thus the vapor pressure of the droplet is calculated to be about 1% higher than that of the bulk water. Although this change is small, the fact that the vapor pressure increases as the drop size decreases has some interesting consequences which we will consider later.

Equation 14–2 can be derived in the following way: Consider a large amount of liquid at equilibrium with its vapor at temperature T and pressure P_o

under such conditions that the liquid surface is planar. Imagine that an infinitesimal number of moles (dn) of the bulk liquid is transferred to a spherical droplet of radius r. As a result of this transfer the radius of the droplet increases from r to $r + dr$. The only change in free energy involved in this process comes from the increase in the surface area, and therefore in the surface free energy of the droplet. The area of the planar surface is unchanged. Since the surface free energy of the droplet is equal to $4\pi r^2 \gamma$, it follows that

$$dG = d(4\pi r^2 \gamma) = 8\pi \gamma r\, dr$$

Now let us carry out the same process in a different way. First let us reversibly evaporate dn moles of the bulk liquid at its vapor pressure, P_o. There is no change in free energy for this step because it is carried out at equilibrium. Then we raise the pressure of the dn moles of vapor from P_o to P, a step for which the free energy change (p. 87) is $dG = RT \ln P/P_o\, dn$. Finally we reversibly condense this amount of vapor at pressure P on the droplet, and again there is no change in free energy.

Because the initial and final states are the same for both paths, the two free energy changes must be equal. Consequently,

$$RT \ln \frac{P}{P_o} dn = 8\pi \gamma r\, dr \qquad (14\text{--}3)$$

Since the number of moles in a spherical droplet is given by

$$n = \frac{\rho}{M} \frac{4}{3} \pi r^3$$

it follows that

$$dn = \frac{4\pi r^2 \rho}{M} dr$$

When this expression for dn is substituted in Equation 14–3, Equation 14–2 is obtained.

Dependence of Vapor Pressure on Total Pressure

We have made the tacit assumption up to now that the only pressure on the liquid under consideration is the pressure of its own vapor. We now consider the effect on the vapor pressure when an external hydrostatic pressure, P_{ext}, is applied to the liquid. From a thermodynamic point of view, how such a pressure is applied to the liquid is unimportant. We can imagine that it is applied by a membrane that is permeable to the vapor but not to the liquid. In a more practical case the system would contain an inert gas at high pressure.[1]

Let us start with a liquid in equilibrium with its vapor; then $\mu_v = \mu_l$. If the external pressure on the liquid is raised at constant temperature by an amount dP_{ext}, the resulting changes in the chemical potentials of vapor and liquid must be equal and thus

$$d\mu_v = d\mu_l$$

Making use of Equation 3–62 for the change of chemical potential with pressure, we have

$$V_v \, dP = V_l \, dP_{ext}$$

in which dP is the change in the vapor pressure and V_v and V_l are the molar volumes of the vapor and liquid, respectively. It follows that

$$\frac{dP}{dP_{ext}} = \frac{V_l}{V_v}$$

Because V_l is much smaller than V_v, the rate of change of P with P_{ext} is small. With the assumption that the vapor behaves ideally, we have

$$\frac{dP}{dP_{ext}} = \frac{PV_l}{RT}$$

and consequently

$$d \ln P = \frac{V_l}{RT} dP_{ext}$$

V_l changes only slightly with P_{ext} at constant T. If we assume that V_l is independent of P_{ext}, by integrating this equation from $P_{ext\,1}$ to $P_{ext\,2}$ we obtain

$$\ln \frac{P_2}{P_1} = \frac{V_l}{RT}(P_{ext\,2} - P_{ext\,1})$$

and

$$\log P_2 = \log P_1 + \frac{V_l}{2.30\,RT} \Delta P_{ext}$$

For example let us calculate the change in the vapor pressure of water at 25°C produced by a change in external pressure of 50 atm. In this case,

$$\log P_2 = \log 23.8 + \frac{18 \times 50}{2.30 \times 82.1 \times 298}$$

$$P_2 = 24.7 \text{ torr}$$

[1]Although solubility of the inert gas in the liquid complicates the interpretation of vapor pressure measurements carried out on this system, we ignore this complication.

Thus we see that this increase in external pressure raises the vapor pressure of water by about 4%.

Most of what has been said above about liquid-vapor equilibria is applicable with little change to the equilibrium between a pure solid and its vapor. Again the 2-phase system is univariant and the vapor pressure of the solid, usually called its sublimation pressure, is a characteristic function of the temperature. This function can be obtained by integrating the Clausius–Clapeyron equation with the heat of sublimation, ΔH_s, substituted for ΔH_v. For most solids the sublimation pressure at room temperature is very small. A few molecular crystals, notably naphthalene and iodine, have appreciable sublimation pressures well below their melting points because the intermolecular forces are relatively weak.

Equilibrium Between Condensed Phases

A pure solid in equilibrium with its melt is another type of univariant system. The melting point of the solid is then a function of the pressure on the system. In order to obtain this function we must use the Clapeyron equation, which we can write in the following form:

$$\frac{dT}{dP} = \frac{T(V_l - V_s)}{\Delta H_f}$$

ΔH_f is the molar heat of fusion, and V_l and V_s are the molar volumes of the liquid and the solid. As mentioned previously (p. 91) over small temperature ranges it is a good approximation to replace the differential coefficient dT/dP by a ratio of finite differences, $\Delta T/\Delta P$, and to assume that the molar volumes of the condensed phases and the heat of fusion are constant. With these assumptions, the change in the melting point resulting from a pressure change ΔP is given by

$$\Delta T = \frac{T(V_l - V_s)}{\Delta H_f} \Delta P$$

Whether a rise in pressure raises or lowers the melting point depends on the relative values of V_l and V_s, since ΔH_f is always positive. For nearly all substances, V_l is larger than V_s (p. 492), and the melting point is raised by rising pressure. Water is one of the very few substances that expand on freezing and whose melting points are lower at higher pressures.

As an example of an application of the equation above let us calculate the change in the melting point of ice produced by a pressure change of 10 atm. The densities of water and ice at 0°C are 1.000 and 0.917 g/cc, respectively, and the value of ΔH_f is 1430 cal/mole. The change in molar volume on melting is $18(1/1.000 - 1/0.917) = -1.63$ cc/mole. Then ΔT is given by

$$\Delta T = -\frac{273 \times 1.63}{1430} \times 0.0242 \frac{\text{cal}}{\text{cc atm}} \times 10 \text{ atm}$$

$$= -0.075°$$

Vapor-Liquid-Solid Equilibria

When three phases of a pure substance are in equilibrium, according to the phase rule $F = 1 - 3 + 2 = 0$ and the system is said to be *invariant*. In this case there are no variables and both the temperature and the pressure have definite numerical values that are characteristic of the substance. An invariant system is represented on a phase diagram (Fig. 14-2) by a point which is the intersection of two curves, each representing a 2-phase equilibrium. Such a point is called a *triple point*.

For water the pressure and the temperature at the triple point are 4.58 torr and 273.16°K, respectively.[2] Because this temperature is independent of all variables, it can be precisely reproduced anywhere that pure water can be obtained. For this reason the triple-point temperature of water is often used as a primary reference temperature in the calibration of thermometers.

Many pure substances can exist in more than one crystalline phase, a phenomenon known as polymorphism. For example, carbon exists in the forms of diamond and graphite, and sulfur has two crystalline forms, one with a rhombic space lattice and one with a monoclinic lattice. Even for polymorphic substances all of the phase relationships can be represented by a 2-dimensional phase diagram. Because the number of degrees of freedom cannot be negative, in a 1-component system not more than three phases of any type can be in equilibrium at the same time.

Let us consider the sulfur system as an example of a somewhat complex 1-component system. The phase diagram is shown schematically in Figure 14-3. The form of sulfur that is stable, that is, has minimum chemical potential, at atmospheric pressure and room temperature is the form with a rhombic lattice. When rhombic sulfur is slowly heated at atmospheric pressure, at 95.5°C it undergoes a reversible solid-state transition to the monoclinic form. With continued heating, the monoclinic form melts at 119.0°C, and the liquid sulfur boils at 444.6°C.

If rhombic sulfur is heated rapidly, it does not undergo the transition to the monoclinic form but instead it melts at 113°C. In this case the rhombic form is metastable above 95.5°C since there is insufficient time for the formation of the monoclinic structure. The dashed line GE represents the melting curve of

---

[2] In an international conference held in 1954, the triple point of water was adopted as the only fixed point in the definition of the absolute (or thermodynamic) temperature scale. The value of 273.16°K for the temperature of the triple point of water is now considered as exact by definition.

metastable rhombic sulfur. Similarly, the line **BG** is the sublimation curve of metastable rhombic sulfur and **GC** is the vapor pressure curve of metastable liquid sulfur.

The sulfur phase diagram has four triple points. At point B sulfur vapor is in equilibrium with both the rhombic and the monoclinic forms. Point C

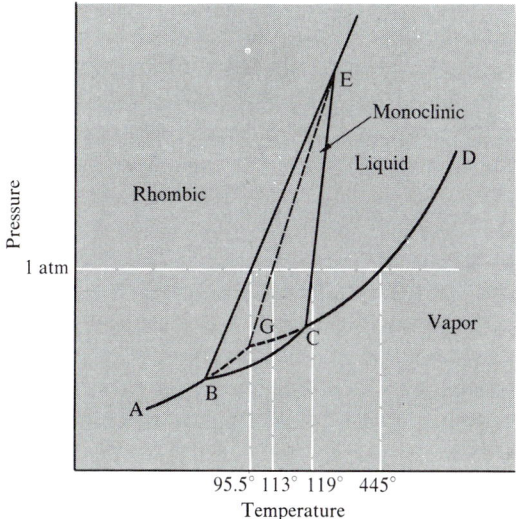

Figure 14-3. Schematic phase diagram for sulfur.

represents the equilibrium of the monoclinic, liquid, and vapor phases, and at point E the liquid is in equilibrium with the two crystalline phases.

The triple point represented by the intersection of the dashed lines at point G is different from the other triple points because it represents a state of metastable equilibrium.

Dynamics of Phase Transitions

Introduction

It has been pointed out previously that thermodynamics does not help us understand processes occurring under nonequilibrium conditions. In particular, it provides no information about the process by which a new phase is formed in a system or the rate at which a new phase will grow. No amount of examination of a phase diagram will yield the details of the transformation of rhombic sulfur to the monoclinic form, discussed in the last section. Neither will it account for the fact that diamonds do not revert spontaneously to graphite at an observable rate, or that liquids can exist at temperatures below their freezing points.

The problem of the initiation and growth of a new phase is important in many subjects other than chemistry. It is of basic importance, for example, to the meteorologist concerned with the conditions under which atmospheric water vapor will condense to form rain drops or fog particles. Likewise, metallurgists are concerned with the growth of desired or undesired solid phases in metallic systems.

We start our discussion of phase transitions with one of the simplest types, the condensation of a vapor to a liquid phase. Many of the ideas introduced in this connection are applicable to other types of phase transitions.

Condensation of Vapor to Form Liquid

Let us consider what happens when a gas below its critical temperature (in other words, a vapor) is isothermally compressed. When the pressure of the system, P, reaches the vapor pressure, P_o, of the substance being compressed, condensation will usually start and the pressure will then remain constant at the value P_o as long as any vapor remains. Under some circumstances, however, the pressure of the system may rise considerably above the vapor pressure

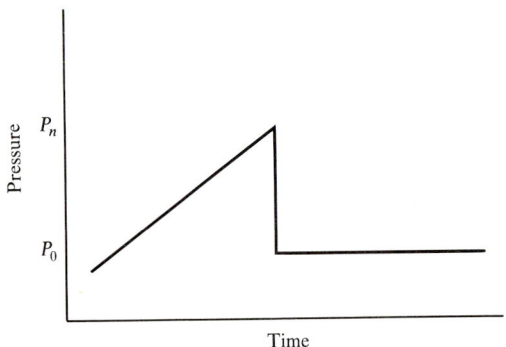

Figure 14-4. Change of pressure with time for the homogeneous nucleation of a vapor.

without any sign of condensation in the vapor phase. When a certain value of the pressure, P_n, is reached condensation will start uniformly throughout the vapor. It then proceeds almost instantaneously and at the same time the pressure of the system drops to the value P_o (Fig. 14-4).

We can interpret these observations qualitatively in the following way: As the pressure of the system rises, the concentration of vapor molecules increases and the rate of molecular collisions, given by Equation 5-20, also increases. At a pressure near P_o randomly distributed molecular clusters, each containing a small number of molecules, begin to form. These clusters are often called *embryos*. An embryo can grow as a result of "sticky" inelastic collisions with vapor molecules. On the other hand, we saw earlier in this chapter that the vapor pressure of very small droplets is significantly higher than the vapor pressure of the bulk liquid. Consequently, even though the system pressure is equal to P_o, the embryos viewed as minute droplets are not in equilibrium with the vapor. When an embryo is formed there is a high probability that it will disappear by vaporization, and for $P < P_n$ the average lifetime of an embryo is very small.

In order for an embryo to continue to grow, it must in some way reach a critical size at which its vapor pressure is less than the system pressure P. Then the rate of growth by gain of vapor molecules is greater than the rate of loss by evaporation. We can say that an embryo whose radius is larger than the critical value, r_c, has become a *nucleus* of the liquid phase. Nuclei, once formed, continue to grow into droplets of liquid which collide and coalesce, thus forming a bulk liquid phase.

The concentration of embryos becomes higher, and their growth into nuclei becomes more probable, as the system pressure increases. The rate of this process increases very rapidly as the pressure approaches a limiting value, P_n, and then in a very short time condensation takes place homogeneously throughout the entire volume occupied by the vapor.

Condensation of a vapor is usually carried out by cooling the vapor. A lowering of the temperature of the system at constant pressure has the same effect as raising the pressure at constant temperature. In this case, both P_o and P_n drop with falling temperature until P_n reaches the system pressure.

It is difficult to make observations of the process just described because condensation by another mechanism for nucleation of a liquid phase usually occurs first. This mechanism, called *heterogeneous nucleation*, results in a rapid condensation when the pressure of the vapor reaches P_o. It may be described in the following way: In the vapor being compressed it is likely that, in addition to the vapor molecules, some suspended minute solid particles are present unless special methods have been used to eliminate them. For lack of a better name we may call these suspended particles "dust" particles. Vapor molecules may become attached to, or adsorbed on, these dust particles. In this case the curvature of the surface and therefore the vapor pressure of the adsorbed molecules is determined by the size of the dust particles, which usually are much larger than the embryos mentioned above. Consequently the vapor pressure of the molecules adsorbed on the dust particles is not much higher than P_o. When the system pressure reaches this value, the nuclei with the dust particles at their centers will grow rapidly with resulting condensation. The walls of the vessel containing the vapor, which are always present, also provide a surface for heterogeneous nucleation. Thus we see that the common mechanism for condensation is heterogeneous nucleation, in which the embryo stage of the vapor molecules is by-passed.

An interesting application of heterogeneous nucleation is the seeding of clouds to induce the precipitation of supersaturated water vapor. It has been found that minute crystals of AgI are particularly effective centers for the nucleation of water vapor.

Another type of particle that has outstanding effectiveness for the heterogeneous nucleation of water vapor is a gaseous ion. This property of ions is the basis for the usefulness of the Wilson cloud chamber in nuclear physics. A rapidly moving charged particle, such as an electron, proton, or α-particle, and also a γ-ray photon, produces ions along its path through a gas. When this gas is supersaturated water vapor resulting from the adiabatic expansion of dust-free air saturated with water, the ions serve as nuclei for the formation of water droplets, which scatter light and make visible the path of the particle or photon.

If the adiabatic expansion of dust-free saturated air is carried out in the absence of ionizing radiation in such a way that at the end of the expansion the partial pressure of the water vapor exceeds the vapor pressure at the existing temperature by a sufficiently large amount, a fog of water droplets will suddenly fill the vessel. This type of condensation is interpreted as the result of homogeneous nucleation. Every liquid has a characteristic value of the *supersaturation ratio*, P_n/P_o, depending on the temperature, at which the rate of homogeneous nucleation is so high that the formation of a fog of droplets is practically instantaneous.

By making several simplifying assumptions, we can calculate the number of molecules, n_c, in an embryo of critical size for a substance in the vapor phase at a pressure P. The major assumptions are: (1) that an embryo which may contain as few as 10 or 20 molecules can be treated as a droplet with a surface tension and density equal to that of the bulk liquid phase; and (2) that thermodynamic equations are applicable to particles this small. The errors resulting from these assumptions are not large except for the smallest embryos which, as mentioned above, are likely to evaporate and thus make a negligible contribution to condensation.

With these assumptions, we can apply Equation 14–2, which was derived for liquid droplets, to embryos. This equation is:

$$\ln \frac{P}{P_o} = \frac{2M\gamma}{RTr\rho}$$

Phase Transitions and Equilibria for Pure Substances 529

Solving this equation for r and replacing $\ln P/P_o$ by x, we obtain

$$r = \frac{2M\gamma}{RTx\rho} \qquad (14\text{-}4)$$

Since this value represents the radius of an embryo that is in equilibrium with vapor molecules at a system pressure P, we can identify it as the critical radius, r_c.

The mass of an embryo is its volume times its density, and this mass divided by the mass of a molecule gives the number of molecules in the embryo. Thus,

$$n_c = \frac{\rho \mathfrak{N}}{M} \frac{4}{3} \pi r_c^3$$

When the value from Equation 14-4 is substituted for r_c the result is

$$n_c = \frac{\rho \mathfrak{N}}{M} \frac{4}{3} \pi \left(\frac{2M\gamma}{RTx\rho}\right)^3$$

$$= \frac{32}{3} \pi \mathfrak{N} \left(\frac{M}{\rho}\right)^2 \left(\frac{\gamma}{RTx}\right)^3$$

In order to get some idea of the magnitude of n_c, we evaluate this expression for water vapor at 275°K. At this temperature the observed value of the supersaturation ratio P_n/P_o when spontaneous nucleation occurs is 4.2. Thus $x = 1.44$ and

$$n_c = \frac{32}{3} \times 3.14 \times 6.02 \times 10^{23} \left(\frac{18.0}{1.00}\right)^2 \left(\frac{75.3}{8.32 \times 10^7 \times 275 \times 1.44}\right)^3$$

$$= 78$$

The corresponding value of the critical radius in this case is 8.2 Å.

For the same system a calculation of the concentration of critical embryos (see p. 533) yields a value of 3.9×10^{-7} cm^{-3} and the rate at which these embryos grow into droplets is about 11 cm^{-3} sec^{-1}. Although these values seem low, they increase very rapidly for even small increases in the supersaturation ratio.

Vaporization

It is a reasonable hypothesis that the maximum rate of vaporization from a liquid surface at a given temperature is equal to the rate at which vapor molecules strike the surface at equilibrium, assuming that all of the molecules striking the surface enter the liquid. This rate in molecules/cm² sec is given by either of the two equivalent equations:

$$R = \frac{1}{4} n\bar{v} \qquad (5\text{–}17)$$

$$R = \frac{P_o}{(2\pi mkT)^{\frac{1}{2}}} \qquad (5\text{–}18)$$

in which $\bar{v}$ is the mean molecular speed and P_o is the vapor pressure at temperature T.

The maximum rate given by the above equations would be expected to be found when a liquid evaporates into a vacuum with no vapor molecules returning to the liquid. The main problem in the experimental determination of vaporization rates that can be used to test Equation 5–18 is the measurement of the surface temperature T. One of the ways in which this problem has been minimized is by allowing liquid drops of known size and temperature to fall through a known distance in an evacuated tube, whose side walls are cooled by liquid air, to a refrigerated container at the bottom. It is necessary to assume that in the short time the drop spends in the evacuated space its temperature is unchanged. The difference in weight between the substance introduced at the top of the evacuated space and the substance collected in the container at the bottom is the amount that evaporated while the drops were falling. From this amount and the time of fall of the drops it is possible to calculate the rate of vaporization at a known temperature.

From measurements on a variety of substances by this and other methods, it has been found that Equation 5–18 is followed quite satisfactorily except for water and other hydrogen-bonded liquids which show a significantly lower rate. Thus, in general, the hypothesis stated above is verified.

Now let us consider vaporization under quite different circumstances. Let us take the case of a liquid that is heated to its boiling point while it is in contact with its vapor and other gases (such as air) at a constant total pressure of 1 atm. When boiling occurs vaporization takes place into bubbles of vapor formed within the body of the liquid. The pressure of the vapor in the bubble is higher than 1 atm as a result of two factors (compare Eq. 13–5): the hydrostatic pressure in the liquid at the location of the bubble ($= \rho gh$) and the pressure due to the surface tension of the liquid ($= 2\gamma/r$). Consequently the temperature of the liquid must be higher than the normal boiling point, at least locally, if the bubble is to form, and the smaller the bubble, the higher is the temperature required.

In this process embryonic bubbles play the same role that embryonic droplets play in condensation, and the theoretical interpretation of the formation and growth of nuclei is basically the same for vaporization as for condensation. An embryonic bubble formed at a temperature just above the boiling point is unstable with respect to a bubble that has the critical radius, and the embryo is almost certain to collapse. If it grows enough to reach the critical radius and thus become a nucleus, it can grow still further, become a visible bubble, and rise through the liquid.

Phase Transitions and Equilibria for Pure Substances

If, as is often the case, heat is supplied through the bottom of the vessel containing the liquid, the layer of liquid at the bottom is superheated with respect to liquid at a higher level. When a bubble that is formed at the bottom rises to a cooler layer it becomes unstable and rapidly collapses. This behavior is responsible for the sometimes violent form of ebullition known as bumping.

The process just described results from homogeneous nucleation. If the liquid contains some dissolved gas whose solubility decreases with rising temperature, the gas forms bubbles in the liquid and near the boiling point these gas bubbles become nuclei for vaporization. Suspended solid particles such as dust also act as nuclei. In this case the main factor promoting nucleation is probably the fact that less energy is required to form the surface of a vapor bubble by detaching a liquid from a solid surface than by separation of the liquid from itself.

Phase Transitions Involving Solids

We can classify the phase transitions of solids into the following types: solid-vapor, solid-liquid, and solid-solid. We briefly consider each of these types in sequence.

1. *Solid-vapor transitions.* Condensation and vaporization are basically the same for solids as for liquids except for complications resulting from the structure of a crystalline solid phase. Not only must a condensation nucleus have a critical size but in addition, the molecules in the nucleus must be arranged in a crystal lattice. Accordingly, the shape of the nucleus instead of being a sphere is some kind of polyhedron for which the total free energy, including the surface energy, is minimized. As a result of the existence of surface energy the sublimation pressure of crystalline nuclei, and crystallites in general, is higher than that of macroscopic crystals, just as the vapor pressure of droplets is higher than that of the bulk liquid.

When a crystal grows by condensation of vapor, the different crystal faces in general grow at different rates, the faces with the lowest surface energy growing at the highest rate. Thus the shape of a crystal can be changed substantially as it grows. As mentioned on p. 485, the growth of a face of a crystal is facilitated by the presence in the growing face of certain types of crystal defects, such as dislocations ending in the surface, which act as surface nuclei. In an analogous way vaporization of a crystal occurs at a higher rate from some faces than from others.

2. *Solid-liquid transitions.* As the temperature of a crystal is slowly raised at constant pressure, the properties of the crystal change only slightly until a temperature is reached at which the structure of the crystal breaks down. Then the crystal disappears completely and is replaced by a liquid with quite different values of some properties, notably heat capacity and viscosity. The sharpness of the melting point, in other words, the narrowness of the temperature range in which melting occurs, is a remarkable phenomenon. It can be attributed to the

fact that a crystal lattice can tolerate only a very limited amount of disorder. With rising temperature the amplitude of the lattice vibrations increases to a magnitude that requires relatively large numbers of molecules to leave their equilibrium positions at the same time. This is an example of what is known as a *cooperative transition*.

The freezing of a pure liquid can take place by either homogeneous or heterogeneous nucleation. Many liquids, when carefully freed of suspended particles, can be cooled well below the freezing point without crystallization and held in this metastable state. Once crystallization starts, however, it proceeds very rapidly. This behavior makes it possible to make an accurate measurement of the freezing point of a liquid, even though some supercooling almost always occurs as the temperature is lowered. When crystallization starts, the rapid release of the heat of fusion raises the temperature of the system, and with adequate stirring the equilibrium temperature is maintained as long as both solid and liquid are present.

When the formation of a large single crystal from a melt is desired, it is necessary to purify the starting liquid as much as possible to minimize the number of nuclei present. Then a small portion of the liquid near one of the walls of the container is brought to a temperature just below the freezing point while the main part of the liquid is held just above this temperature. A seed crystal is introduced into the cooler portion of the liquid to start crystallization at one point. Then the temperature gradient is moved through the liquid at a controlled rate until crystallization is complete.

If a pure liquid can be supercooled without crystallization to a temperature at which its viscosity is extremely high, it forms a glass whose structure, as mentioned earlier, is the same as that of the liquid. On page 509 it was pointed out that this type of behavior is often observed with compounds whose molecules consist of (or contain) long chains, since such molecules do not readily fit into a crystal lattice. It is also observed with some compounds, such as glycerol, that are strongly hydrogen-bonded in the liquid phase.

Another kind of solid-liquid transition is the formation of a solid phase from a solution. (The equilibrium aspects of this transition will be discussed in the next chapter.) Many solutions can be kept in a supersaturated state almost indefinitely, provided that nuclei of the solid solute are absent. When precipitation of the solute occurs homogeneously, the first stage is a suspension of fine crystallites. Because a finely divided solid has a very large surface area per unit mass, it also has a relatively large surface energy. This large surface energy accounts for the fact that a finely divided solid has a higher solubility than the same substance in bulk form. Since crystallites are unstable with respect to the solution in the presence of macroscopic crystals, large crystals will grow in a saturated solution at the expense of the smaller ones.

3. *Solid-solid transitions*. It has been mentioned previously (p. 523) that many substances are polymorphic, that is, they can exist in more than one crys-

talline form. The outstanding fact about transitions between such forms is their very low rate, particularly at temperatures well below the melting point, even though the transitions are favored by a relatively large difference in chemical potentials between the two forms. This low rate is, of course, a consequence of the limited mobility of the particles in a crystal lattice and the necessity for a considerable rearrangement of these particles during a transition from one lattice type to another. The presence of crystal defects markedly increases the rate of such a rearrangement.

Supplementary Material

Rate of Homogeneous Nucleation in the Vapor Phase

In order for a critical embryo to contribute to condensation by homogeneous nucleation, it must gain an additional molecule before it loses a molecule by evaporation. We can represent the process of growth by the following "chemical" equation:

$$A_c + A \rightarrow A_{c+1}$$

in which A_c and A represent a critical embryo and a vapor molecule, respectively.

The rate of formation of nuclei, R, is equal to the number of critical embryos that gain a molecule per second per cubic centimeter. If we consider the growth process as a bimolecular reaction, its rate is given by the rate of collision of vapor molecules with a critical embryo multiplied by the concentration of these embryos, c_c, assuming that every collision is effective.

Let us consider the rate of nucleation in a system containing vapor molecules at a concentration c_1, with temperature T and a pressure P that is higher than P_o, the vapor pressure of the bulk liquid at the same temperature. The number of collisions per second made by vapor molecules with a surface of unit area is, by Equation 5–17, $1/4\ c_1 \bar{v}$ in which $\bar{v}$ is the mean molecular speed. The surface area of a critical embryo with a spherical shape and radius r_c is $4\pi r_c^2$. Thus the rate R is given by the following expression:

$$R = \tfrac{1}{4} c_1 \bar{v} \times 4\pi r_c^2 \times c_c \tag{14-5}$$

We now need to find an expression for c_c. We assume that the concentration, c_r, of embryos with radius r, containing n_r molecules, depends on the free energy change, ΔG_r, in forming an embryo from n_r vapor molecules. We also assume that the distribution function for embryos has the Boltzmann form. Thus we use the following distribution function:

$$c_r = c_1 e^{-\frac{\Delta G_r}{kT}}$$

(This equation implies that the total concentration of all embryos is negligible with respect to the concentration of vapor molecules.)

The expression for ΔG_r consists of two terms. One of these is the free energy change in transferring n_r molecules from the vapor phase at pressure P to the liquid phase. This term can be expressed as the free energy change per unit volume of liquid times the volume of the embryo. The second term is the free energy of formation of the surface of the embryo which is the product of the surface energy (or tension) and the area of the embryo surface. With μ_l and μ_v representing the chemical potentials of the liquid and vapor, respectively, the total free energy change for the formation of a nucleus whose radius is r is then

$$\Delta G_r = \frac{\mu_l - \mu_v}{V_m} \times \frac{4}{3}\pi r^3 + 4\pi r^2 \gamma \tag{14-6}$$

in which V_m is the molar volume of the liquid.

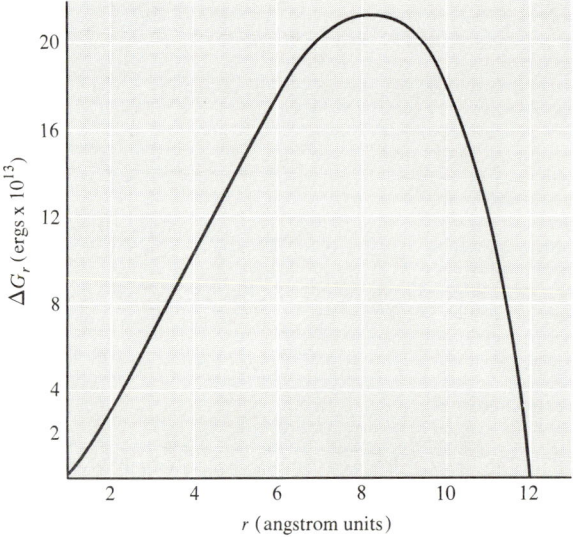

Figure 14-5. Graph of ΔG_r vs radius of embryo r.

ΔG_r as a function of r (Fig. 14–5) has a maximum at a value of r, $r(\Delta G_{\max})$, which can be found from the usual condition, $d\,\Delta G_r/dr = 0$. Thus,

$$\frac{d\,\Delta G_r}{dr} = \frac{\mu_l - \mu_v}{V_m} \times 4\pi r^2 + 8\pi r \gamma = 0$$

and it follows that

$$r(\Delta G_{\max}) = -\frac{2\gamma V_m}{\mu_l - \mu_v} \tag{14-7}$$

Phase Transitions and Equilibria for Pure Substances

This value for $r(\Delta G_{max})$ is necessarily positive, since by hypothesis $\mu_v > \mu_l$.

Whether r increases or decreases from the value just found, ΔG decreases. Therefore $r(\Delta G_{max})$ represents a critical value of the radius for which an embryo is in unstable equilibrium with respect to the vapor, and we conclude that $r(\Delta G_{max})$ must be equal to r_c.

The same conclusion can be reached from the relationship $\mu_v = \mu_v^o + RT \ln P$ by recognizing that μ_l is equal to the chemical potential of vapor at pressure P_o that is in equilibrium with bulk liquid. Thus $\mu_l = \mu_v^o + RT \ln P_o$, and

$$\mu_l - \mu_v = -RT \ln P/P_o = -RTx \tag{14-8}$$

When this substitution is made and V_m is replaced by its equal, M/ρ, Equation 14-7 becomes identical with Equation 14-4.

In order to obtain an expression for ΔG_c for the formation of a critical embryo, we substitute Equation 14-7 into Equation 14-6, as follows:

$$\Delta G_c = \frac{\mu_l - \mu_v}{V_m} \times \frac{4}{3}\pi\left(-\frac{2\gamma V_m}{\mu_l - \mu_v}\right)^3 + 4\pi\gamma\left(-\frac{2\gamma V_m}{\mu_l - \mu_v}\right)^2$$

$$= -\frac{32}{3}\pi\gamma^3\left(\frac{V_m}{\mu_l - \mu_v}\right)^2 + 16\pi\gamma^3\left(\frac{V_m}{\mu_l - \mu_v}\right)^2 \tag{14-9}$$

$$= \frac{16}{3}\pi\gamma^3\left(\frac{V_m}{\mu_l - \mu_v}\right)^2$$

When $\mu_l - \mu_v$ is replaced by its equal from Equation 14-8, the result is[3]

$$\Delta G_c = \frac{16}{3}\pi\gamma^3\left(\frac{V_m}{RT}\right)^2\left(\ln\frac{P}{P_o}\right)^{-2}$$

For the particular case of water at 275°K with a value of P/P_o of 4.20, the value of ΔG_c is

$$\Delta G_c = \frac{16}{3} \times 3.14(75.3)^3 \left(\frac{18.0}{8.32 \times 10^7 \times 275}\right)^2 \left(\frac{1}{1.44}\right)^2$$

$$= 2.12 \times 10^{-12} \text{ ergs}$$

and $\Delta G_c/kT = 55.8$. The concentration of the water molecules is given by P/kT where $P = 4.20 \times 5.29 \times 1.33 \times 10^3 = 2.96 \times 10^4$ dynes/cm². Then

$$c_c = \frac{2.96 \times 10^4}{1.38 \times 10^{-16} \times 275} e^{-55.8}$$

$$= 7.8 \times 10^{17} \times 5.0 \times 10^{-25}$$

$$= 3.9 \times 10^{-7} \text{ cm}^{-3}$$

---

[3] The work, w, done in forming the surface of a critical embryo at constant T and P is equal to the free energy of formation of the surface which is given by the second term of Equation 14-9. Thus it follows that $\Delta G_c = w/3$.

The process of formation of nuclei is opposed by their removal due to evaporation as well as their growth into droplets. Consequently, the net rate of condensation is less than the value given by Equation 14–5. The result of a rather lengthy calculation of the net rate, which is not given here, is the following expression:

$$R_{net} = \frac{1/4 \, c_1 \bar{v} V_m c_1}{r_c \mathfrak{N}} \left(\frac{3 \Delta G_c}{\pi kT}\right)^{\frac{1}{2}} e^{-\frac{\Delta G_c}{kT}}$$

On introduction of the appropriate expressions for c_1, $\bar{v}$, r_c, and ΔG_c into the pre-exponential factor, this equation reduces to

$$R_{net} = (2\gamma)^{\frac{1}{2}} \left(\frac{P}{kT}\right)^2 \frac{V_m}{(\pi M \mathfrak{N})^{\frac{1}{2}}} e^{-\frac{\Delta G_c}{kT}}$$

When the values for water at 275°K and a pressure of 2.96×10^4 dynes/cm² are used in this equation, the result is $R_{net} \sim 11$ cm⁻³ sec⁻¹.

The calculated values of c_c and R_{net} seem rather low, but they increase very rapidly as P increases because of the exponential factor. For a pressure of 3.87×10^4 dynes/cm², corresponding to a value of 5.0 for P/P_o, $\Delta G_c/kT = 44.7$, $c_c = 0.04$ cm⁻³, and $R_{net} \sim 10^6$ cm⁻³ sec⁻¹. Thus as a result of an increase in P of about 30%, c_c and R_{net} have increased by a factor of about 10^5.

Problems

1. What is the number of degrees of freedom of a system that contains equimolar amounts of PCl_3 vapor and Cl_2 gas in equilibrium with PCl_5 vapor?

Ans. 2

2. A system containing four components was found to be invariant. How many phases are present in this system?

Ans. 6

3. Show that for a pure substance the largest number of phases that can coexist in equilibrium is three.

4. The vapor pressure of acetone is 282.7 torr at 30°C and 612.6 torr at 50°C. Calculate (a) ΔH_v for acetone, and (b) its normal boiling point.

Ans. (a) 7.50 kcal/mole; (b) 56°C

5. The normal boiling point of carbon tetrachloride is 76.8°C and its vapor pressure at 50°C is 317 torr. Calculate the value of ΔH_v.

6. Between 10°C and 30°C, the vapor pressure of benzene in torr is given by the following equation: $\log P = -1780/T + 7.960$. Calculate ΔH_v.

Ans. 8.15 kcal/mole

7. In a temperature range near the normal boiling point, the vapor pressure of a certain liquid in atmospheres is given by the equation: $\log P = -3680/T + 10.52$. Calculate (a) the heat of vaporization of this liquid, and (b) its normal boiling point.

8. Calculate the pressure of mercury vapor in equilibrium with mercury droplets with a radius of 0.20μ at 25°C. The vapor pressure of liquid mercury at 25°C is 1.85×10^{-3} torr, its surface tension is 484 dynes/cm, and its density is 13.6 g/cm³. **Ans.** 1.91×10^{-3} torr

9. At 20°C the vapor pressure of CCl_4 is 91 torr and its density is 1.595 g/cm³. How high an external pressure must be applied to CCl_4 at 20°C to double the vapor pressure?

10. The vapor pressure of solid CsI at a series of temperatures is:

T(°K)	767.2	801.8	816.3	830.3	846.8
P(dynes/cm²)	2.03	7.45	12.5	20.5	36.4

Find the heat of sublimation of CsI. **Ans.** 47.2 kcal/mole

11. The melting point of acetone at atmospheric pressure is −94.8°C. At this temperature the densities of the solid and the liquid are 0.987 and 0.929 g/cc, respectively, and the heat of fusion is 23.4 cal/g. Calculate the melting point of acetone at 50 atm.

12. The vapor pressures for solid and liquid UF_6 in torr are given by the equations:

$$\log P_s = 10.648 - 2559.5/T$$
$$\log P_l = 7.540 - 1511.3/T$$

Calculate (a) the heat of fusion of UF_6; (b) its triple point temperature.
Ans. (a) 4.80 kcal/mole; (b) 337°K

13. In the neighborhood of the triple point involving sulfur vapor and the monoclinic and rhombic crystalline forms, the vapor pressure of the monoclinic form in torr is given by the following equation: $\log P_m = -5082/T + 11.364$. The corresponding equation for the rhombic form is: $\log P_r = -5267/T + 11.866$. (a) Calculate the triple point temperature and pressure. (b) Calculate ΔH for the transition $r \to m$ at the triple-point.

14. Calculate the maximum rate of vaporization of toluene (C_7H_8) in moles/cm² sec from a surface maintained at 30°C. The vapor pressure of toluene at 30°C is 36.7 torr. **Ans.** 1.28×10^{-2} mole/cm² sec

15

Phase Equilibria in Multicomponent Systems

Introduction

In this chapter we consider the subject of equilibria in systems containing more than one component. A phase that contains more than one component, and thus has a composition that can be varied, is called a *solution*. It is often convenient to call one of the components of a solution, usually the one at the highest concentration, the solvent, and the others the solutes, but in many cases this distinction has little significance. We first focus on binary systems, that is, those containing two components.

Because all gases are miscible with each other in all proportions, any gaseous mixture can be considered as a solution. Liquid solutions may be composed of a gas and a liquid, two liquids, or a solid and a liquid. The most common type of solid solution contains two solids, although solutions of a gas or a liquid in a solid are known. We devote most attention to liquid solutions.

A solution containing two components has 3 degrees of freedom ($F = 2 - 1 + 2$), which are usually taken to be pressure, temperature, and composition. The phase diagram for such a system is thus a 3-dimensional graph. For convenience it is common to use 2-dimensional sections of this 3-dimensional graph in which either the temperature or the pressure is held constant.

For a binary solution in equilibrium with its vapor, $F = 2 - 2 + 2$ and there are 2 degrees of freedom usually taken to be the composition and the temperature. If the composition of the solution is fixed, the pressure is a function of the temperature only. When a binary solution is in equilibrium with both a solid phase and a vapor phase, there is only 1 degree of freedom which is usually taken to be the temperature.

When a solution is in equilibrium with its vapor, the chemical potential of each component must, as we have seen, be the same in the liquid phase as in the vapor phase. Because the chemical potential of a component in the vapor phase depends on its partial pressure if the vapor behaves ideally, the study of the vapor pressure of a solution as a function of composition provides information about the chemical potentials in the liquid phase.

A wide variety of solutions are known with different types of vapor pressure-composition relationships, and it is difficult to discuss these in general terms. Fortunately there is a type of solution, called the ideal solution, that shows a particularly simple type of behavior, and we consider this type first.

Ideal Solutions Containing a Nonvolatile Solute

Raoult's Law

Let us consider a solution that consists of a solvent A and a solute B with negligible vapor pressure at some definite temperature, T. Many years ago (1887) Raoult found that for dilute solutions of this type, when equilibrium is established between the solution and its vapor, the vapor pressure of the solvent, P_A, is related to its mole fraction in the solution, X_A, by the following equation, now known as *Raoult's law*:

$$P_A = X_A P_A^\circ \tag{15-1}$$

in which P_A° represents the vapor pressure of pure A at temperature T.

An equivalent form of Raoult's law is expressed in terms of ΔP_A, the lowering of the vapor pressure of A in the solution relative to the vapor pressure of pure A, as follows:

$$\Delta P_A = P_A^\circ - P_A$$
$$= P_A^\circ - X_A P_A^\circ$$

On dividing by P_A°, we have

$$\frac{\Delta P_A}{P_A^\circ} = 1 - X_A = X_B \tag{15-2}$$

For example, if $P_A = 0.9 P_A^\circ$, the mole fraction of B is 0.1.

With the assumption that the vapor behaves as an ideal gas, the chemical potential of A in the vapor phase is given by Equation 3–63 which is

$$\mu_A(v) = \mu_A^\circ(v) + RT \ln P_A$$

in which $\mu_A^\circ(v)$ is the chemical potential of pure A in the vapor phase at a pressure of 1 atm. At equilibrium the chemical potential of A in the solution must be given by the same expression, since $\mu_A(v) = \mu_A(\text{soln})$. Substituting P_A by its equal from Raoult's law, Equation 15–1, we have

$$\mu_A(\text{soln}) = \mu_A^\circ(v) + RT \ln X_A P_A^\circ$$

Phase Equilibria in Multicomponent Systems

This equation can be written in the following equivalent forms:

$$\mu_A(\text{soln}) = \mu_A^\circ(v) + RT \ln P_A^\circ + RT \ln X_A$$

and

$$\mu_A(\text{soln}) = \mu_A^\circ(l) + RT \ln X_A \qquad (15\text{-}3)$$

in which $\mu_A^\circ(l)$ is the chemical potential of pure liquid A at temperature T under its own vapor pressure, P_A°. These equations are the basis for most of the treatment of solutions which follows.

Ideal Solutions

A solution for which the chemical potential of each component is related to its mole fraction by Equation 15–3 is said to be an *ideal solution*. No actual solution is ideal over the entire range of compositions but all solutions approach ideality with increasing dilution, that is, as the mole fraction of the solvent approaches 1.

The concept of ideal solution plays a role in the treatment of solutions analogous to that which the concept of ideal gas plays in the treatment of gases. In each of these cases the behavior of the ideal system is relatively simple, and the laws describing this behavior are often adequate approximations when applied to real systems. Deviations of solutions from ideality can be substantial, however, and we will consider them at a later point.

Equation 15–3 implies that when an ideal solution is formed from its pure components, the volume of the solution is the sum of the volumes of the individual components. Likewise, the heat of solution is zero, and consequently the free energy change in forming an ideal solution is due entirely to the entropy change of mixing ($\Delta G = -T \Delta S$).

In order to prove these statements let us first differentiate Equation 15–3 with respect to the pressure at constant composition and temperature. Because the second term on the right-hand side of this equation is independent of the pressure, we have

$$\left(\frac{\partial \mu_A(\text{soln})}{\partial P}\right)_{T,X_A} = \left(\frac{\partial \mu_A^\circ(l)}{\partial P}\right)_T \qquad (15\text{-}4)$$

When we substitute for μ_A in the left-hand side of this equation its equal from Equation 3–57, we obtain

$$\left(\frac{\partial \mu_A(\text{soln})}{\partial P}\right)_{T,X_A} = \left[\frac{\partial}{\partial P}\left(\frac{\partial G(\text{soln})}{\partial n_A}\right)_{T,P,n_B}\right]_{T,X_A}$$

Because the order of the two partial differentiations in the right-hand side of this equation is not significant (Eq. 2–2), we can reverse this order. Thus we obtain

$$\left(\frac{\partial \mu_A(\text{soln})}{\partial P}\right)_{T,X_A} = \left[\frac{\partial}{\partial n_A}\left(\frac{\partial G(\text{soln})}{\partial P}\right)_{T,X_A}\right]_{T,P,n_B}$$

From Equation 3–48, $\left(\frac{\partial G(\text{soln})}{\partial P}\right)_{T,X_A}$ is equal to the molar volume of the solution. It follows that

$$\left(\frac{\partial \mu_A(\text{soln})}{\partial P}\right)_{T,X_A} = \left(\frac{\partial V(\text{soln})}{\partial n_A}\right)_{T,P,n_B}$$

Since $\left(\frac{\partial V(\text{soln})}{\partial n_A}\right)_{T,P,n_B} = \overline{V}_A$, the partial molar volume of A in the solution (p. 92), we have for the left-hand side of Equation 15–4

$$\left(\frac{\partial \mu_A(\text{soln})}{\partial P}\right)_{T,X_A} = \overline{V}_A \tag{15-5}$$

The right-hand side of Equation 15–4 refers to pure A only, and thus it is equal to the molar volume of pure A. It follows that for each component the partial molar volume at any mole fraction is equal to the molar volume of this pure component. Consequently, there is no change in volume when the components are mixed to form an ideal solution.

In a similar way we show that the heat of solution is zero for an ideal solution. Let us divide both sides of Equation 15–3 by T and then differentiate with respect to T at constant P and composition. Since $R \ln X_A$ is independent of the temperature, we obtain

$$\left(\frac{\partial \frac{\mu_A(\text{soln})}{T}}{\partial T}\right)_{P,X_A} = \left(\frac{\partial \frac{\mu_A^\circ(l)}{T}}{\partial T}\right)_P \tag{15-6}$$

Noting that $\mu_A^\circ(l)$ is simply $G_A(l)$, the free energy of 1 mole of pure liquid A, we can carry out the differentiation indicated in the right-hand side of this equation in the following way:

$$\left(\frac{\partial \frac{\mu_A^\circ(l)}{T}}{\partial T}\right)_P = \left(\frac{\partial \frac{G_A(l)}{T}}{\partial T}\right)_P$$

$$= \frac{1}{T}\left(\frac{\partial G_A(l)}{\partial T}\right)_P - \frac{G_A(l)}{T^2}$$

Phase Equilibria in Multicomponent Systems 543

When we substitute for $\left(\dfrac{\partial G_A(l)}{\partial T}\right)_P$ its equal, $-S_A(l)$ (Eq. 3–49), and for $G_A(l)$ its equal from the definition of the Gibbs free energy, the result is

$$\left(\dfrac{\partial \dfrac{\mu_A^\circ(l)}{T}}{\partial T}\right)_P = -\dfrac{S_A(l)}{T} - \dfrac{H_A(l)}{T^2} + \dfrac{S_A(l)}{T}$$

$$= -\dfrac{H_A(l)}{T^2} \qquad (15\text{–}7)$$

in which $H_A(l)$ is the molar enthalpy of pure liquid A. (Compare the derivation of Eq. 3–51.)

We now treat the left-hand side of Equation 15–6 in an analogous manner. Carrying out the indicated differentiation, we obtain

$$\left(\dfrac{\partial \dfrac{\mu_A(\text{soln})}{T}}{\partial T}\right)_{P,X_A} = \dfrac{1}{T}\left(\dfrac{\partial \mu_A(\text{soln})}{\partial T}\right)_{P,X_A} - \dfrac{\mu_A(\text{soln})}{T^2} \qquad (15\text{–}8)$$

When we replace $\mu_A(\text{soln})$ in the first term on the right by its equal, $\left(\dfrac{\partial G(\text{soln})}{\partial n_A}\right)_{T,P,n_B}$, this term becomes

$$\dfrac{1}{T}\left(\dfrac{\partial \mu_A(\text{soln})}{\partial T}\right)_{P,X_A} = \dfrac{1}{T}\left[\dfrac{\partial}{\partial T}\left(\dfrac{\partial G(\text{soln})}{\partial n_A}\right)_{T,P,n_B}\right]_{P,X_A}$$

Reversing the order of partial differentiation, we obtain

$$\dfrac{1}{T}\left(\dfrac{\partial \mu_A(\text{soln})}{\partial T}\right)_{P,X_A} = \dfrac{1}{T}\left[\dfrac{\partial}{\partial n_A}\left(\dfrac{\partial G(\text{soln})}{\partial T}\right)_{P,X_A}\right]_{T,P,n_B}$$

From

$$\left(\dfrac{\partial G(\text{soln})}{\partial T}\right)_{P,X_A} = -S(\text{soln}),$$

and

$$\left(\dfrac{\partial S(\text{soln})}{\partial n_A}\right)_{T,P,n_B} = \overline{S}_A(\text{soln})$$

in which $\overline{S}_A(\text{soln})$ is the partial molar entropy of A in the solution, it follows that

$$\dfrac{1}{T}\left(\dfrac{\partial \mu_A(\text{soln})}{\partial T}\right)_{P,X_A} = -\dfrac{\overline{S}_A}{T}(\text{soln}) \qquad (15\text{–}9)$$

Since

$$\mu_A(\text{soln}) = \left(\dfrac{\partial G(\text{soln})}{\partial n_A}\right)_{T,P,n_B}$$

substitution of $G(\text{soln})$ by its equal leads to

$$\mu_A(\text{soln}) = \left[\frac{\partial}{\partial n_A}\{H(\text{soln}) - TS(\text{soln})\}\right]_{T,P,n_B}$$
$$= \overline{H}_A(\text{soln}) - T\overline{S}_A(\text{soln})$$

When this relation and Equation 15–9 are introduced into the right-hand side of Equation 15–8, this equation reduces to

$$\left(\frac{\partial \frac{\mu_A(\text{soln})}{T}}{\partial T}\right)_{P,X_A} = -\frac{\overline{H}_A(\text{soln})}{T^2}$$

Thus Equation 15–6 becomes

$$-\frac{\overline{H}_A(\text{soln})}{T^2} = -\frac{H_A(l)}{T^2}$$

Because the partial molar enthalpy of A in the solution is equal to the molar enthalpy of pure liquid A, it follows that the enthalpy change of mixing, that is, the heat of solution, is zero.

If an ideal solution contains n_A moles of A and n_B moles of B, its free energy is given by Equation 3–60, as follows:

$$G(\text{soln}) = n_A\mu_A(\text{soln}) + n_B\mu_B(\text{soln})$$

The free energy change in forming the solution from its pure components is

$$\Delta G = n_A(\mu_A(\text{soln}) - \mu_A^\circ(l)) + n_B(\mu_B(\text{soln}) - \mu_B^\circ(l))$$

In view of Equation 15–3, this equation becomes

$$\Delta G = RT(n_A \ln X_A + n_B \ln X_B)$$

(Notice that ΔG is always negative.) To find the corresponding entropy change, we differentiate this equation with respect to T at constant composition and we thus obtain

$$\Delta S = -\left(\frac{\partial \Delta G}{\partial T}\right)_{X_A}$$
$$= -R(n_A \ln X_A + n_B \ln X_B)$$

For 1 mole of solution ($n_A + n_B = 1$), this equation becomes

$$\Delta S = -R(X_A \ln X_A + X_B \ln X_B)$$

This is the same as Equation 3–16 for the molar entropy of mixing of ideal gases.

Because the enthalpy and volume changes of mixing are each equal to zero for an ideal solution, it follows that the energy of mixing ($\Delta E = \Delta H - P \Delta V$) is also equal to zero. In other words, the transfer of a molecule from one of the pure liquids to the solution does not result in an energy change. This implies that the intermolecular forces between A and B molecules must be the same, or nearly so, as those between two A molecules or two B molecules. The volumes occupied by A and by B molecules must also be nearly the same in an ideal solution.

We now demonstrate that if the above requirements for equality of intermolecular forces and volumes are met, a statistical derivation of Raoult's law can be carried out on the basis of the lattice model of a liquid (p. 496). Although this model is overly simplified, the ideas involved in the derivation are significant.

Because of the assumed similarity of their molecules, pure A and B in the liquid phase will have the same type of lattice structure. When n_A molecules of A and n_B molecules of B are mixed, at equilibrium the two kinds of molecules will be randomly distributed among the $n_A + n_B$ lattice sites in the liquid solution. The number of ways such a distribution can be realized is equal to the number of ways that $n_A + n_B$ lattice sites can be divided into two sets with n_A molecules in one set and n_B in the other. According to Equation 9-3, this number is given by

$$w = \frac{(n_A + n_B)!}{n_A! n_B!}$$

Before the components are mixed, there is only one distinguishable way of distributing the n_A molecules of A among the n_A lattice sites and the n_B molecules of B among the n_B lattice sites. Thus the entropy of mixing, as given by the Boltzmann relation (Eq. 9-19), is

$$\Delta S = k \ln \frac{w}{1}$$

$$= k \ln \frac{(n_A + n_B)!}{n_A! n_B!}$$

Making use of the Stirling approximation (p. 341), we obtain

$$\Delta S = k[(n_A + n_B) \ln (n_A + n_B) - (n_A + n_B) - n_A \ln n_A + n_A - n_B \ln n_B + n_B]$$

$$= k \left[n_A \ln \frac{(n_A + n_B)}{n_A} + n_B \ln \frac{(n_A + n_B)}{n_B} \right]$$

$$= -k(n_A \ln X_A + n_B \ln X_B)$$

If we now convert n_A and n_B from numbers of molecules to numbers of moles and replace k by R, the equation above becomes

$$\Delta S = -R(n_A \ln X_A + n_B \ln X_B)$$

The change in free energy for the formation of the solution from the pure components in the liquid phase is $\Delta G = -T\Delta S$, since ΔE and ΔH are both zero. Thus

$$\Delta G = RT(n_A \ln X_A + n_B \ln X_B)$$

The change in the chemical potential of A for the formation of the solution is given by[1]

$$\mu_A - \mu_A^\circ = \left(\frac{\partial \Delta G}{\partial n_A}\right)_{n_B, T, P}$$
$$= RT \ln X_A \quad (15\text{-}10)$$

The values of the chemical potential of A in the vapor phase over the solution and over pure liquid A are given by $\mu_A^\circ + RT \ln P_A$ and $\mu_A^\circ + RT \ln P_A^\circ$, respectively. Because these chemical potentials are equal to the corresponding ones in the liquid phase, it follows that

$$\mu_A - \mu_A^\circ = RT \ln \frac{P_A}{P_A^\circ}$$

A comparison of this equation with Equation 15-10 shows that $X_A = P_A/P_A^\circ$, which is an expression of Raoult's law.

Colligative Properties

It was mentioned earlier that all solutions become more nearly ideal as they become more dilute, and consequently their properties become easier to treat in a quantitative manner. There are four properties of dilute solutions, known as the *colligative properties*, which have had an important role in the development of physical chemistry because they make possible the determination of the molecular weight of nonvolatile substances. These properties are: vapor pressure lowering, boiling-point elevation, and freezing-point depression, each relative to the property of the pure solvent, and osmotic pressure. They are characterized by the fact that their values in dilute solutions depend only on the concentration of the solute and not on its chemical properties. We consider each of the colligative properties in sequence.

1. *Vapor pressure lowering.* For solutions in the concentration range in which Raoult's law is obeyed, we have seen that the relative vapor pressure lowering is given by Equation 15-2. This equation can be written in the following form:

---

[1]Strictly speaking, this equation should be applied only to a process carried out at constant pressure. In our case, the solution is at a pressure equal to the sum of the vapor pressures of both components in the solution, while μ_A° represents the chemical potential of pure liquid A at a pressure equal to its own vapor pressure. The pressure dependence of the chemical potentials of liquids is so small, however, that the effect of a small pressure difference is usually negligible. See Williamson, *J. Chem. Ed.* **43**, 211 (1966).

$$\frac{\Delta P_A}{P_A^\circ} = \frac{w_B/M_B}{\dfrac{w_A}{M_A} + \dfrac{w_B}{M_B}}$$

in which w_A and w_B are the weights of A and B in a solution and M_A and M_B are the corresponding molecular weights. If B is a solute whose molecular weight is to be found, this equation can be solved for M_B. Since the solution is usually dilute, n_B is much less than n_A and it is then an adequate approximation to neglect the second term in the denominator. In this case M_B is given approximately by

$$M_B = \frac{w_B M_A}{w_A} \frac{P_A^\circ}{\Delta P_A}$$

2. *Boiling-point elevation.* Since the boiling-point of a liquid is defined as the temperature at which the vapor pressure of the liquid is equal to the external pressure, a graph of vapor pressure versus temperature (Fig. 15–1)

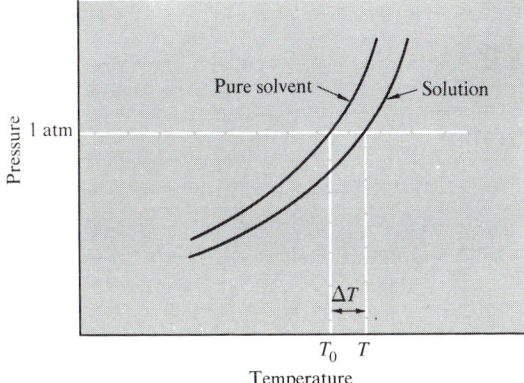

Figure 15–1. Vapor pressure as a function of the temperature, or boiling point as a function of the pressure.

can be interpreted as a graph of boiling point versus external pressure. Because the vapor pressure of a dilute solution of a nonvolatile solute is always less than that of the pure solvent at the same temperature, the solution must be brought to a higher temperature than the pure solvent in order for its vapor pressure to reach the same external pressure. This rise in the boiling point depends on the lowering of the vapor pressure and therefore on the concentration of the solute, according to Raoult's law.

To obtain a relation between the boiling-point elevation and the concentration, let us apply Equation 15–3 to 1 mole of a solution at its boiling point, T, corresponding to some constant external pressure, P. In view of the equality between $\mu_A(v)$ and $\mu_A(\text{soln})$, we have

$$\mu_A(v) = \mu_A^\circ(l) + RT \ln X_A$$

and

$$\frac{\mu_A(v) - \mu_A^\circ(l)}{T} = R \ln X_A$$

Now let us differentiate the equation above with respect to T at constant P. From Equation 15–7,

$$\left(\frac{\partial \frac{\mu_A^\circ(l)}{T}}{\partial T}\right)_P = -\frac{H_A(l)}{T^2}$$

and its analog,

$$\left(\frac{\partial \frac{\mu_A(v)}{T}}{\partial T}\right)_P = -\frac{H_A(v)}{T^2}$$

it follows that

$$\frac{H_A(v) - H_A(l)}{T^2} = -R\left(\frac{\partial \ln X_A}{\partial T}\right)_P \tag{15-11}$$

The quantity $H_A(v) - H_A(l)$ is the enthalpy change in the vaporization of 1 mole of pure liquid A at temperature T. It is essentially the same as ΔH_{vap}, the molar heat of vaporization of pure A at T_o, the boiling point of pure A at pressure P. Making this substitution, we can write Equation 15–11 in the following form:

$$-d \ln X_A = \frac{\Delta H_{A\ \text{vap}}}{RT^2} dT$$

We can obtain a relation between X_A and T by integrating this equation between the limits T_o and $X_A = 1$ for the pure solvent and T and X_A for the solution, assuming that $\Delta H_{A\ \text{vap}}$ is constant in this small temperature range. The result is

$$\ln X_A = \frac{\Delta H_{A\ \text{vap}}}{R}\left(\frac{1}{T} - \frac{1}{T_o}\right)$$

$$= \frac{\Delta H_{A\ \text{vap}}}{R}\left(\frac{T_o - T}{TT_o}\right) \tag{15-12}$$

On substitution of X_A by $1 - X_B$ and $T - T_o$ by ΔT, the boiling-point elevation, this equation becomes

$$\ln(1 - X_B) = -\frac{\Delta H_{A\ \text{vap}}}{R}\frac{\Delta T}{TT_o}$$

Phase Equilibria in Multicomponent Systems

We can convert this equation to a more convenient form by making several approximations. Because the solution under consideration is dilute, X_B is small relative to 1 and $\ln(1 - X_B)$ is nearly equal to $-X_B$. Also T is nearly the same as T_o, and $TT_o \sim T_o^2$. When we make these approximations we have

$$X_B = \frac{\Delta H_{A\ \text{vap}}}{R} \frac{\Delta T}{T_o^2}$$

Again because X_B is small relative to 1, it is approximately equal to n_B/n_A, which is equal to $w_B M_A / w_A M_B$. Thus the boiling-point elevation on the basis of the approximations we have made is given by

$$\Delta T = \frac{w_B M_A}{w_A M_B} \frac{RT_o^2}{\Delta H_{A\ \text{vap}}} \tag{15-13}$$

The equation above can be further simplified by expressing the concentration of the solute in terms of its *molality*, m, defined as the number of moles of solute per 1000 grams of solvent. In this case, since the number of moles of solute per gram of solvent is equal to $w_B / w_A M_B$, the molality is given by

$$m = \frac{w_B}{w_A M_B} \times 1000 \tag{15-14}$$

When this expression is substituted into Equation 15-13 and the numerator and denominator of the right-hand side are divided by M_A, the result is

$$\Delta T = \frac{RT_o^2}{1000\ h_v} m$$

in which h_v is the heat of vaporization per gram of solvent. Because all of the quantities on the right-hand side of this equation except m are constants for a particular solvent, we can write this equation in the simple form:

$$\Delta T = K_b m \tag{15-15}$$

K_b is called the molal boiling-point elevation constant. The values of K_b for a few common solvents are given in Table 15-1.

In view of the approximations made in deriving Equation 15-15 it is not surprising that the proportionality between ΔT and m indicated in this equation is observed only for dilute solutions. Also, only for such solutions is there good agreement between the theoretical value of K_b and the observed value of $\Delta T/m$.

In order to determine the molecular weight of an unknown solute from the boiling-point elevation, we can substitute Equation 15-14 into 15-15 and thus obtain

$$\Delta T = \frac{1000\ K_b w_B}{w_A M_B}$$

TABLE 15-1   BOILING-POINT ELEVATION AND FREEZING-POINT DEPRESSION CONSTANTS

Compound	Boiling point (°C)	K_b	Freezing point (°C)	K_f
Water	100	0.51	0	1.86
Benzene	80.2	2.60	5.5	5.08
Dioxane			11.7	4.71
Camphor			178.5	39.6
Acetic acid	118.3	3.08		
Carbon tetrachloride	76.8	5.03		

When this is solved for M_B the result is

$$M_B = \frac{K_b}{\Delta T} \frac{1000 \; w_B}{w_A}$$

Let us apply these equations to a solution with benzene as a solvent. The value of h_v for benzene at its normal boiling point of 353°K is 94.3 cal/g. Thus the theoretical value of its molal boiling-point constant is

$$K_b = \frac{1.99(353)^2}{1000 \times 94.3}$$
$$= 2.63$$

If 0.450 g of a solute dissolved in 102.5 g of benzene produce a boiling-point elevation of 0.117°, the molecular weight of this solute is given by

$$M_B = \frac{2.63}{0.117} \frac{1000 \times 0.450}{102.5}$$
$$= 98$$

3. *Freezing-point depression.* At equilibrium between a solution and the pure solid solvent, the chemical potentials of the solvent in the two phases must be equal. As shown schematically in Figure 15–2 this requirement implies that the equilibrium temperature is lower than the freezing point of the pure solvent.

Since the derivation of the lowering of the freezing point of a solution as a function of composition is quite similar to the derivation in the preceding section, we present it briefly. For an ideal solution at some definite pressure, for example at atmospheric pressure, the equality of the chemical potentials of the solvent in the liquid and solid phases is expressed by

$$\mu_A^\circ(s) = \mu_A^\circ(l) + RT \ln X_A$$

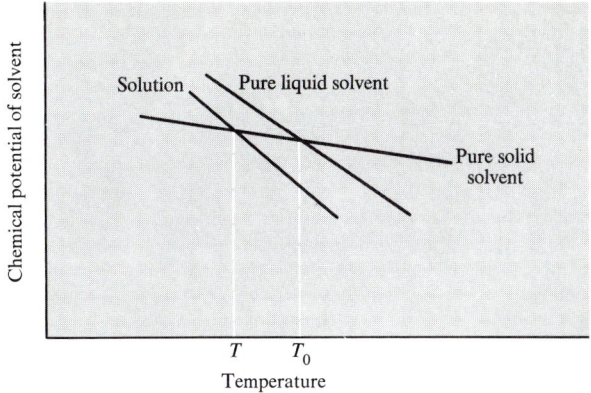

Figure 15–2. Schematic graph of chemical potentials of a solvent in the pure solid phase, the pure liquid phase, and in a solution, as a function of temperature. The slopes of the lines are negative because they are equal to the negative of the corresponding molar entropies, which by the third law are necessarily positive. The intersection of two lines represents an equilibrium state.

and thus

$$\frac{\mu_A^\circ(s) - \mu_A^\circ(l)}{T} = R \ln X_A$$

Differentiating both sides with respect to T at constant P, we obtain

$$R\, d \ln X_A = \frac{H_A(l) - H_A(s)}{T^2} dT$$

This equation can be written in the following form:

$$d \ln X_A = \frac{\Delta H_{A\,f}}{RT^2} dT \qquad (15\text{–}16)$$

in which $\Delta H_{A\,f}$ is the heat of fusion of pure solvent. When this equation is integrated between the limits T_o and $X_A = 1$ for the pure solvent, and T and X_A for the solution, we obtain the following equation which is the analog of Equation 15–12:

$$\ln X_A = \frac{\Delta H_{A\,f}}{R}\left(\frac{T - T_o}{TT_o}\right) \qquad (15\text{–}17)$$

When the same approximations are made as in the preceding section, the result is

$$\Delta T = K_f m \qquad (15\text{–}18)$$

in which $\Delta T\ (= T_o - T)$ is the freezing-point depression, and K_f is the molal freezing-point constant given by

$$K_f = \frac{RT_o^2}{1000\, h_f} \qquad (15\text{–}19)$$

In this equation h_f is the heat of fusion per gram of solvent. The values of K_f for a few common solvents are given in Table 15–1.

The molecular weight of an unknown solute B in a dilute solution is given by

$$M_B = \frac{1000\, K_f}{\Delta T} \frac{w_B}{w_A} \qquad (15\text{–}20)$$

If the solute in the solution dissociates into molecules or ions, or if it associates into dimers or polymers, the molecular weight obtained in this way is the number average molecular weight, defined as the sum for all species present in the solution of the products of molecular weight and mole fraction

$$(\overline{M} = \sum_i M_i X_i)$$

The lowering of the freezing point of a sample of a substance relative to the freezing point of the pure substance is often used as a criterion of the purity of the sample. If no solid solution is formed, the observed depression of the freezing point depends only on the total molar concentration of the impurities, regardless of the nature or the number of the various impurities present. It should be remembered, however, that the validity of Equation 15–18 or Equation 15–20, even for dilute solutions, is based on the restriction that the solid phase formed on freezing is pure solvent. As we will see later, the formation of solid solutions destroys the validity of these equations.

4. *Osmotic pressure.* Certain types of films have been found to be permeable to some substances but not to others, which are usually those substances having relatively large molecules. Because the first such films to be discovered were naturally-occurring animal membranes, films with this property are known as semipermeable membranes. An example of a semipermeable membrane is a thin layer of collodion, which is permeable to water but not to sucrose in an aqueous solution.

Let us imagine that a membrane which is permeable only to solvent molecules separates two phases consisting of an ideal solution and the corresponding pure solvent in a vessel diagrammed in Figure 15–3. Initially the liquid levels in the vertical tubes a and b are the same. As time passes, the level in tube a rises spontaneously and that in tube b falls until equilibrium is reached and the process stops. It can then be observed that the solution has become more dilute, which implies that solvent molecules have passed through the membrane from the pure solvent into the solution. This transfer of solvent

Phase Equilibria in Multicomponent Systems

through a semipermeable membrane as a result of a difference in concentration is called *osmosis*. At equilibrium, the pressure on the solution is higher than that on the solvent phase. The difference between the pressure on the solvent and on the solution at equilibrium is known as the osmotic pressure of this solution, represented by π.

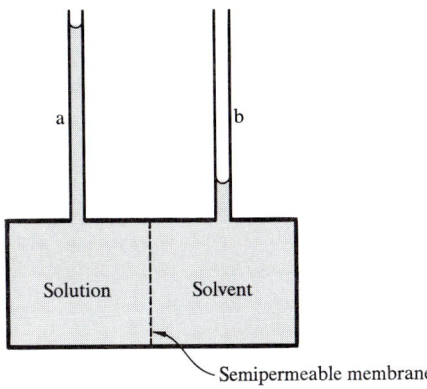

Figure 15-3. Diagram of apparatus for demonstrating osmosis.

The criterion for osmotic equilibrium is the same one we have used previously, namely, the chemical potential of the solvent must have the same value on both sides of the membrane. Initially, $\mu_A(\text{soln})$ is less than $\mu_A^\circ(\text{l})$ because of the presence of the solute. Consequently $\mu_A(\text{soln})$ tends to approach $\mu_A^\circ(\text{l})$ spontaneously by transfer of solvent into the solution. At the same time the value of $\mu_A(\text{soln})$ rises also because the pressure on the solution rises. As a result of the two processes of dilution and pressure rise during osmosis, the values of $\mu_A(\text{soln})$ and $\mu_A^\circ(\text{l})$ eventually become equal and equilibrium is established. The osmotic pressure can thus be seen to depend on the composition of the solution.

We now express these relationships more precisely. Let us assume that the solution in question is ideal, and that at equilibrium at temperature T the mole fraction of the solvent is X_A. The condition for osmotic equilibrium is

$$\mu_A(\text{soln})(\pi + P) = \mu_A^\circ(\text{l})(P) \tag{15-21}$$

The relationship between $\mu_A(\text{soln})$ at a pressure $\pi + P$ and μ_A in the same solution at a pressure P is given by

$$\mu_A(\text{soln})(\pi + P) = \mu_A(\text{soln})(P) + \int_P^{\pi+P} \left(\frac{\partial \mu_A(\text{soln})}{\partial P}\right)_{T, X_A} dP \tag{15-22}$$

To evaluate the integral we can make use of Equation 15–5 and the fact shown previously that for an ideal solution $\bar{V}_A$ equals V_A°, the molar volume of pure solvent. It follows that

$$\int_P^{\pi+P} \left(\frac{\partial \mu_A(\text{soln})}{\partial P}\right)_{T,X_A} dP = \int_P^{\pi+P} V_A^\circ \, dP$$

With the assumption that the solvent is incompressible, the integral on the right-hand side of the above equation has the value πV_A°. Substituting this value in Equation 15–22 we obtain

$$\mu_A(\text{soln})(\pi + P) = \pi_A(\text{soln})(P) + \pi V_A^\circ$$

From Equation 15–21 and Equation 15–3, it follows that

$$\mu_A^\circ(l)(P) = \mu_A^\circ(l)(P) + RT \ln X_A + \pi V_A^\circ$$

and consequently,

$$\pi V_A^\circ = -RT \ln X_A = -RT \ln(1 - X_B)$$

When the solution is so dilute that X_B is much less than 1, as before we can use the approximation $\ln(1 - X_B) \sim -X_B$ and thus we obtain

$$\pi V_A^\circ = RT \, X_B$$

For very dilute solutions, $X_B \sim n_B/n_A$ and the above equation can be written in the following form:

$$\pi = \frac{n_B RT}{n_A V_A^\circ}$$

For these dilute solutions the denominator is essentially the volume of the solution, V. Thus we have

$$\pi = \frac{n_B RT}{V} = RTc_B \tag{15-23}$$

in which c_B is the concentration of the solute in moles/liter.

This interesting equation has the same form as the ideal gas law, a fact which was pointed out by van't Hoff in 1885. It is misleading, however, to picture osmotic pressure as the result of the impacts of gas-like solute molecules on one side of a semipermeable membrane. Although little is known about the transport mechanism in semipermeable membranes or the molecular origin of osmotic pressure, it is likely that a variety of transport mechanisms exists. Thermodynamics throws no light on this problem, but here again it yields useful relationships that are independent of mechanisms.

Equation 15–23 provides the basis for the application of osmotic pressure measurements to the determination of molecular weights. With the solute concentration expressed in the form $c_B = w_B/M_B V$, where w_B is the weight of solute contained in a volume V of solution, the molecular weight is given by

$$M_B = \frac{w_B RT}{\pi V}$$

This use of osmotic pressure measurements is particularly advantageous for the determination of the molecular weights of relatively insoluble substances or those with high molecular weights because in either of these cases the available molar concentrations are very low. Even at these low concentrations, however, the osmotic pressure is so large that it can be measured with a precision which is higher than the precision with which measurements of the boiling-point elevation or the freezing-point depression can be made. For example, the osmotic pressure at 300°K of a solution containing 2.0 g of a polymer with a molecular weight of 20,000 dissolved in 100 cc of solution is given by

$$\pi = 82.1 \times 300 \times \frac{2.0}{20,000 \times 100}$$

$$= 0.0246 \text{ atm} = 18.7 \text{ torr}$$

A pressure of this magnitude can easily be measured with an uncertainty of less than 1%.

To minimize the effect of deviations from solution ideality in the determination of molecular weights, osmotic pressure measurements are usually made at several different concentrations of solute. Then the ratio of the osmotic pressure to the concentration is plotted against the concentration, the graph line is extrapolated to zero concentration, and the molecular weight of the solute is calculated from the limiting value of π/c.

As mentioned above, the measurement of osmotic pressures can be carried out by allowing transfer of solvent through the semipermeable membrane and measuring the resulting pressure difference at equilibrium. An alternative method consists in the application of a higher pressure on the solution than on the solvent and a measurement of the pressure difference that is just sufficient to prevent transfer of solvent through the membrane. If the pressure applied to the solution is larger than the osmotic pressure, solvent is transferred spontaneously through the membrane from the solution to the pure solvent. This process of "reverse osmosis" is one of the processes being investigated for use in the desalinization of salt water.

Ideal Solutions Containing Two Volatile Components

Vapor Pressure

When both components, A and B, in a binary solution are volatile and their vapors behave as ideal gases, the total pressure of the vapor in equilibrium with the solution is the sum of contributions from both components.

When, in addition, the solution is ideal, both components obey Raoult's law and the total vapor pressure is given by the following equation:

$$P_{tot} = X_A P_A^\circ + (1 - X_A) P_B^\circ \qquad (15\text{–}24)$$

in which $1 - X_A$ has been substituted for X_B. A graph of P_{tot} as a function of X_A is shown in Figure 15–4.

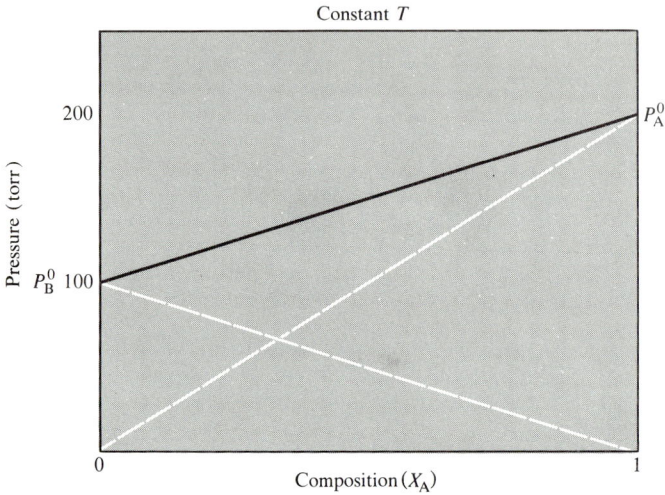

Figure 15–4. Total vapor pressure and partial vapor pressures for an ideal solution as a function of the composition.

In the vapor phase the partial pressure of component A is related to its mole fraction in the vapor phase, Y_A, by Dalton's law, which is $P_A = Y_A P_{tot}$. Equating this expression for P_A with that in Equation 15–1, we have

$$Y_A P_{tot} = X_A P_A^\circ$$

When the expression for P_{tot} in Equation 15–24 is substituted in this equation, the result is

$$Y_A = \frac{X_A P_A^\circ}{X_A P_A^\circ + (1 - X_A) P_B^\circ} \qquad (15\text{–}25)$$

We thus see that in general the composition of the vapor phase is different from that of the solution with which it is in equilibrium.

In the phase diagram for a 2-component system shown in Figure 15–5 the total pressure at some constant temperature is plotted against the mole fraction of A in both the liquid and vapor phases. The values of both X_A and Y_A are read on the same scale. When X_A is given the total pressure is found from the straight line ab which represents the sum of the partial vapor pressures of A and B. (The line ab is the same as the solid line in Figure 15-4.) When Y_A is given, the total pressure is found from the curved line acb, which is the graph of Equation 15-25.

The region between these two lines represents the range of compositions and pressures in which liquid and vapor can coexist at the given temperature.

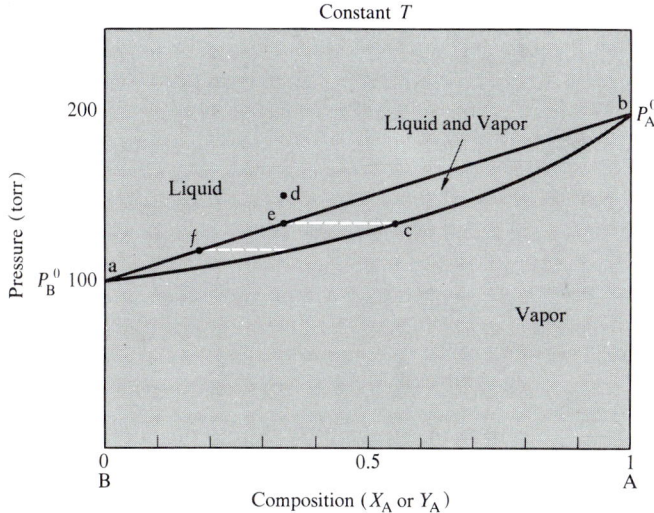

Figure 15-5. Phase diagram for an ideal solution at constant temperature. The composition coordinate is X_A (liquid phase) for points along line aeb or at higher pressures, and is Y_A (vapor phase) for points along acb or at lower pressures.

In the region above the line ab only the liquid phase exists. Similarly, in the region below the line acb only the vapor phase exists.

To illustrate the significance of this diagram, let us take as a specific example two liquids, A and B, with $P_A^\circ = 200$ torr and $P_B^\circ = 100$ torr at temperature T, and let us say that a solution of these liquids contains 1 mole of A and 2 moles of B. The total vapor pressure is $P_{tot} = 0.33 \times 200 + 0.67 \times 100 = 133$ torr. If the pressure on the system is higher than this value, no vapor phase can exist. The system may then be represented by point d in Figure 15-5. Now imagine that the pressure on the system is lowered to a value just less than 133 torr (point e). At this pressure a bubble of vapor will appear whose composition can be read from the phase diagram or calculated from Equation 15-25, as follows:

$$Y_A = \frac{0.33 \times 200}{133} = 0.50$$

With the appearance of a second phase the number of degrees of freedom of this system drops from 3 to 2. In this case, when the temperature and the total pressure of the system are fixed, the compositions of the two phases are also fixed. At each value of the total pressure the points of the phase diagram, such as c and e, representing the mole fractions in the liquid and the vapor phases can be connected by a horizontal line known as a *tie-line*.

As the system pressure is lowered, more and more of the liquid will evaporate and the vapor will become richer in B. As the last drop of liquid evaporates the composition of the vapor approaches the composition of the entire system, and thus $Y_A = 0.33$. (This value can also be read from the phase diagram.) When the value of Y_A is introduced into Equation 15-25, the calculated value of X_A is 0.20, and from Equation 15-24 the total pressure of the system when evaporation is just complete is 120 torr. At pressures lower than this, the system consists entirely of vapor and again has 3 degrees of freedom.

For any given pressure between 120 and 133 torr we can calculate the amount of each component in each phase because the composition of each phase is fixed. When Equation 15-24 is solved for X_A we obtain

$$X_A = \frac{P_{\text{tot}} - P_B^\circ}{P_A^\circ - P_B^\circ}$$

We can also find the corresponding value of Y_A. For example, if $P_{\text{tot}} = 125$ torr, $X_A = (125 - 100)/(200 - 100) = 0.25$, and $Y_A = (0.25 \times 200)/125 = 0.40$. Because the number of moles of A in the system is given as 1, we have

$$X_A n_l + Y_A n_v = 1$$

in which n_l and n_v represent the total number of moles of liquid and vapor, respectively. Also, since the total number of moles in the system is given as 3, we have

$$n_l + n_v = 3$$

When n_v is eliminated between these two simultaneous equations, we obtain

$$0.25 n_l + 0.40(3 - n_l) = 1$$

It follows that $n_l = 1.33$ moles. Thus when $P_{\text{tot}} = 125$ torr, the liquid phase of our system consists of 0.33 moles of A and 1 mole of B, while the vapor phase consists of 0.67 moles of A and 1 mole of B.

Whenever a 2-component system consists of two phases of known composition at equilibrium, the ratio of the numbers of moles in the two phases can be found from a general relationship known as the *lever rule*. Let X_{A1}, X_{A2},

X_{A3} represent the mole fraction of component A in phase 1, phase 2, and the total system, respectively, and let n_1 and n_2 represent the total number of moles of A and B in phase 1 and in phase 2, respectively. Since all of A is either in phase 1 or phase 2, we have the following identity:

$$X_{A1}n_1 + X_{A2}n_2 = X_{A3}(n_1 + n_2)$$

It follows that

$$(X_{A3} - X_{A1})n_1 = (X_{A2} - X_{A3})n_2$$

and consequently

$$\frac{n_1}{n_2} = \frac{X_{A2} - X_{A3}}{X_{A3} - X_{A1}}$$

The distances on the tie-line, represented by the differences of mole fractions on the right-hand side of this equation, can be considered to represent the lengths of lever arms.

Applying the lever rule to the example discussed above, we find the following result:

$$\frac{n_l}{n_v} = \frac{0.40 - 0.33}{0.33 - 0.25} = 0.80$$

This value of the liquid-to-vapor ratio is consistent with the values of the numbers of moles found above.

Fractional Distillation

As we have seen, when a solution is in equilibrium with its vapor the compositions of the two phases differ from each other. If the vapor is separated from the liquid and then condensed, the resulting condensate is richer in the more volatile component than the original liquid. If the condensate is then partially revaporized by reducing the pressure on the system, the vapor is richer in the more volatile component than the condensate. By repeated vaporization and condensations resulting from varying the pressure at constant temperature, it is possible to obtain samples that consist predominantly of each of the individual components. This process is called *isothermal fractional distillation*.

A much more common process, however, is fractional distillation carried out at constant pressure with temperature as a variable. To follow this process it is convenient to use a phase diagram in which the boiling point of the solution at a pressure of 1 atm is plotted against its composition (Fig. 15-6). This kind of diagram can be drawn for any solution that can be considered as ideal when the vapor pressures of the two components are known as a function of the temperature. At any particular temperature, X_A can be calculated from Equation 15–24, since P_{tot}, P_A°, and P_B° are known, and then Y_A can be calculated from Equation 15–25. Qualitatively, it can be seen that the component with the higher vapor pressure has the lower boiling point.

If we have a solution in which the mole fraction of component A is X_1, the liquid phase only is present in the system at atmospheric pressure and a temperature below the normal boiling point. When the solution is heated at

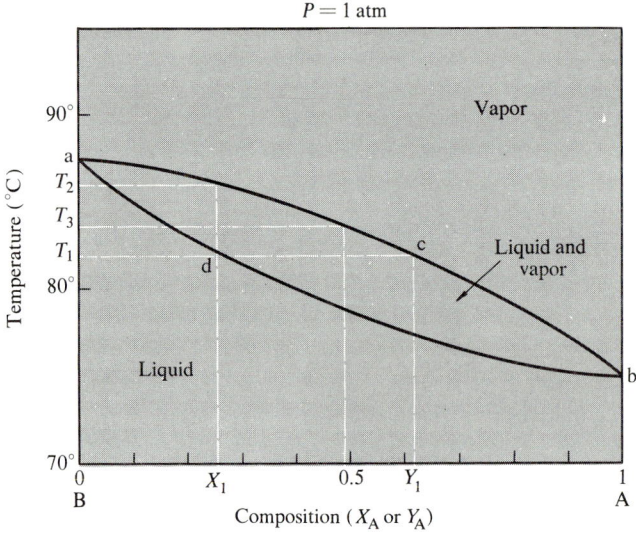

Figure 15–6. Boiling point-composition diagram for an ideal solution.

constant pressure without superheating, a bubble of vapor will first appear at temperature T_1 which is the normal boiling point. The composition of this vapor is given by the other end of the tie-line, namely the mole fraction X_2.

As the temperature continues to rise and more and more of the liquid vaporizes, the point representing the vapor composition moves along the line acb while the point for the liquid residue at the other end of the tie-line moves along the line adb. Ultimately, at temperature T_2 all of the liquid will have vaporized and the mole fraction of component A in the vapor phase is equal to X_1.

If at some temperature T_3 between T_1 and T_2 the vapor is separated from the liquid residue and is then condensed, the condensate is richer in the lower-boiling component (A, in this case) than the original solution. If the condensate is now reheated to a temperature somewhat lower than T_3 and the equilibrium vapor is again separated and condensed, the new condensate is still richer in component A. By repetition of the steps of vapor-liquid equilibration, sepa-

Phase Equilibria in Multicomponent Systems 561

ration of vapor, and condensation, eventually the initial solution can be separated into two parts consisting of almost pure A and almost pure B. In principle, an infinite number of repetitions of these steps would be required for complete fractionation because each repetition becomes successively less effective as the limits of pure A and pure B are approached.

Fractional distillation is usually carried by means of a distillation column in which many cycles of the evaporation-condensation steps are carried out continuously. The fractionating ability of such a column is often expressed by the number of *theoretical plates*, the term theoretical plate referring to the degree of fractionation produced by a single evaporation-condensation cycle in which equilibrium between liquid and vapor is achieved.

One type of distillation column, known as a bubble-cap column, is shown in Figure 15-7. This column contains a series of plates or trays, each of which

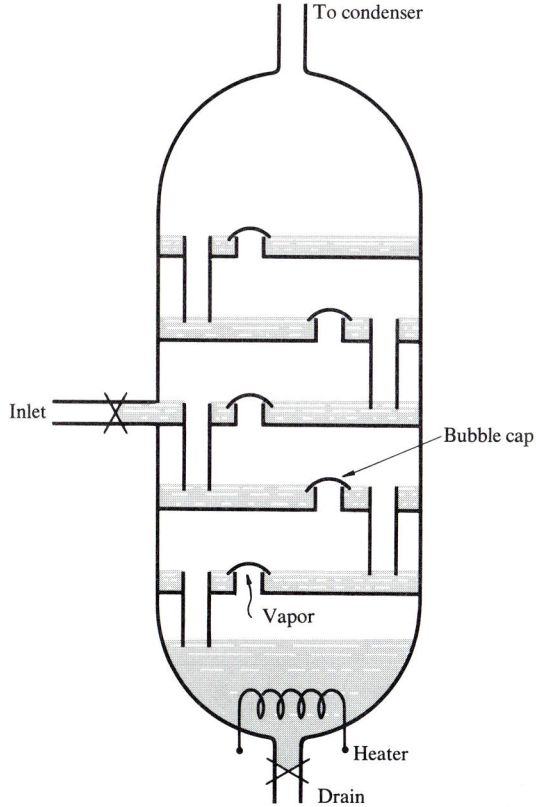

Figure 15-7. Diagram of bubble-cap distillation column.

holds a thin layer of liquid. Hot vapor ascending from the boiler at the bottom bubbles through the liquid layer and is partially condensed while the portion that is richer in the more volatile components continues upwards. The condensate on each tray overflows to the tray below where it is equilibrated with rising vapor. Eventually the most volatile component in a nearly ideal starting solution can be obtained at the top of the column in almost pure form, while the least volatile component is concentrated at the bottom of the column.

Some types of distillation columns are packed with glass beads or other types of packing, but the basic fractionation process is the same as that described above.

Henry's Law

A special case of a solution containing two volatile components is a solution of a gas in a liquid. Near the beginning of the nineteenth century Henry observed that the weight of a gas which dissolves in a unit volume of liquid is directly proportional to the partial pressure of the gas above the liquid. A more useful form of Henry's law is given by the following equation:

$$P_B = kX_B$$

in which X_B is the mole fraction of the gas B in the solution when its partial pressure is P_B, and k is a constant whose value is characteristic for the two components of the solution.

Component B can be thought of either as a gaseous solute, as mentioned above, or as a volatile component of a solution. If the solution is ideal, the partial pressure of component B in the vapor phase is given by Raoult's law; thus $P_B = P_B^\circ X_B$. By comparing this equation with the one above we see that for an ideal solution the constant in Henry's law is simply the vapor pressure of the more volatile component at the temperature of the solution, and Henry's law is the same as Raoult's law. This is *not* the case for solutions that are not ideal.

Solutions which Deviate from Ideality

Effect of Intermolecular Forces

As mentioned earlier, very few solutions obey Raoult's law within experimental error over the whole range of concentrations. In general, the total pressure of the vapor above a solution is larger or smaller than the value calculated from Equation 15-24 and the vapor pressures of the pure components. Likewise, significant heat effects and volume changes are usually observed when the components of a solution are mixed. On the other hand, it is an experimental fact that all solutions approach ideality as the solute concentration approaches zero.

In the following discussion of nonideal solutions we consider only those solutions that are composed of electrically neutral molecules. The subject of electrolytic solutions is deferred until Chapter 16.

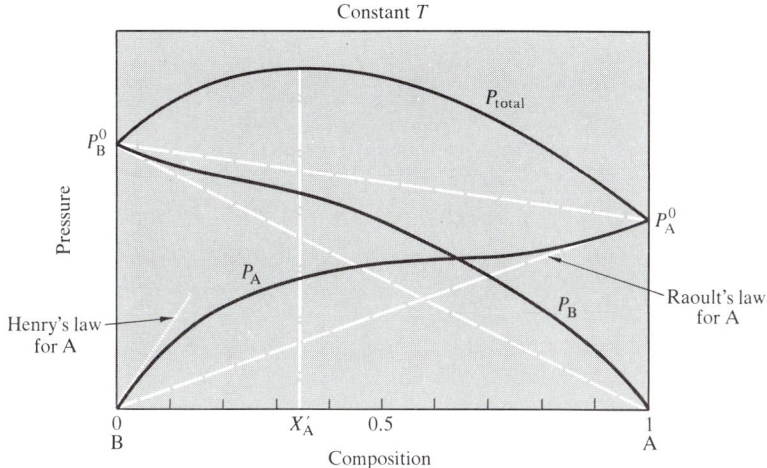

Figure 15-8. Diagram for a solution that shows a positive deviation from Raoult's law. P_A and P_B are the partial vapor pressures of the components and P_{total} is the total vapor pressure of the solution. The dashed lines show the values these quantities would have if the solution were ideal. The dotted line shows Henry's law for A considered as a solute.

From a molecular point of view, deviations of solutions from ideality are directly related to the relative values of the forces between the molecules in the solution. Positive deviations from Raoult's law, for which the observed total vapor pressure is higher than that calculated from Raoult's law (Fig. 15-8), are associated with a smaller attractive force between A and B molecules than exists between A molecules or B molecules by themselves. When the force between A and B molecules is quite weak, at equilibrium the system containing A and B will consist of two liquid phases called *conjugate solutions*, one having a higher concentration of A and the other a higher concentration of B. In such cases one speaks of limited miscibility. In extreme cases (paraffin oil and water, for example), the two components are practically mutually insoluble, and the liquid phases consist of the almost pure components.

In contrast, negative deviations from Raoult's law are associated with forces between A and B molecules that are larger than those existing between A molecules or between B molecules. Negative deviations are usually observed when the component molecules have relatively large dipole moments or when hydrogen bonds are formed between the two kinds of molecules in the solution.

Raoult's and Henry's Laws as Limiting Laws

As solutions become more dilute, an average solvent molecule is increasingly likely to have other solvent molecules as nearest neighbors. In addition, the average distance between solute molecules becomes greater and the forces between these molecules become less significant. Thus it can be understood that the solvent obeys Raoult's law more closely with increasing dilution.

At the same time, the solute molecules tend to be more completely surrounded by solvent molecules, and thus they are in an increasingly uniform environment. As a consequence, the solute obeys Henry's law more closely the lower its concentration becomes. The Henry's law constant in this case, however, is not the vapor pressure of the pure solute, but instead depends on the nature of the solute-solvent interaction.

From a thermodynamic point of view we can take into consideration the deviation of a solution from ideality by expressing the chemical potential of a component in terms of its activity (Eq. 3–65) rather than its mole fraction, as follows:

$$\mu_A = \mu_A^\circ + RT \ln a_A$$

This equation is exact by definition, since all of the effects of the deviation from ideality are included in the value of the activity. Any of the equations we have obtained for the ideal solution which involve component A can be made applicable to the nonideal solution by replacing X_A by a_A. For example, when A is taken to be the solvent, Equation 15–17 for the boiling-point elevation becomes

$$\ln a_A = \frac{\Delta H_{A\ vap}}{R}\left(\frac{T_o - T}{TT_o}\right)$$

This relationship and the analogous ones for the other colligative properties can be used to determine activities from experimental data.

If A is the solvent, with decreasing concentration of the solute the chemical potential of A in the solution approaches that of pure liquid A and consequently a_A approaches 1. Thus Equation 3–65 implies that the standard state of A is the pure liquid.

In discussing deviations from ideality it is often convenient to introduce the activity coefficient γ_A, defined by a_A/X_A. Then Equation 3–65 can be written in the following form:

$$\mu_A = \mu_A^\circ + RT \ln X_A + RT \ln \gamma_A$$

Because solutions approach ideality as they become more dilute, we can say that $\gamma_A \to 1$ as $X_A \to 1$.

If A is a minor component in a solution, and thus is considered to be a solute, then the solution approaches ideality, and consequently $\gamma_A \to 1$, as $X_A \to 0$. This implies that the standard state of a solute is a hypothetical state in which the solute has a mole fraction of 1 while its properties have the same values that they have at infinite dilution. The fact that this state cannot be actually realized does not cause any difficulty in using it as a reference state in calculations.

Let us now consider a solution in equilibrium with its vapor when the solution deviates from ideality but the vapor behaves as an ideal gas. Equating chemical potentials of component A in the vapor and in the solution, we have

$$\mu_A^\circ(v) + RT \ln P_A = \mu_A^\circ(\text{soln}) + RT \ln a_A$$

and

$$\frac{\mu_A^\circ(\text{soln}) - \mu_A^\circ(v)}{RT} = \ln \frac{P_A}{a_A}$$

$$= \ln \frac{P_A}{\gamma_A X_A}$$

The left-hand side of this equation includes only quantities that are constant for particular components at a definite temperature. Consequently,

$$\frac{P_A}{\gamma_A X_A} = K$$

If A is considered as the solvent, since $\gamma_A \to 1$ as $X_A \to 1$, in the limit, $K = P_A^\circ$. This expresses Raoult's law as a limiting law.

If, in contrast, A is considered as a solute, then as we have just seen $\gamma_A \to 1$ as $X_A \to 0$. For dilute solutions $P_A/X_A \to K$, and this is Henry's law as a limiting law. These results suggest that in any solution the solute must obey Henry's law in the same concentration range in which the solvent obeys Raoult's law (Fig. 15–8).

In order to prove that this is indeed the case, let us start with the Gibbs–Duhem equation (Eq. 3–61) which is completely general for a process at constant T and P:

$$n_A \, d\mu_A = -n_B \, d\mu_B$$

When both sides are divided by $n_A + n_B$, this equation becomes

$$X_A\, d\mu_A = -X_B\, d\mu_B \qquad (15\text{-}26)$$

Since $X_A = 1 - X_B$, it follows that $dX_A = -dX_B$. On dividing the left-hand side of Equation 15–26 by dX_A and the right-hand side by $-dX_B$ and indicating the constancy of T and P, we have

$$X_A \left(\frac{\partial \mu_A}{\partial X_A}\right)_{T,P} = X_B \left(\frac{\partial \mu_B}{\partial X_B}\right)_{T,P} \qquad (15\text{-}27)$$

Considering component A as a solvent that obeys Raoult's law, from

$$\mu_A = \mu_A^\circ + RT \ln P_A$$

we obtain

$$\mu_A = \mu_A^\circ + RT \ln P_A^\circ + RT \ln X_A$$

Partial differentiation with respect to X_A leads to

$$\left(\frac{\partial \mu_A}{\partial X_A}\right)_{T,P} = RT \left(\frac{\partial \ln X_A}{\partial X_A}\right)_{T,P}$$

$$= RT \frac{1}{X_A} \qquad (15\text{-}28)$$

Turning now to the other component, from $\mu_B = \mu_B^\circ + RT \ln P_B$ we obtain

$$\left(\frac{\partial \mu_B}{\partial X_B}\right)_{T,P} = RT \left(\frac{\partial \ln P_B}{\partial X_B}\right)_{T,P} \qquad (15\text{-}29)$$

When Equations 15–28 and 15–29 are introduced into Equation 15–27, the result is

$$X_B \left(\frac{\partial \ln P_B}{\partial X_B}\right)_{T,P} = 1$$

This equation may be written as

$$d \ln P_B = \frac{dX_B}{X_B} = d \ln X_B \qquad \text{(constant } T \text{ and } P)$$

When this equation is integrated to obtain the relation between P_B and X_B, the result is

$$\ln P_B = \ln X_B + K'$$
$$= \ln K X_B$$

The constant of integration K (or K') is independent of X_B but it is a function of T and P. Taking the antilogarithm of both sides, we obtain

$$P_B = K X_B$$

Phase Equilibria in Multicomponent Systems

which is a statement of Henry's law. We thus have shown that whenever one component of a solution obeys Raoult's law the other component must obey Henry's law.

Up to this point we have assumed that all deviations from ideality can be associated with the liquid phase. When it is not an adequate approximation to assume that the vapors obey the ideal gas law, use is made of Equation 3-64 to express the chemical potential of a vapor in terms of its fugacity. The only effect of this change on the equations in this chapter is the replacement of each partial pressure by the corresponding fugacity. The introduction of the fugacity of a component implies that sufficient P,V,T data are available for this component to permit calculation of the appropriate fugacity values.

Azeotropes and Azeotropic Distillation

When the deviations from ideality of a solution are sufficiently large, in some concentration range the total vapor pressure of the solution may be higher than that of the more volatile component or lower than that of the less volatile component at the same temperature. It follows that the boiling point of such a solution may be lower than that of the lower-boiling component or higher than that of the higher-boiling component.

Let us first consider a solution for which the total vapor pressure has a maximum value at some particular mole fraction, X'_A (Fig. 15-8). At this composition the boiling point of the solution is at a minimum. The composition-boiling point diagram of the system has the form shown in Figure 15-9. It can be seen that the vapor and liquid lines are tangent to each other at X'_A. An important consequence of this tangency is the fact that a solution with this composition is in equilibrium with vapor of the same composition. Such a solution can be distilled without any change in composition, and thus its boiling point remains constant. A constant-boiling solution is called an *azeotropic solution*, or *azeotrope*. At a particular value of the total pressure, the composition of an azeotrope depends only on the components present in the solution, and consequently it is highly reproducible like the composition of a pure compound. The composition of the azeotrope, however, is changed by a change in the pressure on the system.

If a solution whose mole fraction of A is less than X'_A is fractionally distilled, the distillate is richer in A than the residue. The mole fraction of A in the distillate, however, cannot exceed X'_A. It follows that on exhaustive fractional distillation, the solution can be separated into pure B, obtained as a residue, and the azeotrope. If the initial mole fraction of A in the solution is greater than X'_A, exhaustive fractional distillation will lead to pure A as a residue and the azeotrope as a distillate.

For a solution for which negative deviations from ideality result in a minimum in the vapor pressure-composition curve, the boiling point will be a maximum at the azeotropic composition. In this case, exhaustive fractional distilla-

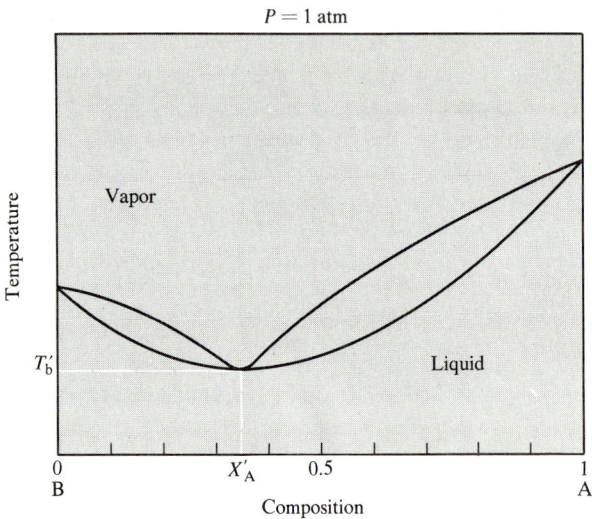

Figure 15-9. Boiling point-composition diagram for a solution with a minimum boiling point. X'_A is the mole fraction of A in the azeotrope and T'_b is its boiling point.

tion will lead to one or the other component as a pure distillate, depending on the composition of the initial solution, and the azeotrope will be obtained as a residue.

Distillation of Immiscible Liquids

As mentioned before, when deviations from ideality are extreme the components are immiscible and no solution at all is formed. The system under consideration in this case consists of two liquid phases, each containing only one component, and a vapor phase. With two components and three phases, the system has 1 degree of freedom according to the phase rule. This degree of freedom may be the temperature, the pressure, or the composition of the vapor. When such a system is distilled at constant pressure, the boiling point and the composition of the distillate are fixed as long as both components are present as liquid phases.

Phase Equilibria in Multicomponent Systems

In the vapor phase the partial pressure of each liquid is equal to its own vapor pressure, and the boiling point of the solution is the temperature at which the sum of the vapor pressures equals the external pressure. In addition, the molar ratio of the components in the vapor phase, and thus in the distillate, is equal to the ratio of the vapor pressures at this temperature. Since $P_{tot} = P_A^\circ + P_B^\circ$ and $n_A/n_{tot} = P_A^\circ/P_{tot}$, it follows that

$$\frac{n_A}{n_B} = \frac{P_A^\circ}{P_B^\circ}$$

Because $n_A = w_A/M_A$, the weight ratio of the components in the distillate is given by

$$\frac{w_A}{w_B} = \frac{P_A^\circ M_A}{P_B^\circ M_B}$$

In most distillations of immiscible liquids one of the components is water. Such a process is known as a *steam distillation*. The molecular weight of an unknown volatile substance that is insoluble in water can be found from the equation above when the composition of the distillate and the distillation temperature are known, since the vapor pressure of water is known quite accurately as a function of the temperature.

A more important reason for carrying out a steam distillation is the fact that in this way a high-boiling substance can be distilled at a temperature well below its normal boiling point. This process is particularly useful for the distillation of those compounds which decompose at their normal boiling points.

Equilibria in Condensed Systems

In the remainder of this chapter we consider equilibria in systems that do not contain a vapor phase. If the system is in a closed container, absence of a vapor phase implies that the external pressure is higher than the vapor pressure or sublimation pressure of the system. If the system is in a container open to the atmosphere (for example, water in an open beaker) because of the diffusion of the vapor its partial pressure in the air is far below the equilibrium value and consequently the presence of the vapor can usually be neglected.

Liquid-Liquid Systems with Partial Miscibility

It is often observed that as one liquid, A, is added to another, B, at some definite temperature, the system at first is homogeneous, but at some point in the process a second liquid phase is formed. As the addition of B is continued, the amount of the second liquid phase increases while that of the original phase decreases, until ultimately the system is homogeneous again. As mentioned previously, two liquid phases in equilibrium with each other are known as conjugate solutions.

In the composition range in which conjugate solutions are present the system has 2 degrees of freedom. It follows that at a definite temperature and pressure, the compositions of both conjugate solutions are fixed. The formation of conjugate solutions is attributed, as indicated earlier, to deviations from solution ideality that are large enough to lead to phase separation over a range of compositions of the system, but not large enough to result in a complete segregation of the components.

If a vapor phase is also present, the system is univariant. In this case as long as the system contains two conjugate solutions at a definite temperature, both the pressure and the composition of the vapor phase are fixed at equilibrium. Since both solutions are in equilibrium with the same vapor phase, it follows that the partial vapor pressure of each component is the same for both conjugate solutions, although their compositions may be widely different. For example, in a system containing benzene and water at equilibrium at 25°C, even though one of the liquid phases contains about 0.08 mole % water while the conjugate phase contains about 99.9 mole % water, the partial pressure of water vapor is the same for both solutions.

The temperature-composition phase diagrams for various types of liquid systems with partial miscibility are shown in Figure 15-10. In the first case, which is the most common one, the two components become more soluble in each other as the temperature is raised, until a certain temperature is reached above which the two components are completely miscible. This temperature is known as the *upper critical solution*, or *consolute, temperature*. The composition corresponding to the maximum of the curve in Figure 15-10a is known as the critical composition. When a 2-phase system having this composition is slowly heated, little seems to happen until the critical solution temperature is reached, when the interface between the liquid phases disappears and the system becomes homogeneous.

With a slight drop in temperature, the interface becomes reestablished. In this process an opalescence is often observed which is the result of the scattering of light by molecular clusters formed by growth of nuclei during the formation of a new phase. There is a close analogy with the formation of a liquid from a gas under critical conditions.

There are some systems which separate into conjugate solutions when the temperature is raised, and thus have a lower critical solution temperature, as shown in Figure 15–10b. There are a very few systems which have both an upper and a lower critical solution temperature (Fig. 15–10c).

For a system that contains two conjugate solutions the relative amounts of each phase can be calculated from the compositions of the two phases and the over-all composition of the system by use of the lever rule (p. 558). For example, at 59°C perfluoromethylcyclohexane and benzene form conjugate solutions containing 31.0 and 95.5 mole % benzene, respectively. When a

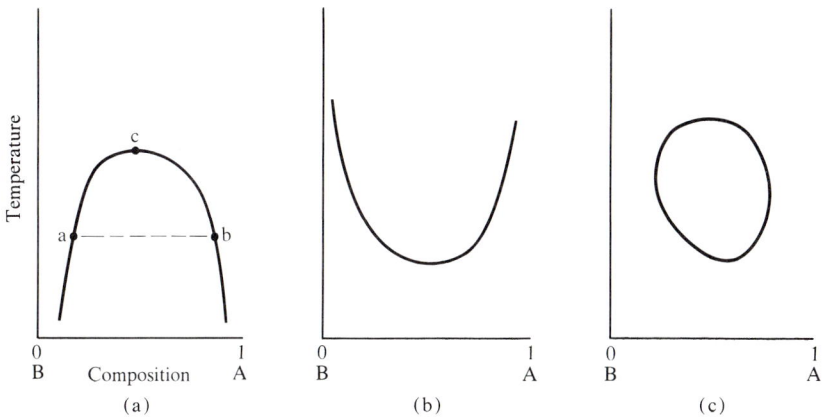

Figure 15–10. Phase diagrams for solutions with (a) upper consolute temperature, (b) lower consolute temperature, and (c) both upper and lower consolute temperatures. The areas within the concave sides of the curves represent the coexistence regions.

system containing 2.00 moles of benzene and 1.00 mole of perfluoromethylcyclohexane is equilibrated at 59°C, the ratio of the number of moles of the benzene-poor phase, n_p, to the number of moles of the benzene-rich phase, n_r, is given by the lever rule as follows:

$$\frac{n_p}{n_r} = \frac{0.955 - 0.667}{0.667 - 0.310} = 0.807$$

Since

$$\frac{n_p}{n_r} + 1 = \frac{n_p + n_r}{n_r} = 1.81$$

and $n_p + n_r = 3.00$, we obtain

$$n_r = \frac{3.00}{1.81} = 1.65$$

The benzene-rich phase contains $1.65 \times 0.955 = 1.58$ moles of benzene. The other phase then contains the remaining 0.42 mole of benzene at a mole fraction of 0.310. Consequently the number of moles in this phase is given by $0.42/0.310 = 1.35$, which is equal to $3.00 - 1.65$.

Solid-Liquid Equilibria

Measurements of temperature and composition have been made for a large number of binary condensed systems containing liquids and solids, and a wide variety of phase diagrams have been developed. We will not attempt to discuss

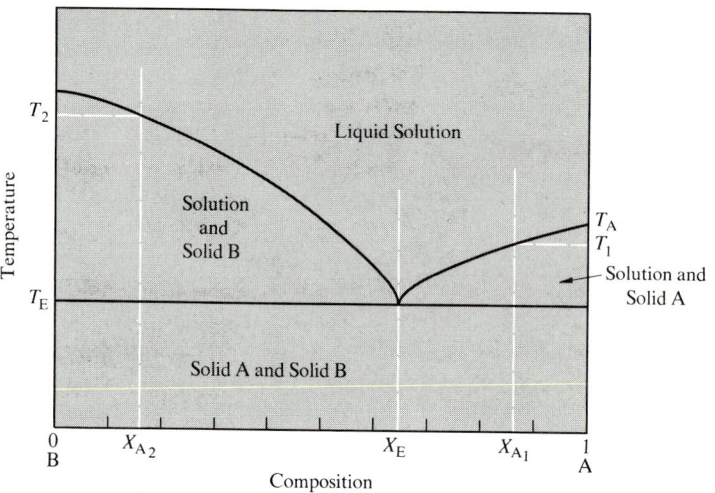

Figure 15–11. Phase diagram for solid-liquid system. A and B are completely miscible in the liquid phase but the solid phases are completely immiscible.

these completely but confine our attention to a few of the simpler and more common types of phase equilibria.

1. The simplest type of solid-liquid phase equilibrium is found when the two components are completely miscible in the liquid phase and the corresponding solids are completely immiscible. A typical phase diagram for this case is shown in Figure 15–11, in which the pressure is assumed to be constant. When a solution in which the mole fraction of component A is X_{A1} is slowly cooled, a solid phase consisting of pure A first appears at temperature T_1. As the temperature is lowered more of A crystallizes and the remaining solution becomes richer in B until it reaches the mole fraction X_E when the temperature is T_E. At this temperature the solid phase of B starts to form. As the system

is cooled still more, both A and B crystallize in the same proportion as exists in the liquid phase until the system is completely solid.

If the initial solution has the composition X_{A2}, the first solid to appear is pure B at temperature T_2. This is the only solid phase until the solution reaches the mole fraction X_E, when A starts to crystallize also. If the initial solution has the mole fraction X_E, no change occurs on cooling until the temperature T_E is reached. Then both A and B crystallize simultaneously with no change in the composition of the solution until the system is completely solidified. The solid that is formed at this temperature is a 2-phase system that consists of an intimate mixture of crystallites of A and of B which usually can be distinguished only by means of a microscope.

The temperature T_E and the mole fraction X_E are known as the *eutectic temperature* and *composition*, respectively. The point on the phase diagram representing the values of T_E and X_E is fixed by the nature of the system. This follows from the fact that when a liquid phase is in equilibrium with two solid phases at constant pressure, the number of degrees of freedom is given by $F = 2 - 3 + 1 = 0$. Thus the system is invariant at constant P and both the eutectic temperature and the eutectic composition are properties of the system.

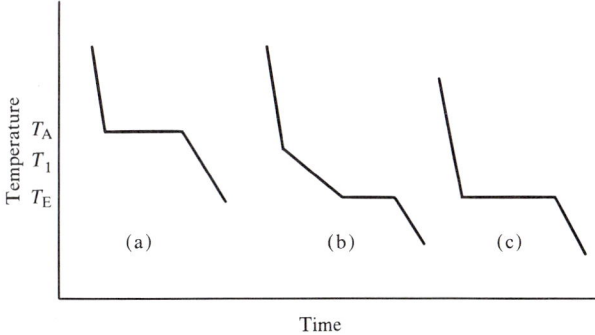

Figure 15-12. Cooling curves for (a) pure A, (b) solution with composition X_{A1}, and (c) solution with eutectic composition.

One common method for obtaining data from which phase diagrams can be drawn is the determination of cooling curves for a series of samples having a range of compositions. This method consists in the measurement of the temperature of an initially molten sample as a function of time while a constant temperature difference is maintained between the sample and its surroundings.

The idealized cooling curve[2] shown in Figure 15-12b corresponds to the

[2] These cooling curves do not show the supercooling of the liquid that often occurs before the start of crystallization.

composition X_{A1} in Figure 15–11. The temperature of the sample drops more or less linearly until it reaches the value T_1. When crystallization of pure A commences, the release of the heat of fusion decreases the rate of temperature lowering. The temperature, however, continues to drop at the lower rate because the composition of the melt changes as crystallization proceeds. When the eutectic temperature is reached, the composition of the melt remains constant at the eutectic composition, and consequently the temperature remains constant until complete crystallization has occurred. Then a rapid temperature drop follows. The length of the flat portion of the cooling curve is proportional to the amount of the sample that crystallizes at the eutectic composition. The cooling curve for a melt that has the eutectic composition (Fig. 15–12c) has the same appearance as the curve for a 1-component sample.

As we have just seen, the two curves in Figure 15–11 represent the freezing point of a 2-component liquid as a function of composition. The addition of some A to pure B, or B to pure A, lowers the freezing point of the pure major component. In fact, if the solution is ideal, these curves must be graphs of Equation 15–17, which gives the relationship between composition and freezing point of a solution in terms of the freezing point and heat of fusion of the component that is present in the solid phase.

Looking at these two curves from another point of view, we can say that they give the composition of the liquid phase (the solution) that is in equilibrium with one solid component or the other (the solute) as a function of the temperature. In other words, these curves represent the solubility of A in B, or B in A. According to Equation 15–17, the mole fraction of a substance in any saturated solution that can be considered as an ideal solution depends only on the properties of this substance. It follows that at a given temperature, a substance has the same solubility expressed in terms of mole fraction in all solvents with which it forms an ideal solution.

For example, naphthalene forms nearly ideal solutions with chlorobenzene, nitrobenzene, benzene, and toluene. The heat of fusion of naphthalene is 4.60 kcal/mole at its melting point of 80°C. When these values are substituted into Equation 15–17, the ideal solubility of naphthalene at 20°C is given by

$$\log X = \frac{4600}{2.30 \times 1.99}\left(-\frac{60}{353 \times 293}\right) = -0.583$$

$$X = 0.261$$

The observed solubility of naphthalene at 20°C, in terms of its mole fraction in the solvents mentioned above, is 0.256, 0.243, 0.241, and 0.224, respectively.

2. It often happens that the components in a binary system interact with each other so strongly that a loosely-bound compound is formed in the solid phase. Such compound formation can result in substantial complications in the phase relationships. The phase diagram in Figure 15–13 applies to the case

in which the two components are completely miscible in the liquid phase, and a compound with a molar ratio of 1:1 is formed. When this compound melts a liquid having the same composition is formed. This process is called *congruent melting*.

The phase diagram for this system can be considered to be divided into two halves, each of which is similar to Figure 15-11. There are two eutectic points, one for each component and the compound. If a liquid phase with a mole fraction of 0.5 is cooled, the cooling curve is exactly the same as that for a

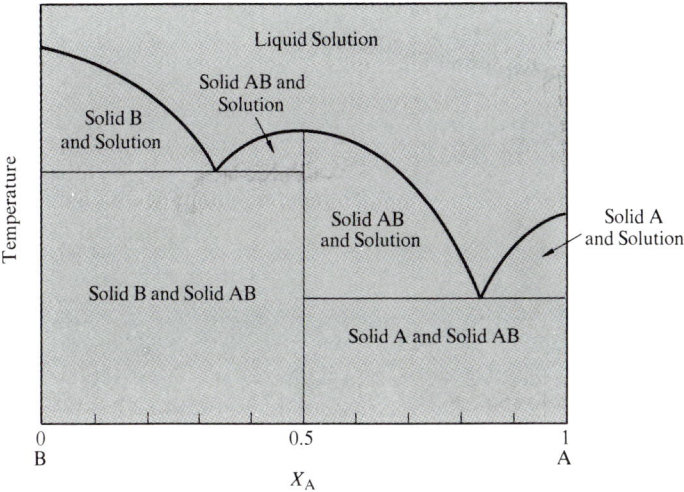

Figure 15-13. Phase diagram for solid-liquid system with the formation of a solid compound, AB.

single component. When either component is added to this liquid, its freezing point is lowered, as for a pure substance.

In some systems in which a solid compound is formed this compound is so unstable that, as its temperature is raised, before it could melt it decomposes to form a liquid and one of the solid components. The liquid phase formed in this process has a composition that is different from that of the original solid compound. The process of forming a liquid in this way is called *incongruent melting*, a term which is misleading because the process is actually decomposition, and not melting. The temperature at which it occurs is called an incongruent melting point, or a *peritectic point*. A phase diagram which includes a peritectic point is shown in Figure 15-14.

3. In the systems discussed so far, the solid phases have been immiscible. In many systems, however, the two components involved form a solid solution, in some cases over the whole range of compositions, in others over limited

ranges. So far as phase relationships and thermodynamics are concerned, a solid solution is just like any other solution, namely, a single phase of variable composition. The phase diagram, Figure 15-15, for a system in which there is complete miscibility in the solid phase is quite similar to Figure 15-6 representing the equilibrium between a liquid-phase solution and its vapor. In the 2-phase

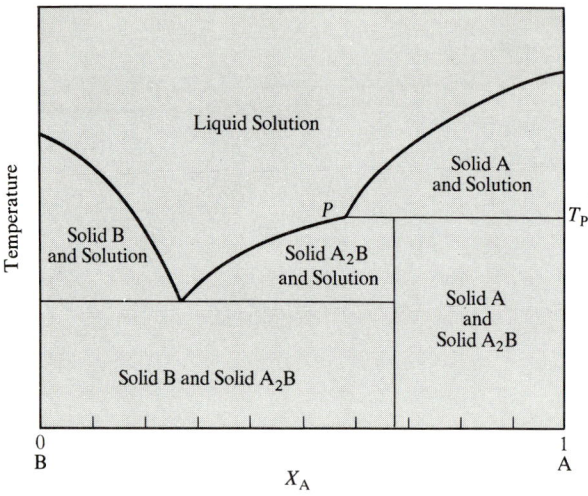

Figure 15-14. Phase diagram showing peritectic point, *P*, representing equilibrium between solid A_2B, solid A, and solution. When solid A_2B is heated above temperature T_p, it decomposes to form solid A and a solution.

region between the freezing and the melting curves a liquid solution is in equilibrium with a solid solution of a different composition. In this case the addition of a small amount of one component to the other may raise the melting point of the major component. Thus we see that the use of the freezing-point depression for the determination of molecular weight or impurity concentration must be limited to those systems in which solid solutions are not formed (see p. 552).

Since the two components that form a solid solution must be uniformly mixed on a molecular scale, they must be present in the same crystal lattice. When the size and shape of the molecules[3] are closely similar for the two components, one kind of molecule can be substituted for the other in the crystal lattice. Many solid solutions of metallic elements are of this substitutional type, which includes a number of alloys of practical importance. Solid solutions can also be formed when the molecules of one component are small

[3]The molecule is the same as the atom for monatomic elements, such as the metals.

compared to those of the other component. In such a solid solution the smaller molecules are distributed in the interstices of the solvent.

Since a change in the composition of a solid solution requires a migration of one kind of molecule into or out of a crystal lattice, this process is relatively slow. Consequently, although the composition of a solid solution is a variable

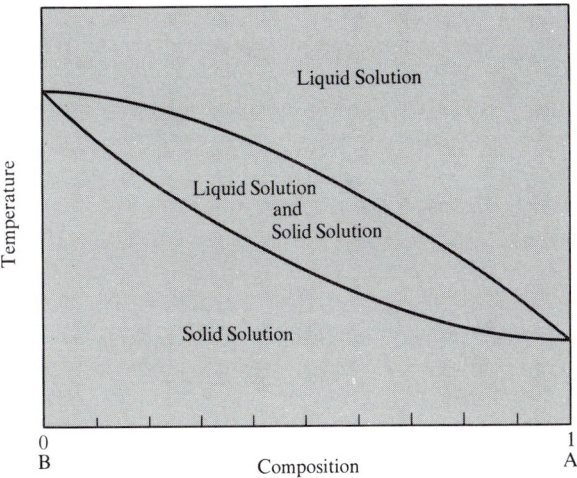

Figure 15-15. Phase diagram for a system in which there is complete miscibility in the solid phase. The presence of B dissolved in solid A raises its melting point.

from a thermodynamic point of view, a solid solution may take a long time to reach equilibrium unless it is formed from a melt.

Three-Component Systems

As we have seen, a system that contains only two components can have a rather complex phase diagram involving, for example, combinations of limited miscibility in both liquid and solid phases with the possible presence of several compounds and peritectic points. A 3-component system can have a much greater variety and complexity. Even the representation of the simplest type of phase diagram for a ternary system presents some difficulties, since three components in a single phase have 4 degrees of freedom. If, however, the temperature and pressure of the system are fixed, then the two compositional variables can be plotted in a plane.

A very convenient and widely used method for plotting the composition of a ternary system makes use of the following geometrical property of the equilateral triangle: If from any point within the triangle lines are drawn perpendicular to the three sides, the sum of the lengths of these three lines is equal to the height of the triangle and this sum is therefore independent of the location of the point. This property is utilized by letting each vertex of the triangle represent one of the pure components and by setting up a scale for this component, either in mole % or weight %, with its zero on the side opposite the vertex (Fig. 15–16). Thus each point within the triangle represents a possible composition of a ternary system, and a point on a side represents a binary

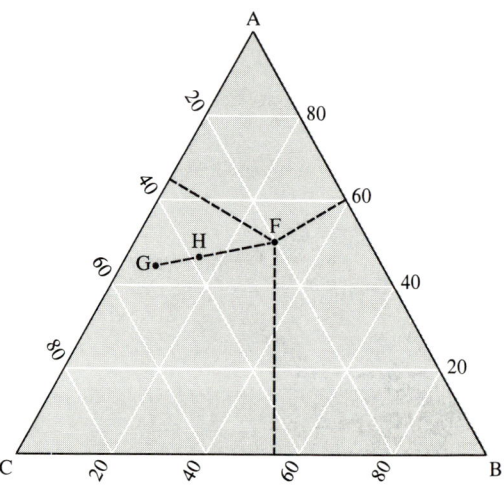

Figure 15–16. Phase diagram for ternary system at constant *T* and *P*. Point F represents a phase whose composition in mole percent is 50% A, 30% B, and 20% C.

system. The temperature (or some other variable) can be plotted in a direction perpendicular to the plane of the triangle, or this variable may be plotted as a set of contour lines in the plane.

There are several other geometrical properties of the triangular phase diagram that are sometimes useful. If two 3-component systems, represented by points F and G in Figure 15–16, are mixed, the point H representing the composition of the resulting mixture lies along the line FG with the distance ratio FH/HG equal to the ratio of the amount of G to the amount of F. Another property leads to the rule that if a straight line is drawn through any vertex, A for example, the points on this line represent systems with various amounts of A in which the ratio of the amounts of B and C is constant.

There is an extensive literature on the subject of the phase equilibria in ternary systems, and many of these systems are of great practical importance. We consider here only a relatively simple system as an example. This system contains water (W), acetone (A), and chloroform (C), at atmospheric pressure and 25°C. The binary systems W–A and C–A are completely miscible while W and C are almost completely immiscible. The phase diagram is shown in Figure 15–17. The region inside the curve, known as a *binodal curve*, is a 2-phase region. A system whose composition is represented by point b separates

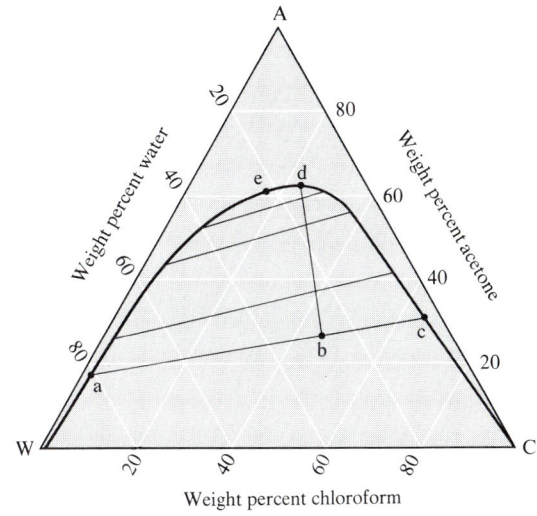

Figure 15–17. Phase diagram of water-chloroform-acetone ternary system.

into two conjugate solutions, a W-rich phase, a, and a C-rich phase, c. The points representing these ternary conjugate solutions are connected by the tie-line abc. The relative amounts of each phase can be found by use of the lever rule.

If more of component A is added to the system, the composition point moves in the direction of vertex A. The W-rich phase grows at the expense of the C-rich phase until at point d the system becomes homogeneous. If the composition of the system is changed in such a way that the composition point approaches the point e, the relative amounts of the two phases remain nearly the same, while their compositions approach each other as the tie-lines become shorter. When the composition point reaches e, the two compositions are the same and the system becomes homogeneous. The point e is called the *plait point* of the system.

Distribution Law

A type of system that is of considerable importance is one which consists of a solute A that is distributed between two mutually immiscible solvents B and C. At equilibrium, the chemical potential of A must be the same in both solvents. It follows that at a definite temperature T,

$$\mu^\circ_{A(B)} + RT \ln a_{A(B)} = \mu^\circ_{A(C)} + RT \ln a_{A(C)}$$

and

$$\ln \frac{a_{A(B)}}{a_{A(C)}} = \frac{\mu^\circ_{A(C)} - \mu^\circ_{A(B)}}{RT}$$

where A(B) and A(C) refer to the solution of A in B and in C, respectively. Because the standard state chemical potentials are independent of composition, the right-hand side of this equation is constant, and thus the left-hand side is also constant; consequently,

$$\frac{a_{A(B)}}{a_{A(C)}} = K$$

If the solutions under consideration are so dilute that they can be considered as ideal,

$$\frac{X_{A(B)}}{X_{A(C)}} = K$$

In dilute solutions the mole fraction is approximately proportional to the concentration. Making this substitution, we have

$$\frac{c_{A(B)}}{c_{A(C)}} \sim K'$$

Expressed in words, this means that if a solute is distributed between two immiscible solvents the ratio of its equilibrium concentrations in the two solvents is approximately constant, and independent of the amount of solute. This statement is known as the *Nernst distribution law*, and the constant K (or K') is called the *distribution coefficient*.

◊◊◊◊◊◊◊

In some cases the solute exists as a dimer or a polymer (*n*-mer) in one of the solvents and as single molecules in the other solvent. The process of transferring solute from a solvent in which it is an *n*-mer to the other solvent can be represented by

$$A_{n(B)} \rightarrow nA_{(C)}$$

Then the equality of chemical potentials in the two solvents at equilibrium leads to

$$\mu^\circ_{A_n(B)} + RT \ln X_{A_n(B)} = n\mu^\circ_{A(C)} + nRT \ln X_{A(C)}$$

It follows that

$$\frac{X^n_{A(C)}}{X_{A(B)}} = K$$

The Nernst distribution law is the basis for the important process of solvent extraction which is widely used for the separation or purification of dissolved substances. For most effective use of the solvent, it is advantageous to carry out this process in a series of successive stages. When the distribution coefficient for the system is known, the weight of solute remaining unextracted (w_n) after n successive extractions can be calculated in the following way: Let the initial solution contain w grams of solute in a volume V_1 of solvent 1. When this solution is equilibrated with a volume V_2 of solvent 2, according to the distribution law

$$\frac{c_1}{c_2} = \frac{w_1/V_1}{(w-w_1)/V_2} = K$$

in which w_1 is the weight of solute remaining in solvent 1. When this equation is solved for w_1 the result is

$$w_1 = \frac{KV_1}{KV_1 + V_2} w$$

w_1/w is the fraction of solute remaining in solvent 1.

If now the solution of concentration c_1 is equilibrated with a second volume V_2 of fresh solvent 2, the weight of solute remaining in solvent 1 is given by

$$w_2 = \frac{KV_1}{KV_1 + V_2} w_1$$

and the fraction of solute remaining is

$$\frac{w_2}{w} = \left(\frac{KV_1}{KV_1 + V_2}\right)^2$$

After n successive extractions, each with volume V_2, the fraction of solute remaining in solvent 1 is

$$\frac{w_n}{w} = \left(\frac{KV_1}{KV_1 + V_2}\right)^n \tag{15-30}$$

582 Physical Chemistry

In contrast, if the same total volume of solvent 2, namely nV_2, is used in a single extraction step the fraction of solute remaining is

$$\frac{w'_1}{w} = \frac{KV_1}{KV_1 + nV_2} \qquad (15\text{--}31)$$

It can be shown that w_n is always less than w'_1, and thus repeated extractions, each with a small amount of fresh solvent, are more effective than a single extraction with a large amount of solvent.

◈◈◈◈◈◈◈

To show that this is so we can write the reciprocal of Equation 15–30 in the form

$$\frac{w}{w_n} = \left(1 + \frac{V_2}{KV_1}\right)^n$$

When this binomial is expanded, we have

$$\left(1 + \frac{V_2}{KV_1}\right)^n = 1 + \frac{nV_2}{KV_1} + \frac{n(n-1)}{2}\left(\frac{V_2}{KV_1}\right)^2 + \cdots$$

The reciprocal of Equation 15–31 can be written in the form

$$\frac{w}{w'_1} = 1 + \frac{nV_2}{KV_1}$$

Since $\dfrac{w}{w_n} > \dfrac{w}{w'_1}$, it follows that $w_n < w'_1$.

◈◈◈◈◈◈◈

Supplementary Material

Surfaces of Two-Component Systems: Adsorption

In our discussion of the phase equilibria of multicomponent systems we have not mentioned the surfaces (or interfaces) that separate phases from each other. As in the case of 1-component systems (p. 519), this neglect of surfaces is justified in most cases by the fact that the amount of matter contained in the surfaces is so small relative to the amount in the bulk phases that it is negligible. There are systems, however, in which the effect of the nature and extent of surfaces is significant, or even dominant, and then the properties of the surface must be considered. The physical chemistry of surfaces is a very broad subject, and only a few of its aspects can be treated here.

Gibbs Adsorption Equation. We first consider a surface between two fluid phases, assuming for simplicity that it is planar. From a molecular point of view, this surface has a finite thickness because it consists of a transition region between phases that extends over at least several molecular diameters. Since the properties of this transition region vary from their values in one phase to those in the adjoining phase, the exact definition of the boundary between the two phases appears somewhat arbitrary. Gibbs showed that a consistent

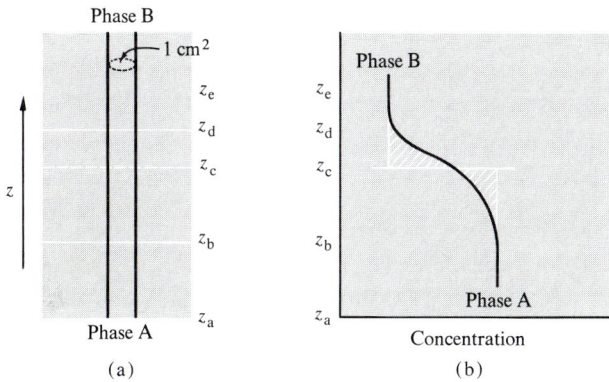

Figure 15–18. (a) Cross-section of the surface between two phases, A and B, with a cylindrical volume of unit cross-section area perpendicular to the surface; z_b and z_d represent the limits of the transition region and z_c is the location of the theoretical boundary. (b) Graph of the concentration of a component in the system represented in (a) as a function of z. The cross-hatched areas are obtained by extending the concentration lines for this component in the bulk phases to the theoretical boundary. These areas are proportional to the number of moles of this component in the corresponding cylindrical volume.

thermodynamic treatment of surfaces can be developed if the actual physical surface is considered to be replaced by a mathematical boundary of infinitesimal thickness whose location is specified in a suitable way. We imagine that the boundary is located in the transition region by the requirement that the number of moles (or molecules) of a component in a cylindrical volume of unit cross-section (Fig. 15–18a) in the transition region is the same on one side of the boundary as on the other side. The properties of each phase are then considered to have their bulk values up to this theoretical boundary.

This choice of boundary is illustrated in Figure 15–18b which represents the concentration of a component in the transition region $z_b - z_d$ between phases A and B as a function of the coordinate z. Since the volume in the cylinder of unit cross-section is proportional to the magnitude of z, the area under the curve between two values of z is proportional to the number of moles in the corresponding volume. The value of z_c is determined by the requirement that the cross-hatched area between z_b and z_c is equal to the cross-hatched area between z_c and z_d.

For a 1-component system consisting of a pure liquid in equilibrium with its vapor, we then have the relationship $n = c_A V_A + c_B V_B$ in which n is the total number of moles in the system, c_A and c_B are the molar concentrations in the two phases, and V_A and V_B are the volumes of the two phases up to the theoretical boundary. For a multicomponent system the situation is more complicated because the concentrations of the components are additional independent variables.

Let us restrict our attention to a 2-component system with components 1 and 2 considered as the solvent and the solute, respectively. We define the location of the boundary between two phases in such a way that in the transition region the numbers of moles of component 1 on the two sides of the boundary are equal to each other. From the discussion above it follows that $n_1 = c_{1A} V_A + c_{1B} V_B$. With the boundary defined in terms of component 1, however, in general the numbers of moles of component 2 in the transition region on the two sides of the boundary are not equal. This can be expressed by the equation $n_2 = c_{2A} V_A + c_{2B} V_B + n_2^s$, in which n_2^s is the difference between the total number of moles of component 2 in the system and the number of moles that would be present if the concentrations of component 2 in the two bulk phases remained constant up to the defined phase boundary. (The superscript s in n_2^s refers to the surface.) Depending on the nature of components 1 and 2, n_2^s can have either positive or negative values.

The ratio of n_2^s to the area of the surface is a measure of the amount of component 2 that is adsorbed at the surface. This ratio, represented by Γ_2, is interpreted as the surface concentration in moles/cm² of component 2 adsorbed at the surface; that is, its concentration in a 2-dimensional phase lying between the bulk phases.

The importance of the definition of the boundary between bulk phases as carried out above is that it permits the definition in a consistent way of all of the thermodynamic properties of the components in a surface phase. The treatment can be generalized to include the case of curved surfaces. As one might expect, the surface free energy plays a predominant role in the thermodynamic treatment of surfaces.

We have seen (p. 503) that the formation of an element of surface area (da^s) of a pure liquid requires an amount of work γda^s, in which γ is the surface energy or the surface tension. From a thermodynamic point of view, we can equate the work done in forming a fresh surface at constant temperature and pressure to the free energy[4] of the surface; thus $dG^s = \gamma da^s$. For a solution,

[4]There is no need to distinguish between the Helmholtz and the Gibbs free energies in this case. They have the same value because $G^s = A^s + PV$, and $V = 0$ for a surface.

Phase Equilibria in Multicomponent Systems

the free energy of a surface is changed also by adsorption; that is, by a change in the composition of the surface. Thus the total change in the surface free energy (at constant T and P) for an infinitesimal process is given by

$$dG^s = \gamma da^s + \mu_2 \, dn_2^s \tag{15-32}$$

in which dn_2^s is the number of moles of component 2 transferred to the surface from a bulk phase in which its chemical potential is μ_2.

When we integrate the equation above keeping the composition of the surface constant (compare the derivation of Eq. 3-60), we obtain

$$G^s = \gamma a^s + \mu_2 n_2^s$$

Now we can differentiate this equation in general, noting that γ is a function of the composition; thus

$$dG^s = \gamma \, da^s + a^s \, d\gamma + \mu_2 \, dn_2^s + n_2^s \, d\mu_2$$

On subtracting Equation 15-32 from the equation above, we obtain the following equation:

$$a^s \, d\gamma + n_2^s \, d\mu_2 = 0$$

When this equation is divided by a^s, it becomes

$$d\gamma + \Gamma_2 \, d\mu_2 = 0$$

which is called the *Gibbs adsorption equation* or the *Gibbs adsorption isotherm*. This equation can be written in several closely-related ways. For example, when we introduce the relationship $d\mu_2 = RT \, d \ln a_2$, we obtain

$$d\gamma = -\Gamma_2 \, RT \, d \ln a_2$$

or

$$\Gamma_2 = -\frac{1}{RT} \frac{d\gamma}{d \ln a_2}$$

For a solution so dilute that it can be considered as ideal, $d \ln a_2 = d \ln X_2$, and it follows that

$$\Gamma_2 = -\frac{1}{RT} \frac{d\gamma}{d \ln X_2}$$

For a solution so dilute that $d \ln X_2 \sim d \ln c_2$, with c_2 the concentration of component 2,

$$\Gamma_2 = -\frac{1}{RT} \frac{d\gamma}{d \ln c_2}$$

$$= -\frac{c_2}{RT} \frac{d\gamma}{dc_2}$$

This equation says that when a solute is adsorbed positively (that is, Γ_2 is a positive number), the surface tension of the solution is lower than that of the pure solvent. A striking verification of this conclusion was carried out by McBain. He was able to scoop up a thin surface layer of an aqueous solution with a rapidly moving sharp knife. Analysis of the surface material collected in this way showed reasonable agreement with the surface concentration predicted by the Gibbs adsorption equation.

A number of solutes are known to lower the surface tension of water markedly even though their bulk concentration is very low. Among these solutes, known as surface active agents, are soaps and long-chain fatty acids and alcohols. Electrolytes raise the surface tension of water, but the magnitude of the effect is small.

Adsorption of Gases on Solids. The only other type of surface to be discussed here is the surface between a solid phase and a gas or vapor phase of a different substance. It has proved to be difficult to treat the surface behavior of solids quantitatively, mainly because the surfaces of nearly all solids are nonuniform and heterogeneous in structure and composition on a molecular scale. Even the determination of surface area is associated with considerable uncertainty as shown by the fact that different methods often lead to significantly different results.

The molecules in the surface of a solid, like the molecules in the surface of a liquid, are in an environment that is different from the interior. Unbalanced intermolecular forces exist at the surface and as a consequence a solid can adsorb measurable quantities of a gas, particularly if the solid is porous and finely divided which is the usual case in the study of adsorption by solids.[5] The amount of gas adsorbed by a solid adsorbent at equilibrium is a function of the gas pressure and the temperature. The amount of adsorbed gas is usually expressed in terms of the weight, or volume at 0°C and 1 atm, of gas per gram of adsorbent. A function relating this amount to the gas pressure at constant temperature, or a graph representing such a function, is called an *adsorption isotherm*. Two common forms of adsorption isotherm are shown in Figure 15-19. These two forms are associated with two limiting types of adsorption known as *physical adsorption* and *chemisorption*.

1. *Physical adsorption* is characterized by reversibility; that is, the adsorbed vapor may be removed unchanged from the adsorbent by reducing the pressure with a vacuum pump. Also, the heat of adsorption per mole of adsorbed vapor has the same order of magnitude as the heat of condensation of the vapor to a liquid. Thus it can be concluded that a layer of adsorbed vapor molecules is held to the surface of the solid adsorbent by relatively weak van der Waals

[5] If a solid sample initially in the form of a cube with smooth sides and an edge length of 1 cm is divided into cubes with an edge length of 1μ (10^{-4} cm), the area of the surface increases from 6 cm² to 6×10^4 cm². If the solid contains pores and fissures, the available surface area may be much greater than this.

forces. As the pressure of the vapor is raised, a second layer of adsorbed molecules is formed and then a third layer, and so on, until the vapor pressure is reached, when condensation occurs.

By treating multilayer physical adsorption on the basis of simplifying assumptions, Brunauer, Emmett, and Teller were able to derive an adsorption isotherm which is in fair agreement with a variety of experimental data. Their equation makes it possible to estimate the amount of gas adsorbed on a particular adsorbent when the first layer of adsorbed molecules has been completed.

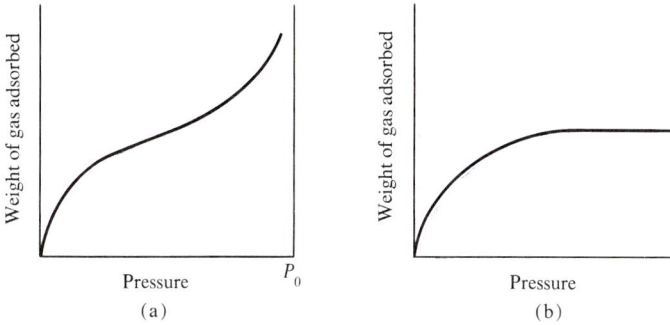

Figure 15-19. Typical adsorption isotherms for (a) physical adsorption, (b) chemisorption. P_0 is the equilibrium vapor pressure of the substance whose vapor is adsorbed.

By assuming an effective cross-section area of the adsorbed molecule, one can then calculate the area of the adsorbent surface. This BET method, with nitrogen as the adsorbed gas, is widely used as a method for the semiquantitative determination of the areas of adsorbent surfaces.

2. *Chemisorption* is often associated with adsorption isotherms of the form shown in Figure 15-19b. The amount of gas adsorbed increases at low gas pressures to a certain value, and then is essentially independent of the gas pressure. In contrast to physical adsorption, chemisorption is irreversible; that is, the adsorbed gas cannot be desorbed by merely lowering the gas pressure, but pumping at an elevated temperature is required. The gas that is desorbed is often different from the gas that was adsorbed. For example, when oxygen is adsorbed on charcoal the gas that is desorbed is carbon monoxide. In addition, the heat of chemisorption is relatively high, of the same order of magnitude as the heat of reaction of the gas with the adsorbent. All of these facts are consistent with the interpretation of chemisorption as the result of the formation on the adsorbent surface of a single layer of adsorbed molecules which are held to the surface by chemical bonds. There is then little tendency in general to form a second layer of adsorbed gas molecules on top of the first layer.

In 1918 Langmuir derived an adsorption isotherm whose graph resembles Figure 15–19b. Although this equation was derived on the basis of an unrealistically simple model of an adsorbent surface, it provides a fairly good representation of the experimental data for some systems, particularly at low pressures. Langmuir assumed that the adsorbent surface, with area a, is uniform and that at equilibrium the rate of adsorption is equal to the rate of desorption. The rate of adsorption is proportional to the gas pressure and to the available area of "bare" surface. The rate of desorption is proportional to the area covered by adsorbed molecules. If θ represents the fraction of the surface area that is covered, we can express the equality of rates in the following way:

$$k_1 P(1 - \theta)a = k_2 \theta a$$

Solving this equation for θ, we obtain

$$\theta = \frac{k_1 P}{k_2 + k_1 P}$$
$$= \frac{bP}{1 + bP}$$

in which $b = k_1/k_2$. If V is the volume of gas adsorbed and V' is the volume needed to form a single adsorbed layer, $\theta = V/V'$ and thus

$$V = \frac{V'bP}{1 + bP}$$

At pressures so low that $bP \ll 1$, $V = V'bP$ and the volume adsorbed is directly proportional to the gas pressure. At pressures so high that $bP \gg 1$, $V = V'$.

It is possible to derive the Langmuir adsorption isotherm on a statistical basis without making the assumptions needed for this kinetic derivation. It turns out, however, that the assumptions needed to evaluate the 2-dimensional partition functions are also rather drastic.

Supplementary References

A. W. Adamson, *The Physical Chemistry of Surfaces*, 2nd edition, Interscience, New York, 1967.

G. N. Lewis and M. Randall, 2nd edition revised by K. S. Pitzer and L. Brewer, *Thermodynamics*, McGraw-Hill, New York, 1961. Chapters 17–21 inclusive; Chapter 29.

J. E. Ricci, *The Phase Rule and Heterogeneous Equilibrium*, Van Nostrand, Princeton, 1951.

S. Ross and J. P. Oliver, *On Physical Adsorption*, Interscience, New York, 1964.

Problems

1. The vapor pressure at 100°C of a solution of 9.75 g of a nonvolatile substance in 148.5 g of water is 749 torr. Calculate the molecular weight of the substance. **Ans.** 82

2. When 3.60 g of a nonvolatile hydrocarbon containing 93.50 weight % carbon was dissolved in 100 g of benzene at 30°C, the vapor pressure dropped from 119.6 torr to 117.4 torr. Obtain the molecular formula of the hydrocarbon.

3. A solution of 0.925 g of a nonvolatile solute in 150.0 g of carbon tetrachloride has a boiling point 0.210°C above that of the pure solvent. Calculate the molecular weight of the solute. **Ans.** 148

4. When 640 mg of naphthalene are dissolved in 40.0 g of chloroform the boiling point of the solution is 0.455°C higher than that of the pure solvent (61.2°C). Calculate (a) the molal boiling-point elevation constant; (b) the molar heat of vaporization of chloroform. **Ans.** (a) 3.64; (b) 7.29 kcal/mole

5. For a solution containing 50.0 g of sucrose ($M = 342$) in 1000 g of water, calculate (a) the vapor pressure at 25°C; (b) the freezing point; (c) the normal boiling point. The vapor pressure of pure water at 25°C is 23.76 torr.

6. What weight of ethylene glycol ($C_2H_6O_2$) must be added to a kilogram of water to lower its freezing point by 5°C? **Ans.** 167 g

7. When 1.750 g of a certain substance was dissolved in 100 g of water the freezing point of the water was depressed by 0.357°C. A solution containing 2.710 g of this substance in 100 g of benzene had a freezing point 0.770°C lower than that of pure benzene. What conclusion can be reached about the state of this substance when it is dissolved in benzene?

8. A sample of benzene has a freezing point of 4.20°C. (a) What is the purity of the benzene in mole % assuming that no solid solution is formed? (b) Calculate the normal boiling point of the sample of benzene.
Ans. (a) 98.0 mole %; (b) 80.9°C

9. An aqueous solution of a solute whose concentration is 8.52 g/liter has an osmotic pressure of 1140 torr at 300°K. (a) What is the molecular weight of the solute? (b) Calculate the osmotic pressure of a solution at 75°C that contains 1.12 g of this solute in 400 ml of solution. **Ans.** (a) 140; (b) 434 torr

10. In an apparatus for the measurement of the osmotic pressure of an aqueous solution the water side of the membrane is maintained at atmospheric pressure while the pressure on the solution side is measured with an open-end water manometer. A solution containing 1.45 g of a polymer in 300 ml of solution at 50°C produced a manometer reading of 28.8 mm of water at equilibrium. Calculate the molecular weight of the polymer.

11. Two liquids, A and B, form an ideal solution at temperature T. When the total vapor pressure above the solution is 600 torr the mole fraction of A in the vapor phase is 0.35 and in the liquid phase is 0.70. What are the vapor pressures of pure A and pure B at temperature T?

Ans. $P_A^\circ = 300$ torr; $P_B^\circ = 1300$ torr

12. At 50°C the vapor pressures of pure n-hexane and n-heptane are 408 and 141 torr, respectively. Assuming ideality: (a) What is the mole fraction of n-hexane in a solution in equilibrium with a vapor in which the mole fraction of n-hexane is 0.85? (b) What is the molar entropy change when this solution is formed from pure components at 50°C?

13. At 30°C the vapor pressures of pure toluene and pure benzene are 36.7 and 118.2 torr, respectively, and the two liquids form a nearly ideal solution. (a) For a solution containing 50.0 weight % of toluene, calculate the total vapor pressure and the mole fraction of each compound in the vapor phase. (b) What is the composition of a solution of benzene and toluene that will boil at 30°C at a pressure of 50.0 torr?   Ans. (a) $P_{tot} = 80.9$ torr, $Y_{Bz} = 0.793$; (b) $X_{Bz} = 0.163$

14. Plot P_{tot} versus the composition of the liquid and vapor phases for the benzene-toluene system at 30°C, assuming ideality. Use data from the preceding problem.

15. At 80°C the vapor pressures of ethylene bromide and propylene bromide are 172 and 127 torr, respectively, and these compounds form a nearly ideal solution. 3 moles of ethylene bromide and 2 moles of propylene bromide are equilibrated at 80°C and a total pressure of 153 torr. (a) What is the composition of the liquid phase? (b) How many moles of each compound are present in the vapor phase?

Ans. (a) 58 mole % ethylene bromide; (b) 0.91 moles ethylene bromide, 0.49 moles propylene bromide

16. For the preceding problem, use the lever rule to obtain the number of moles of each compound in the vapor phase.

17. The vapor pressures of two pure liquids, A and B, that form an ideal solution are 300 and 800 torr, respectively, at temperature T. A mixture of the vapors of A and B for which the mole fraction of A is 0.25 is slowly compressed at temperature T. Calculate: (a) the composition of the first drop of condensate; (b) the total pressure when this drop is formed; (c) the composition of the solution whose normal boiling point is T; (d) the pressure in the system when only the last bubble of vapor remains; and (e) the composition of this bubble.

Ans. (a) $X_A = 0.471$; (b) 565 torr; (c) $X_A = 0.080$; (d) 675 torr; (e) $Y_A = 0.11$

18. The vapor pressures of two pure liquids, A and B, that form an ideal solution are 300 and 800 torr, respectively, at temperature T. A liquid solution of A and B for which the mole fraction of A is 0.60 is contained in a cylinder closed by a piston on which the pressure can be varied. The solution is slowly vaporized at temperature T by decreasing the applied pressure, starting with a pressure of about 1 atm. Calculate: (a) The pressure at which the first bubble of vapor is formed; (b) the composition of the vapor in this bubble; (c) the composition of the last droplet, and (d) the pressure when only this last droplet of liquid remains.

19. The vapor pressures in torr of two liquids, A and B, that form an ideal solution are given by the equations:

$$\log P_A = -5100/T + 16.24$$
$$\log P_B = -4530/T + 13.38$$

Draw a temperature-composition phase diagram for solutions of A and B at a total pressure of 1 atm.

20. The Henry's-law constants, as defined on p. 562, for oxygen and nitrogen dissolved in water at 0°C are 1.9×10^7 torr and 4.1×10^7 torr, respectively. A sample of water at a temperature just above 0°C was equilibrated with air (20% oxygen and 80% nitrogen) at 1 atm. (a) The dissolved gas was separated from a sample of this water and then dried. What was the composition of this gas? (b) Calculate the freezing point of the saturated water.

Ans. (a) 35 mole % O_2; (b) $-0.0024°C$

21. The following vapor pressure-composition data were obtained for acetone(A)–carbon disulfide(B) solutions at 29.2°C:

P_{tot}(torr)	490.4	523.5	526.4	530.8	534.0	528.7	518.0
X_B	0.295	0.340	0.500	0.568	0.670	0.770	0.850
Y_B	0.525	0.600	0.615	0.635	0.665	0.700	0.755

Plot the portion of the pressure-composition diagram covered by the data and estimate the composition of the azeotrope. In addition, plot the partial pressures of each of the components.

22. The following data were obtained for benzene(A)–ethanol(B) solutions at atmospheric pressure:

Boiling point (°C)	78	75	70	70	75	80
X_A	0	0.04	0.21	0.86	0.96	1
Y_A	0	0.18	0.42	0.66	0.83	1

(a) Plot the temperature-composition diagram for this system and estimate the composition and boiling point of the azeotrope. (b) For a solution containing 20 g of benzene and 50 g of ethanol, estimate the boiling point and the composition of the vapor first formed at this temperature. (c) If the solution described in (b) is exhaustively fractionated, which component will be obtained as a pure substance? Is this component the distillate or the residue? How much of this component will be obtained as a pure substance?

23. The steam distillation of chlorobenzene is observed to occur at a temperature of 90.6°C when the total pressure is 1.00 atmosphere. Assuming complete immiscibility of these liquids, calculate the weight of chlorobenzene in 100 g of distillate. The vapor pressure of water at 90.6°C is 538.9 torr.

Ans. 72.0 g

24. When a liquid that is immiscible with water was steam-distilled at 95.2°C at a total pressure of 747.3 torr, the distillate contained 1.27 g of the liquid per gram of water. Calculate the molecular weight of the liquid. The vapor pressure of water is 638.6 torr at 95.2°C.

25. The vapor pressures at 50°C of the immiscible liquids n-hexane and water are 401 and 92.5 torr, respectively. When a system containing equal numbers of moles of these substances was equilibrated at 50°C, the total pressure of the vapor phase was found to be 275 torr. (a) What phases were present in the system? (b) What was the composition of the vapor phase? (c) What was the ratio of the number of moles of vapor to the number of moles of liquid?

Ans. (a) Vapor and liquid water; (b) 66.3 mole % hexane; (c) 3.08

26. Biphenyl has a melting point of 71°C and a heat of fusion of 4.05 kcal/mole. Assuming that solutions of biphenyl in benzene are ideal, calculate the solubility of biphenyl in benzene at 50°C.

27. The following data, in mole %, were obtained for the ternary system: aniline, toluene, n-hexane at 25°C.

Aniline-rich phase		Heptane-rich phase	
aniline	n-heptane	aniline	n-heptane
95.6	4.4	8.4	91.6
89.1	7.2	12.2	81.3
74.0	13.2	21.0	60.7
62.6	19.3	32.5	45.4

(a) Plot these data in a triangular graph and indicate the 1- and 2-phase regions. (b) Using this graph, indicate the changes that occur when successive increments of toluene are added to a system containing equal numbers of moles of aniline and n-heptane.

28. When different amounts of phenol were equilibrated with a mixture of water (W) and chloroform (C) at 25°C the concentrations of the two solutions formed in each case had the following values in moles/liter:

C_W	7.85×10^{-4}	1.73×10^{-3}	2.64×10^{-3}	4.65×10^{-3}
C_C	1.64×10^{-3}	8.10×10^{-3}	1.97×10^{-2}	5.77×10^{-2}

Phenol dissolved in water is almost completely monomeric. (a) Is this also true of phenol dissolved in chloroform? (b) Calculate the value of the distribution coefficient. **Ans.** (a) No; (b) 3.7×10^{-4}

29. When different amounts of succinic acid were equilibrated with a mixture of water (W) and diethylether (E) at 25°C, the two solutions formed in each case had the following compositions expressed as moles of acid per mole of solution times 10^3:

C_W	3.64	7.20	10.88	15.13
C_E	2.48	4.85	7.27	10.14

Verify the validity of the Nernst distribution law and calculate the value of the distribution coefficient.

30. A solute, A, is distributed between two immiscible solvents, B and C, with the value of the distribution coefficient, $C_{A(C)}/C_{A(B)}$, equal to 10 when expressed in terms of grams of solute per liter of solvent. It is desired to remove 99% of the amount of A from a solution containing 1 gram of A in 1 liter of B by extraction with successive 100 ml portions of solvent C. Approximately what volume of solvent C will be required? **Ans.** 700 ml

16

Electrochemistry

Introduction

Today we know that all of chemistry is concerned with the interaction of electrically charged particles. From this point of view the term "electrochemistry" is too inclusive. In the present context this term refers to the study of the production and behavior of charged particles (ions) in certain solutions, usually with water as the solvent, that are distinguished by their ability to conduct an electric current. These solutions are known as electrolytic solutions. Their conductivity differs from metallic conductivity in that *electrolytic conductivity* always involves both the transport of matter and chemical changes at the electrodes, the locations where the electric current enters and leaves the electrolyte. Neither of these phenomena is observed with metallic conductivity.

The observation that electric currents can produce chemical reactions and that, conversely, chemical reactions can produce electric currents, has had a strong influence on the development of chemistry. In the introduction to Chapter 7 it was pointed out that electrochemical studies led to the first even partially successful theory of chemical binding — the dualistic theory of Berzelius. In the latter part of the nineteenth century, advances in electrochemistry, both theoretical and experimental, led to the recognition of physical chemistry as a distinct branch of chemistry.

The earliest quantitative work in electrochemistry was the study by Faraday of the relation between amounts of electricity transferred through solutions and the amounts of chemical change observed during electrolysis. In an investigation which even today is a model of care and thoroughness, he laid the basis for his two laws of electrolysis (1834) which can be stated as follows:

1. The amount of a substance formed or consumed during electrolysis is proportional to the amount of electric charge passed through the electrolyte as given by the product of current and time.

2. When equal amounts of electric charge are passed through electrolytes,

the amounts of different substances formed or consumed are proportional to their equivalent weights.

These two laws can be expressed more concisely by the following equation:

$$m = \frac{It}{\mathcal{F}} \frac{M}{z}$$

in which m is the mass of a substance (an element, in the simplest cases) deposited on or removed from an electrode, M is the atomic weight of the element and z is the magnitude of the charge on its ions in terms of the electronic charge as a unit, I is the electric current in amperes, t is the time in seconds, and $\mathcal{F}$ is a constant which is called the *faraday*. The product It gives the amount of charge passed through the electrolyte in coulombs, and M/z is the equivalent weight of the ion.

If the ion in question carries a unit charge, which is numerically equal to the charge on an electron, then 1 faraday is the amount of charge that is transferred when 1 mole of substance, containing Avogadro's number of atoms, is deposited or removed. It follows that the faraday is the amount of charge carried by Avogadro's number of electrons. Since the charge on the electron is 1.6021×10^{-19} coulombs, the value of the faraday is given by

$$\mathcal{F} = 1.6021 \times 10^{-19} \times 6.0225 \times 10^{23}$$
$$= 9.6487 \times 10^4 \text{ coulombs/equivalent}$$

It will be sufficient for our purposes here to round off this number to three significant figures; then $\mathcal{F} = 9.65 \times 10^4$ coulombs/equivalent. It should be noted that the term "faraday" is also used as the name of a unit quantity of electricity equal to 9.65×10^4 coulombs.

Faraday's laws are the basis for a useful device, called a *coulometer*, for determining the amount of charge that has passed through a solution during a certain time. A coulometer is a small electrolytic cell that is connected in series with the cell containing the solution under study. Coulometers often consist of silver electrodes dipping in a solution of $AgNO_3$, although copper coulometers are sometimes used. The number of coulombs that have passed through the main electrolysis cell can be calculated by dividing the gain in weight of the negative electrode (the cathode), or the loss in weight of the positive electrode (the anode), by the equivalent weight of the ion formed or discharged, and then multiplying by the value of the faraday. In effect, the coulometer acts as an analog computer which carries out the integration of the equation

$$Q = \int_{t_1}^{t_2} I\, dt$$

even though the current may vary during the time interval.

We have referred to the existence of ions in solution, but it should be noted that the theoretical concept that electrolytes in solution are dissociated more or less completely into ions was not proposed for about half a century after Faraday's work on electrolysis. At that time van't Hoff was interested in applications of the then newly-derived equations for the colligative properties to electrolytic solutions. He found that when the molecular weight of a salt or some acids and bases was calculated from measured values of any of the colligative properties of a conducting solution, an abnormally low value was found that was inconsistent with values obtained in other ways. In the case of NaCl, for example, the value found was near 30, which is about half the expected value.

Soon after van't Hoff's work on the properties of electrolytic solutions Arrhenius arrived at a theoretical interpretation of electrolytic conductivity. In 1887 he pointed out that just those solutions which conducted electricity also showed abnormal values of the colligative properties, and that these values could be understood if there were more particles in electrolytic solutions than the number of electrolyte molecules — twice as many in the case of NaCl solutions.

On the basis of these and related observations, Arrhenius suggested that in electrolytic solutions a spontaneous dissociation of the electrolyte into ions occurs and that an equilibrium exists between undissociated molecules and their ions, whose motion in an electric field carries the current. He also proposed that the degree of dissociation could be calculated from measurements of either conductivities or colligative properties. He found good agreement between the degrees of dissociation calculated from both kinds of data for a number of electrolytes.

Arrhenius' suggestion of spontaneous dissociation into ions met with great resistance at first. The idea that positively and negatively charged particles could exist in solutions in close proximity without neutralizing their charges was repugnant to many chemists and physicists at that time. Also most chemists found it difficult to believe that the charges carried by ions could drastically affect their chemical properties. Consequently they thought that NaCl could not be dissociated in aqueous solution because in that case sodium ions would react with the water as sodium atoms do and the "chlorine" ions would be evident by their color and odor.

Eventually Arrhenius' basic concept of ionization won universal acceptance. Today we know that many substances are completely ionized, even in the solid state. It is also recognized that the effect of the solvent on solutions of electrolytes, which Arrhenius did not consider, is far from negligible. Although this fact has led to some modification of Arrhenius' theory, as we will see later, this theory is still the foundation of electrochemistry.

Electrochemistry deals with the behavior of ions from two points of view. One is the study of the time-dependent processes of ionic conductance and reaction carried out under nonequilibrium conditions; the other is the study of reversible processes and reactions at equilibrium. For the latter, the methods of thermodynamics are particularly appropriate. We first consider the time-dependent and irreversible process of electrolytic conductance.

Electrolytic Conductance

Equivalent Conductivity

It has been known since the work of Ohm in 1827 that the electrical resistance of a metallic conductor is given by Ohm's law, $R = \mathcal{E}/I$, where $\mathcal{E}$ is the difference in electrical potential between the ends of the conductor and I is the current passing through it. With $\mathcal{E}$ expressed in volts and I in amperes, R is expressed in ohms.

The resistance of a particular conductor, such as a piece of copper, depends not only on its composition but also on its size and shape. The resistance of a conductor with a uniform cross-section is directly proportional to its length, L, and inversely proportional to its cross-sectional area, A. Thus

$$R = \frac{L}{A} r$$

in which the constant of proportionality, r, is a measure of the characteristic property of the conducting substance called its *resistivity*. For a conductor in the shape of a cube with an edge length of 1 cm, $R = r$. The unit of resistivity is the ohm-cm.

For many years there was uncertainty about the applicability of Ohm's law

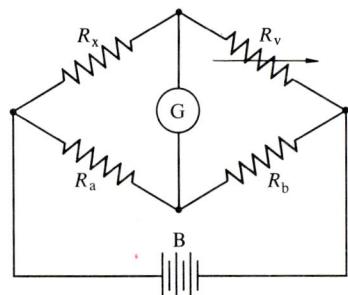

Figure 16–1. Diagram of a Wheatstone bridge. R_x is an unknown resistance, R_v is a variable known resistance, R_a and R_b are resistances whose ratio is known, G is a galvanometer, and B is a battery. When the bridge is balanced and the galvanometer reading is zero, $R_x = \dfrac{R_a}{R_b} R_v$.

Electrochemistry

to electrolytic solutions because of inconsistent experimental data. Eventually, measurements under carefully controlled conditions established the fact that electrolytic solutions do indeed obey Ohm's law.

In principle, electrical resistances can be determined by making simultaneous measurements of potential difference and current and then calculating the ratio of these quantities. It is usually more convenient, however, to employ a comparative method such as the use of a Wheatstone bridge (Fig. 16-1). To minimize the effect of chemical changes at the electrodes in electrolytic solutions, low-frequency alternating current, rather than direct current, is used for resistance measurements.

In electrochemistry it is simpler to express relationships in terms of the reciprocals of resistance and resistivity instead of these quantities themselves. These reciprocals are called the *conductance* (unit, ohm^{-1}) and the *conductivity*, κ, (unit, ohm^{-1} cm^{-1}), respectively. In terms of the resistance, κ is given by $\kappa = \frac{1}{R}\left(\frac{L}{A}\right)$. The ratio L/A cm^{-1} is called the *cell constant*.

A typical conductance cell is shown in Figure 16-2. For the calculation of the conductivity of a solution from a measurement of its resistance, it would

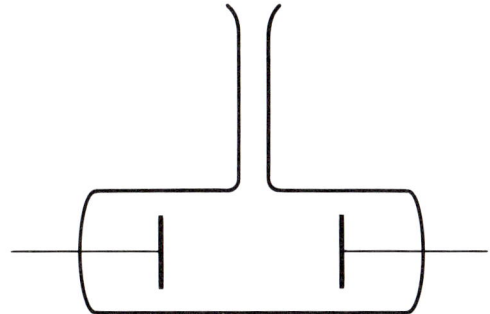

Figure 16-2. A conductance cell.

be possible to use measured values of the area of the electrodes and the distance between them. A more accurate method, however, involves an indirect determination of the cell constant for a particular conductance cell by measurement of the resistance of the cell when it is filled with a solution of KCl at known concentration and temperature. The cell constant is then found from a tabulated value of the conductivity of the KCl solution which has been accurately determined as a function of the concentration.

For example, a conductance cell is found to have a resistance of 35.42 ohms when it is filled with 0.100 molar KCl at 25°C. The conductivity of this solution is 0.01288 ohm^{-1} cm^{-1}. Consequently the value of the cell constant is

$35.42 \times 0.01288 = 0.4562$ cm^{-1}. If a solution of some other electrolyte has a resistance of 112.7 ohms in this cell, its conductivity is $0.4562/112.8 = 4.050 \times 10^{-3}$ ohm^{-1} cm^{-1}.

The conductivity of a particular electrolytic solution depends not only on the nature of the solute but also on its concentration. To take the concentration into consideration, Kohlrausch, who was a pioneer in the precise measurement of electrolytic conductance, defined the *equivalent conductivity* (Λ) in the following way:

$$\Lambda = \frac{1000\kappa}{c} \tag{16-1}$$

in which c is the concentration of the solution in equivalents per liter and the units of Λ are cm^2 ohm^{-1} equiv^{-1}. The equivalent conductivity is equal to the conductance of enough solution to contain 1 equivalent when this solution is placed between parallel electrodes of sufficiently large area placed 1 cm apart.

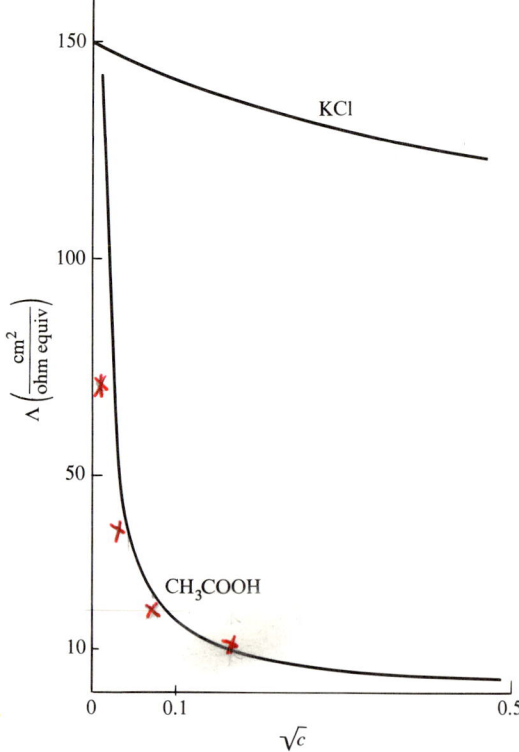

Figure 16-3. Equivalent conductivity as a function of $\sqrt{c}$ for a strong and a weak electrolyte.

Kohlrausch found empirically from measurements made on a large number of aqueous solutions that for many electrolytes, of which KCl is a typical example, the equivalent conductivity is a nearly linear function of the square root of the concentration. This function, which fits the data most closely at low concentrations (Fig. 16-3), can be represented by the equation

$$\Lambda = \Lambda_o - \beta c^{\frac{1}{2}} \qquad (16\text{-}2)$$

in which β is a constant and Λ_o is the limiting value of the equivalent conductivity extrapolated to infinite dilution ($c = 0$). Electrolytes in this class are called *strong*. For other electrolytes, such as acetic acid, the value of Λ increases greatly at low concentrations. Such electrolytes are called *weak*. There are very few electrolytes in an intermediate class. It can be seen that reliable values of Λ_o can be obtained for strong electrolytes, but not for weak electrolytes.

Values of Λ_o are critically dependent on the reliability of the measurements made on solutions whose concentrations are low. In this concentration range, the conductance due to the ions from the water or other solvent may be a significant fraction of the total conductance. In that case an appropriate correction must be applied to the determined values to obtain the contribution due to the electrolyte. Specially prepared water must be used whose resistivity is as high as possible ($\sim 10^7$ ohm cm), in contrast to ordinary distilled water whose resistivity is usually about 10^5 ohm cm because of dissolved CO_2 from the atmosphere and traces of electrolytes from glass containers.

When Kohlrausch had obtained the limiting equivalent conductivities at infinite dilution for a number of strong electrolytes, he found empirically that these values could be expressed as sums of *limiting ionic conductivities*, λ_o^+ and λ_o^-, each of which is independent of the nature of the other ion; thus

$$\Lambda_o = \lambda_o^+ + \lambda_o^-$$

This fact implies that the positive and negative ions move independently of each other when an electric current is passed through an electrolytic solution of very low concentration. This statement is called Kohlrausch's law of the independent migration of ions. Since the λ_o values for a given ion depend only on the solvent and the temperature, they can be tabulated as in Table 16-1.

The additivity of limiting ionic conductivities makes it possible to calculate the limiting equivalent conductivity of a weak electrolyte, a quantity that, as mentioned above, cannot be evaluated directly. For this purpose the Λ_o values of strong electrolytes can be added and subtracted to yield the desired value. For example, given the following Λ_o values: 91.0 for sodium acetate, 426.1 for HCl, and 126.4 for NaCl, Λ_o for acetic acid is given by

$$\Lambda_o(HAc) = \Lambda_o(NaAc) + \Lambda_o(HCl) - \Lambda_o(NaCl)$$
$$= \lambda_{o\,Na^+} + \lambda_{o\,Ac^-} + \lambda_{o\,H^+} + \lambda_{o\,Cl^-} - \lambda_{o\,Na^+} - \lambda_{o\,Cl^-}$$
$$= 91.0 + 426.1 - 126.4$$
$$= 390.7 \text{ cm}^2 \text{ ohm}^{-1} \text{ equiv}^{-1}$$

Alternatively, if the λ_o values for the ions of the weak electrolyte are available,

TABLE 16-1   LIMITING IONIC EQUIVALENT CONDUCTIVITIES AT 25°C
($\text{cm}^2 \text{ ohm}^{-1} \text{ equiv}^{-1}$)

Ion	λ_o	Ion	λ_o
H^+	349.8	OH^-	197.8
Na^+	50.11	Cl^-	76.35
NH_4^+	73.4	Br^-	78.20
K^+	73.52	NO_3^-	71.44
Ca^{2+}	59.50	CH_3COO^-	40.9

*A more complete tabulation can be found in H. S. Harned and B. B. Owen, *The Physical Chemistry of Electrolytic Solutions*, Reinhold Publishing Corp., New York, 1958.

these values may simply be added. Thus from the λ_o values of H^+ and Ac^- from Table 16-1, Λ_o of HAc is $349.8 + 40.9 = 390.7$.

In many cases it is an adequate approximation to use the limiting values of conductivities for ions that are at moderately low concentrations, although these values apply only at infinite dilution. Unless otherwise stated, we will assume that this approximation is justified.

Transference Number and Ionic Mobility

Early investigators of electrolysis observed that when the solution being electrolyzed was undisturbed the concentration changes that took place around the anode and the cathode were different from each other, even though according to Faraday's law the numbers of equivalents of ions chemically changed were the same at both electrodes. Hittorf interpreted this result to mean that the positive and negative ions migrated at different velocities in the electric field between the electrodes. This, in turn, implied that these ions carried different fractions of the total current through the electrolyte. He defined the *transference number*, τ, of an ion in a particular electrolytic solution as the fraction of the total current that is carried by that ion. Thus, $\tau_+ + \tau_- = 1$ and

$$\tau_+ = \frac{I_+}{I_+ + I_-} = \frac{Q_+}{Q_+ + Q_-}$$

in which I_+ is the current carried by the positive ion and $Q_+ (= I_+ t)$ is the amount of charge transported by the positive ion during a particular experiment. Since the current carried by an ion is proportional to its conductivity, it follows that

$$\tau_+ = \frac{\lambda_+}{\lambda_+ + \lambda_-} = \frac{\lambda_+}{\Lambda} \qquad (16\text{-}3)$$

Hittorf showed how the transference number could be calculated from quantitative analytical data for the changes in concentration that occurred in the anode and cathode portions of the electrolyte as a result of electrolysis. A typical Hittorf cell is shown in Figure 16-4. This design makes it possible

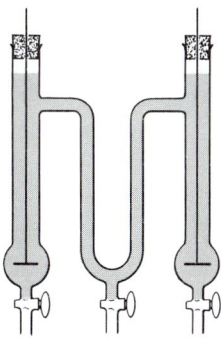

Figure 16-4. Hittorf cell for the determination of transference numbers.

to separate the anode, center, and cathode portions after the electrolysis is completed, and it minimizes the tendency of these portions to mix as a result of diffusion or convection. If the electrolysis is not continued too long, the concentration of the electrolyte in the center compartment remains unchanged because whatever ions leave on one side of the compartment will be replaced by an equal number entering from the other side.

To illustrate the principle of the Hittorf method, let us say that the electrolyte is M^+X^-, that the velocity of X^- is twice that of M^+, and that the electrolysis is carried out between inert electrodes. Assume that the 3 compartments of the Hittorf cell contain initially 3 equivalents of M^+X^- each, and that 3 faradays of electric charge are passed through the cell. We can diagram the situation before and after the electrolysis in the following way:

$$\text{before } +\left|\begin{array}{c|c|c} +++ & +++ & +++ \\ --- & --- & --- \end{array}\right| - \qquad \text{after } +\left|\begin{array}{c|c|c} ++ & +++ & + \\ -- & --- & - \end{array}\right| -$$

In this diagram $+$ and $-$ represent one equivalent of positive and negative ions, respectively.

As a result of the electrolysis, 3 equivalents of X^- are discharged at the anode, 2 equivalents of X^- move into the anode compartment, and 1 equivalent of M^+ moves out, leaving a net amount of $2M^+X^-$. At the cathode, 3 equivalents of M^+ are discharged, 1 equivalent moves in, and 2 equivalents of X^- move out, leaving a net amount of one equivalent of M^+X^- in the cathode compartment. At no time has there been a detectable separation of charge; the solution is electrically neutral in all portions at all times. The concentration changes at the two electrodes, however, are clearly different from each other.

The transference number of either ion can be obtained from the concentration change in either electrode compartment. To obtain τ_+ from data for the cathode compartment we need to find the number of equivalents, n_m, of M^+ that have migrated into the cathode compartment. The final number of equivalents in this compartment is given by

$$n_f = n_i - n_e + n_m$$

in which n_i is the initial number of equivalents of M^+ and n_e is the number of equivalents deposited on the cathode. We thus obtain

$$n_m = n_f - n_i + n_e$$

Since n_e is equal to the number of faradays passed through the cell and n_m is equal to the number of faradays carried by M^+, we can calculate τ_+ as follows:

$$\tau_+ = \frac{n_m}{n_e} = \frac{n_f - n_i + n_e}{n_e}$$

From the data for the cathode compartment in the example above we obtain

$$\tau_+ = \frac{1 - 3 + 3}{3} = 0.33$$

We can obtain the same value of τ_+ from the data for the anode compartment. In this compartment the only change in the concentration of M^+ is due to migration; thus $n_f = n_i - n_m$, and

$$\tau_+ = \frac{n_i - n_f}{n_e}$$

$$= \frac{3-2}{3} = 0.33$$

When τ_+ is known, τ_-, equal to $1 - \tau_+$, is also known.

The determination of a transference number by the Hittorf method is basically a matter of bookkeeping. The reactions taking place at the electrodes must be known and all processes involving the ions must be taken into consideration because with some electrolytes, Na_2SO_4 for example, the ions that carry the current through the solution are not the ions that are formed or discharged. In some way the number of coulombs passed through the cell must also be known.

We have seen that the transference number of an ion is closely related to the rate of migration of this ion relative to its partner of opposite sign. In fact, the ratio of the transference numbers is equal to the ratio of the ionic velocities; thus

$$\frac{\tau_+}{\tau_-} = \frac{v_+}{v_-}$$

and it follows that

$$\tau_+ = \frac{v_+}{v_+ + v_-} \tag{16-4}$$

Because the velocity of an ion depends on the strength of the electric field in which the ion is located, it is preferable to use a related quantity, the *ionic mobility*, defined as the velocity of an ion in a unit electric field. The strength of the field between parallel electrodes is given by the ratio of the potential difference to the distance between the electrodes. (More precisely, the field strength is given by $d\mathcal{E}/dL$.) Thus the ionic mobility, u, is given by $u = v/(\mathcal{E}/L)$ with units of cm/sec per volt/cm, or cm²/volt sec. Since all ions in a solution being electrolyzed are in the same electric field, $v_+/v_- = u_+/u_-$ and from Equation 16-4 we obtain

$$\tau_+ = \frac{u_+}{u_+ + u_-}$$

It is possible to measure the velocity of an ion directly, and thus to determine its mobility and its transference number, by the moving boundary method. This method involves the use of two different electrolytic solutions with one ion in common, which are placed in a vertical tube in such a way that as sharp a boundary as possible is formed between the solutions (Fig. 16-5). The location of the boundary may be observed by a difference in color if one of the ions other than the common ion absorbs light, or by a difference in some property, such as the refractive index, of the two solutions. When a current is passed through these solutions the boundary between them will move up or down, depending on the ions present and the polarity of the electrodes.

To be more specific, let us assume that the ion of interest is M^+ in the electrolyte MX and that the boundary will move upwards. Then the solution

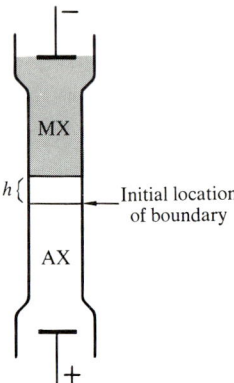

Figure 16–5. Moving boundary method for the determination of ionic mobilities and transference numbers.

of MX will be made the upper solution and the electrode immersed in it will be made the cathode. The lower solution will contain X^- as the negative ion and a positive ion, A^+, whose mobility is less than that of M^+.

The need for this restriction arises from the fact that as the M^+ ions above the boundary move out of a certain volume of solution their place must be taken by A^+ ions. If the velocity of the A^+ ions is greater than that of the M^+ ions, the A^+ ions will overtake the M^+ ions and thus blur the boundary. If, in contrast, the A^+ ions have the lower velocity and if they should lag behind the M^+ ions, the layer of solution just below the boundary would have a lowered electrolyte concentration and therefore a relatively high resistance. By Ohm's law, the potential drop across this layer would be increased proportionately, and the resulting higher electric field would accelerate the A^+ ions. Thus the A^+ ions must keep up with the moving boundary, which remains sharp. The effect of convection is minimized by making the lower solution be the one of higher density.

Let us say that when a current of I amperes has passed for t seconds through a solution of concentration c in MX in a tube with cross-sectional area of A cm², the boundary has moved a distance of h cm. The velocity of the M⁺ ions is given by

$$v_+ = \frac{h}{t}$$

The electric field strength in a layer of solution of thickness dL is given by $d\mathcal{E}/dL$ which from Ohm's law, is equal to $I\,dR/dL$. Expressed in terms of the conductivity $dR/dL = 1/\kappa A$. Then the mobility of the M⁺ ions is given by

$$\begin{aligned}
u_+ &= \frac{h/t}{d\mathcal{E}/dL} \\
&= \frac{h}{t}\frac{A\kappa}{I} \\
&= \frac{V\kappa}{Q}
\end{aligned}$$

in which V is the volume swept out by the moving boundary when Q coulombs have passed through the cell.

To obtain the transference number of M⁺ we note that the number of equivalents of M⁺ in 1 ml of solution is $c/1000$ and the number of equivalents that have migrated when the boundary moves a distance h is $cAh/1000$. This number of equivalents of M⁺ carries $\mathcal{F}cAh/1000$ coulombs of charge. The ratio of this number to Q, the total amount of charge transported, is the transference number. Thus we obtain the following relationship:

$$\tau_+ = \frac{\mathcal{F}cAh}{1000\,Q}$$

We can find the relation between ionic mobility and equivalent ionic conductivity by starting from Equation 16-3 in the form $\lambda_+ = \tau_+ \Lambda$. On substituting for τ_+ its equal from the equation above and using Equation 16-1 for Λ, the result is

$$\begin{aligned}
\lambda_+ &= \frac{\mathcal{F}cAh}{1000\,Q}\frac{1000\kappa}{c} \\
&= \frac{\mathcal{F}\kappa Ah}{Q} \\
&= \mathcal{F}u_+
\end{aligned}$$

As an example of the application of these equations let us consider a solution of LiCl whose molarity is 0.10 and whose conductivity is found to be 1.06×10^{-2} ohm⁻¹ cm⁻¹. When this solution was placed in a moving boundary cell of cross-sectional area 1.17 cm² and it was electrolyzed for 131 minutes with a constant current of 9.42 milliamperes, the Li⁺ boundary was observed to move a distance of 2.08 cm. What are the velocity, mobility, transference number, and equivalent conductivity of the Li^+ ions in this solution?

The velocity is $2.08/(131 \times 60) = 2.65 \times 10^{-4}$ cm/sec. The mobility is

$$u_+ = \frac{2.08 \times 1.17 \times 1.06 \times 10^{-2}}{131 \times 60 \times 9.42 \times 10^{-3}}$$

$$= 3.48 \times 10^{-4} \text{ cm}^2/\text{volt sec}$$

The transference number is

$$\tau_+ = \frac{96{,}500 \times 0.10 \times 2.08 \times 1.17}{10^3 \times 9.42 \times 10^{-3} \times 131 \times 60}$$

$$= 0.318$$

The equivalent ionic conductivity is

$$\lambda_+ = 96{,}500 \times 3.48 \times 10^{-4}$$

$$= 33.7 \text{ cm}^2/\text{ohm equiv}$$

Alternatively, λ_+ can be found from τ_+ and Λ as follows:

$$\lambda_+ = \tau_+ \frac{1000\kappa}{c}$$

$$= \frac{0.318 \times 10^3 \times 1.06 \times 10^{-2}}{0.10}$$

$$= 33.7 \text{ cm}^2/\text{ohm equiv}$$

A representative set of ionic mobilities at infinite dilution in water is presented in Table 16-2. All of these values are of the same order of magnitude except those for the H^+ and OH^- ions. The mobilities of these two ions are so large that the mechanism by which they transport charge through the solution must be different than that of the other ions. This mechanism is undoubtedly related to the facts that these two ions are produced by the dissocia-

TABLE 16-2   LIMITING IONIC MOBILITIES AT 25°C
(cm^2 $volt^{-1}$ sec^{-1})

Ion	Mobility x 10^4	Ion	Mobility x 10^4
H^+	36.3	OH^-	20.5
Li^+	4.01	Cl^-	7.90
Na^+	5.20	Br^-	8.12
K^+	7.62	NO_3^-	7.40
Ca^{2+}	6.16	SO_4^{2-}	8.27

tion of solvent water and that water has a relatively dense hydrogen-bonded structure (p. 260).

There is ample evidence that an H^+ ion in water does not exist as a "bare" proton but rather as a proton associated with at least one molecule of water (H_3O^+, hydronium ion). We can imagine that a proton can "jump" from an H_3O^+ ion to a neighboring water molecule. The resulting H_3O^+ ion can, in turn, transfer a different proton to a neighboring water molecule, since all protons are alike. Thus the net result of a series of steps like these is a transfer of a proton through the solution, although any particular proton needs to travel only a very small distance.

A similar mechanism accounts for the high mobility of the OH^- ion, since a proton can be transferred from an H_2O molecule to the OH^- ion. This is equivalent to a movement of the OH^- ion in the opposite direction.

Another point of interest in Table 16–2 is the inverse dependence of the mobility of the alkali metal ions on their ionic radius, as determined from X-ray diffraction data for crystals. One might expect that the K^+ ion with 3 filled electron shells would have a lower mobility than the Li^+ ion with just 1 filled shell, but the reverse order is observed. The accepted explanation for this fact is based on the idea that the electric field strength near a Li^+ ion is relatively large because its unit charge is concentrated on a very small sphere. Consequently it can polarize and attract a relatively large number of water molecules by ion-dipole interaction. When it migrates it must carry all of these water molecules with it, and thus its motion is retarded. In contrast, the Na^+ ion and still more, the K^+ ion, have their unit charges distributed over a larger volume. Because of the lower field strength in their neighborhood, these ions are less extensively hydrated than the Li^+ ion and thus have a higher mobility.

It should be noted that the discussion of the transference number and ionic mobility in this section has been based on the simplest situation, namely, a single electrolyte which dissociates completely into only two kinds of ions. Many electrolytes, particularly at relatively high concentrations, dissociate into a variety of ions. In this case the definition of transference number given above is not adequate.[1] An example of the type of complication which can be found is the observation that in the electrolysis of some salts, such as $CuCl_2$, some of the metal constituent migrates toward the positive electrode. This is a consequence of the presence of negative complex ions, such as $CuCl_4^=$, which contain the metallic element.

Equivalent Conductivity as a Function of Concentration

As we have seen, most electrolytes can be divided into two classes, strong and weak, which differ from each other in the dependence of their equivalent

[1]M. Spiro, *J. Chem. Ed.* **33**, 464 (1956).

conductivities on the square roots of their concentrations. We now try to interpret these two types of behavior.

We can begin by noting that if the conductivity of an electrolyte were directly proportional to its concentration, the equivalent conductivity would be independent of concentration. This is clearly not the case. To account for the observed relations between equivalent conductivities and concentrations of electrolyte, two theoretical approaches are used: (1) The electrolyte is dissociated into ions to a degree, α, which varies with the concentration; or (2) the degree of dissociation is constant but the ionic mobilities depend on the concentration. These alternatives are not mutually exclusive but in most cases one or the other is predominant. We examine each of them in turn.

1. It was pointed out previously that as part of his theory of ionization Arrhenius assumed that in all cases an equilibrium was established between an undissociated electrolyte in solution and its ions. For the simplest case we can represent this process by the equation

$$MX \rightleftharpoons M^+ + X^-$$

He further assumed that the degree of dissociation, α, is equal to Λ/Λ_o, which yielded values of α consistent with those calculated by van't Hoff from the colligative properties. Ostwald treated this equilibrium like that for a chemical reaction and calculated the equilibrium constant in the following way:

$$K_{eq} = \frac{[M^+][X^-]}{[MX]}$$

If the total concentration of the electrolyte is represented by c, then $[M^+] = [X^-] = \alpha c$ and $[MX] = c - \alpha c$. Substitution of these expressions in the equation above results in

$$K_{eq} = \frac{\alpha^2 c^2}{c(1-\alpha)} = \frac{\alpha^2 c}{1-\alpha}$$

Ostwald then replaced α by its equal, Λ/Λ_o according to the theory of Arrhenius and thus obtained the following equation which is known as the *Ostwald dilution law*:

$$K_{eq} = \frac{\Lambda^2 c}{\Lambda_o(\Lambda_o - \Lambda)} \tag{16-5}$$

This relationship accounts fairly well for the observed concentration dependence of Λ for weak electrolytes at low values of the concentration, since in these cases K_{eq} is indeed nearly constant as illustrated by Table 16-3. The implicit assumption in this approach is that the ionic mobilities are independent of concentration. This is a good approximation at low concentrations of weak electrolytes but not at higher concentrations.

2. The success of the Arrhenius theory for weak electrolytes led to efforts

to apply the same theory to strong electrolytes which were thought to be characterized by values of α close to 1. As good experimental data for strong electrolytes accumulated, however, numerous inconsistencies were found. For example, the values of α obtained by different experimental methods were not in agreement, and transference numbers were found to vary with concentration. It gradually became clear that the assumption of constant ionic mobilities is not valid and that another theoretical approach was needed for strong electrolytes.

TABLE 16-3 IONIZATION CONSTANT OF ACETIC ACID CALCULATED FROM THE OSTWALD DILUTION LAW*

$c \times 10^3$ (equiv/liter)	Λ ($cm^2 \, ohm^{-1} \, equiv^{-1}$)	K_{eq}
0.0280	210.4	1.760
0.1532	112.1	1.767
1.0283	48.15	1.781
2.4140	32.22	1.789

*D. A. MacInnes, *The Principles of Electrochemistry*, Reinhold Publishing Corp., New York, 1939; republished by Dover Publications, New York, 1961, p. 56.

This approach was provided in 1923 when Debye and Hückel carried out a detailed calculation of the deviations from ideality of solutions of strong electrolytes. They started from the assumption that these electrolytes are completely dissociated into ions, in consistency with X-ray diffraction data for the solid electrolyte. The Debye-Hückel theory is outlined on p. 652. For the present we merely refer to some aspects of this theory, together with its extension to the motion of ions in solution, carried out by Onsager.

Let us focus our attention on a single ion in solution in the absence of an applied electric field. As a result of collisions with nearby solvent molecules, this ion moves about randomly with a very small mean free path. On the basis of a hole theory of liquids (p. 496), we can imagine that an ion, like a solvent molecule, moves from one hole in the liquid to another.

There are, however, at least two ways in which an ion is different from a molecule. In the first place, an ion carries an electric charge that interacts with the charges of other ions by Coulomb's law. In the second place, there are always at least two kinds of ions in a solution, with charges of opposite sign.

The fact that the force between any two ions varies, according to Coulomb's law, as r^{-2} with r the interionic distance, is of the greatest importance in the behavior of ions in solution. When we compare the interionic force to intermolecular forces, which vary as r^{-7} (p. 386), we see that at values of r for which the intermolecular forces are completely negligible, the interionic force still has a considerable magnitude. In other words, as we saw in the treatment of ionic crystals (p. 481), the coulombic forces (and the related interaction potential) extend over relatively large distances on a molecular scale.

A direct consequence of Coulomb's law is the conclusion that the distribution of ions in the solution is not completely random. Debye and Hückel pointed out that on the average each ion tends to be surrounded symmetrically by an excess of ions of the opposite sign, the ionic atmosphere. The formation of this atmosphere is a compromise between the tendency of the long-range coulombic forces to produce order, as in an ionic crystal, and the disordering effect of thermal motion. Thus the extent of the ionic atmosphere is a function of the temperature of the solution. It is also a function of the dielectric constant of the solvent because the force between two charges at a fixed distance is inversely proportional to the dielectric constant of the medium in which the charges are immersed.

When an electrical potential difference is applied to electrodes in the solution, each ion experiences a force which superimposes on the otherwise random motion a drift toward the electrode of opposite sign. As a result of this drift each ion tends to leave its atmosphere lagging behind, and since this atmosphere has a net charge opposite to that of the ion in question, it acts as a drag and thus decreases the velocity of the ion and also its mobility. Onsager showed how this *atmosphere effect* can be calculated as a function of the dielectric constant of the solvent and the temperature.

Onsager also showed that another factor, called the *electrophoretic* effect, must be taken into consideration. This is a consequence of the solvation of the ions which was mentioned previously. As a particular ion moves toward its electrode, the ions of opposite charge in its atmosphere move in the opposite direction. Since these ions carry solvent molecules along with them, this reverse flow of solvent, which depends on the viscosity of the solvent, retards the ion in question. By considering the ion as a sphere whose viscous drag is given by Stokes' law (Eq. 13–2) as a function of the viscosity coefficient, η, Onsager calculated the magnitude of the electrophoretic effect.

By inclusion of both these effects, Onsager derived the following equation for the equivalent conductivity of a completely dissociated uni-univalent electrolyte in a dilute solution:[2]

$$\Lambda = \Lambda_o - \left[\frac{82.48}{(DT)^{\frac{1}{2}}\eta} + \frac{8.20 \times 10^5}{(DT)^{\frac{3}{2}}}\Lambda_o\right]c^{\frac{1}{2}} \qquad (16\text{–}6)$$

[2] A somewhat more complex form of this equation is applicable to electrolytes with bivalent and trivalent ions.

This equation says that for a given solvent and temperature the equivalent conductivity of a strong electrolyte is a linear function of $c^{\frac{1}{2}}$. This is just the relationship found by Kohlrausch (Eq. 16-2) many years before Onsager's theoretical equation was derived, with the quantity in square brackets equal to β. Unfortunately, efforts to extend the range of validity of this equation into the higher concentration range have not as yet been successful. One of the complications affecting the relationship between conductivity and concentration in this range is undoubtedly the association of ions of opposite sign to form ion-pairs which do not contribute to conductance. The study of ion-pair formation is a difficult problem which is still a subject of research.

A few additional points concerning ionic mobility may be of interest:

1. Debye and Falkenhagen predicted, and Sack found, that when the conductance of a strong electrolyte is measured with radio-frequency currents ($\nu \sim 10^6$ sec^{-1}), the conductance increases with increasing frequency. This result is interpreted to mean that when an ion oscillates in an electric field at a sufficiently high frequency, its atmosphere does not have enough time in each half-cycle to reform completely. Consequently the effect of the atmosphere diminishes with increasing frequency and thus the conductance increases.

2. Wien found that when he measured the conductance of a strong electrolyte as a function of the strength of the applied electric field, at very high field strengths the conductance increased. This implies that when the velocity of an ion becomes very high, there is insufficient time for its atmosphere to form.

3. There is a serious question as to whether one is justified in substituting in the Onsager equation values of the viscosity coefficient and dielectric constant which are obtained with macroscopic samples of solvent and thus applying these values to interactions on a molecular scale. There is, however, no alternative at present, and the Onsager equation does give the correct limiting slope of the plot of Λ versus $c^{\frac{1}{2}}$.

Conductance in Nonaqueous Systems

According to Coulomb's law, the force acting between two ions at a fixed distance apart is inversely proportional to the dielectric constant of the solvent. Since the values of D for water and, for example, dioxane are 78.5 and 2.2, respectively, it follows that for a given interionic distance the force between ions in dioxane is approximately 35 times the force in water. One of the consequences of a lower dielectric constant of a solvent is a greater tendency for ion-pair formation. Thus, in general, electrolytes that are strong in aqueous solutions are weaker, with lower equivalent conductivities, in a solvent of low dielectric constant.

Another type of nonaqueous electrolytic system is a fused salt. Although many salts, such as NaCl, are completely ionized in the crystalline state (p. 467), they are good insulators because the mobility of the ions is essentially zero. Fused salts are excellent conductors because when the crystal lattice breaks down on melting, the ions have relatively high mobilities.

Ionic Equilibria

Deviations from Ideality, Activity Coefficients

In Chapter 15 we started the discussion of solution equilibria by defining an ideal solution as one in which the chemical potential of each component is related to its mole fraction by the following equation:

$$\mu_i = \mu_i^\circ + RT \ln X_i$$

Essentially on the basis of this equation, we can treat all of the equilibrium properties of ideal solutions. It was pointed out that although no solution is ideal over the whole range of mole fractions, many solutions can be considered as ideal if the mole fraction of one component, the solute, is not too large. We attributed deviations from ideality to the effect of intermolecular forces.

When we come to a consideration of equilibria in electrolytic solutions we find that deviations from ideality become apparent at concentrations much below those needed for ideality of unionized solutes. This fact is a direct consequence of the long range of coulombic forces which was mentioned before. Because significant deviations from ideality are practically always present with ionized solutes, from the outset of this section we express the chemical potentials of these solutes in terms of activities and activity coefficients.

Now let us recall that we have expressed the composition of a solution in three different ways, namely, as the mole fraction, the molarity, and the molality. Corresponding to each of these there is a scale of activity coefficients. (In Chapter 15 we used the activity coefficient based on mole fractions, see page 565.) In applications of thermodynamics to electrochemistry it is customary to express composition in terms of molality. Accordingly, the *molal activity coefficient*, which is used exclusively in this chapter, is defined by the equation $\gamma_i = a_i/m_i$, in which m_i is the molality of component i. The basic thermodynamic equation relating chemical potential and composition,

$$\mu_i = \mu_i^\circ + RT \ln a_i$$

then becomes for a solute of molality m_i

$$\mu_i = \mu_i^\circ + RT \ln \gamma_i m_i$$

This equation implies that the standard state for the solute is a hypothetical state in which both $m_i = 1$ and $\gamma_i = 1$. In actual solutions, as $m_i \to 0$, $\gamma_i \to 1$ and 1 and $a_i \to m_i$.

The chemical potential, μ_e, of a strong uni-univalent electrolyte, M^+X^-, in solution can be considered to be the sum of the chemical potentials, μ_+ and μ_-, of its ions and we would like to find the values of these quantities. Unfortunately, because positive and negative ions always occur together in solutions, we cannot add just one kind of ion to a solution. It follows that we cannot obtain the individual values of ionic activities or ionic activity coefficients. The best that can be done is to employ some kind of average of the individual values.

In order to see what kind of average would be useful, let us notice that the equality $\mu_e = \mu_+ + \mu_-$ implies that $\mu_e = (\mu_+^\circ + \mu_-^\circ) + RT \ln a_+ a_-$, and thus the activity of the electrolyte, a_e, is equal to $a_+ a_-$. This relation suggests that a *mean activity* for the two kinds of ions, $a_\pm$, can be defined as their geometric mean;[3] that is, $a_\pm = (a_+ a_-)^{\frac{1}{2}}$. It follows that $a_e = a_\pm^2$. By analogy with this definition we define a mean molality as $m_\pm = (m_+ m_-)^{\frac{1}{2}}$, and a mean ionic activity coefficient as $\gamma_\pm = (\gamma_+ \gamma_-)^{\frac{1}{2}}$. Then

$$a_\pm^2 = \gamma_\pm^2 m_\pm^2$$

and

$$\gamma_\pm = \frac{a_\pm}{m_\pm} \qquad (16\text{-}7)$$

Now let us consider the general case of any strong electrolyte, $M_{\nu_+} X_{\nu_-}$, which dissociates to yield ν_+ moles of positive ions and ν_- moles of negative ions per mole of electrolyte. The total number of moles of ions formed, ν, is then $\nu_+ + \nu_-$, and the chemical potential of the electrolyte is equal to $\nu_+ \mu_+ + \nu_- \mu_-$. The activity of the electrolyte is equal to $a_+^{\nu_+} a_-^{\nu_-}$.

As an extension of the above definition of mean activity, we now write

$$a_\pm = (a_+^{\nu_+} a_-^{\nu_-})^{\frac{1}{\nu}}$$

Correspondingly, the mean ionic molality, $m_\pm$, is equal to $(m_+^{\nu_+} m_-^{\nu_-})^{\frac{1}{\nu}}$ and the mean activity coefficient, $\gamma_\pm$, is equal to $(\gamma_+^{\nu_+} \gamma_-^{\nu_-})^{\frac{1}{\nu}}$.

When the electrolyte in question is the only solute in a solution and its molality is m, then $m_+ = \nu_+ m$ and $m_- = \nu_- m$. It follows that

$$m_\pm^\nu = (\nu_+ m)^{\nu_+} (\nu_- m)^{\nu_-}$$

and

$$m_\pm = m(\nu_+^{\nu_+} \nu_-^{\nu_-})^{\frac{1}{\nu}}$$

From $a_\pm = m_\pm \gamma_\pm$ for the electrolyte we obtain

$$a_\pm = (\nu_+^{\nu_+} \nu_-^{\nu_-})^{\frac{1}{\nu}} m \gamma_\pm$$

[3] This is equivalent to the definition of $\ln a_\pm$ as $\frac{1}{2}(\ln a_+ + \ln a_-)$.

If, for example, we have a solution of $CrCl_3$ whose molality is 0.10, $a_{CrCl_3} = a_{\pm}^4$, and $a_{\pm} = 0.10(1^1 \times 3^3)^{\frac{1}{4}}\gamma_{\pm}$, or $0.228\gamma_{\pm}$.

Determination of Activity Coefficients

When either the mean activity or the mean activity coefficient can be found for a solution of known molality, then the other of these two is also known. A number of different experimental methods for the determination of activity coefficients have been employed. Some of these methods depend on the measurement of one of the colligative properties of electrolytic solutions, while others depend on kinds of measurements (solubility, emf of galvanic cells) we will discuss at a later point in this chapter.

As an example of the determination of activity coefficients from measurements of a colligative property, we describe the freezing-point depression method. When the freezing-point depression was discussed in Chapter 15, we arrived at an equation (Eq. 15–16) whose general form, applicable to non-ideal solutions, is

$$d \ln a_A = \frac{\Delta H_{Af}}{RT^2} dT \tag{16-8}$$

In this equation, a_A is the activity of the solvent. In order to obtain the activity of the solute, a_B, we once more make use of the Gibbs-Duhem equation (Eq. 3–61) which is

$$n_A \, d\mu_A = -n_B \, d\mu_B$$

From $\mu_i = \mu_i^\circ + RT \ln a_i$, it follows that at constant temperature $d\mu_i = d \ln a_i$. This relationship is independent of the choice of standard state or the particular activity scale being used. Thus we can write the Gibbs-Duhem equation in the following way:

$$n_A \, d \ln a_A = -n_B \, d \ln a_B$$

Since we want to express composition in terms of molalities rather than numbers of moles, let us take as our system 1 kilogram of solvent containing m moles of electrolyte. In this case $n_B = m$, $n_A = 1000/M_A$, and the equation above becomes

$$\frac{1000}{M_A} d \ln a_A = -m \, d \ln a_B$$

Electrochemistry

Because the activity of the electrolyte, a_B, is related to the mean activity of its ions by $a_B = a_\pm^\nu$, it follows that $d \ln a_B = \nu \, d \ln a_\pm$. Substituting this expression for $d \ln a_B$ in the equation above and solving for $d \ln a_\pm$, we obtain

$$d \ln a_\pm = -\frac{1000}{M_A m \nu} d \ln a_A$$

On replacing $d \ln a_A$ by its equal from Equation 16-8, the result is

$$d \ln a_\pm = -\frac{1000}{M_A m \nu} \frac{\Delta H_{Af}}{RT^2} dT$$

We now limit our treatment to dilute solutions; then we can set $T = T_o$, the freezing point of pure solvent. We can simplify the equation above by introducing the molal freezing-point constant (Eq. 15-19) defined by

$$K_f = \frac{M_A R T_o^2}{1000 \, \Delta H_{Af}}$$

and thus we obtain

$$d \ln a_\pm = \frac{-1}{K_f m \nu} dT$$

This relationship can be expressed in terms of the freezing-point depression, $\Delta T \, (= T_o - T)$, by noting that $dT = -d(\Delta T)$. Thus

$$d \ln a_\pm = \frac{1}{K_f m \nu} d(\Delta T)$$

In order to obtain the mean activity coefficient at a definite molality, we replace $a_\pm$ by $\gamma_\pm m_\pm$. Because $m_\pm$ differs from m by a constant factor, $d \ln m_\pm = d \ln m$, and we have

$$d \ln \gamma_\pm = \frac{1}{K_f m \nu} d(\Delta T) - d \ln m \qquad (16\text{-}9)$$

In principle, if we know ΔT as a function of m we can integrate this equation. For mathematical reasons, it turns out to be preferable to introduce a new variable before undertaking the integration. First let us note that if the solution were ideal, the following generalization of Equation 15-18 would be satisfied:

$$\frac{\Delta T}{K_f m \nu} = 1$$

Since the solution is not ideal we can express the deviation from ideality in terms of a variable j, defined by

$$j = 1 - \frac{\Delta T}{K_f m \nu} \qquad (16\text{-}10)$$

It follows that $\Delta T = K_f m \nu(1 - j)$, and the result of differentiating this equation, considering both m and j as variables, is

$$d(\Delta T) = K_f \nu[(1 - j)\, dm - m\, dj]$$

When this expression for $d(\Delta T)$ is introduced into Equation 16–9 we obtain

$$d \ln \gamma_\pm = (1 - j)\frac{dm}{m} - dj - d \ln m$$

$$= -\frac{j}{m} dm - dj$$

Now we are ready to integrate this equation. The lower limit is the infinitely dilute solution, for which $m \to 0$, $\gamma_\pm \to 1$, and $j \to 0$. From

$$\int_1^{\gamma_\pm} d \ln \gamma_\pm = -\int_0^m \frac{j}{m} dm - \int_0^j dj$$

we obtain

$$\ln \gamma_\pm = -\int_0^m \frac{j}{m} dm - j$$

The remaining integral must be evaluated graphically by finding ΔT at a series of values of m extended to as low a value of m as is experimentally feasible, calculating a value of j for each ΔT and m from Equation 16–10, plotting j/m versus m, extrapolating the graph line to $m = 0$, and finding the area under the curve from $m = 0$ to whatever value of m is desired.

TABLE 16-4   MEAN IONIC ACTIVITY COEFFICIENTS*

Electrolyte molality	0.001	0.002	0.005	0.01	0.05	0.1	0.5	1.0
HCl	0.966	0.952	0.928	0.904	0.830	0.796	0.758	0.809
NaCl	0.966	0.953	0.929	0.904	0.823	0.780	0.68	0.66
H_2SO_4	0.830	0.757	0.639	0.544	0.340	0.265	0.154	0.130
$CaCl_2$	0.89	0.85	0.79	0.73	0.57	0.52	0.52	0.71
$CuSO_4$	0.74	—	0.53	0.41	0.21	0.16	0.068	0.047
$ZnSO_4$	—	—	0.48	0.39	0.20	0.15	0.063	0.044

*A more complete table can be found in W. M. Latimer, *The Oxidation States of the Elements and Their Potentials in Aqueous Solutions*, Prentice-Hall, Englewood Cliffs, N. J., 1952.

Electrochemistry

Some representative values of the mean ionic activity coefficient determined by a variety of experimental methods for a number of electrolytes in a range of molalities are given in Table 16-4. A few of these values are graphed in Figure 16-6. It can be seen that the characteristic behavior of $\gamma_\pm$ is a relatively sharp drop from a value of 1 at very low molalities to a minimum, followed by increasing values at higher molalities.

There is at present no theory on the basis of which activity coefficients can be calculated over the whole range of concentrations. This fact is basically a consequence of the complexity of the electrical interactions of the ions among

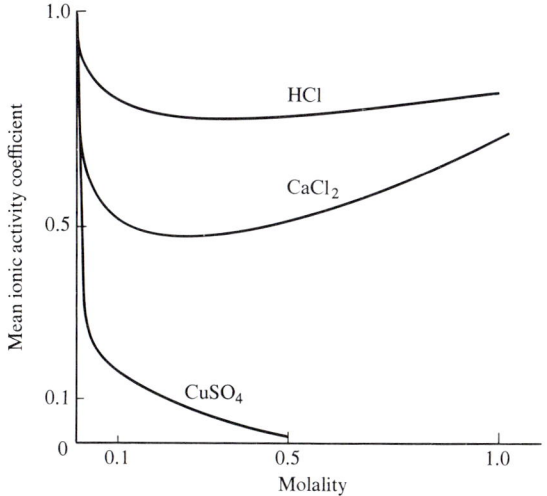

Figure 16-6. Mean ionic activity coefficient versus molality.

themselves. There is, however, a theoretical treatment of the relationship between the mean activity coefficient and the ionic concentration in very dilute solutions. This is the Debye-Hückel theory, mentioned above, which we now consider.

Theoretical Calculation of Activity Coefficients

The Debye-Hückel theory has undergone little change or extension since it was published in 1923. The basic idea behind this theory is that deviations of dilute electrolytic solutions from ideality can be attributed to the electrical interaction of each ion with the ions in its immediate neighborhood, its ionic atmosphere. Strong electrolytes are assumed to be completely dissociated into ions. For weak electrolytes, it is assumed that undissociated molecules at low concentrations can be neglected, and only the presence of ions need be considered.

The existence of an ionic atmosphere about any ion, i, and the resulting electrostatic interaction cause the chemical potential of this ion to have a larger magnitude than it would have in an ideal solution. Thus for a single ion we can write the following expression:

$$\frac{\mu_i}{\mathfrak{N}} = \frac{\mu_i}{\mathfrak{N}}(\text{ideal}) + \frac{\mu_i}{\mathfrak{N}}(\text{electrical})$$

Alternatively, we can express μ_i for an ion in terms of the activity coefficient as follows:

$$\frac{\mu_i}{\mathfrak{N}} = \frac{\mu_i^\circ}{\mathfrak{N}} + kT \ln \gamma_i m_i$$

$$= \frac{\mu_i}{\mathfrak{N}}(\text{ideal}) + kT \ln \gamma_i$$

It follows that

$$kT \ln \gamma_i = \frac{\mu_i}{\mathfrak{N}}(\text{electrical})$$

Debye and Hückel carried out a derivation of a theoretical equation for μ_i (electrical), presented on p. 652, from which an equation for $\ln \gamma_\pm$ can be obtained. A simplified form of this equation that is applicable to very dilute solutions of an electrolyte is the following equation which is usually called the *Debye-Hückel limiting law*

$$\log \gamma_\pm = -1.825 \times 10^6 \left(\frac{\rho_s}{D^3 T^3}\right)^{\frac{1}{2}} z_+ z_- I^{\frac{1}{2}} \qquad (16\text{-}11)$$

In this equation ρ_s and D are the density and dielectric constant of the solvent, T is the temperature of the solution, and z_+ and z_- are positive integers (valences) representing the charges on the ions as multiples of the charge on the electron. The quantity I, which is called the *ionic strength*, is defined by

$$I = \tfrac{1}{2} \Sigma \, m_i z_i^2 \qquad (16\text{-}12)$$

in which m_i is the molality of the ion of kind i with charge z_i, and the summation is carried out over all of the kinds of ions present in the solution.[4]

For a single uni-univalent electrolyte, such as $NaNO_3$, the ionic strength is equal to the molality. For a 0.1 molal solution of $CrCl_3$, for example, $I = \tfrac{1}{2}(0.1 \times 9 + 0.3 \times 1) = 0.6$. If this solution is also 0.2 molal in $Ba(NO_3)_2$, $I = \tfrac{1}{2}(0.1 \times 9 + 0.3 \times 1 + 0.2 \times 4 + 0.4 \times 1) = 1.2$.

[4] It is interesting to note that Lewis and Randall found empirically that many properties of an electrolyte could be expressed as functions of concentration more simply in terms of ionic strength than in terms of total molality or molarity. They defined this quantity and used it before Debye and Hückel gave it a theoretical basis.

Electrochemistry

When the solvent is water and the temperature is 25°C, the Debye-Hückel limiting law becomes

$$\log \gamma_\pm = -0.509 z_+ z_- I^{\frac{1}{2}} \tag{16-13}$$

The agreement of values of the mean activity coefficient calculated by use of this equation with experimental values is shown graphically in Figure 16-7.

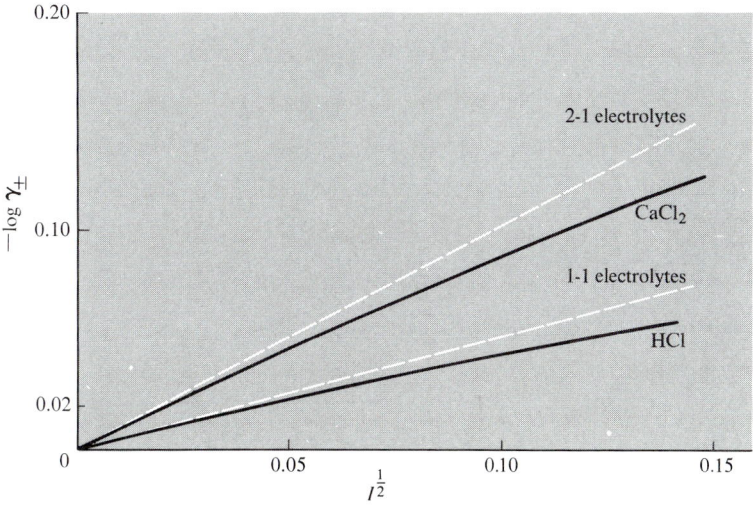

Figure 16-7. Negative of the logarithm of the mean activity coefficient as a function of the square root of the ionic strength. The dashed lines represent the Debye-Hückel limiting law for 1-1 and 2-1 electrolytes.

It is apparent that the range of validity is indeed limited to very low values of the ionic strength. The extension of the theory to somewhat larger values of I is discussed on p. 659.

Now we are ready to consider some specific examples of ionic equilibria. One of the simpler cases is the equilibrium of a slightly soluble electrolyte with a solution of its ions.

Equilibria of Slightly Soluble Electrolytes

We can represent the equilibrium of a slightly soluble salt, such as $BaSO_4$, and a saturated aqueous solution of its ions in the following way:

$$BaSO_4(s) \rightleftarrows Ba^{++} + SO_4^=$$

The equilibrium constant is

$$K = \frac{a_{Ba^{++}} a_{SO_4^=}}{a_{BaSO_4}}$$

We can say that either the solid $BaSO_4$ is in its standard state, which means that $a_{BaSO_4} = 1$, or the activity of the solid $BaSO_4$ is a constant that can be combined with the equilibrium constant. From either point of view we have

$$K_{sp} = a_{Ba^{++}} a_{SO_4^=}$$

in which K_{sp} is the thermodynamic solubility product constant.

With concentrations expressed in terms of molalities, the equation above becomes

$$K_{sp} = m_{Ba^{++}} m_{SO_4^=} \gamma_{Ba^{++}} \gamma_{SO_4^=}$$
$$= m_s^2 \gamma_\pm^2$$

in which m_s is the solubility expressed in moles of salt per 1000 g of water. (For aqueous solutions as dilute as a saturated solution of $BaSO_4$, the molality is equal to the molarity within experimental error.) If the solution were ideal, the solubility, m_o, would be given by $\sqrt{K_{sp}}$. Consequently,

$$m_s = \frac{m_o}{\gamma_\pm}$$

Thus it can be seen that the effect of the deviation from ideality which lowers the activity coefficient is to increase the solubility of a slightly soluble salt above the value it would have in a hypothetical ideal solution.

In order to calculate the value of K_{sp} from an experimental value of m_s it is necessary to know the value of $\gamma_\pm$. This value can be obtained in either of the following two ways:

1. The value of $\gamma_\pm$ can be calculated from the Debye-Hückel limiting law. Let us take 25°C as the temperature of a $BaSO_4$ solution. Then $m_s = 1.06 \times 10^{-5}$ moles per 1000 g of water. For a 2-2 salt, such as $BaSO_4$, $I = \frac{1}{2}(4m + 4m) = 4m$. Consequently $I^{\frac{1}{2}} = 2(1.06 \times 10^{-5})^{\frac{1}{2}} = 6.51 \times 10^{-3}$. Then from Equation 16-13

$$\log \gamma_\pm = -0.509 \times 2 \times 2 \times 6.51 \times 10^{-3}$$
$$= -1.32 \times 10^{-2}$$
$$\gamma_\pm = 0.970$$

The value of K_{sp} is then given by

$$K_{sp} = (1.06 \times 10^{-5})^2 (0.970)^2$$
$$= 1.06 \times 10^{-10}$$

2. Since $\gamma_\pm = 1$ when $I = 0$, if m_s could be measured at a series of values of I and plotted against $I^{\frac{1}{2}}$, by extrapolation to $I = 0$ the value of m_o could be found from which $\gamma_\pm$ could be calculated. Of course, I is a constant when $BaSO_4$ is the only electrolyte in the saturated solution. Its value, however, can be changed by adding other electrolytes to the solution which do not yield either Ba^{++} or $SO_4^=$ ions. Thus a series of m_s values can be determined at

Acid-Base Equilibria

various values of I, the extrapolation to $I = 0$ can be carried out to obtain the value of m_o, and then $\gamma_\pm$ can be calculated.

If the added electrolyte has an ion in common with $BaSO_4$, the effect of this common ion on the equilibrium far outweighs the effect of changing ionic strength.

Acid-Base Equilibria

One of the interesting aspects of the development of chemistry is the change with time of the definition of acid and base. Initially, the term seems to have referred to solutions that "attacked" certain metals, changed the colors of certain dyes, and had a sour taste. Likewise, a base was a solution that "neutralized" an acid. It was recognized early that a "strong" acid could displace a "weak" acid from its salts. By 1790, Lavoisier, who is sometimes called the father of chemistry, thought that the then newly discovered element oxygen was the essential element in both acids and bases. In fact, the word "oxygen," which he invented, means acid-former.

With improvements in the techniques of chemical analysis, it became recognized that hydrogen was the essential element in acids. After the theory of ionization was established, an acid became a substance which yielded hydrogen ions in aqueous solution, while a base yielded hydroxide ions, these two kinds of ions having a unique relationship with water as a solvent.

In this century there have been two additional definitions of acid and base, both being applicable in differing degrees to nonaqueous solutions. One of these definitions was supplied by J. N. Brönsted and M. Lowry, the other, by G. N. Lewis. We discuss these briefly in sequence.

1. Brönsted and Lowry defined an acid as any substance that can donate protons to any other substance, called the base. Removal of a proton from an acid forms a conjugate base. From this point of view, when acetic acid, HAc, is dissolved in water, the water acts as a base according to the following equation:

$$HAc + H_2O \rightleftarrows H_3O^+ + Ac^- \tag{16-14}$$

The hydronium ion is the acid conjugate to H_2O and the acetate ion is the base conjugate to HAc. When ammonia is dissolved in water, the water acts as an acid according to the equation

$$NH_3 + H_2O \rightleftarrows NH_4^+ + OH^-$$

In the reverse reaction the ammonium ion is the acid and the hydroxide ion is the base conjugate to water. In liquid ammonia, the NH_4^+ ion is an acid and the NH_2^- ion is a base. The neutralization reaction in this case is

$$NH_4^+ + NH_2^- \rightarrow 2NH_3$$

2. To include the case of systems that do not contain any protons or, at least, any exchangeable protons, Lewis defined an acid as a substance that can accept a pair of electrons from a base, which he defined as an electron donor. NH_3, for example, is a Lewis base because of the lone pair of electrons on the nitrogen atom, and BF_3 is a Lewis acid because the boron atom can accept a pair of electrons. From this point of view, the reaction represented by the following equation in Lewis' notation:

$$\begin{array}{cc} H & F \\ \ddot{} & \ddot{} \\ H:\ddot{N}: + B:\ddot{F} \\ \ddot{} & \ddot{} \\ H & F \end{array} \rightarrow \begin{array}{c} H\ F \\ \ddot{}\ \ddot{} \\ H:\ddot{N}:B:\ddot{F} \\ \ddot{}\ \ddot{} \\ H\ F \end{array}$$

is considered to be an acid-base reaction, although no ions are involved.

The dissociation of water into ions, from the Brönsted-Lowry point of view, is an acid-base reaction which we might represent by the following equation:

$$H_2O + H_2O \rightleftarrows H_3O^+ + OH^-$$

The number of water molecules associated with a proton is not accurately known, however. Consequently we represent the hydrogen ion, whatever its state of hydration might be, simply by the symbol H^+. If this is done consistently there is no loss in definiteness. The dissociation of water is then represented by

$$H_2O \rightleftarrows H^+ + OH^-$$

The equilibrium constant for this dissociation is given by

$$K_w = \frac{a_{H^+} a_{OH^-}}{a_{H_2O}}$$

Since the water is in its standard state, $a_{H_2O} = 1$, and thus we have

$$K_w = a_{H^+} a_{OH^-}$$

When the mean ionic activity coefficient is introduced, this equation becomes

$$K_w = m_{H^+} m_{OH^-} \gamma_\pm^2$$

In order to evaluate K_w let us make the assumption that the concentrations of the ions in water are so low that $\gamma_\pm = 1$, and that molality can be replaced by molarity. Then the molarity of the ionized water may be determined from conductivity data by use of Equation 16–1. Solving this equation for c, we have

$$c = \frac{1000\kappa}{\Lambda}$$

Kohlrausch and Heydweiller[5] found that the conductivity of the purest sample of water they could prepare was 5.8×10^{-8} ohm^{-1} cm^{-1} at 25°C. Since the

[5] *Z. physik. Chemie*, **14**, 317 (1894).

ionic concentrations are very low, $\Lambda \sim \Lambda_o$, and we can obtain the value of Λ_o from the conductivities of the H$^+$ and OH$^-$ ions listed in Table 16-2. Thus $\Lambda_o = 349.8 + 198.0 = 547.8$ cm^2 ohm^{-1} equiv^{-1}, and

$$c = \frac{10^3 \times 5.8 \times 10^{-8}}{547.8}$$

$$= 1.06 \times 10^{-7} \text{ equiv/liter}$$

It follows that $K_w = 1.12 \times 10^{-14}$. The presently accepted value of this quantity, determined by a more accurate method to be mentioned later, is 1.008×10^{-14}.

Let us now consider the ionization of the typical weak acid, acetic acid, which is represented according to Brönsted and Lowry by Equation 16-14. The simplified equation for this "reaction" is

$$HAc \rightleftarrows H^+ + Ac^-$$

The equilibrium constant K_a, which is usually called the acid dissociation constant, is given by the following expression:

$$K_a = \frac{a_{H^+} a_{Ac^-}}{a_{HAc}}$$

Replacing the activity by the product of molarity and activity coefficient, we have

$$K_a = \frac{m_{H^+} m_{Ac^-}}{m_{HAc}} \frac{\gamma_{H^+} \gamma_{Ac^-}}{\gamma_{HAc}} \tag{16-15}$$

Because γ_{HAc} is the activity coefficient of an electrically neutral molecule, its value is essentially 1 in dilute solutions and it can be dropped from the equation above. The product $\gamma_{H^+}\gamma_{Ac^-}$, equal to $\gamma_\pm^2$, can be calculated from the Debye-Hückel equation if the molalities are low enough, or it can be determined experimentally by one of the methods previously mentioned. The ionic molalities can be determined in several different ways, one of the most direct being conductivity measurements. Since HAc is a weak electrolyte, the degree of dissociation is given by the Arrhenius relationship, $\alpha = \Lambda/\Lambda_o$. Then K_a can be calculated by use of the corrected Ostwald dilution law, Equation 16-5, in the following form:

$$K_a = \frac{\alpha^2 m}{1 - \alpha} \gamma_\pm^2$$

in which m, the total molality of the acetic acid, is equal to $m_{HAc} + m_{Ac^-}$. Values of K_a at 25°C for a range of molalities are listed in Table 16-5. It can be seen that in contrast to the values in Table 16-3, K_a is indeed constant within narrow limits for low molalities. The dissociation constant for a weak base, K_b, is obtained in a completely analogous way.

TABLE 16-5 IONIZATION CONSTANT OF ACETIC ACID CALCULATED FROM THE OSTWALD DILUTION LAW AND THE DEBYE-HÜCKEL EQUATION*

$c \times 10^3$ (equiv/liter)	K_{eq}
0.0280	1.752
0.1532	1.750
1.0283	1.751
2.4140	1.750
5.9115	1.749
9.8421	1.747

*D. A. MacInnes, *The Principles of Electrochemistry*, Reinhold Publishing Corp., New York, 1939; republished by Dover Publications, New York, 1961, page 345.

Since K_a and K_b are the thermodynamic equilibrium constants for the dissociation of acids and bases, respectively, the standard free energy of dissociation can be calculated from Equation 4-5. For example, from the accepted value of K_a for HAc at 25°C, which is 1.754×10^{-5}, we can calculate $\Delta G°$ for its dissociation as follows:

$$\Delta G° = -RT \ln K_a$$
$$= -1.99 \times 298 \times 2.30(-4.756) = 6.50 \text{ kcal}$$

In dilute solutions of acids and bases in water, a quantity that is often of interest is the hydrogen ion concentration, [H⁺]. To calculate this quantity it is a common approximation to assume that all activity coefficients are unity; thus $a \sim m$. Furthermore, for dilute solutions the density is nearly the same as that of pure water, and then $m \sim c$, where c is the concentration in moles/liter.

With the above approximations, let us calculate the value of [H⁺] in a solution of HAc at a given total concentration, c. In this solution there are four unknown concentrations, namely, [H⁺], [OH⁻], [Ac⁻], and [HAc]. It is possible to find the values of each of these because they are related to each other by four simultaneous equations, all of which must be satisfied. One of these equations is essentially an expression of the law of conservation of matter. Since the HAc molecules added to the water are either ionized or unionized, we have

Electrochemistry

$$c = [HAc] + [Ac^-] \tag{a}$$

The solution must be electrically neutral; consequently,

$$[H^+] = [Ac^-] + [OH^-] \tag{b}$$

Because undissociated HAc must be in equilibrium with its ions, the following approximate version of Equation 16–15 must be applicable:

$$K'_a = \frac{[H^+][Ac^-]}{[HAc]} \tag{c}$$

in which K'_a is an approximate dissociation constant whose value is 1.8×10^{-5}. Finally, water must be in equilibrium with its ions, and thus

$$K_w = [H^+][OH^-] \tag{d}$$

The relationship between $[H^+]$ and c can be found by solving this set of equations for $[H^+]$. One method of solution is the successive elimination of variables, starting with [HAc]. From Equation (c), [HAc] is given by

$$[HAc] = \frac{[H^+][Ac^-]}{K'_a}$$

When this expression is substituted in Equation (a), the result is

$$c = \frac{[H^+][Ac^-]}{K'_a} + [Ac^-]$$

$$= \left(\frac{[H^+]}{K'_a} + 1\right)[Ac^-] \tag{e}$$

We now eliminate $[Ac^-]$. From Equation (b) we have

$$[Ac^-] = [H^+] - [OH^-] \tag{f}$$

and from Equation (d)

$$[OH^-] = \frac{K_w}{[H^+]} \tag{g}$$

When Equation (g) is substituted into Equation (f), and the result is substituted into Equation (e), we obtain

$$c = \left(\frac{[H^+]}{K'_a} + 1\right)\left([H^+] - \frac{K_w}{[H^+]}\right)$$

This is a cubic equation in $[H^+]$ which on expansion has the following form:

$$[H^+]^3 + K'_a[H^+]^2 - (K_w + cK'_a)[H^+] - K'_aK_w = 0 \tag{h}$$

Although this equation could be solved exactly, there is little point in doing so because Equations (c) and (d) are only approximations. Consequently, approximate solutions will suffice, but some judgment is needed as to what approximations are appropriate. K_w is very small relative to cK'_a and so can be neglected. Likewise, $K'_a K_w \sim 10^{-19}$ and this term can also be neglected. Then on division by [H⁺] Equation (h) reduces to the following quadratic:

$$[H^+]^2 + K'_a[H^+] - cK'_a = 0 \tag{i}$$

The equation above can also be obtained by ignoring the ionization of water and using the following approximations in Equation (c):

$$[H^+] \sim [Ac^-] \quad \text{and} \quad [HAc] \sim c - [H^+]$$

Equation (i) can be written in the form:

$$[H^+]^2 + ([H^+] - c)K'_a = 0$$

For a weak acid such as acetic acid, $[H^+] \ll c$, and consequently

$$[H^+] \sim \sqrt{cK'_a}$$

For example, with $c = 0.1$, we find that

$$[H^+] \sim (0.1 \times 1.8 \times 10^{-5})^{\frac{1}{2}} \sim 1.3 \times 10^{-3}$$

A check of the validity of this result can be obtained by substituting it in Equation (h).

We now extend this treatment to the reaction represented by

$$Ac^- + H_2O \rightleftarrows HAc + OH^-$$

which is called *hydrolysis* or, in the case of the analogous reaction in nonaqueous solutions, *solvolysis*. Its thermodynamic equilibrium constant is given by

$$K_h = \frac{a_{HAc} a_{OH^-}}{a_{Ac^-} a_{H_2O}}$$

The approximate form of this equation is

$$K'_h = \frac{[HAc][OH^-]}{[Ac^-]} \tag{j}$$

It can be seen from Equations (c) and (j) that $K'_a K'_h \sim K_w$. Thus the value of K'_h can be readily found from the value of K'_a. Then Equation (j) can be used to calculate the approximate value of [OH⁻], or of [H⁺], in a solution of known concentration of acetate ions. The calculation of [OH⁻] and [H⁺] in aqueous solutions of salts of weak bases follows a similar pattern.

The concepts applied here are also adequate for the treatment of ionic equilibria more complex than the ones mentioned above. In each case, as many equations must be found as the number of unknowns by consideration of the various equilibria involved in the solution as well as the requirements of electrical neutrality and mass balance. The exact solutions of such a set of equations are usually quite complicated, but they can always be simplified by making suitable approximations.

Hydrogen Ion Concentration: pH

The concentration of hydrogen ions in an aqueous solution is an important quantity in many branches of chemistry, particularly in their practical applications. Because its value is usually a very small number, Sorensen in 1909 introduced the symbol pH as a matter of convenience of notation to represent $-\log [H^+]$. The pH value of pure water, for example, is 7.0.

We saw in the last section that for solutions of weak acids and bases and their salts hydrogen ion concentrations, and thus the pH values, can be calculated approximately on the assumption that activities can be replaced by concentrations. The values of pH can also be determined approximately by several different methods. One of these methods, which was mentioned previously, depends on the measurement of the conductivity of a solution of an acid or a base and the assumption of the Arrhenius relation, $\alpha = \Lambda/\Lambda_o$. Another makes use of the fact that for many weak acids and bases, known as *indicators*, the absorption spectrum in the visible region of the undissociated molecules is different from that of the corresponding ions. This difference results in the familiar change in color with changing pH that is characteristic of the various indicators. The most precise methods depend on measurements made on electrochemical cells, which we consider in the next section.

The approximate values of pH obtained by these methods are adequate and useful for many purposes. When an attempt is made, however, to obtain more accurate values by the use of equations or experimental methods that do not involve approximations, a fundamental difficulty is encountered. The thermodynamic equations all depend on activities. If we try to refine Sorensen's definition of pH by defining pH as $-\log a_{H^+}$, we are blocked by the fact that the activity of a single kind of ion cannot be determined (p. 615). Only mean ionic activities can be found. We will return to this topic in a later section.

By analogy with Sorensen's definition of pH, the dissociation constants of weak acids and bases are often expressed in terms of pK values, with $pK_a = -\log K_a$ and $pK_b = -\log K_b$. Thus, for example, from the value of K_a for acetic acid at 25°C, 1.754×10^{-5} (p. 626), we obtain $pK_a = -\log 1.754 \times 10^{-5} = 4.756$.

Electrochemical Cells

Galvanic Cells

The subject to be considered in this section started with discoveries made by Luigi Galvani (1786) and Alessandro Volta (1796). Galvani, a professor of anatomy, found that dissected frogs' legs hanging on a copper hook underwent a muscular twitch when exposed nerve ends were touched with a piece of iron. He thought that this effect was necessarily associated with animals. Volta observed that the nature of the metals used in this experiment was significant and he also observed that if disks of dissimilar metals were separated by a piece of cloth soaked in acidified water, he could obtain an electric current when the end disks were connected by a wire. Stacks of such separated metallic disks in alternating sequence became known as a *voltaic pile*.

Neither of these men realized that the effects he observed were associated with chemical reactions. Their discoveries, however, stimulated much research, and it gradually came to be recognized that in a *galvanic*, or *voltaic*, *cell* the energy released by a chemical reaction is transformed into electrical energy, and that this process is the reverse of the one that occurs in an *electrolytic cell*. Galvanic cells have proved to be of great value, both as a practical source of electrical energy and as a source of a large amount of information of interest to chemists.

A typical galvanic cell is diagrammed in Figure 16–8. In this cell, which is known as a Daniell cell, a piece of metallic zinc dips into an aqueous solution

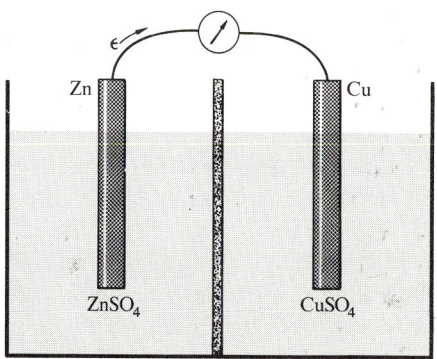

Figure 16–8. Diagram of a Daniell cell.

of $ZnSO_4$ and a piece of copper dips into a solution of $CuSO_4$. The two solutions are separated by a piece of porous ceramic which retards the interdiffusion of the electrolytes while it does not prevent completely the movement of the ions between the solutions. The Daniell cell is represented schematically in the following way:

$$Zn/ZnSO_4(m_1) \vdots CuSO_4(m_2)/Cu$$

The solid lines represent the boundaries between phases and the dotted line represents the boundary, called the *liquid junction*, between the two different solutions in electrical contact in the ceramic separator; m_1 and m_2 are the molalities of the respective electrolytes.

Nothing appears to happen in this cell until the metallic electrodes are connected by an electrical conductor, such as a piece of copper wire. Then an electric current flows through the wire and the cell can be used to provide heat, light, or mechanical power. At the same time it is found that the zinc electrode appears to dissolve and that the concentration of the $ZnSO_4$ solutions increases, while the mass of the copper electrode increases and the concentration of the $CuSO_4$ solution decreases. These changes mean, of course, that the following chemical reaction is taking place:

$$Zn + Cu^{++} \rightarrow Zn^{++} + Cu \qquad (16\text{-}16)$$

This is the same reaction which occurs almost instantaneously when a piece of zinc wire is dipped into a $CuSO_4$ solution. Although no electrical energy is produced in this case, we now recognize the basic process as the transfer of electrons, in one case directly from zinc atoms to copper ions, and in the other by the flow of electrons in the copper wire from the zinc electrode to the copper electrode, and then to copper ions.

At this point attention should be called to one of the possible sources of confusion in this subject. An electric current was recognized to be a flow of "something" before the something was identified. Purely as a matter of convention, the current was defined by Benjamin Franklin as the flow of positive charge. Long afterwards, it became evident that an electric current is a flow of negative charge carried by electrons. Nevertheless, the earlier definition of current was retained by physicists. Consequently, the direction of the conventional current is the opposite of the direction of the flow of electrons. In the treatment here we emphasize the flow of electrons.

If the wire that connects the zinc and copper electrodes is cut, the chemical reaction stops but a difference in electrical potential still exists between the two pieces of wire. This potential difference is usually called the *electromotive force* (emf) of the cell, represented by $\mathcal{E}$, and its unit is the *volt*. When Q coulombs of charge flow between the electrodes of a galvanic cell whose emf is $\mathcal{E}$ volts, the amount of chemical energy converted into electrical energy is $Q\mathcal{E}$ joules.

When a current is drawn from a galvanic cell its emf is always less than the "open-circuit" value, which is the quantity of primary interest in chemistry, due to irreversible processes. Consequently, cell emfs are not measured with a voltmeter which, because of its finite resistance, allows a small current to pass, but with a potentiometer (Fig. 16–9) which is a device for opposing the cell emf with a variable known potential difference from an external source. When the external potential difference is varied until it is equal and opposite to the cell emf, no current is drawn from the cell and the magnitude of the external potential difference is equal to that of the open-circuit emf of the cell.

If the external potential difference is larger than the cell emf, the cell operates as an electrolytic cell in which electrical energy from the external source is converted into chemical energy at the electrodes. The chemical reaction in this case is the reverse of the reaction carried out when the cell operates as a galvanic cell.

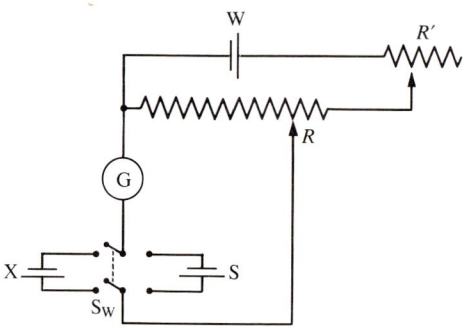

Figure 16-9. Diagram of a potentiometer. X is the cell whose emf is to be measured, S is a standard cell supplying a reference emf, G is a galvanometer, W is a working cell producing a potential difference across the variable resistance R, R' is a variable resistance, and SW is a double-pole, double-throw switch for connecting either X or S into the circuit.

It is found that cell emfs depend on the nature of the electrodes, the concentrations of the electrolytes, the cell temperature, and, in some cases, on the external pressure. We will see later how these dependences can be understood.

Kinds of Electrodes

The electrodes that are useful in electrochemistry are those at which the transfer of electrons can occur in a thermodynamically reversible manner. A cell both of whose electrodes are reversible is called a *reversible cell*. Reversibility in this case means that when the cell emf is nearly balanced by an external potential difference, a minute change in this external potential difference reverses the direction of the current and thus the direction of the chemical reaction that takes place in the cell. The emf of a reversible cell has the same value whether the balance point is approached from one direction or the other. A reversible electrode is said to be reversible with respect to the kind of ion or molecule which reacts by electron transfer at the electrode.

A number of different types of reversible electrodes are known. Among the more common of these types are the following:

1. *Metallic.* A metallic electrode dipping into a solution of one of its

Electrochemistry

salts is usually reversible with respect to its ions. For example, the electrodes of the Daniell cell described above are reversible.

2. *Metal, insoluble salt.* This electrode consists of a piece of metal coated with a layer of an insoluble salt of the metal and dipping into a solution that contains the negative ion of the salt. One of the most widely used electrodes of this type is the Ag/AgCl electrode in a solution containing Cl^- ions, with respect to which it is reversible.

3. *Gas.* To circumvent the difficulty of making electrical contact with a gas, this kind of electrode consists of a piece of metal around which a gas is

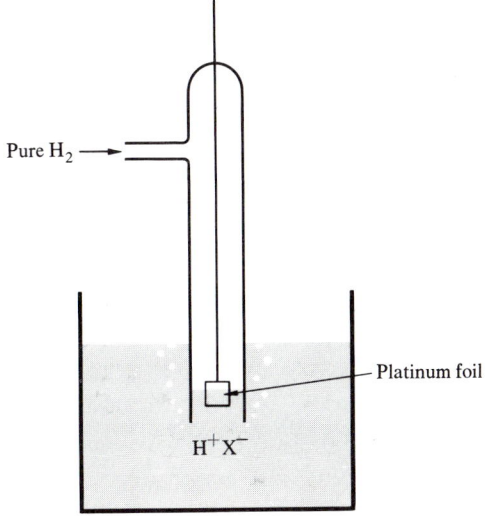

Figure 16–10. Hydrogen electrode dipping into an acid solution.

allowed to bubble while the metal dips into a solution containing the ion formed by the loss or gain of electrons by the gas (Fig. 16–10). The metal is almost always a piece of platinum foil covered with a coating of finely divided platinum, known as platinum black. The most common type of gas electrode is the hydrogen electrode dipping into a solution containing H^+ ions.

4. *Oxidation-reduction, redox.* Although oxidation or reduction, in the sense of electron transfer, occurs at every electrode, this term is used here to apply to an inert metal which dips into a solution containing two ions in different oxidation states, such as Sn^{2+} and Sn^{4+}, or two interconvertible compounds, such as quinone and hydroquinone. The electrode which makes use of these two compounds is referred to as the quinhydrone electrode because these compounds combine reversibly to form a complex known as quinhydrone.

Electrode Potentials and Cell Emf

A wide variety of cells can be constructed from electrodes of the types just mentioned. While it is possible to study each of these cells individually, it is more useful to consider that each cell is a combination of two half-cells, each containing one electrode. By focusing on the half-cells we can understand the behavior of the much larger number of cells made from the different combinations of the half-cells. A half-cell reaction is associated with each half-cell. The over-all reaction is the sum of the half-cell reactions.

Let us again use the Daniell cell as an example. One of the half-cells in this case consists of a Zn electrode dipping into a solution of $ZnSO_4$; the other consists of a Cu electrode dipping into a solution of $CuSO_4$. When the Daniell cell operates as a galvanic cell the half-cell reactions are:

$$Zn \rightarrow Zn^{++} + 2\epsilon$$

and

$$Cu^{++} + 2\epsilon \rightarrow Cu$$

and the sum of these reactions is $Zn + Cu^{++} \rightarrow Zn^{++} + Cu$, which is the same as Equation 16–16.

Associated with each of these half-cells and its reaction there is an electrode potential. In accordance with an international convention, the electrode potential related to any half-cell reaction is formulated as a reduction, that is, a gain of electrons by the reactant. Thus the potential of the copper electrode, $\mathcal{E}_{Cu}$, is associated with the reaction as given above:

$$Cu^{++} + 2\epsilon \rightarrow Cu$$

In contrast, the potential for the zinc half-cell, $\mathcal{E}_{Zn}$, is associated with the reverse of the reaction given above, namely:

$$Zn^{++} + 2\epsilon \rightarrow Zn$$

It can be seen that if this equation for the Zn half-cell is subtracted from the equation for the Cu half-cell, the equation for the over-all reaction is obtained. It also follows that the emf of the complete cell is equal to

$$\mathcal{E}_{Cu} - \mathcal{E}_{Zn}$$

When any cell is represented by a schematic diagram, as is the Daniell cell on page 630, it is conventional to equate the cell emf to the potential of the electrode written to the right in the diagram, minus the potential of the electrode written to the left:

$$\mathcal{E}_{cell} = \mathcal{E}_{right} - \mathcal{E}_{left}$$

If the reaction taking place at the right-hand electrode is actually a reduction, then the cell emf is positive, and the cell reaction represented by the sum of

Electrochemistry

the equations for the half-cell reactions takes place spontaneously. In this case electrons travel from left to right in the external part of the circuit. The electrode at which reduction occurs is often referred to as the cathode and the electrode at which oxidation takes place, as the anode.

We now have the problem of finding numerical values for single electrode potentials, such as $\mathcal{E}_{Zn}$. Before we can find these values we must consider a basic difficulty. When a piece of metal dips into an electrolytic solution, there is a certain tendency for electrons to leave the metal, in rough analogy with the tendency of molecules to leave a liquid and enter the vapor phase. This tendency is in principle measured by the potential difference between the metal and the solution into which it dips. There is, however, no method for measuring this potential difference because such a measurement would require that electrical contact be made with the solution by another piece of metal. Then the measured potential difference would be that of the whole system and it could not be attributed to just a single electrode.

It follows that although we can measure the emf of pairs of electrodes, there is no absolute way of dividing the measured value into contributions from the individual electrodes. This fact is not a serious difficulty, however, because only differences of potential are of interest in chemistry and we do not need absolute values of electrode potentials. Consequently, we can do what has been done several times previously, namely, arbitrarily define a zero of the quantity of interest and then determine values relative to this defined zero.

In the present case, the zero of electrode potential is defined by selecting a carefully specified electrode and assigning it a value of zero by convention. This reference electrode has been chosen to be a hydrogen electrode in which hydrogen at a pressure of one atmosphere bubbles around a piece of platinum foil coated with platinum black which dips into an acid solution containing hydrogen ions at unit activity; more precisely, the mean ionic activity is equal to 1, since single-ion activities cannot be determined. This electrode is called the *standard hydrogen electrode* (SHE).

If a complete cell is made of a SHE and any other half-cell, a measurement of the emf of this cell gives the potential of the other half-cell. For example, we may take a cell consisting of a SHE and a Ag/AgCl electrode dipping into a solution of HCl whose concentration is such that $a_{\pm} = 1$. This cell is represented by

$$Pt/H_2(1 \text{ atm})/H^+Cl^-(a_{\pm} = 1)/AgCl(s)/Ag$$

The emf of this cell is equal to the potential of the half-cell containing the Ag/AgCl electrode.

Although, as mentioned above, the SHE is the primary reference electrode, it is not a convenient electrode to use in the laboratory. It requires a source of H_2 which must be carefully purified because the surface of the platinum foil is easily contaminated or "poisoned," and the pressure of the H_2

must be maintained at 1 atm. Consequently it is common to use secondary reference electrodes whose half-cell potentials have been carefully measured relative to the SHE.

We mention here only the most commonly used secondary reference electrode, the calomel electrode. The half-cell containing this electrode consists of mercury in contact with a saturated solution of mercurous chloride (calomel) in which there is some specified concentration of KCl. This half-cell can be represented in the following way:

$$Hg/Hg_2Cl_2(s)/Cl^-$$

and the half-cell reaction is

$$Hg_2Cl_2 + 2\epsilon \rightarrow 2\ Hg(l) + 2\ Cl^-$$

The potential of the calomel electrode depends on the concentration of the KCl solution used to construct it. The concentrations, expressed as molarity or normality, that are used in reference electrodes are 0.1 M, 1.0 M, and a saturated solution. The corresponding half-cell potentials are given as functions of the temperature in °C by the following expressions:

0.1 M KCl	$0.334 - 0.7 \times 10^{-4}(t - 25)$ volts
1.0 M KCl	$0.280 - 2.4 \times 10^{-4}(t - 25)$
Saturated KCl	$0.242 - 7.6 \times 10^{-4}(t - 25)$

When the calomel electrode is used as part of a cell, a tube extending from the half-cell is dipped into the solution of the other half-cell (Fig. 16-11). At

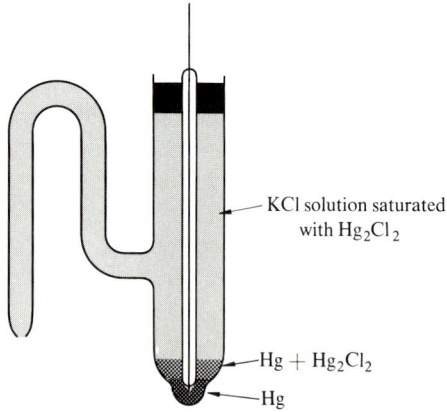

Figure 16–11. One form of calomel electrode.

the end of this tube two electrolytic solutions are in contact and form a liquid junction. The effect of a liquid junction presents a problem which we now consider.

In any cell, such as the Daniell cell, that includes a liquid junction there is a potential difference at the liquid junction due to differences in mobility between the various ionic species present. This junction potential is a disturbing factor when accurate values of electrode potentials are desired. It cannot be measured by itself because electrodes would have to be introduced for this purpose. Neither can its value be calculated accurately, in spite of much effort in this direction, because it is the result of irreversible processes, such as ionic diffusion and transference, to which a thermodynamic treatment is not applicable.

Consequently, for accurate measurements cells with liquid junctions are avoided by use of a single electrolytic solution whenever this is possible. If a liquid junction must be used, its effect is minimized by the use of a *salt bridge* between the electrolytic solutions. In most cases this consists of a U-tube containing a solution of KCl with the ends of the tube plugged with glass wool to retard diffusion. Sometimes an agar gel or gelatin saturated with KCl solution is used in the tube. The extension of the calomel electrode, mentioned above, is essentially a salt bridge.

The reason for using a KCl solution in a salt bridge is the fact that the mobilities of the K⁺ and Cl⁻ ions are nearly the same. Thus there is a partial compensation for the effects of the two ends of the tube.

Thermodynamics and the Nernst Equation

It was pointed out previously that the emf, $\mathcal{E}$, of a reversible galvanic cell has a maximum value when it is opposed by an external potential difference that is at most infinitesimally less than $\mathcal{E}$. If under these conditions a certain amount of charge, say 1 faraday, flows through the external circuit, the electrical work done by the cell reaction is the maximum amount of work that can be done in moving this amount of charge; then $w_{max} = \mathcal{F}\mathcal{E}$ joules.

The fact that an infinite amount of time would be required to move a finite amount of charge, since the current is infinitesimal by hypothesis, is of no consequence here because we are concerned only with thermodynamic relationships. There is a close analogy between the reversible transfer of charge and the reversible expansion of a gas. The cell emf is analogous to the pressure of the gas, and the work done is a maximum in each case when the process is carried out reversibly.

When the transfer of charge is carried out reversibly at constant temperature and pressure, the electrical work done by the cell reaction is equal to the decrease in the Gibbs free energy of the system resulting from carrying out the cell reaction (p. 85). Let us take as an example the reaction of the Daniell cell with specified molalities of the electrolytes:

$$Zn + CuSO_4 \,(m = 0.10) \to Cu + ZnSO_4 \,(m = 0.50) \qquad (16\text{-}17)$$

The free energy change implied by this equation is the free energy change for the complete reaction of 1 mole of Zn with 1 mole of Cu^{2+} ions at a molality of 0.10 in which 2 faradays of charge are transferred from Zn to Cu^{2+} ions and 1 mole of Zn^{2+} ions is formed at a molality of 0.50. The electrical work done when this process is carried out reversibly is $2 \times 96{,}500\,\mathcal{E}_D$, where $\mathcal{E}_D$ is the emf of the Daniell cell. Consequently, $\Delta G = -2 \times 96{,}500\,\mathcal{E}_D$ joules.

In general, for a cell reaction in which n faradays of charge are transferred reversibly, the corresponding free energy change is given by the following fundamental equation:

$$\Delta G = -n\mathcal{F}\mathcal{E} \qquad (16\text{-}18)$$

In connection with the use of this equation, it should be noted that ΔG and n have definite values only when a specific chemical equation is written, or at least clearly understood. If all quantities in the chemical equation for a cell reaction are multiplied by the same factor, then n must be multiplied by this factor and the equation above is still valid. Since $\mathcal{E}$ depends on $\Delta G/n$, it is an intensive property, independent of the size of the cell or the amount of chemical change.

Having found the relationship between free energy change and cell emf, we can now make use of the relationship between free energy change and composition that was developed in Chapter 4 for the general gas reaction

$$aA + bB = eE + fF$$

At that point we used the partial pressure of an ideal gas as a measure of its thermodynamic activity. From a more general point of view we can rewrite Equation 4–3 in terms of activities in the following way:

$$\Delta G = \Delta G^\circ + RT \ln \frac{a_E^e a_F^f}{a_A^a a_B^b}$$

When this expression for ΔG is substituted in Equation 16–18 and the resulting equation is solved for $\mathcal{E}$, we have

$$\mathcal{E} = -\frac{\Delta G^\circ}{n\mathcal{F}} - \frac{RT}{n\mathcal{F}} \ln \frac{a_E^e a_F^f}{a_A^a a_B^b} \qquad (16\text{-}19)$$

From the equation above it can be seen that the standard emf, $\mathcal{E}^\circ$, for a cell in which all reactants and products are at unit activity is equal to $-\Delta G^\circ/n\mathcal{F}$. It follows that

$$\mathcal{E} = \mathcal{E}^\circ - \frac{RT}{n\mathcal{F}} \ln \frac{a_E^e a_F^f}{a_A^a a_B^b}$$

This equation is known as the *Nernst equation*. For 25°C, the temperature at which most measurements of cell emfs are made, $2.303\,RT/\mathcal{F} = 0.0591$, and the Nernst equation becomes

$$\mathcal{E} = \mathcal{E}^\circ - \frac{0.0591}{n} \log \frac{a_E^e a_F^f}{a_A^a a_B^b}$$

The standard emf of a cell can be considered to be the difference between the standard potentials of the two half-cells that comprise the cell; thus, $\mathcal{E}^\circ = \mathcal{E}_r^\circ - \mathcal{E}_l^\circ$. The standard potential for a half-cell (or an electrode) is defined as the emf of a cell in which the left-hand half-cell is a SHE and the right-hand half-cell consists of the electrode under consideration with all reactants and products at unit activity.

The standard potentials of most electrodes depend on the temperature only; for gas electrodes the pressure has the standard value of 1 atm. It follows that standard electrode potentials can be tabulated for the reference temperature of 25°C. Such a compilation is presented in Table 16–6. In conformity with the set of conventions mentioned above, the electrode reactions in this table are written as reductions, and the tabulated values are standard reduction potentials. The greater the tendency of the electrode to supply electrons to the external circuit, the more negative is its reduction potential. (In some references the opposite sign convention is used and oxidation potentials are tabulated.)

Let us now apply the Nernst equation to the calculation of the emf at 25°C of the Daniell cell previously mentioned. From Table 16–6 we find that $\mathcal{E}^\circ$ for this cell is $0.337 - (-0.763)$, or 1.100 v. Turning now to the second term in the Nernst equation, we see that because the solid Zn and Cu electrodes are in their standard states, $a = 1$. For the reaction as written in Equation 16–17 the transfer of 2 faradays is implied and consequently $n = 2$. Thus the cell emf is given by

$$\mathcal{E} = 1.100 - \frac{0.0591}{2} \log \frac{a_{Zn^{2+}}}{a_{Cu^{2+}}}$$

Since we do not know the activity values for single ions, we must make use of the mean ionic activity and its relation to the mean molality and the mean activity coefficient. Thus the equation above becomes

$$\mathcal{E} = 1.100 - 0.0295 \log \frac{0.50 \gamma_\pm(ZnSO_4)}{0.10 \gamma_\pm(CuSO_4)}$$

On introducing the values of the mean activity coefficients from Table 16–4, we have

$$\mathcal{E} = 1.100 - 0.0295 \log \frac{0.50 \times 0.063}{0.10 \times 0.16}$$
$$= 1.100 - 0.009$$
$$= 1.091 \text{ v}$$

TABLE 16-6   STANDARD ELECTRODE POTENTIALS*

Electrode	$\mathscr{E}°$ (volts)	Half-Cell Reaction
Li^+; Li	−3.045	$Li^+ + e = Li$
K^+; K	−2.925	$K^+ + e = K$
Rb^+; Rb	−2.925	$Rb^+ + e = Rb$
Na^+; Na	−2.714	$Na^+ + e = Na$
Mg^{2+}; Mg	−2.37	$\frac{1}{2}Mg^{2+} + e = \frac{1}{2}Mg$
Al^{3+}; Al	−1.66	$\frac{1}{3}Al^{3+} + e = \frac{1}{3}Al$
Zn^{2+}; Zn	−0.763	$\frac{1}{2}Zn^{2+} + e = \frac{1}{2}Zn$
Fe^{2+}; Fe	−0.440	$\frac{1}{2}Fe^{2+} + e = \frac{1}{2}Fe$
Cd^{2+}; Cd	−0.403	$\frac{1}{2}Cd^{2+} + e = \frac{1}{2}Cd$
Tl^+; Tl	−0.336	$Tl^+ + e = Tl$
Br^-; $PbBr_2(s)$, Pb	−0.280	$\frac{1}{2}PbBr_2 + e = \frac{1}{2}Pb + Br^-$
Co^{2+}; Co	−0.277	$\frac{1}{2}Co^{2+} + e = \frac{1}{2}Co$
Ni^{2+}; Ni	−0.250	$\frac{1}{2}Ni^{2+} + e = \frac{1}{2}Ni$
I^-; $AgI(s)$, Ag	−0.151	$AgI + e = Ag + I^-$
Sn^{2+}; Sn	−0.140	$\frac{1}{2}Sn^{2+} + e = \frac{1}{2}Sn$
Pb^{2+}; Pb	−0.126	$\frac{1}{2}Pb^{2+} + e = \frac{1}{2}Pb$
H^+; H_2, Pt	0.0000	$H^+ + e = \frac{1}{2}H_2$
Br^-; $AgBr(s)$, Ag	0.095	$AgBr + e = Ag + Br^-$
Sn^{4+}, Sn^{2+}; Pt	0.15	$\frac{1}{2}Sn^{4+} + e = \frac{1}{2}Sn^{2+}$
Cu^{2+}, Cu^+; Pt	0.153	$Cu^{2+} + e = Cu^+$
Cl^-; $AgCl(s)$, Ag	0.2224	$AgCl + e = Ag + Cl^-$
Cl^-; $Hg_2Cl_2(s)$, Hg	0.268	$\frac{1}{2}Hg_2Cl_2 + e = Hg + Cl^-$
Cu^{2+}; Cu	0.337	$\frac{1}{2}Cu^{2+} + e = \frac{1}{2}Cu$
Cu^+; Cu	0.521	$Cu^+ + e = Cu$
I^-; $I_2(s)$, Pt	0.5355	$\frac{1}{2}I_2 + e = I^-$
H^+, quinhydrone(s); Pt	0.6996	$\frac{1}{2}C_6H_4O_2 + H^+ + e = \frac{1}{2}C_6H_6O_2$
Fe^{3+}, Fe^{2+}; Pt	0.771	$Fe^{3+} + e = Fe^{2+}$
Hg_2^{2+}; Hg	0.789	$\frac{1}{2}Hg_2^{2+} + e = Hg$
Ag^+; Ag	0.7991	$Ag^+ + e = Ag$
Hg^{2+}, Hg_2^{2+}; Pt	0.920	$Hg^{2+} + e = \frac{1}{2}Hg_2^{2+}$
Br^-; $Br_2(l)$, Pt	1.0562	$\frac{1}{2}Br_2(l) + e = Br^-$
Tl^{3+}, Tl^+; Pt	1.250	$\frac{1}{2}Tl^{3+} + e = \frac{1}{2}Tl^+$
Cl^-; $Cl_2(g)$, Pt	1.3595	$\frac{1}{2}Cl_2(g) + e = Cl^-$
Pb^{2+}; PbO_2, Pb	1.455	$\frac{1}{2}PbO_2 + 2H^+ + e = \frac{1}{2}Pb^{2+} + H_2O$
Au^{3+}; Au	1.50	$\frac{1}{3}Au^{3+} + e = \frac{1}{3}Au$
Ce^{4+}, Ce^{3+}; Pt	1.61	$Ce^{4+} + e = Ce^{3+}$
Co^{3+}, Co^{2+}; Pt	1.82	$Co^{3+} + e = Co^{2+}$

*A more complete tabulation can be found in W. M. Latimer, *loc. cit.*

Electrochemistry

It is possible to carry out the calculation of the cell emf in a different but equivalent way by obtaining the two electrode potentials individually, and then substituting these values in $\mathcal{E}_{cell} = \mathcal{E}_r - \mathcal{E}_l$. For this purpose we can write a separate Nernst equation for each half-cell. Thus for the Daniell cell just considered the half-cell reaction for the copper electrode is $Cu^{2+} + 2\epsilon \rightarrow Cu$, and the corresponding electrode reduction potential is given by

$$\mathcal{E}_{Cu} = \mathcal{E}_{Cu}^{\circ} - \frac{0.0591}{2} \log \frac{a_{Cu}}{a_{Cu^{2+}}}$$
$$= \mathcal{E}_{Cu}^{\circ} + 0.0295 \log 0.10\gamma_{\pm}(CuSO_4)$$
$$= 0.337 + 0.0295 \log 0.016$$
$$= 0.284 \text{ v}$$

For the left-hand half-cell we have $Zn^{2+} + 2\epsilon \rightarrow Zn$, and the zinc reduction potential is given by

$$\mathcal{E}_{Zn} = \mathcal{E}_{Zn}^{\circ} - 0.0295 \log \frac{a_{Zn}}{a_{Zn^{2+}}}$$
$$= -0.763 + 0.0295 \log 0.0315$$
$$= -0.807 \text{ v}$$

It follows that the cell emf is given by

$$\mathcal{E}_{cell} = 0.284 - (-0.807) = 1.091 \text{ v}$$

Having found the value of the cell emf, we can now readily find the free energy change for the cell reaction, Equation 16–17. From Equation 16–18,

$$\Delta G = -2 \times 96,500 \times 1.091$$
$$= -211,000 \text{ joules} = -50.5 \text{ kcal}$$

Additional information about the cell reaction can be obtained if the cell emf has been measured at several temperatures and the temperature coefficient of $\mathcal{E}$ is known. When Equation 16–18 is differentiated with respect to T at constant P, we have

$$\left(\frac{\partial \Delta G}{\partial T}\right)_P = -n\mathcal{F}\left(\frac{\partial \mathcal{E}}{\partial T}\right)_P$$

Since by Equation 3–49 $\left(\frac{\partial \Delta G}{\partial T}\right)_P = -\Delta S$, the entropy change in the cell reaction in units of joules/deg is given by

$$\Delta S = n\mathcal{F}\left(\frac{\partial \mathcal{E}}{\partial T}\right)_P$$

From the relationship $\Delta H = \Delta G + T \Delta S$, it follows that the enthalpy change in the cell reaction at temperature T is given by

$$\Delta H = -n\mathcal{F}\mathcal{E} + Tn\mathcal{F}\left(\frac{\partial \mathcal{E}}{\partial T}\right)_P$$

This equation is one form of the Gibbs-Helmholtz equation, Equation 3–50.

Concentration Cells

The type of galvanic cell we have discussed up to this point consists of two half-cells containing different electrodes. Such a cell is often called a *chemical cell* because its operation depends on the carrying out of a chemical reaction. There is another type of galvanic cell, called a *concentration cell*, in which the two half-cells contain the same electrodes but the concentrations, and therefore the activities, of a reactant involved in the electrode reaction are different. The process producing electrical energy in a concentration cell is essentially a transfer of a reactant from a higher activity to a lower one.

Concentration cells can be divided into two types: (1) electrode-concentration cells, in which the reactant whose concentration differs in the two half-cells is a component of the electrodes; and (2) electrolyte-concentration cells, in which this reactant is the electrolyte. The electrolyte-concentration cells are further divided into (2a), those without a liquid junction; and (2b), those with a liquid junction at which transfer of ions between solutions occurs. We consider each of these types in sequence.

1. *Electrode-concentration cells.* The simplest example of this type of cell is one containing two gas electrodes with the same gas at different pressures, such as the following cell:

$$\text{Pt}/\text{H}_2(P_1)/\text{HCl}(a_\pm)/\text{H}_2(P_2)/\text{Pt}$$

In this cell the reaction occurring at the left-hand electrode can be represented by

$$\text{H}_2(P_1) \rightarrow 2\text{H}^+(a_\pm) + 2\epsilon$$

The corresponding reaction at the right-hand electrode is

$$2\text{H}^+(a_\pm) + 2\epsilon \rightarrow \text{H}_2(P_2)$$

The over-all "reaction," which is independent of the composition of the electrolyte, is

$$\text{H}_2(P_1) \rightarrow \text{H}_2(P_2)$$

Since H_2 at moderate pressures can be considered to be an ideal gas, the ratio of the pressures is equal to the ratio of the hydrogen activities. Thus the Nernst equation reduces to the following (with $n = 2$):

$$\mathcal{E} = -\frac{RT}{2\mathcal{F}} \ln \frac{P_2}{P_1}$$

When P_2 is less than P_1, the process is spontaneous because it is equivalent to an expansion of the hydrogen and $\mathcal{E}$ then has a positive value.

Another type of electrode-concentration cell consists of electrodes composed of a metallic solution or alloy. For example, electrodes of Na or K dissolved in Hg (amalgam electrodes) are sometimes useful. The mercury is an inert carrier for the alkali metal whose concentrations can be varied over a considerable range.

2a. *Electrolyte-concentration cells without transference.* Let us consider the following cell, which actually consists of two cells connected back to back by a metallic conductor (Fig. 16–12):

$$\text{Pt}/H_2(1 \text{ atm})/\text{HCl}(a_1)/\text{AgCl}/\text{Ag}/\text{AgCl}/\text{HCl}(a_2)/H_2(1 \text{ atm})/\text{Pt}$$

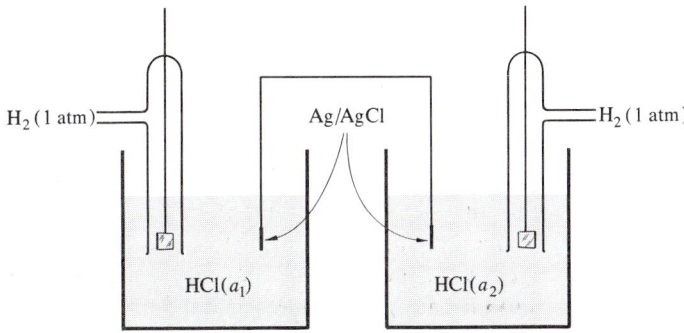

Figure 16–12. HCl concentration cell without transference.

The cell reaction for the left-hand cell is

$$\tfrac{1}{2}H_2(1 \text{ atm}) + \text{AgCl} \rightarrow H^+Cl^-(a_1) + \text{Ag}$$

The reaction for the right-hand cell as written is

$$\text{Ag} + H^+Cl^-(a_2) \rightarrow \text{AgCl} + \tfrac{1}{2}H_2(1 \text{ atm})$$

The over-all "reaction" is

$$H^+Cl^-(a_2) \rightarrow H^+Cl^-(a_1)$$

The emf of this cell is given by

$$\mathcal{E} = -\frac{RT}{\mathcal{F}} \ln \frac{a_1}{a_2} \qquad (16\text{--}20)$$

When a_1 is less than a_2, HCl is formed in the left-hand cell and consumed in the right-hand cell. The spontaneous process which takes place is, in effect, a dilution of HCl from an activity a_2 to an activity a_1, even though the two solutions are not in contact. $\mathcal{E}$ then has a positive value. If, for example, $a_2 = 10a_1$, at 25°C $\mathcal{E} = -0.0591 \log(0.1) = 0.059$ v. The molar free energy of dilution is given in this case by

$$\Delta G = -n\mathcal{F}\mathcal{E} = -5700 \text{ joules}$$
$$= -1.36 \text{ kcal}$$

Since by definition $a_1 = (a_\pm)_1^2$, Equation 16–20 can also be written in the following equivalent forms:

$$\mathcal{E} = -\frac{2RT}{\mathcal{F}} \ln \frac{(a_\pm)_1}{(a_\pm)_2} = \frac{2RT}{\mathcal{F}} \ln \frac{(a_\pm)_2}{(a_\pm)_1} \qquad (16\text{–}21)$$

2b. *Electrolyte-concentration cell with transference.* In this cell there is a liquid junction and ions are transferred from one of the solutions to the other during the operation of the cell. Such a cell containing two HCl solutions in electrical contact can be represented in the following way:

$$\text{Pt}/\text{H}_2(1 \text{ atm})/\text{HCl}(a_1) \vdots \text{HCl}(a_2)/\text{H}_2(1 \text{ atm})\text{Pt}$$

The operation of this cell, as in the case just considered, results in a transfer of HCl from a solution of higher activity to one of lower activity. H^+ ions are formed at the left-hand electrode (the anode). In order to maintain electrical neutrality Cl^- ions must migrate into this half-cell from the other one. At the same time some H^+ ions leave the left-hand half-cell to enter the right-hand one.

Let us now imagine that our cell contains so much electrolyte that the passage of 1 faraday through the cell has a negligible effect on the composition of the solutions. As a result of the passage of 1 faraday, 1 mole of H^+ ions with the mean activity $(a_\pm)_1$ has been formed in the left-hand cell, while τ_- moles of Cl^- ions have migrated in and τ_+ moles of H^+ ions have migrated out, where τ_- and τ_+ are the respective transference numbers. Thus a net amount of $1 - \tau_+ (= \tau_-)$ moles of H^+ ions and τ_- moles of Cl^- ions have been added to this half-cell. At the same time 1 mole of H^+ ions at a mean activity of $(a_\pm)_2$ has reacted at the right-hand electrode, and τ_+ moles of H^+ ions have migrated into, and τ_- moles of Cl^- ions have migrated out of this half-cell. Thus the net change at the right-hand electrode is a removal of τ_- moles of H^+ ions and τ_- moles of Cl^- ions.

The result of all of these changes produced by the passage of 1 faraday through the cell is that τ_- moles of H^+ and of Cl^-, each at a mean activity of $(a_\pm)_2$, have been removed from the right-hand half-cell, and the same amounts of H^+ and Cl^- ions, each at a mean activity of $(a_\pm)_1$, have been added to the left-hand half-cell. We can write the following equation for this process:

$$\tau_- H^+(a_\pm)_2 + \tau_- Cl^-(a_\pm)_2 \rightarrow \tau_- H^+(a_\pm)_1 + \tau_- Cl^-(a_\pm)_1$$

Electrochemistry

The corresponding free energy change is given by

$$\Delta G = 2\tau_- RT \ln \frac{(a_\pm)_1}{(a_\pm)_2}$$

The factor of 2 results from the inclusion of changes for both kinds of ions. Since $n = 1$ in this case, the cell emf is given by

$$\mathcal{E} = 2\tau_- \frac{RT}{\mathcal{F}} \ln \frac{(a_\pm)_2}{(a_\pm)_1}$$

Notice that the transference number that appears in this equation is the one for the ion which is *not* involved in the electrode reactions.

The equation above for the cell emf is the same as Equation 16–21, except for the factor τ_-. It follows that

$$\tau_- = \frac{\mathcal{E}_w}{\mathcal{E}_{wo}} \qquad (16\text{–}22)$$

in which $\mathcal{E}_w$ and $\mathcal{E}_{wo}$ are the cell emfs for concentration cells with and without transference, respectively. This assumes, of course, that the same process is carried out in both cells and that the transference numbers are essentially independent of the electrolyte concentration. Then Equation 16–22 can be used for the determination of transference number values.

Applications of Galvanic Cells

Because measurements of cell emfs can be carried out with excellent precision and accuracy, such measurements have been applied wherever feasible to obtain data of chemical interest. One of these applications, the determination of transference numbers, has been mentioned in the preceding section. Among the other applications are the following:

1. *Determination of equilibrium constants.* From Equation 16–19 we saw that the standard emf of a cell is related to the standard free energy change of the cell reaction by the equation $\mathcal{E}° = -\Delta G°/n\mathcal{F}$. Introducing Equation 4–5 for the relation between the standard free energy change and the equilibrium constant, we obtain

$$\mathcal{E}° = \frac{RT}{n\mathcal{F}} \ln K_{eq}$$

Thus by determination or calculation of $\mathcal{E}°$ the equilibrium constant for the cell reaction can be found.

Let us take as an example the following reaction:

$$Ag^+ + Fe^{2+} \rightleftarrows Ag + Fe^{3+}$$

whose equilibrium constant is given by

$$K_{eq} = \frac{a_{Fe^{3+}}}{a_{Fe^{2+}}a_{Ag^+}}$$

This reaction is the cell reaction for the following cell:

$$Pt/Fe^{2+}, Fe^{3+} \vdots Ag^+/Ag$$

From the $\mathcal{E}^\circ$ values in Table 16-6 for the Ag/Ag$^+$ and the Pt/Fe^{2+}, Fe^{3+} electrodes, we find that

$$\mathcal{E}^\circ = \mathcal{E}_r^\circ - \mathcal{E}_l^\circ$$
$$= 0.799 - 0.771 = 0.028 \text{ v}$$

It follows that at 25°C, since $n = 1$,

$$\ln K_{eq} = \frac{0.028 \times 96{,}500}{8.32 \times 298} = 1.09$$

and

$$K_{eq} = 2.97$$

2. *Determination of solubility products.* This topic is closely related to the previous one because the solubility product is a kind of equilibrium constant. Let us suppose that a slightly soluble salt, MX, is in equilibrium with its ions; thus $MX \rightleftarrows M^+ + X^-$ and $K_{sp} = a_+ a_-$. This dissociation is the cell reaction for the following cell:

$$M/M^+X^-(\text{satd.})/MX(s)/M$$

since the reaction at the left-hand electrode is

$$M \rightarrow M^+ + \epsilon$$

and the reaction at the right-hand electrode is

$$\epsilon + MX(s) \rightarrow M + X^-$$

Thus the over-all reaction is

$$MX(s) \rightarrow M^+ + X^-$$

From the value of $\mathcal{E}^\circ$ for the cell under consideration the solubility constant is obtained by use of the equation

$$\ln K_{sp} = \frac{n\mathcal{F}\mathcal{E}^\circ}{RT}$$

For the particular case of AgCl as the slightly soluble salt, $\mathcal{E}_r^\circ$ is the standard potential of the Ag/AgCl electrode and $\mathcal{E}_l^\circ$ is the corresponding value for the

Ag/Ag$^+$ electrode. Referring to Table 16–6 for these standard electrode potentials, we obtain $\mathcal{E}° = 0.222 - 0.799 = -0.577$ v. Then

$$\ln K_{sp} = -\frac{96{,}500 \times 0.577}{8.32 \times 298} = -22.5$$

and

$$K_{sp} = 1.7 \times 10^{-10}$$

With the assumption that a saturated solution of AgCl is so dilute that $\gamma_\pm = 1$ and the molality is practically equal to the molarity, it follows that the solubility is given by $\sqrt{K_{sp}} = 1.3 \times 10^{-5}$ moles/liter.

3. *Determination of activity coefficients.* Since the emf of a cell depends on the activities of the ions involved in the electrode reactions according to the Nernst equation, measurements of cell emfs can be combined with data for the composition of the electrolyte solutions to yield values of activity coefficients. For an illustration of this procedure we return to the following cell:

$$\text{Pt/H}_2(1 \text{ atm})/\text{HCl}(m)/\text{AgCl/Ag}$$

whose emf at 25°C is given by

$$\mathcal{E} = \mathcal{E}° - 0.0591 \log a_{H^+} a_{Cl^-}$$

From Equation 16–7, $a_{H^+} a_{Cl^-} = \gamma_\pm^2 m_\pm^2$ and in this case $m_\pm = m$, the molality of the HCl solution. It follows that we can express the equation for the cell emf as follows:

$$\mathcal{E} = \mathcal{E}° - 0.0591 \log (\gamma_\pm m)^2$$
$$= \mathcal{E}° - 0.1182 \log m - 0.1182 \log \gamma_\pm$$

If $\mathcal{E}°$ can be obtained from a table of standard electrode potentials and m is known for a particular solution by analysis or otherwise, a measurement of $\mathcal{E}$ yields a value of $\gamma_\pm$.

If $\mathcal{E}°$ is not known, it can be determined from measurements of $\mathcal{E}$ at a series of values of m. At low values of m, according to the Debye-Hückel limiting law, $\log \gamma_\pm$ is proportional to $\sqrt{m}$. It follows that when values of $\mathcal{E} + 0.1182 \log m$ are plotted against $\sqrt{m}$, the graph line can be extrapolated linearly to $m = 0$. The intercept on the ordinate is the value of $\mathcal{E}°$ because $\log \gamma_\pm$ is then equal to zero.

Some of the most reliable activity coefficient values have been obtained in this way.

4. *Determination of pH. The glass electrode.* The difficulty with the accurate definition of pH, mentioned previously (p. 629), results in the fact that the pH value of a solution depends to some extent on the method used to determine it. The differences between the values obtained with different methods, however, can be neglected for most purposes.

A widely used method for obtaining pH values involves the measurement of the emf of a galvanic cell containing an electrode reversible with respect to hydrogen ions and a reference electrode of known potential. A cell that can be used for this purpose is the combination of a hydrogen electrode and a "normal" calomel electrode represented by

$$\text{Pt}/\text{H}_2(1 \text{ atm})/\text{H}^+(a_{\text{H}^+}) \vdots \text{Cl}^-(1 \text{ molar})/\text{Hg}_2\text{Cl}_2/\text{Hg}$$

The potential of the hydrogen electrode can be represented by ($\mathcal{E}^\circ = 0$):

$$\mathcal{E}_+ = -0.0591 \log a_{\text{H}^+}$$

When the problem of the measurement of single ion activities is ignored and pH is assumed to be equal to $-\log a_{\text{H}^+}$, this equation becomes

$$\mathcal{E}_+ = 0.0591 \text{ pH}$$

On page 636 the potential of the normal calomel electrode was given as 0.280 v at 25°C. Thus the Nernst equation for the cell emf is[6]

$$\mathcal{E}_{\text{cell}} = 0.280 + 0.059 \text{ pH}$$

and it follows that

$$\text{pH} = \frac{\mathcal{E}_{\text{cell}} - 0.280}{0.059}$$

As mentioned earlier (p. 635) the hydrogen electrode is not convenient in practice, and it is customary to employ other electrodes reversible with respect to hydrogen ions. One of these is the quinhydrone electrode (p. 633). The electrode most widely used for pH determination is the glass electrode, one form of which is shown in Figure 16-13.

The essential feature of the glass electrode is a very thin bubble or membrane made of a special type of glass that is permeable to hydrogen ions. When the pH of a solution on one side of this glass membrane is different from that on the other side, a potential difference develops between the two sides which is a measure of the difference in the two pH values. Inside the bubble, there is a solution of constant pH (buffer solution) containing Cl⁻ ions, and a Ag/AgCl electrode is immersed in this solution. The glass electrode as a unit is immersed in the solution whose pH is to be measured and a reference electrode, usually a calomel electrode, is also immersed in this solution. Then the potential difference between these two electrodes is measured[7] and converted to a pH value. The relation between pH and potential difference is found from a calibration with solutions whose pH values are given by the National Bureau of Standards.

---

[6]This equation does not include the contribution of the junction potential which is usually ignored.

[7]Because a glass membrane has a very high electrical resistance, some form of vacuum tube voltmeter is used for this measurement.

5. *Conversion of chemical to electrical energy.* A type of application of galvanic cells that is completely different from those mentioned previously is their employment as sources for usable amounts of electrical energy. A galvanic cell whose electrical output is derived from the reaction of substances used in the construction of the cell is often called a primary cell. Probably the most familiar primary cell is the so-called dry cell which consists essentially of a zinc anode in contact with a moist paste of $ZnCl_2$ and NH_4Cl in which a carbon cathode surrounded by MnO_2 is supported.

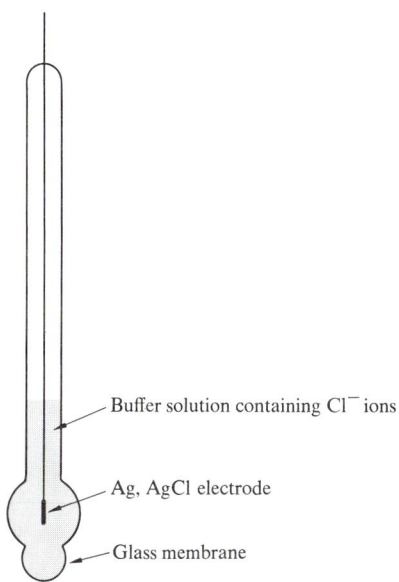

Figure 16–13. Schematic diagram of a glass electrode.

In contrast to primary cells, secondary cells require an external source of electrical energy to charge them. A number of these cells connected together is usually called a storage battery. During the charging period the secondary cell operates as an electrolytic cell and chemical changes are produced at the electrodes. The products of this reaction then react when the cell is used as a galvanic cell and electrical energy is produced until these products are exhausted. The most common type of storage battery is the lead-sulfuric acid battery in which the cell reaction is

$$Pb + PbO_2 + 2H_2SO_4 \underset{\text{discharge}}{\overset{\text{charge}}{\rightleftarrows}} 2PbSO_4 + 2H_2O$$

During the charging period spongy lead is formed on one electrode and lead dioxide is formed on the other, and the sulfuric acid concentration increases. This reaction is reversed when the cell acts as a source of electrical energy.

A type of primary cell that has received much attention recently is a cell constructed in such a way that reactants can be continuously supplied and products can be continuously removed at appropriate rates. If one of the reactants is oxygen, such a cell is called a fuel cell. For example, a fuel cell used in some practical applications is based on the oxidation of methanol by atmospheric oxygen at an electrode that is suitable for this reaction and H_2–O_2 fuel cells are used as sources of power in space ships.

One of the incentives in research on fuel cells is the fact that in principle the direct conversion of chemical energy into electrical energy is more efficient than the alternate route of burning the fuel under a steam boiler and converting the thermal energy to the mechanical energy of a dynamo, which then converts the mechanical energy into electrical energy. The heat of combustion of methanol is about 171 kcal/mole. If this heat is used to form superheated steam at 150°C which drives a dynamo and is discharged at 20°C, the maximum obtainable efficiency according to the second law of thermodynamics (p. 63), is $(423 - 293)/423$, or 31%, even with all losses ignored. Thus a maximum of 53 kcal is obtainable from a mole of methanol for conversion into electrical energy, while the rest of the heat of combustion must be wasted.

If the combustion reaction is carried out reversibly and isothermally in a galvanic cell, the entire free energy change of about 169 kcal/mole is available for conversion into electrical energy. This figure is unrealistic, of course, for as soon as appreciable amounts of power are drawn from the cell its emf drops, and the energy actually available is less than ΔG. Nevertheless, the over-all efficiency of the direct conversion of chemical energy into electrical energy is still substantially greater than that for the indirect thermal process.

A major economic advantage would result if mixtures of hydrocarbons, such as those present in gasoline, could be used as the fuel in a fuel cell. Unfortunately, the development of an electrode-electrolyte combination which would make this type of cell possible has presented serious problems that have not been solved up to the present time.

Electrolysis

It was mentioned earlier that a cell with reversible electrodes acts as an electrolytic cell even when the applied potential difference is only slightly larger than the cell emf. In this case, the electrode potential at which an ion is discharged, the *decomposition potential*, is essentially the same as that at which the ion is formed. Thus the decomposition potential at a reversible electrode is related to the concentration of the ion by the Nernst equation for this electrode.

When a direct current is passed through a cell, a number of time-dependent and irreversible processes occur.[8] We consider here only the two most important of these processes.

[8]For this reason, a small alternating current is used for the measurement of electrolytic conductance (p. 599) in order to minimize the effect of these irreversible processes.

1. *Concentration polarization.* The passage of a direct current through an electrolytic cell results in a lowering of the concentration of an ion in the neighborhood of the electrode at which this ion is discharged because ions cannot diffuse through the solution fast enough to replace those removed by discharge. This irreversible effect, which is known as *concentration polarization*, increases the effective resistance of the cell and thus reduces the current resulting from a given applied potential difference. In addition, the difference in the ionic concentrations at the two electrodes produces a back emf opposing the applied potential difference because the cell operates to some extent as a concentration cell. Concentration polarization can be minimized by rapid stirring of the electrolyte.

2. *Overvoltage.* With some types of electrodes, particularly those at which a gas is evolved, passage of a direct current results in an irreversible change at the electrode surface. In this case, the decomposition potential is significantly higher than the reversible electrode potential. Consequently the minimum potential difference which must be applied to a cell containing an irreversible (or polarized) electrode for electrolysis to occur is larger than that calculated by use of the Nernst equation. The difference between these values is known as *overvoltage*.

TABLE 16-7   HYDROGEN (BUBBLE) OVERVOLTAGE ON VARIOUS METALS*

Platinized platinum	0.00 v
Smooth platinum	0.09
Silver	0.15
Nickel	0.21
Copper	0.23
Lead	0.64
Zinc	0.70
Mercury	0.78

*Data quoted in MacInnes, *loc. cit.*

For the discharge of the ions of most metals the overvoltage is essentially zero. For the evolution of gases the magnitude of the overvoltage depends not only on the nature of the gas, but also on the current density at the electrode and on the nature of the electrode surface, as may be seen from the values for hydrogen listed in Table 16-7. This behavior, which is not understood completely, may be due to the effect of the electrode surface on (a) the rate of combination of hydrogen atoms, formed by discharge of H^+ ions, to form H_2 molecules; or

(b) the rate of desorption of H_2 molecules from the surface to form bubbles of H_2 gas. In either case, a relatively slow rate-determining step must be involved.

The existence of overvoltage makes it possible to plate some metals electrolytically even though the evolution of hydrogen would be expected from a consideration of reversible electrode potentials. For example, if a solution of a zinc salt is electrolyzed in a cell with inert electrodes, metallic zinc is deposited on the cathode although zinc is above hydrogen in the electromotive series; that is, the standard potential of a zinc electrode is negative with respect to the SHE. Hydrogen will not be evolved until the concentration of Zn^{2+} ions has reached such a low value that the decomposition potential of the Zn^{2+} ions has a larger magnitude than the sum of the reversible decomposition potential of the H^+ ions and the hydrogen overvoltage.

In order to see what the maximum concentration of Zn^{2+} ions must be before hydrogen is evolved from a neutral solution of a zinc salt, let us carry out a rough calculation using concentration instead of activity and a value of 0.70 v for the hydrogen overvoltage on a zinc electrode. The reversible potential for the discharge of H^+ ions is given by $\varepsilon = 0.059 \log c_{H^+} = -0.059$ pH. Thus in a neutral solution the zinc cathode must reach a potential of $-0.70 - 0.059 \times 7 \sim -1.11$ v before hydrogen is evolved. The value of $c_{Zn^{2+}}$ that corresponds to this potential is given by

$$-1.11 \sim -0.76 + \frac{0.059}{2} \log c_{Zn^{2+}}$$

$$\log c_{Zn^{2+}} \sim -12$$

$$c_{Zn^{2+}} \sim 10^{-12} \text{ moles/liter}$$

Thus we find that practically all of the Zn^{2+} ions must be discharged before hydrogen gas is evolved.

Supplementary Material

The Debye-Hückel Theory

The derivation of the Debye-Hückel limiting law (Equation 16–11) starts with a statistical calculation of the average distribution of ions around any particular ion in a solution. Then the electrical potential associated with this charge distribution, which we called the ionic atmosphere, is found. The potential energy that results from the interaction of this ion with its atmosphere is identified with the deviation of the ionic chemical potential from its ideal value due to the electrical charge on this ion. This deviation from ideality is then expressed in terms of the activity coefficient.

Electrochemistry

The average distribution of ions around any particular ion, designated as α, is assumed to be given by the Boltzmann distribution law, Equation 5-25. Ion α produces an electric field in its neighborhood. Any other ion with charge q in this neighborhood has a potential energy $q\psi$, where ψ is the potential of the field due to ion α at the position of the other ion.[9] If q is given in terms of the electronic charge and ψ is measured in volts, the potential energy is given in electron-volts.

With N_i representing the number of ions per unit volume of solution which are of kind i and carry a charge q_i, and $N_{i\psi}$ representing the number of these ions per unit volume located in an electric field with potential ψ, the Boltzmann equation takes the following form:

$$N_{i\psi} = N_i e^{-q_i\psi/kT}$$

If, for example, ion α is a positive ion, in its neighborhood the distributions of ions with charges $+q$ and $-q$, respectively, are given by

$$N_{+\psi} = N_+ e^{-q\psi/kT} \quad \text{and} \quad N_{-\psi} = N_- e^{-(-q\psi/kT)}$$

It can be seen that in the neighborhood of α the average concentration of the positive ions is lower than their over-all concentration, and likewise the average concentration of the negative ions is higher than their over-all concentration.

The charge density, ρ_c, associated with a set of charges is simply the total charge of the set per unit volume. The total charge density due to all of the kinds of ions that surround ion α is given as a function of ψ by

$$\rho_c = \sum_i q_i N_{i\psi}$$

$$= \sum_i q_i N_i e^{-q_i\psi/kT} \tag{16-23}$$

Debye and Hückel assumed that this charge density could be treated as though it resulted from a continuous distribution of charges; that is, they ignored the fact that the charge around ion α is carried by individual ions with discrete charges. To reduce mathematical complications in the manipulation of the equation above, they limited the application of their theory to very dilute solutions, for which $q_i\psi$ is much smaller than kT. In this case, it is a good approximation to write

$$e^{-q_i\psi/kT} \sim 1 - \frac{q_i\psi}{kT}$$

---

[9]The potential, ψ, at a point in an electric field may be defined as the potential energy of a unit positive charge located at that point. In our case, it is the work that must be done on an ion that carries a unit positive charge in order to bring it from an infinite distance, that is, zero potential, to the point of interest in the neighborhood of ion α.

Then Equation 16–23 for the charge density becomes

$$\rho_c = \sum q_i N_i \left(1 - \frac{q_i \psi}{kT}\right)$$

$$= \sum q_i N_i - \frac{\psi}{kT} \sum N_i q_i^2 \qquad (16\text{–}24)$$

The first term in this equation is the sum over all ionic charges in the unit volume of solution, and it vanishes because the solution is electrically neutral.

Now we must make use of one of the important differential equations in electrostatics, known as Poisson's equation, which gives the relation between ρ_c and ψ, expressed as a function of the coordinates of the charge distribution. This equation is

$$\nabla^2 \psi = -\frac{4\pi}{D} \rho_c$$

in which ∇^2 is the Laplacian operator in terms of polar coordinates (compare Equation 6–28), and D is the dielectric constant of the medium in which the charges are immersed. In the case of an ion, the potential is spherically symmetrical and it therefore depends only on the radial distance, r, from the center of ion α. Thus in this case Poisson's equation reduces to

$$\frac{1}{r^2} \frac{d}{dr}\left(r^2 \frac{d\psi}{dr}\right) = -\frac{4\pi}{D} \rho_c$$

Substituting the expression for ρ_c from Equation 16–24, we have

$$\frac{1}{r^2} \frac{d}{dr}\left(r^2 \frac{d\psi}{dr}\right) = \frac{4\pi}{D} \frac{\psi}{kT} \sum N_i q_i^2 \qquad (16\text{–}25)$$

We can write this equation in a simpler form by combining all of the constants in a single quantity κ, defined by

$$\kappa = \left(\frac{4\pi}{DkT} \sum N_i q_i^2\right)^{\frac{1}{2}} \qquad (16\text{–}26)$$

(κ has the dimensions of r^{-1}). At the same time we replace the left-hand side of Equation 16–25 by a mathematically identical expression. With these changes, Equation 16–25 becomes

$$\frac{d^2\psi}{dr^2} + \frac{2}{r} \frac{d\psi}{dr} = \kappa^2 \psi$$

When both sides of this equation are multiplied by r, it can be put into the following form, which we have seen a number of times previously, for example, Equation 6–30:

$$\frac{d^2(r\psi)}{dr^2} = \kappa^2(r\psi)$$

(This step may be verified by carrying out the indicated differentiation.)

Electrochemistry

The solution of this differential equation can be expressed in terms of either trigonometric functions or exponentials. For our purpose here the exponential form is more appropriate. Thus we write the general solution as follows:

$$r\psi = Ae^{-\kappa r} + Be^{\kappa r}$$

As r becomes very large, ψ must approach zero; consequently, $B = 0$ and the equation above reduces to

$$\psi = \frac{A}{r} e^{-\kappa r} \tag{16-27}$$

In order to evaluate the constant of integration A, we take as our model of ion α a sphere of radius a_α with all of the charge q_α concentrated at the center of the sphere. No other ion can penetrate within this sphere. For electrical neutrality, the charge carried by ion α must be equal and opposite in sign to the net charge carried by all other ions in its ionic atmosphere. We can obtain this net charge by integrating the charge density in a spherical shell of volume $4\pi r^2\, dr$ from $r = a_\alpha$ to $r = \infty$. Thus

$$q_\alpha = -\int_{a_\alpha}^{\infty} \rho_c\, 4\pi r^2\, dr \tag{16-28}$$

To carry out this integration we must have ρ_c as a function of r. We can find this function by substituting Equation 16–27 into Equation 16–24 which yields

$$\rho_c = -\frac{A}{r} e^{-\kappa r} \frac{\sum N_i q_i^2}{kT} \tag{16-29}$$

From the definition of κ, $\sum N_i q_i^2 / kT = D\kappa^2 / 4\pi$. On introducing this relation into Equation 16–29 we obtain

$$\rho_c = -\frac{D}{4\pi} \kappa^2 \frac{A}{r} e^{-\kappa r} \tag{16-30}$$

Then Equation 16–28 becomes

$$q_\alpha = D\kappa^2 A \int_{a_\alpha}^{\infty} r e^{-\kappa r}\, dr \tag{16-31}$$

On carrying out the integration by parts or by consulting a table of integrals we obtain

$$\int_{a_\alpha}^{\infty} r e^{-\kappa r}\, dr = \left| -\frac{r e^{-\kappa r}}{\kappa} - \frac{e^{-\kappa r}}{\kappa^2} \right|_{r=a_\alpha}^{r=\infty}$$

$$= \frac{e^{-\kappa a_\alpha}}{\kappa^2} (a_\alpha \kappa + 1)$$

When this value of the integral is introduced into Equation 16–31, and the resulting equation is solved for A, we obtain

$$A = \frac{q_\alpha e^{\kappa a_\alpha}}{D(a_\alpha \kappa + 1)}$$

Now we substitute this expression for A in Equation 16–27 to obtain the electrical potential as a function of r in the following form:

$$\psi = \frac{q_\alpha e^{\kappa a_\alpha} e^{-\kappa r}}{Dr(a_\alpha \kappa + 1)}$$

The value of the potential at the "surface" of the ion ($r = a_\alpha$) is given by

$$\psi_{r=a_\alpha} = \frac{q_\alpha}{Da_\alpha(a_\alpha \kappa + 1)}$$

This value of the potential can be considered to be made up of two contributions, one from ion α itself, and the other, ψ_κ, from the charge distribution around ion α, that is, the ionic atmosphere. To obtain the contribution due to the ionic atmosphere, which depends on κ and therefore on the ionic concentration, we subtract the contribution of ion α, which is equal to q_α/Da_α. Thus

$$\psi_\kappa = \frac{q_\alpha}{Da_\alpha(a_\alpha \kappa + 1)} - \frac{q_\alpha}{Da_\alpha}$$

$$= -\frac{q_\alpha \kappa}{D(a_\alpha \kappa + 1)} \qquad (16\text{–}32)$$

The theoretical development up to this point has been based on electrostatic theory only. Now we come to the important step of relating the equation above to the chemical potential of ion α. In order to accomplish this, let us imagine that the charge on ion α can be varied continuously from 0 to its actual value, q_α. Since our electrolytic solution is assumed to be very dilute, when $q_\alpha = 0$ the ion α can be considered to be a neutral molecule in a solution so dilute that it is ideal. Consequently its chemical potential can be expressed by

$$\frac{\mu_\alpha(\text{ideal})}{\mathfrak{N}} = \frac{\mu_\alpha^\circ}{\mathfrak{N}} + kT \ln m_\alpha$$

in which $\mathfrak{N}$ is Avogadro's number and m_α is the molality of species α. Notice that this equation refers to the chemical potential per particle.

When the ion is fully charged, the solution is not ideal and an additional term due to the electrostatic interaction must be added to the chemical potential. Thus we have

$$\mu_\alpha = \mu_\alpha(\text{ideal}) + \mu_\alpha(\text{electrical})$$

Electrochemistry

As the charge on ion α is increased continuously from 0 to q_α at constant temperature and pressure, electrical work must be done against the field of the charges in the ionic atmosphere. This work is given by

$$w = \int_0^{q_\alpha} \psi_\kappa \, dq_\alpha$$

On substitution of the expression for ψ_κ from Equation 16-32, the result is

$$w = -\int_0^{q_\alpha} \frac{\kappa}{D(a_\alpha \kappa + 1)} q_\alpha \, dq_\alpha$$

$$= -\frac{q_\alpha^2 \kappa}{2D(a_\alpha \kappa + 1)}$$

The work that is done in this process lowers the potential energy of the ion. We now identify this change in potential energy with the change in the free energy of the ion, and thus in its chemical potential. Consequently,

$$\frac{\mu_\alpha}{\mathfrak{N}} \text{(electrical)} = -\frac{q_\alpha^2 \kappa}{2D(a_\alpha \kappa + 1)}$$

Expressing the deviation from ideality indicated in the equation above in terms of the activity coefficient, we have

$$\frac{\mu_\alpha}{\mathfrak{N}} \text{(electrical)} = kT \ln \gamma_\alpha$$

It follows that

$$\ln \gamma_\alpha = -\frac{q_\alpha^2 \kappa}{2kTD(a_\alpha \kappa + 1)} \quad (16\text{-}33)$$

In order to obtain the Debye-Hückel limiting law (Equation 16-11) which is applicable only to very dilute solutions, we note that the quantity represented by κ depends on the ionic concentrations, N_i. For concentrations so low[10] that $a_\alpha \kappa$ is much less than 1, the equation above reduces to

$$\ln \gamma_\alpha = -\frac{q_\alpha^2 \kappa}{2kTD} \quad (16\text{-}34)$$

For application of this equation to a particular electrolyte, we now need to take into consideration the fact γ_α is the activity coefficient of a single kind of ion. Since we have seen previously that such a quantity has no experimental significance, we must introduce the mean activity coefficient, defined by $\gamma_\pm^\nu = \gamma_+^{\nu_+} \gamma_-^{\nu_-}$. Thus

$$\nu \ln \gamma_\pm = \nu_+ \ln \gamma_+ + \nu_- \ln \gamma_-$$

---

[10] For uni-univalent electrolytes, the molality must be below about 0.01. For other electrolytes, this limiting molality is even lower.

Because Equation 16–34 does not depend on the sign of the charge on ion α, we can apply this equation to both the positive and the negative ions and add the resulting equations. Accordingly, we then have

$$\nu_+ \ln \gamma_+ + \nu_- \ln \gamma_- = -(\nu_+ q_+^2 + \nu_- q_-^2) \frac{\kappa}{2kTD}$$

For electrical neutrality, on dissociation of the electrolyte the amount of charge on the positive ions must equal the amount of charge on the negative ions; thus

$$\nu_+ q_+ = \nu_- q_-$$

When this equation is first multiplied by q_+ and then by q_-, and the two resulting equations are added, we obtain

$$\nu_+ q_+^2 + \nu_- q_-^2 = \nu_+ q_+ q_- + \nu_- q_+ q_-$$
$$= \nu q_+ q_-$$

It follows that

$$\ln \gamma_\pm = - \frac{q_+ q_- \kappa}{2kTD}$$

We can express the magnitudes of the charges on the ions in terms of the charge on the electron as a unit by $q_+ = z_+ \epsilon$ and $q_- = z_- \epsilon$ where z_+ and z_- are integers. Then

$$\ln \gamma_\pm = - \frac{z_+ z_- \epsilon^2 \kappa}{2kTD} \tag{16-35}$$

Our next step will be the expression of κ, defined in Equation 16–26, in terms of more useful quantities. It contains $\Sigma N_i q_i^2$ which is equal to $\Sigma N_i z_i^2 \epsilon^2$ with N_i representing the number of ions of kind i per cm³ of solution. The molarity is given by $1000 N_i / \mathfrak{N}$ and in very dilute solutions such as those under consideration, for which the density of the solution is essentially the density of the solvent, ρ_s, the molality is $1000 N_i / \rho_s \mathfrak{N}$. Consequently,

$$\sum N_i z_i^2 = \frac{\rho_s \mathfrak{N}}{1000} \sum m_i z_i^2$$

and κ can be expressed in the following way:

$$\kappa = \left(\frac{4\pi \epsilon^2}{kTD} \frac{\rho_s \mathfrak{N}}{1000} \sum m_i z_i^2 \right)^{\frac{1}{2}}$$

In terms of the ionic strength (Eq. 16–12) this equation becomes

$$\kappa = \left(\frac{8\pi \epsilon^2}{kTD} \frac{\rho_s \mathfrak{N}}{1000} \right)^{\frac{1}{2}} I^{\frac{1}{2}} \tag{16-36}$$

Electrochemistry

On substituting this expression for κ into Equation 16–35, we obtain

$$\ln \gamma_\pm = -\frac{z_+ z_- \epsilon^2}{2kTD}\left(\frac{8\pi\epsilon^2}{kTD}\frac{\rho_s \mathfrak{N}}{1000}\right)^{\frac{1}{2}} I^{\frac{1}{2}}$$

When the values of the constants are introduced into this equation and a change from natural to common logarithms is made, the result is Equation 16–11.

In order to extend the range of ionic strengths for which this equation is applicable, we must return to Equation 16–33 because in the more general case $a_\alpha\kappa$ is not small relative to 1. For the calculation of the mean activity coefficient of an electrolyte when the factor $(a_\alpha\kappa + 1)^{-1}$ is retained in the equation, it is necessary to broaden the interpretation of the quantity a. This quantity was defined on page 655 as the effective size of an ion, but since the mean activity coefficient involves two kinds of ions, a now represents an average ionic radius which cannot as yet be obtained theoretically. Consequently it is either treated as an adjustable parameter for each electrolyte, or assigned a constant value for all ions of 3.0×10^{-8} cm, which is sufficiently accurate for almost all purposes.

With the average ionic radius represented by a, the more general form of Equation 16–11 for the mean activity coefficient is

$$\log \gamma_\pm = -1.825 \times 10^6 \left(\frac{\rho_s}{D^3 T^3}\right)^{\frac{1}{2}} z_+ z_- \frac{I^{\frac{1}{2}}}{a\kappa + 1}$$

When κ is replaced by its value from Equation 16–36, the result is

$$\log \gamma_\pm = -1.825 \times 10^6 \left(\frac{\rho_s}{D^3 T^3}\right)^{\frac{1}{2}} z_+ z_- \frac{I^{\frac{1}{2}}}{1 + a\left(\dfrac{8\pi\epsilon^2}{kTD}\dfrac{\rho_s \mathfrak{N}}{1000}\right)^{\frac{1}{2}} I^{\frac{1}{2}}}$$

For aqueous solutions at 25°C this equation reduces to

$$\log \gamma_\pm = -0.51 z_+ z_- \frac{I^{\frac{1}{2}}}{1 + 3.3 \times 10^7 a I^{\frac{1}{2}}}$$

With a expressed in angstrom units, we have

$$\log \gamma_\pm = -0.51 z_+ z_- \frac{I^{\frac{1}{2}}}{1 + (a/3.0)I^{\frac{1}{2}}}$$

Finally, with $a = 3\text{Å}$, this equation reduces to

$$\log \gamma_\pm = -0.51 z_+ z_- \frac{I^{\frac{1}{2}}}{1 + I^{\frac{1}{2}}}$$

In order to get some idea of the physical significance of the parameter κ, we return to Equation 16–30 for the charge density of the ionic atmosphere:

$$\rho_c = -\frac{DA}{4\pi r}\kappa^2 e^{-\kappa r}$$

The radial distribution function for the charge density is obtained by multiplying ρ_c by the volume of a spherical shell of radius r, $4\pi r^2\, dr$. Thus we find

$$4\pi r^2 \rho_c\, dr = D\kappa^2 A r e^{-\kappa r}\, dr$$

Now let us obtain the maximum of the radial distribution function in the usual way by differentiating it and setting the derivative equal to zero, as follows:

$$\frac{d}{dr} 4\pi r^2 \rho_c = D\kappa^2 A e^{-\kappa r}(1 - r\kappa) = 0$$

It follows that

$$\kappa = \frac{1}{r_{\max}}$$

and we see that κ is the reciprocal of the distance from the center of an ion to a point at which the charge density of the ionic atmosphere is a maximum. Since κ increases with increasing ionic strength, the mean thickness of the ionic atmosphere decreases as the ionic strength increases.

Supplementary References

R. G. Bates, *Determination of pH — Theory and Practice*, Wiley, New York, 1964.

R. M. Fuoss and F. Accascina, *Electrolytic Conductance*, Interscience, New York, 1959.

H. S. Harned and B. B. Owen, *Physical Chemistry of Electrolytic Solutions*, Reinhold, New York, 1958.

G. Kortüm, Treatise on Electrochemistry, Elsevier, New York, 1965.

W. M. Latimer, *Oxidation States of the Elements and Their Potentials in Aqueous Solution*, 2nd edition, Prentice-Hall, Englewood Cliffs, New Jersey, 1952.

R. M. Noyes, "Conventions Defining Thermodynamic Properties of Aqueous Ions and Other Chemical Species." *J. Chem. Ed.* **40**, 2 (1963).

Problems

1. When a varying electric current was passed for 75.0 min through a copper coulometer containing $CuCl_2$, 12.6 mg of copper was deposited. Calculate (a) the number of faradays passed and (b) the average current.

Ans. (a) 3.96×10^{-4}; (b) 8.50 milliamperes

2. The resistance of a conductance cell containing 0.100 molar KCl was found to be 60.50 ohms at 25°C. When this cell was filled with a solution containing 2.380 g of $MgCl_2$ per liter, its resistance was 151.0 ohms. Calculate (a) the cell constant; (b) the conductivity of the $MgCl_2$ solution; (c) the equivalent conductivity of the solution.

Ans. (a) 0.780 cm^{-1}; (b) 5.16 × 10^{-3} ohm^{-1} cm^{-1}; (c) 103 cm^2/ohm equiv

3. Obtain the value of Λ_o at 25°C for $CaCl_2$ from the following data taken at this temperature:

c (equiv/liter) × 10^4	5.00	10.00	50.00	100.00
Λ (cm^2/ohm equiv)	131.93	130.36	124.25	120.36

4. The conductivity of a saturated solution of $BaSO_4$ in water was found to be 4.23 × 10^{-6} ohm^{-1} cm^{-1} at 25°C. The conductivity of the water used as solvent was 1.25 × 10^{-6} ohm^{-1} cm^{-1}. The value of Λ_o for $BaSO_4$ at 25°C is 143.4 cm^2/ohm equiv. Calculate the solubility in moles/liter of $BaSO_4$ in water at 25°C.
Ans. 1.04 × 10^{-5} mole/liter

5. Calculate Λ_o for NH_4OH from the following data: Λ_o (cm^2/ohm equiv) = 149.7 for NH_4Cl, 126.4 for NaCl, and 248.0 for NaOH.
Ans. 271.3 cm^2/ohm equiv

6. Draw a rough graph that describes qualitatively the variation in the conductance of a solution of a strong acid as it is titrated with (a) a strong base; (b) a weak base.

7. After a 2 molal solution of $FeCl_3$ was electrolyzed between Pt electrodes, the cathode portion of the electrolyte was found to be 1.91 molal in $FeCl_3$ and 0.50 molal in $FeCl_2$. Calculate the transference numbers of the Fe^{3+} and Cl^- ions.
Ans. $\tau_+ = 0.46$

8. A solution of $AgNO_3$ containing 17.480 g of $AgNO_3$ per 100.0 g of water was electrolyzed between silver electrodes in a Hittorf transference cell. After the electrolysis the anode portion of the electrolyte was found to weigh 75.951 g and to contain 12.449 g of $AgNO_3$. The cathode portion weighed 90.856 g and contained 12.351 g of $AgNO_3$. The initial and final weights of the anode were 20.455 g and 18.835 g, respectively. Calculate the transference numbers of the Ag^+ and NO_3^- ions from the data for each portion of the electrolyte.

9. The transference numbers of the ions in a 1.00 molal solution of $AgNO_3$ were determined by the moving boundary method, using a solution of 0.60 molal $Cd(NO_3)_2$ as the following solution. When a current of 15.0 milliamperes was used, the boundary swept through a volume of 0.145 ml in 33.0 min. Calculate the transference numbers of the Ag^+ and NO_3^- ions. Ans. $\tau_+ = 0.471$

10. In an application of the moving boundary method, it is observed that a boundary formed by H^+ ions and Cd^{2+} ions moves a distance of 2.05 cm in 12.52 min. The distance between the electrodes is 9.68 cm and the potential difference is 8.01 v. Calculate the mobility and the ionic conductivity of the H^+ ions.

11. From the data in Table 16–2 calculate (a) the transference number of the Br^- ions in a dilute LiBr solution; (b) the equivalent conductivity of the Li^+ ions in this solution.

12. Obtain an expression for the mean ionic activity of a 3–2 electrolyte ($Fe_2(SO_4)_3$ for example) in terms of the molality of the electrolyte and the mean ionic activity coefficient.   **Ans.** $2.55 m \gamma_\pm$

13. The mean ionic activity coefficient of H_2SO_4 at a molality of 0.002 is 0.757. Calculate the corresponding value of the activity of the H_2SO_4 in this solution.

14. Calculate the ionic strength of 0.10 molal solutions of Na_2SO_4; $CuCl_2$; $La_2(SO_4)_3$.

15. The solubility of $Pb(IO_3)_2$ in water at 25°C is 4.15×10^{-5} moles/liter. Using the Debye-Hückel limiting law, calculate (a) its solubility product; (b) $\Delta G°$ for the process $Pb(IO_3)_2(s) \rightarrow Pb^{2+} + 2IO_3^-$ (sat. soln.).
Ans. (a) 2.63×10^{-13}; (b) 17.1 kcal

16. The solubility of $AgBrO_3$ at 25°C in solutions containing $LiClO_4$ is given in the following table:

Molarity of $LiClO_4$	0	0.025	0.050	0.075
Molarity of $AgBrO_3 \times 10^3$	8.09	8.73	9.09	9.39

(a) Using the graphical method, obtain the thermodynamic solubility product of $AgBrO_3$ at 25°C. (b) Calculate the mean ionic activity coefficient of $AgBrO_3$ in a saturated solution of this salt.

17. The solubility product of $BaSO_4$ at 25°C is 1.06×10^{-10}. Calculate the solubility of $BaSO_4$ in (a) 0.05 molal NaCl; (b) 0.05 molal Na_2SO_4.
Ans. (a) 1.34×10^{-5} mole/liter; (b) 2.12×10^{-9} mole/liter

18. The dissociation constant of propionic acid at 25°C is 1.33×10^{-5}. Calculate the pH of (a) a 0.1 molar solution of propionic acid; (b) a solution 0.1 molar in propionic acid and 0.1 molar in sodium propionate; (c) a 0.1 molar solution of sodium propionate.

19. The dissociation constant of ammonium hydroxide is 1.8×10^{-5}. Calculate the pH of a 0.2 molar solution of NH_4Cl.

Electrochemistry

20. The pH of a saturated solution of phenol in water is 5.49 at 25°C. At this temperature the dissociation constant of phenol is 1.3×10^{-10}. Calculate the solubility of phenol in water at 25°C.

21. For the following galvanic cell:

$$\text{Zn}/\text{Zn}^{2+}(a = 0.01) \vdots \text{Fe}^{2+}(a = 0.001), \text{Fe}^{3+}(a = 0.1)/\text{Pt}$$

(a) Write the equation for the cell reaction; (b) calculate the cell potential at 25°C; (c) calculate the free energy change for the cell reaction.
    Ans. (a) $\text{Zn} + 2\text{Fe}^{3+} \rightarrow \text{Zn}^{2+} + 2\text{Fe}^{2+}$; (b) 1.711 v; (c) -68.4 kcal

22. From the standard potentials in Table 16–6 for the Fe/Fe^{2+} and the $\text{Pt}/\text{Fe}^{2+},\text{Fe}^{3+}$ electrodes, calculate the standard potential of the Fe/Fe^{3+} electrode.
    Ans. -0.036 v

23. From the data in Table 16–6, calculate the reduction potential at 25°C for each of the following half-cells: (a) $\text{Cl}^-(a = 0.001)/\text{AgCl(s)}/\text{Ag}$; (b) $\text{Sn}^{2+}(a = 0.01)/\text{Sn}$; (c) $\text{Cu}^{2+}(a = 0.005)/\text{Cu}$.

24. (a) For the combination of half-cell (a) with half-cell (b) in the preceding problem, calculate the cell potential and state which is the electrode at which oxidation occurs spontaneously. (b) Do the same for the combination of half-cell (a) with half-cell (c).

25. Calculate the potential at 25°C for the following cell:

$$\text{Pt}/\text{H}_2(0.30 \text{ atm})/\text{HCl}(m = 0.010)/\text{Cl}_2(0.60 \text{ atm})/\text{Pt}$$

(For HCl at a molality of 0.010, $\gamma_\pm = 0.904$).   **Ans.** 1.580 v

26. For the following cell:

$$\text{Ag}/\text{AgCl(s)}/\text{HCl}(0.1m)/\text{Hg}_2\text{Cl}_2(s)/\text{Hg(l)}$$

the potential at 25°C is 0.046 v.
    (a) Write the equation for the cell reaction. (b) What is the cell potential if the molality of the HCl solution is 0.5? (c) What is the standard potential of the calomel electrode?

27. For the following cell:

$$\text{Pb}/\text{PbCl}_2(s)/\text{PbCl}_2(\text{soln.})/\text{AgCl(s)}/\text{Ag}$$

the potential at 25°C is 0.490 volts and the temperature coefficient of the cell potential is -1.86×10^{-4} volts/degree.
    (a) Write the equation for the cell reaction. (b) Calculate ΔH and ΔS for this reaction at 25°C.   **Ans.** $\Delta H = -25.2$ kcal, $\Delta S = -8.58$ cal/deg

28. For the following cell:

$$Hg(l)/Hg_2Cl_2(s)/HCl(m = 1.0)/Cl_2(1\ atm)/Pt$$

the cell potential at 25°C is 1.090 volts and its temperature coefficient is -9.45×10^{-4} volts/degree.

(a) Write the equation for the cell reaction. (b) Calculate ΔH and ΔS for this reaction at 25°C. (c) Calculate the standard free energy of formation of $Hg_2Cl_2(s)$.

29. Write the equation for the cell reaction and calculate the potential at 25°C of the following cell:

$$Pt/Cl_2(P = 1.2\ atm)/NaCl\ solution/Cl_2(P = 0.4\ atm)/Pt$$

Is the cell reaction spontaneous?   Ans. $\mathcal{E} = -0.014$ v

30. Calculate the cell potential at 25°C of the following cell:

$$Pt/H_2(1\ atm)/HCl(m = 0.005)/Hg_2Cl_2(s)/Hg(l)/Hg_2Cl_2(s)/HCl(m = 0.10)/H_2(1\ atm)/Pt$$

Use activity coefficients obtained from Table 16–4.

31. Calculate the potential at 25°C of the following cell:

$$Zn/ZnCl_2(m = 0.010,\ \gamma_\pm = 0.708)/AgCl/Ag/AgCl/ZnCl_2(m = 0.10,\ \gamma_\pm = 0.502)/Zn$$

Ans. 0.0755 v

32. The potential of the following concentration cell with transference is 0.0430 v at 25°C:

$$Ag/AgCl/NaCl(m = 0.10) \quad NaCl(m = 0.01)/AgCl/Ag$$

Using data from Table 16–4, calculate the value of the transference number of the chloride ion.

33. From the data in Table 16–6, calculate the equilibrium constant at 25°C for the reaction represented by $FeCl_2 + Cd = CdCl_2 + Fe$.   Ans. 0.083

34. Finely divided tin is added to a solution in which the molality of Pb^{2+} is 0.1. What are the molalities of Pb^{2+} and Sn^{2+} ions when equilibrium is established? Assume that molality is equal to activity for this problem.

35. From the data in Table 16–6, calculate the solubility product of $PbBr_2$.   Ans. 6.3×10^{-6}

36. The standard electrode potential for the half-cell $I^-/AgI(s)/Ag$ is -0.152 v. Calculate the solubility of AgI.

37. A hydrogen electrode in an acid solution had a potential difference of 0.529 v relative to a normal calomel electrode. Calculate the hydrogen ion concentration in the solution.   Ans. 6.0×10^{-5} mole/liter

38. 100 ml of 0.04 molar NaOH is titrated with 0.05 molar HCl, and a hydrogen electrode and a normal calomel electrode are immersed in the solution. Calculate the potential of the cell (a) when 1 ml less acid has been added than the volume needed for neutralization; (b) when neutralization is complete; (c) when 1 ml more of the acid is added.

39. The concentration of Ni^{2+} ions in a solution of a nickel salt is to be reduced to a value of 10^{-3} moles/liter without the evolution of hydrogen. What is the lowest value of the pH of this solution for which this process is possible?

Ans. 2.2

40. A solution that contains $Ag^+(a = 0.01)$, $Ni^{2+}(a = 0.05)$, and $Cu^{2+}(a = 0.10)$ is electrolyzed between inert electrodes. Describe the processes occurring at the cathode as the magnitude of the applied potential difference is slowly increased from zero.

Glossary of Symbols

a	van der Waals constant, activity, lattice parameter
a_0	radius of first Bohr orbit
A	Helmholtz free energy, surface area, pre-exponential factor in Arrhenius equation, absorbance, Madelung number, electron affinity, parameter in Debye-Hückel equation
b	van der Waals constant, absorption cell length, lattice parameter
B	rotational constant, second virial coefficient
c	speed of light, wave speed, concentration, lattice parameter
c_V	specific heat at constant volume
C	capacitance, number of components
C_P, C_V	molar heat capacities at constant pressure or volume
d	differential
d_{hkl}	interplanar distance
D	diffusion constant, dielectric constant, Debye unit
e	base of natural logarithms
E	molar energy, electric field strength
E_a	activation energy
ε	approximate energy, electrical potential difference, cell emf
f	force, fugacity
F	rotational term value, number of degrees of freedom
$\mathcal{F}$	Faraday constant
g	gas, acceleration of gravity, degree of degeneracy, nuclear splitting factor
G	Gibbs free energy, vibrational term value
h	height, Planck's constant, Miller index
h_v	specific heat of vaporization
H	molar enthalpy, Hamiltonian function, magnetic field strength
H_{op}	Hamiltonian operator
i	$\sqrt{-1}$
I	electric current, moment of inertia, beam intensity, ionization potential, ionic strength, nuclear spin quantum number

j	deviation of electrolytic solution from ideality
J	rotational quantum number
k	Boltzmann's constant, force constant, Miller index
k_r	rate constant
K	equilibrium constant
K_f	freezing-point constant
K_b	boiling-point constant
l	liquid, quantum number, Miller index
L	length
m	molecular mass, induced moment, molality, quantum number
M	molecular weight, intensity of magnetization, nuclear mass number
n	number of moles, number of molecules, quantum number
N	number of molecules per cm,^{3} normalization constant
$\mathfrak{N}$	Avogadro's number
p	momentum, quantum number
P	pressure, probability
P_m	molar polarization
$\mathcal{P}$	dielectric polarization, number of phases
q	amount of heat, molecular partition function, generalized coordinate
Q	reaction quotient, molar partition function, electric charge
r	interatomic distance, radius, electrical resistivity
R	gas constant, Rydberg constant, electrical resistance, rate of nucleation
R_m	molar refraction
s	solid, quantum number, electron diffraction parameter
S	molar entropy, overlap integral, quantum number, area
t	time, temperature in °C
T	temperature in °K
u	ionic mobility
U	potential energy
v	velocity, vapor
V	volume
V_m	molar volume
w	work, weight
W	thermodynamic probability
x	distance
X	mole fraction
Y	mole fraction in vapor phase
z	ionic charge
Z	compressibility factor, collision number, atomic number

Glossary of Symbols

α	temperature coefficient of expansion, degree of dissociation, Coulomb integral, molecular polarizability, Lagrange multiplier, spin function, parameter in Debye T^3 law
β	exchange integral, Lagrange multiplier, spin function
γ	heat capacity ratio, γ-ray, activity coefficient, surface tension, magnetogyric ratio
δ	variation, NMR chemical shift, length of pass through energy barrier
Δ	finite difference
ϵ	charge of the electron, molecular energy
ϵ_λ	molar absorptivity
η	viscosity coefficient
θ	angle, reduced temperature
Θ	angular factor of eigenfunction
κ	compressibility coefficient, thermal conductivity, Debye-Hückel parameter, electrical conductivity, transmission coefficient
λ	mean free path, wavelength, ionic equivalent conductivity
Λ	equivalent conductivity
μ	chemical potential, Joule-Thomson coefficient, reduced mass, dipole moment, nuclear magnetic moment
μ_0	nuclear magneton
ν	frequency, number of ions per molecule of electrolyte
ξ	degree of advancement of reaction
π	quantum number, osmotic pressure
Π	reduced pressure, product
ρ	density, charge density
σ	collision diameter, quantum number, symmetry number, reflection through mirror plane
Σ	sum
τ	transference number, $d\tau$, volume element
ϕ	reduced volume, angle, wave function, fluidity, quantum yield
Φ	angular factor of eigenfunction
χ	magnetic susceptibility
ψ	eigenfunction, electrical potential
Ψ	complete eigenfunction
ω	wavenumber, angular frequency
Ω	solid angle, ohm

Index

Absolute entropy, 112
  reaction rate theory, 424
  temperature, 7
Absorbance, 438
Absorption coefficient, 437
Absorptivity, molar, 439
Acid, 623
  Brönsted-Lowry definition, 623
  dissociation constant, 625
  Lewis definition, 624
Activated complex, 421
Activation energy of
  chemical reaction, 413
  viscous flow, 502
Activity, 96
Activity coefficient
  based on molality, 614
  based on mole fraction, 565
  from cell emf, 647
  from Debye-Hückel theory, 619
  mean ionic, 615
Adiabatic process, 36
Adsorption, 434, 582
  equation (Gibbs), 585
  isotherm, 586
    BET, 587
    Langmuir, 588
Alpha particle, 161
Amorphous solid, 508
Angstrom unit, 146
Angular momentum
  atomic, 173, 197
  molecular, 287
  nuclear spin, 315

Anharmonic oscillator, 297
Anode, 158, 635
Antibonding molecular orbital, 222
Antisymmetric wave function, 211
Arrhenius
  equation for reaction rate constant, 413
  theory of electrolytic dissociation, 597
Asymmetric top, 292
Atom, united, 230
Atomic
  electron configuration, 204
  energy levels, 176
  number, 162
  orbital, 197
  spectra, 167
Average values from a distribution function, 132
Avogadro's number from X-ray diffraction by crystals, 462
Azeotrope, 567

Balmer equation, 168
Band model for crystal energy levels, 477
Band spectrum, 309
Barometric equation, 149
Base, 623
  dissociation constant, 625
Battery, 649
Beer's law, 437
Benzene
  molecular orbital theory of, 255
  valence bond theory of, 257
BET adsorption isotherm, 587

Beta-ray spectroscopy, 166
Bimolecular
    collisions, 142, 415
    reactions, 414
Binary systems, 539
Binodal curve, 579
Bivariant system, 514
Black-body radiation, 169
Body-centered cubic lattice, 465
Bohr theory of the hydrogen atom, 173
Boiling-point elevation, 547
Boltzmann
    constant, 126
    distribution law, 150, 343
    equation for entropy, 346
Bond
    covalent, 216
    hydrogen, 259
    ionic, 481
    orbital, 240
    order, 252
Bonding molecular orbital, 222
Born-Haber cycle, 483
Born-Oppenheimer approximation, 217
Born theory of cohesive energy of ionic crystals, 485
Box, particle in a, 187
Boyle's law, 5
    point, 391
Bragg equation, 454
Branching-chain reaction, 411
Bravais lattice, 448
Brönsted-Lowry theory of acids, 623
Bubble-cap distillation column, 561
Bubble-pressure method for surface tension, 504
Butadiene, molecular orbital theory of, 248

Cailletet-Mathias law, 19
Calomel electrode, 636
Calorie, 10
Canonical ensemble (Gibbs), 377
Capillary-rise method for surface tension, 505
Carnot cycle, 61
Catalysis, 433
Cathode, 158
    ray, 158
Cell
    concentration, 642
    conductance, 598
    fuel, 650

Cell *continued*
    galvanic, 629
    primary, 649
Celsius temperature scale, 6
Center of symmetry, 270
Chain reaction, 408
Charles' law, 6
Chemical potential, 92
Chemical shift (NMR), 322
Chemisorption, 587
Clapeyron equation, 89
Clausius-Clapeyron equation, 89
Clausius-Mosotti equation, 274
Closed system, 2
Close-packing of spheres, 465
Cohesive energy of crystals, 481
Colligative properties, 546
Collision
    diameter, 141
    rate, bimolecular, 142, 415
    theory of reaction rates, 414
    with wall, 138
Combustion, heat of, 50
Compressibility
    coefficient, 487
    factor, 13
Compton effect, 172
Concentration
    cell, 642
    polarization, 651
Condensation, 526
Conductance, electrolytic, 598
Conduction band of crystals, 478
Conductivity
    electrolytic, 595
    equivalent, 598
    metallic, 478
    thermal, 146
Configuration integral, 382
Congruent melting point, 575
Conjugate
    acid and base, 623
    solutions, 563
Consecutive reactions, 405
Conservation of energy, 34
Consolute temperature, 570
Contact angle, 505
Continuity of states, 20
Cooling curves, 573
Cooperative transition, 532
Correlation diagram, 231
Corresponding states, 22

Coulomb integral, 221
Coulomb's law, 173
Coulometer, 596
Covalent
  bond, 216
  crystal, 479
Critical
  density, 19
  opalescence, 570
  pressure, 18
  solution temperature, 570
  volume, 18
Cross-section, collision, 141
Crystal
  classes, 450
  defects, 485
  heat capacity, 470
  lattice, 446
  plane, 450
  structure, 464
  symmetry, 448
  systems, 447
  types, 476
Cubic
  lattices, 464
  close-packing, 466
Curie's law, 314
Cyclic integral, 30

Dalton's law of partial pressures, 8
Daniell cell, 630
De Broglie wavelength, 177
Debye (unit), 277
Debye equation for
  heat capacity of crystals, 112, 473
  molar polarization, 277
Debye-Falkenhagen effect, 613
Debye-Hückel
  limiting law, 620
  theory, 619, 652
Debye-Hückel-Onsager equation, 612
Debye-Scherrer powder method, 455
Decomposition potential, 650
Degeneracy of wave functions, 192
Degrees of freedom (phase rule), 514
Delocalization energy, 252
Density
  critical, 18
  orthobaric, 18
Diamagnetism, 313
Diamond structure, 468, 479

Dielectric
  constant, 273
  polarization, 273
Differential
  exact, 30
  total, 29
Differential heat of solution, 52
Diffraction
  electron, 326
  neutron, 464
  X-ray, 452
Diffusion coefficient, 146
Dilution
  heat of, 51
  Ostwald's law of, 610
Dipole-dipole force, 386
Dipole moment
  electric
    induced, 272
    permanent, 274
  magnetic, 313
Dispersion force, 386
Dissociation
  degree of, 108
  electrolytic, 597
Distillation
  azeotropic, 567
  fractional, 559
  of immiscible liquids, 568
Distribution
  coefficient, 580
  function and calculation of averages, 132
    Boltzmann, 150, 344
    Maxwell, 136
    Maxwell-Boltzmann, 150
Dulong and Petit law, 470

Efficiency, heat engine, 62
Effusion, 140
Eigenfunction, 183
Eigenvalue, 185
Einstein
  photochemical law, 436
  relation between mass and energy, 29, 172
  theory of crystal heat capacity, 471
  theory of photoelectric effect, 171
  unit, 436
Electrical
  conductivity
    electrolytic, 595
    metallic, 478

Electrical *continued*
  potential, 653
Electrochemical cells, 629
Electrode
  calomel, 636
  gas, 633
  glass, 648
  hydrogen, 633
  redox, 633
Electrode potential, 634
Electrolysis, 595, 650
Electrolyte, ionic dissociation of, 597
Electromagnetic spectrum, 165
Electromotive force (emf) of galvanic cells, 631
Electron
  affinity, 483
  diffraction by gases, 326
Electronic
  charge, 161
  mass, 161
  spectra, 309
  spin, 202
Electrophoretic effect, 612
Embryo, nucleation, 527
Emf (electromotive force), 631
Energy
  activation of
    chemical reaction, 413
    viscous flow, 502
  band of crystals, 477
  conservation, 34
  delocalization, 252
  equipartition of, 126, 153, 379
  level
    atomic, 176
    electronic, 309
    rotational, 287
    translational, 190
    vibrational, 295
  zero-point, 193, 296
Ensemble, Gibbs, 374
Enthalpy, 37
  of activation, 431
  change of cell reaction, 642
  of formation, 49
  of reaction, 48
Entropy, 68
  absolute, 112
  of activation, 431
  change (cell reaction), 641
  of fusion, 81

Entropy *continued*
  of mixing, 75, 544
  and probability, 346
  statistical calculation of, 346, 360
  of vaporization, 81
Equation of state, 10
  thermodynamic, 78
Equilibrium, 3
  acid-base, 623
  chemical, 101
  criteria for, 81
  phase, 513
Equilibrium constant, 101
  from cell emf, 645
  from partition functions, 366
  temperature dependence of, 115
Equipartition of energy, 126, 153, 379
Equivalent conductivity, 600
Euler criterion for exactness of differentials, 30
Eutectic, 573
Exact differential, 30
Exclusion principle, Pauli, 204, 212
Exothermic reaction, 48
Expansion, temperature coefficient of, 6
Extensive property, 3
Extraction, solvent, 581

Face-centered cubic lattice, 466
Faraday constant, 596
Faraday's laws of electrolysis, 595
Fick's law of diffusion, 146
Flow
  laminar, 143, 498
  molecular, 140
Fluid, 4
Fluidity, 501
Fluorescence, 436
Force constant, Hooke's law, 152, 294
Force, intermolecular, 386
Fractional distillation, 559
Franck-Condon principle, 311
Free energy
  of activation, 431
  of formation, 111
  Gibbs, 85, 347
  Helmholtz, 83, 347
Free energy function, 117
Free path, mean, 141
Freezing-point depression, 550
Fuel cell, 650

Fugacity, 95
Function of the state, 35
Fundamental vibrational frequency, 297

Galvanic cell, 629
Gamma ray, 166
Gas constant, 8
Gases
　effusion of, 140
　heat capacity of, 39, 150, 345
　ideal, 9
　liquefaction of, 16, 526
　thermal conductivity of, 146
　viscosity of, 142
Gay-Lussac's law, 6
Gibbs
　adsorption equation, 585
　ensemble, 374
　free energy, 85, 347
　phase rule, 515
Gibbs-Duhem equation, 93
Gibbs-Helmholtz equation, 88, 642
Glass, 508
　electrode, 648
Graham's law of effusion, 140
Group, 271
　theory, 325
Gyromagnetic ratio, 317

Half-cell potential, 634
Half-life, 397
Hamiltonian
　function, 184
　operator, 186
Harmonic
　oscillator, 153, 184, 293, 374
　waves, 164
Hartree self-consistent field method, 203
Heat of, 28
　activation, 413
　adsorption, 586
　combustion, 50
　dilution, 51
　formation, 49
　fusion, 91
　reaction, 48
　solution, 51
　vaporization, 47, 90
Heat capacity, 38
　of crystals

Heat capacity *continued*
　Debye theory, 472
　Einstein theory, 471
　of gases, empirical equations for, 39
　from partition function, 345
Heisenberg uncertainty principle, 191
Heitler-London treatment of H_2, 235
Helium atom, 209
Helmholtz free energy, 83
　from partition function, 347
Henry's law, 562
Hess' law, 48
Heterogeneous system, 5
Hexagonal close packing, 466
Hole theory of liquids, 496
Homogeneous system, 3
Hooke's law, 152, 294
Hückel approximation, 250
Hund's rule, 204
Hybrid orbital, 241
Hydration of ions, 609
Hydrogen
　atom
　　Bohr theory, 173
　　orbitals, 197
　　Schrödinger equation, 194
　　spectrum, 168
　bond, 259
　bonded crystals, 484
　electrode, 633
　ion concentration, 625, 647
　molecular ion, 219
　molecule
　　Heitler-London theory, 235
　　molecular orbital theory, 225
　　valence bond theory, 235
Hydrolysis, 628

Ideal gas law, 8
　statistical derivation, 352
Ideal solutions, 541
Immiscible liquids, 568
Induction period, 407
Infrared absorption spectrum, 285
Insulator, electrical, 478
Integral
　coulomb, 221
　cyclic, 30
　overlap, 221
　phase, 376
　resonance, 221

Integral heat of
  dilution, 51
  solution, 51
Intensive property, 3
Interatomic distance from
  electron diffraction, 326
  rotation spectra, 285
Interference of light, 164
Interionic attraction, 611
Intermolecular forces, 386
Internuclear distance, 285, 326
Invariant system, 523
Inversion
  operation, 269
  temperature, 47
Ionic
  activity coefficient, 615
  atmosphere, 612
  conductivity, 601
  crystals, 480
  discharge, 650
  equilibrium, 614
  migration, 602
  mobility, 605
  strength, 620
Ionization
  degree of, 610
  potential, 207
Ion pair formation, 613
Isobaric process, 36
Isochoric process, 36
Isolated system, 2
Isothermal process, 36
  fractional distillation, 559
Isotopic substitution, effect of, 291, 298

Joule
  gas expansion experiment, 40
  paddle wheel experiment, 28
Joule-Thomson
  coefficient, 46, 80
  expansion, 45
Junction potential, 637

Kelvin
  equation, 520
  temperature scale, 65
Kinetics, reaction, 393
Kinetic theory of gases, 123
Kirchhoff equation, 54
Kohlrausch's law, 601

Lagrange's method of undetermined multipliers, 343
Langevin equation, 276, 314
Langmuir adsorption isotherm, 588
Laplacian operator, 182
Larmor precession, 320
Lattice, crystal, 446
LCAO-MO method, 220
LeChatelier's rule, 104
Lennard-Jones potential energy function, 388
Lever rule, 558
Lewis acid, 624
Light, nature of, 164
Limiting densities method, 13
Lindemann mechanism for unimolecular reactions, 419
Line integral, 31
Liquefaction of gases, 16, 47, 526
Liquid
  crystal, 509
  junction, 630, 637
Liquids
  structure of, 491
  surface tension of, 503
  viscosity of, 498
  X-ray diffraction pattern of, 492
London dispersion force, 386
Lorentz-Lorenz equation, 281
Lowry-Brönsted acids, 623

Madelung number, 486
Magnetic
  moment
    molecular, 314
    nuclear, 315
  quantum number, 195
  susceptibility, 313
Magneton, nuclear, 316
Mass action law, 104, 404
Mass-energy relation, 29
Mass, reduced, 194, 287
Mass spectroscopy, 166
Maximum work, 43
Maxwell
  -Boltzmann distribution law, 150
  distribution law, 136
  electromagnetic theory, 165, 280
McLeod equation, 507
Mean
  free path, 141
  ionic activity coefficient, 615

Mechanism, reaction, 406
Melting, 531
Melting-point change with pressure, 90, 522
Mesomorphic phase, 509
Metallic
 conductance, 478
 crystals, 476
Microwave spectroscopy, 291
Miller indices, 450
Millikan oil-drop experiment, 160
Mirror plane, 270
Mobility, ionic, 605
Molality, 549
Molar
 absorptivity, 439
 partition function, 349
 polarization, 274
 refraction, 281
Molecular
 collision diameter, 141
 crystals, 483
 effusion, 140
 electron configuration, 230
 kinetic energy, distribution of, 138
 orbital method, 219
 partition function, 344
 polarizability, 272
 spectra, 285
 speed, distribution of, 136
 symmetry, 268
 weight, 12, 546
  number average, 552
Molecularity, 414
Mole fraction, 8
Moment of inertia, 152, 286
Morse potential function, 298
Moving boundary method, 605

Nernst
 distribution law, 580
 equation, 639
Neutron diffraction, 464
Newton's viscosity equation, 143
Node, 180
Nonbonding molecular orbital, 235
Normalization constant, 191
Normalized function, 184
Normal modes of vibration, 301
Nuclear magnetic
 moment, 315
 resonance spectroscopy, 317

Nuclear magneton, 315
Nuclear spin, 315
Nucleation
 in phase transitions, 527
 rate, 533
Nucleus
 atomic, 161, 315
 condensation, 527

Ohm's law, 598
Onsager equation, 612
Open system, 2
Operator, 185
 Hamiltonian, 186
 Hermitian, 220
Orbital, 197
 antibonding, 222
 bonding, 222
 hybrid, 241
 molecular, 219
 symmetry of, 229
Order of reaction, 395
Orientation polarization, 274
Orthobaric density, 18
Orthogonal functions, 193
Oscillator
 anharmonic, 297
 harmonic, 153, 184, 293, 374
Osmotic pressure, 552
Ostwald dilution law, 610
Overlap integral, 221
Overvoltage, 652
Oxidation potential, 639

Parallel reactions, 403
Paramagnetism, 313
Partial molar
 quantities, 92
 volume, 542
Partial pressure, 8
Particle in a box, 187
Partition function
 electronic, 352
 molar, 349
 molecular, 344
 rotational, 353
 translational, 349
 vibrational, 355
Pauli exclusion principle, 204, 212
Peritectic point, 575

Perturbation method of approximation, 208
pH, 629, 647
Phase, 5
   boundary, 583
   diagram, 11
   equilibrium, 88, 513
   integral, 376
   rule, Gibbs, 515
   space, 374
   transitions, dynamics of, 525
Photochemistry, 435
Photoelectric effect, 165
Photon, 171
Physical adsorption, 586
Pi
   bond, 233
   molecular orbital, 228
Plait point, 579
Planck
   constant, 171
   law, 172
Point group, 271
Poise (unit), 143
Poiseuille's equation, 498
Poisson's equation, 654
Polar molecules, 274
Polarizability, molecular, 272
Polarization
   concentration, 651
   dielectric, 273
   molar, 274
Potential
   decomposition, 650
   electrical, 653
   electrode, 634
   ionization, 207
   oxidation, 639
   reduction, 634
   standard electrode, 639
Potential energy diagram for diatomic molecule, 218, 297, 310
Potentiometer, 631
Powder pattern, X-ray, 455
Precession, nuclear, 319
Pre-exponential factor (reaction kinetics), 413
Pressure
   critical, 18
   internal, 39
   from kinetic theory, 125, 140
   from partition function, 352
   reduced, 22
Pressure-volume work, 27

Primitive unit cell, 447
Principal quantum number, 196
Probability, 336
   entropy and, 346
   thermodynamic, 341
   from wave function, 183
Proton, 163
   magnetic resonance of, 317

Quantization, space, 316
Quantum
   numbers, 174
      angular momentum, 196
      magnetic, 195
      molecular, 228
      principal, 196
      rotational, 268
      spin, 202
      translational, 190
      vibrational, 295
   theory, 171
   yield, 436
Quantum mechanics, postulates of, 185
Quasi-static process, 36

Radial distribution function, 200, 330, 493
Radiation
   black-body, 169
   electromagnetic, 165
Raman spectrum, 304
Raoult's law, 540
Rate of
   bimolecular collisions in gases, 142, 415
   chemical reactions, 393
   fluid flow, 498
   nucleation in vapor phase, 533
Rate constant, 395
   temperature dependence of, 412
Rational indices, law of, 450
Rayleigh scattering, 305
Reaction
   activation energy, 413
   chain, 409
   coordinate, 422
   heat, 48
   molecularity, 414
   order, 395
   quotient, 100
   rate, 393
Rectilinear diameter, law of, 19

Reduced
    mass, 194, 287
    pressure, 22
    temperature, 22
    volume, 22
Reduction potential, 634
Reference electrode, 635
Reflection plane, 269
Refraction
    index of, 167, 280
    molar, 281
Relaxation time, 279, 319
Resistance
    electrical, 598
    of metals, 479
Resistivity, 598
Resonance, 238
    energy, 258
    integral, 221
    nuclear magnetic, 317
Reversible process, 36
Root-mean-square speed, 127
Rotational
    constant, 288
    energy, 151, 287
    partition function, 353
    quantum number, 288
    spectra, 285
Rotation axis, 269
Rotation-vibration spectrum, 299
Rutherford model of atom, 162
Rydberg constant, 169

Sackur-Tetrode equation, 351
Salt bridge, 637
Scattering
    light, 305
    X-ray, 453
Schönfliess symbols, 272
Schrödinger equation, 182
Second law of thermodynamics, 60
Second virial coefficient, 384
Secular equation, 221
Selection rule
    nuclear, 316
    rotational, 289
    vibrational, 296
Self-consistent field method, 203
Semiconductors, 480
Semipermeable membranes, 552
Separation of variables, 130, 194

Sigma bond, 233
    molecular orbital, 228
Significant structure theory of liquids, 496
Solid
    solution, 575
    surface, 586
Solids, 445
Solubility of
    gases, 562
    solids, 574
Solubility product constant, 622
    and cell emf, 646
Solution
    heat of, 51
    ideal, 541
    solid, 575
    supersaturated, 532
Solvation, ionic, 609
Solvent extraction, 581
Space
    group, 450
    lattice, 447
    quantization, 316
Spectroscopy, 166
Spectrum
    absorption, 167
    atomic, 167
    band, 167, 310
    black-body, 169
    emission, 168
    infrared, 285
    microwave, 291
    molecular
        electronic, 309
        rotational, 285
        vibrational, 293
    nuclear magnetic resonance, 317
    Raman, 304
    X-ray, 163
Speed
    molecular
        average, 137
        distribution of, 136
        root-mean-square, 127
Spherical top, 292
Spin
    electron, 202
    nuclear, 315
Spin-spin splitting (NMR), 323
Spontaneous process, 59
Standard hydrogen electrode, 635
Standard state, 93

Standard state *continued*
  in solutions, 564
Standing waves, 180
State, 3
  critical, 16
  equation of, 10
  function of, 35
  stationary, 173
Statistical
  mechanics, 335, 373
  weight, 344
Steady state approximation (reaction kinetics), 408
Steam distillation, 569
Stefan-Boltzmann law, 170
Steric factor, 418
Stirling approximation, 341
Stokes law of viscous flow, 501
Sublimation pressure, 90, 522
Sulfur phase diagram, 524
Supercooling, 532
Superheating, 531
Supersaturated solution, 532
Supersaturation ratio, 528
Surface
  area, 586
  concentration, 584
  energy, 503, 520, 584
  tension, 503
Symmetric
  function, 211
  top, 292
Symmetry
  center of, 270
  crystal, 445
  elements, 270
  group, 271
  number, 353
  operations, 269
  of wave functions, 212
System, 2

Temperature
  absolute, 7
  coefficient of expansion, 6
  consolute, 570
  critical, 17
  Debye characteristic, 475
  eutectic, 573
  inversion, 47
  peritectic, 575

Temperature *continued*
  reduced, 22
  scale
    Celsius, 6
    Kelvin, 7, 65
    thermodynamic, 65, 523
Term value
  rotational, 288
  vibrational, 295
Ternary system, 577
Theoretical plate, 561
Thermal conductivity, 146
Thermochemistry, 48
Thermodynamic
  criterion of equilibrium, 81
  equation of state, 78
  probability, 341
Thermodynamics
  first law of, 34
  second law of, 60
    statistical interpretation, 363
  third law of, 112
    statistical interpretation, 360
Tie-line, 558
Top
  asymmetric, 292
  spherical, 292
  symmetric, 292
Transference number, 603, 645
Transition state theory, 421
Translational
  energy, 126, 190
  partition function, 349
Transmission coefficient, 425
Transport properties of gases, 142
Triple point, 523
Triplet state, 212
Trouton's rule, 81

Uncertainty principle, Heisenberg, 191
Undetermined multipliers, Lagrange, 343
Unimolecular reaction, 418
Unit cell, 447
United atom, 230
Univariant system, 515

Valence band of crystals, 478
Valence bond method, 235
Van der Waals
  equation, 14
  forces, 386

Van't Hoff equation for osmotic pressure, 554
Vapor, 18
  pressure, 88, 517
    effect of drop size on, 520
    effect of total pressure on, 521
    of ideal solutions, 540
    lowering of, 546
Vaporization, 47, 529
Variance, 514
Variation method of approximation, 208, 220
Velocity
  molecular, 124
    distribution function, 134
  space, 135
Vibration
  molecular, 293
  normal mode of, 301
Vibrational
  energy, 153, 295
  partition function, 355
  spectrum, 293
  transitions, 296
Vibration-rotation spectrum, 299
Virial
  coefficients, 16, 384
  equation of state, 15, 384
Viscosity
  gas, 142
  liquid, 498
Voltaic cell, 629

Volume
  critical, 18
  free, of liquids, 496
  reduced, 22
Volume element in spherical polar coordinates, 135

Wave, de Broglie, 176
Wave equation
  classical, 177
  Schrödinger, 182
Wave function, 178
Wavenumber, 169
Wave-particle duality, 176
Wheatstone bridge, 598
Wien effect (conductance), 613
Wien's displacement law, 170
Wierl equation, 328
Work, 27
  function, 83
    photoelectric, 171

X-ray
  diffraction, 452
  powder method, 455
  spectrum, 163

Zero-point energy, 193, 296, 356

M. P. PERIASAMY
DEPT OF CHEMISTRY
FSU · TALLAHASSEE.
FLA · 32306 · USA ·
SEPT · 1970.

LOGARITHMS

Natural Numbers	0	1	2	3	4	5	6	7	8	9	\multicolumn{9}{c}{Proportional Parts}
											1	2	3	4	5	6	7	8	9
10	0000	0043	0086	0128	0170	0212	0253	0294	0334	0374	4	8	12	17	21	25	29	33	37
11	0414	0453	0492	0531	0569	0607	0645	0682	0719	0755	4	8	11	15	19	23	26	30	34
12	0792	0828	0864	0899	0934	0969	1004	1038	1072	1106	3	7	10	14	17	21	24	28	31
13	1139	1173	1206	1239	1271	1303	1335	1367	1399	1430	3	6	10	13	16	19	23	26	29
14	1461	1492	1523	1553	1584	1614	1644	1673	1703	1732	3	6	9	12	15	18	21	24	27
15	1761	1790	1818	1847	1875	1903	1931	1959	1987	2014	3	6	8	11	14	17	20	22	25
16	2041	2068	2095	2122	2148	2175	2201	2227	2253	2279	3	5	8	11	13	16	18	21	24
17	2304	2330	2355	2380	2405	2430	2455	2480	2504	2529	2	5	7	10	12	15	17	20	22
18	2553	2577	2601	2625	2648	2672	2695	2718	2742	2765	2	5	7	9	12	14	16	19	21
19	2788	2810	2833	2856	2878	2900	2923	2945	2967	2989	2	4	7	9	11	13	16	18	20
20	3010	3032	3054	3075	3096	3118	3139	3160	3181	3201	2	4	6	8	11	13	15	17	19
21	3222	3243	3263	3284	3304	3324	3345	3365	3385	3404	2	4	6	8	10	12	14	16	18
22	3424	3444	3464	3483	3502	3522	3541	3560	3579	3598	2	4	6	8	10	12	14	15	17
23	3617	3636	3655	3674	3692	3711	3729	3747	3766	3784	2	4	6	7	9	11	13	15	17
24	3802	3820	3838	3856	3874	3892	3909	3927	3945	3962	2	4	5	7	9	11	12	14	16
25	3979	3997	4014	4031	4048	4065	4082	4099	4116	4133	2	3	5	7	9	10	12	14	15
26	4150	4166	4183	4200	4216	4232	4249	4265	4281	4298	2	3	5	7	8	10	11	13	15
27	4314	4330	4346	4362	4378	4393	4409	4425	4440	4456	2	3	5	6	8	9	11	13	14
28	4472	4487	4502	4518	4533	4548	4564	4579	4594	4609	2	3	5	6	8	9	11	12	14
29	4624	4639	4654	4669	4683	4698	4713	4728	4742	4757	1	3	4	6	7	9	10	12	13
30	4771	4786	4800	4814	4829	4843	4857	4871	4886	4900	1	3	4	6	7	9	10	11	13
31	4914	4928	4942	4955	4969	4983	4997	5011	5024	5038	1	3	4	6	7	8	10	11	12
32	5051	5065	5079	5092	5105	5119	5132	5145	5159	5172	1	3	4	5	7	8	9	11	12
33	5185	5198	5211	5224	5237	5250	5263	5276	5289	5302	1	3	4	5	6	8	9	10	12
34	5315	5328	5340	5353	5366	5378	5391	5403	5416	5428	1	3	4	5	6	8	9	10	11
35	5441	5453	5465	5478	5490	5502	5514	5527	5539	5551	1	2	4	5	6	7	9	10	11
36	5563	5575	5587	5599	5611	5623	5635	5647	5658	5670	1	2	4	5	6	7	8	10	11
37	5682	5694	5705	5717	5729	5740	5752	5763	5775	5786	1	2	3	5	6	7	8	9	10
38	5798	5809	5821	5832	5843	5855	5866	5877	5888	5899	1	2	3	5	6	7	8	9	10
39	5911	5922	5933	5944	5955	5966	5977	5988	5999	6010	1	2	3	4	5	7	8	9	10
40	6021	6031	6042	6053	6064	6075	6085	6096	6107	6117	1	2	3	4	5	6	8	9	10
41	6128	6138	6149	6160	6170	6180	6191	6201	6212	6222	1	2	3	4	5	6	7	8	9
42	6232	6243	6253	6263	6274	6284	6294	6304	6314	6325	1	2	3	4	5	6	7	8	9
43	6335	6345	6355	6365	6375	6385	6395	6405	6415	6425	1	2	3	4	5	6	7	8	9
44	6435	6444	6454	6464	6474	6484	6493	6503	6513	6522	1	2	3	4	5	6	7	8	9
45	6532	6542	6551	6561	6571	6580	6590	6599	6609	6618	1	2	3	4	5	6	7	8	9
46	6628	6637	6646	6656	6665	6675	6684	6693	6702	6712	1	2	3	4	5	6	7	7	8
47	6721	6730	6739	6749	6758	6767	6776	6785	6794	6803	1	2	3	4	5	5	6	7	8
48	6812	6821	6830	6839	6848	6857	6866	6875	6884	6893	1	2	3	4	4	5	6	7	8
49	6902	6911	6920	6928	6937	6946	6955	6964	6972	6981	1	2	3	4	4	5	6	7	8
50	6990	6998	7007	7016	7024	7033	7042	7050	7059	7067	1	2	3	3	4	5	6	7	8
51	7076	7084	7093	7101	7110	7118	7126	7135	7143	7152	1	2	3	3	4	5	6	7	8
52	7160	7168	7177	7185	7193	7202	7210	7218	7226	7235	1	2	2	3	4	5	6	7	7
53	7243	7251	7259	7267	7275	7284	7292	7300	7308	7316	1	2	2	3	4	5	6	6	7
54	7324	7332	7340	7348	7356	7364	7372	7380	7388	7396	1	2	2	3	4	5	6	6	7